Polymer and Ceramic Composite Materials

Polymer and Ceramic Composite Materials

Emergent Properties and Applications

By

Noureddine Ramdani

CRC Press
Taylor & Francis Group
Boca Raton London New York

CRC Press is an imprint of the
Taylor & Francis Group, an **informa** business

CRC Press
Taylor & Francis Group
6000 Broken Sound Parkway NW, Suite 300
Boca Raton, FL 33487-2742

First issued in paperback 2021

ISBN-13: 978-0-367-78030-2 (pbk)
ISBN-13: 978-1-138-30221-1 (hbk)

Library of Congress Cataloging-in-Publication Data

Names: Ramdani, Noureddine, 1986- author.
Title: Polymer and ceramic composite materials : emergent properties and applications / authored by Noureddine Ramdani.
Description: Boca Raton : Taylor & Francis, CRC Press, 2019. | Includes bibliographical references and index.
Identifiers: LCCN 2018046108| ISBN 9781138302211 (hardback : alk. paper) | ISBN 9780203731857 (e-book)
Subjects: LCSH: Composite materials.
Classification: LCC TA418.9.C6 R27 2019 | DDC 620.1/18--dc23
LC record available at https://lccn.loc.gov/2018046108

Visit the Taylor & Francis Web site at
http://www.taylorandfrancis.com

and the CRC Press Web site at
http://www.crcpress.com

Dedication

It gives me immense pleasure to thank CRC Press for their kind acceptance of this monograph. I dedicate this book to the memory of my mother, Daoudi Aicha, who always believed in my ability to be successful in the academic arena. I express heartfelt thanks to my wife for her continuous support in writing this book. I wish to acknowledge my colleagues, Khaled Oukil, Hamza Zaimeche, and Nacereddine Abid for their kind help to publish this work.

Contents

Preface

During the last few decades, few polymer/inorganic hybrid composites that have been produced are able to attain the commercialization phase. Polymer/ceramic composite materials are one example. Their ease of processing and outstanding thermal, electrical, and mechanical properties allow for their use in a wide range of industrial applications, including electronic packaging, automotive, energy management, and aerospace.

The need for high-performance multifunctional polymeric composites based on the various thermoplastic, thermoset, and elastomeric matrices, reinforced with ceramic micro- and nanofillers continues to be increasingly developed and investigated to satisfy the requirements of different industrial domains. This book is comprised of twelve chapters, providing an introduction to the unique and striking properties of composite materials produced from different kinds of polymeric matrices and ceramic fillers. In this monograph, the advantageous characteristics of these materials, such as thermal properties, mechanical performance, piezoelectric and ferroelectric, barrier properties, and electrical properties, are discussed and summarized. In addition, the modeling of these properties, the properties of polymer/fiber/ceramic hybrid composites, and the various potential applications of polymer/ceramic composites are reported.

Chapter 1 of this book presents an introduction to different types of polymeric matrices, their structures, processing techniques, and their various inherent properties. Chapter 2 presents the various processing methods, enhanced properties, and applications of ceramic fillers, and Chapter 3 reviews and discusses the fundamental functionalization techniques used for modifying the outer surface of ceramic particles. The various processing methods used for producing polymer/ceramic composite materials, such as sol–gel, solution blending, in situ polymerization, and melt-processing, will be reviewed in Chapter 4. Chapter 5 explains the mechanisms of various improvements in the mechanical properties of polymer/ceramic composites, such as tensile properties, flexural, tribological properties, microhardness, and impact properties. The kinetics of cure, and the change in glass-transition temperature, thermal stability, and coefficient of thermal expansion, thermal conductivity, and the thermomechanical properties of polymer composites reinforced with various types of ceramic fillers, are demonstrated in Chapter 6. Chapter 7 summarizes the piezoelectric and ferroelectric properties of polymer/ceramic composite materials showing the different mechanisms of improvements. The role of ceramic micro- and nanofillers in enhancing the electrical conductivity, dielectric constant, and dielectric loss of polymers are presented in Chapter 8. Chapter 9 provides a good overview on the various theoretical and semi-empirical equation models used for predicting thermal, electrical, and mechanical parameters of polymer/ceramic composites. The barrier properties of polymer/ceramic composites, such as water-uptake properties, corrosion-resistant properties, and gas barrier properties are studied in Chapter 10. Chapter 11 describes how ceramic fillers can improve the thermal, barrier, and mechanical properties of polymer/fiber-laminated composites resulting in multifunctional composite systems suitable for high-performance applications. The last chapter of this book (Chapter 12) reports on and compares several potential applications of polymer/ceramic composites in many industrial sectors, including microelectronics, energy management, aerospace, automotive, biomedical, ballistic protections, and others.

This monograph can be used as a research book for both undergraduate and postgraduate students working on polymer composites from several majors, such as plastic engineering, ceramic technology, nanomaterials, and material science, and engineering. Also, this book can serve as a useful reference for academicians, researchers, materials scientists, and industrial engineers in ceramics and related polymer industries. Thus, my hope is that this book will inspire and enthuse readers to undertake research in the field of polymer/ceramic hybrid materials.

Author Biography

Dr. Noureddine Ramdani achieved his undergraduate degree in multidisciplinary science from the National School Preparatory for Engineering Studies, Algiers, Algeria. He earned his degree in chemical engineering from the National Polytechnic School, Algiers, Algeria. He obtained his doctoral degree in material science from Harbin Engineering University, Harbin, China. His research involved synthesizing stabilizers for studying the stability of propellants and gunpowders. In addition to his works on polymer composites, thermosets, and biomaterials, he is the author of several chapters on polymer composites for electronic packaging materials, application of polymer ceramic composites, polybenzoxazine ceramic composites, and polybenzoxazine/bio-filler composites. He has also served as reviewer for many reputable scientific journals and magazines such as *Composite Science and Technology, Polymer for Advanced Technologies, Applied Polymer Science, IGI Global Publisher.* In 2016, he served as Director of Analytical Chemistry Lab in the Research and Development Center, Blida, Algeria. He is currently working in the Research and Development Institute in Industry and Defense Technologies in the department of Advanced Material Researches. His research work in the recent years focuses on polymeric materials, nanotechnology, detection of hazardous materials, and engineering ceramics.

Introduction

Polymer/ceramic composite materials have been widely explored for a long time. When nanoceramic reinforcing agents are used to prepare such hybrid materials, they are called nanocomposites. These composite materials combine the advantages of the ceramic fillers (e.g., high rigidity, thermal stability, good mechanical strength) and those of organic polymers (e.g., flexibility, low dielectric, improved ductility, and the ease of processability). A special characteristic of polymer/ceramic nanocomposites is that the small size of the ceramic fillers leads to a substantial enhancement in the interfacial area compared to traditional microsized ceramic based polymer composites. This increased interfacial area provokes a higher volume fraction of interfacial polymer exhibiting different properties compared with those of the bulky polymers even at low contents [1-5].

Inorganic fillers, such as carbon nanotubes, graphene, clays, metallic fillers, semiconductors, and so forth, are widely used for reinforcing the various thermoplastics, thermosets, and elastomeric matrices; among them, ceramics are considered the most prominent reinforcing agents. Therefore, polymer/ceramic nanocomposites have received extensive academic and industrial interest. In fact, among the numerous polymer/ceramic composite materials, polymer/silica hybrids are the most commonly studied in the literature. They have attracted much attention in recent decades and have been utilized in several high-performance engineering applications.

Polymer/ceramic micro- and nanocomposites can be produced using different processing techniques, thanks to the ability to join in different ways to incorporate each phase. The polymeric matrix can be introduced as (1) a precursor, in a form of monomer or an oligomer, (2) a preformed linear polymer (in molten, solution, or emulsion states), or (3) a polymer network constituted from a semi-crystalline linear polymer or chemically cross-linked thermosets or elastomers. On the other hand, the ceramic phase can be added as (1) a precursor (e.g., TEOS) or (2) solid particles. In the case where at least one of the starting moieties is a precursor, organic or inorganic polymerization are generally required.

This results in three different methods for the elaboration of polymer/ceramic composites according to the raw materials and the synthesizing techniques: solution blending, sol–gel technique, and in situ polymerization. The solution blending is used for enhancing the degree of mixing of ceramic nanoparticles into the polymeric matrix; a sol–gel method can be achieved in situ in the presence of a preformed polymeric matrix or simultaneously during the polymerization of a monomer(s); and the technique of in situ polymerization, which is based on the dispersion of ceramic fillers into the monomer(s) first, followed by polymerization. In addition, in recent years, great efforts have been engaged in designing and controlling the production of polymer/ceramic colloidal composite particles with tailored morphologies [6, 7].

The processing, properties, and applications of polymer/ceramic micro- and nanocomposites have become a rapidly expanding area of research and have attracted both academic and industrial interests. However, the majority of published books and review articles are partly concentrated on the development of specific types of polymer/ceramic nanocomposites, without investigating this subject in detail [8].

The aim of this book is to explore the synthetic methods, the properties, and the applications of polymer/ceramic composite materials by providing the recent developments in this field by scrutinizing the wide literature published in the last two decades. Due to the numerous papers published on polymer/ceramic composites, it was too difficult to totally describe this theme for each example of composite. Thus, this book will afford a general overview of the techniques and strategies used for producing the polymer/ceramic micro- and nanocomposites, followed by a detailed discussion of their thermal, mechanical, electrical properties, and cite some relevant engineering applications of these hybrid materials, along with selected examples taken from the published literature.

Polymer/ceramic composites are inorganic–organic hybrids composed from ceramic micro- or nanofillers and a matrix of organic polymers. The formation of polymer/ceramic networks is based on thermal curing of functionalized thermosets being able to provide ceramic-like structures due to heat treatment above 200°C [9]. Polymer/ceramics can also be produced via a variety of thermoplastic-forming techniques such as high-pressure injection molding and extrusion technique. Polymer/ceramic composites exhibited high thermal stability (the service temperature exceeded 400°C), low shrinkage, high stability of form, and good dimensional stability.

The addition of ceramic fillers resulted in a substantial improvements in the thermal properties, (1) such as thermal conductivity, thermal stability, glass transition temperature (T_g), thermal expansion coefficient (CTE); electrical properties (2) like electrical conductivity, dielectric constant, dielectric breakdown strength; and mechanical properties (3) including elastic modulus, tensile strength, impact and flexural properties, and wear properties. These enhancements in the properties of polymer/ceramic composites can be tailored by selecting suitable functional ceramic fillers, the type of polymeric matrix, the filler ratios, and type of surface treatment involved. Different mechanisms are responsible for these improvements, which are discussed and compared in detail in the book.

The potential applications of polymer/ceramic composite materials are diverse owing to their low-cost production, easy processing, improved mechanical and dialectical properties, and heat and wear resistance of these material. They are widely applied in development gas sensors and actuators, armor systems, gas-separation membranes, anticorrosion coatings, microwave absorptions, weather-resistant coatings, electronic packaging, biomedical applications, etc.

REFERENCES

1. Koo, J. H. 2006. *Polymer Nanocomposites: Processing, Characterization, and Applications.* Pennsylvania: McGraw-Hill.
2. Thomas, S., Joseph, K., Malhotra, S. K. et al. 2013. *Polymer Composites, Nanocomposites.* Singapore: Wiley-VCH Verlag GmbH & Co.
3. Caseri, W. R. 2006. Nanocomposites of polymers and inorganic particles: Preparation, structure and properties. *Materials Science and Technology* 22: 807.
4. Schadler, L. S., Kumar, S. K., Benicewicz, B. C. et al. 2007. Designed interfaces in polymer nanocomposites: A fundamental viewpoint. *MRS Bulletin* 32: 335–340.

5. Colombo, P., Raj, R. 2010. *Advances in Polymer Derived Ceramics and Composites.* Hoboken, NJ: Wiley & Sons.
6. Bhargava, A. K. 2012. *Engineering Materials: Polymers, Ceramics and Composites.* New Delhi: PHI learning Private Limited.
7. Colombo, P. 2010. *Polymer Derived Ceramics: From Nano-structure to Applications.* Pennsylvania, USA: DEStech Publication.
8. Barsoum, M., Barsoum, M. W. 2003. *Fundamentals of Ceramics.* Boca Raton: CRC Press/Taylor & Francis.
9. Richerson, D., Richerson, D. W., Lee, W. E. 2005. *Modern Ceramic Engineering: Properties, Processing, and Use in Design*, 3rd ed. Boca Raton: CRC Press/Taylor & Francis Group.

1 Structure and Properties of Polymer Matrix

1.1 INTRODUCTION

A polymer is defined as a long-chain organic molecule constituted from many repeated elemental units, known as monomers. Polymers can be synthesized by chemical reactions or generated by nature. Today, these materials can be found everywhere, such as in plastics (bottles, toys, vinyl siding, packaging), cosmetics, shampoos and other haircare products, contact lenses, nature (crab shells, amber), food (proteins, gelatin, and gum), fabric, polysaccharides, sneakers, and even in human DNA. Polymers' widespread nature makes them useful in many industries and products, such as aerospace, sports equipment, microelectronic packaging, 3D printing plastics, holography, molecular recognition, bulletproof vests and fire-resistant jackets, and water purification. Examples of engineering applications of some polymers are illustrated in Figure 1.1.

Polymer matrix–based composites (PMCs) are developed to reduce the heavy weight of metals in various industrial applications. These materials exhibit a light weight, good corrosion-resistance, and low dielectrics constant. However, prolonged exposure of these materials to UV light and some solvents can degrade them. Figure 1.2 compares the strength characteristics for polymers, elastomers, ceramics, and metals [1].

1.2 POLYMER STRUCTURE TYPES

As illustrated in Figure 1.3, there are different types of polymer structures, which include linear, branched, and cross-linked. Linear polymers, such as polyethylene, polyvinyl chloride, and polymethyl methacrylate, consist of a long chain, often bent, of atoms with attached side groups. Branched polymers consist of side-branching of atomic chains. In cross-linked polymers, molecules of one chain are bonded (cross-linked) with those of another, making the polymer strong and rigid with a three-dimensional network. Ladder polymers, in turns, join several linear polymers in a regular manner, which results in a more rigid structure compared to linear polymers.

1.3 THERMAL BEHAVIOR OF POLYMERS

Compared to pure metals, which melt at a fixed temperature, polymers generally reveal a range of temperatures where the crystallinity vanishes when heated. On cooling, polymer liquids contract like metals do. In the case of amorphous polymers, this contraction phenomena continues below the melting point, T_m, of crystalline polymer to a temperature called the glass transition temperature (T_g), at which the

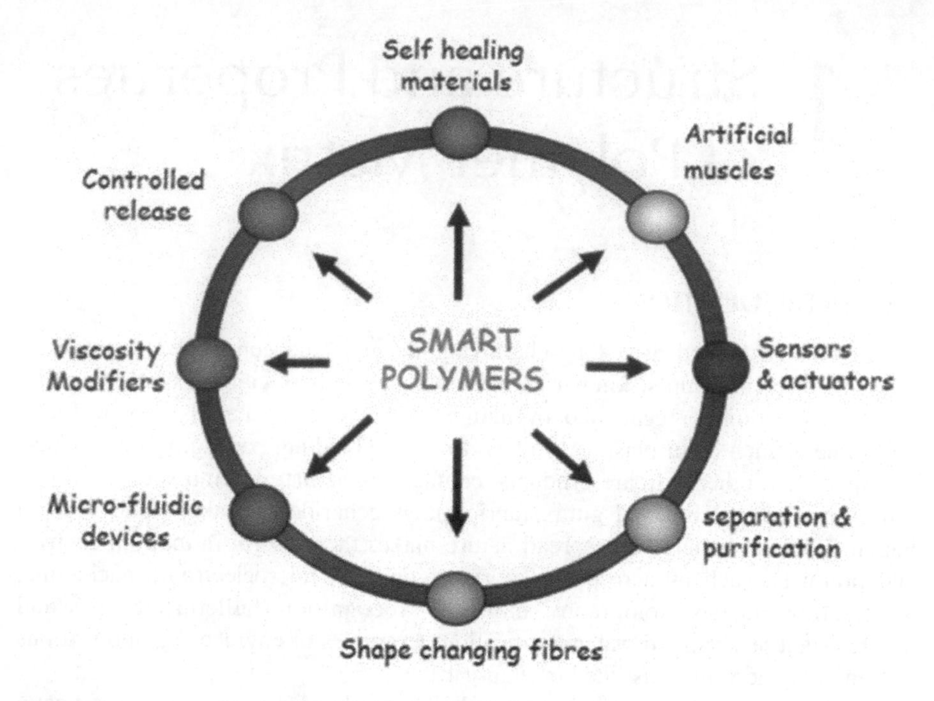

FIGURE 1.1 Engineering Applications of Some Polymers.

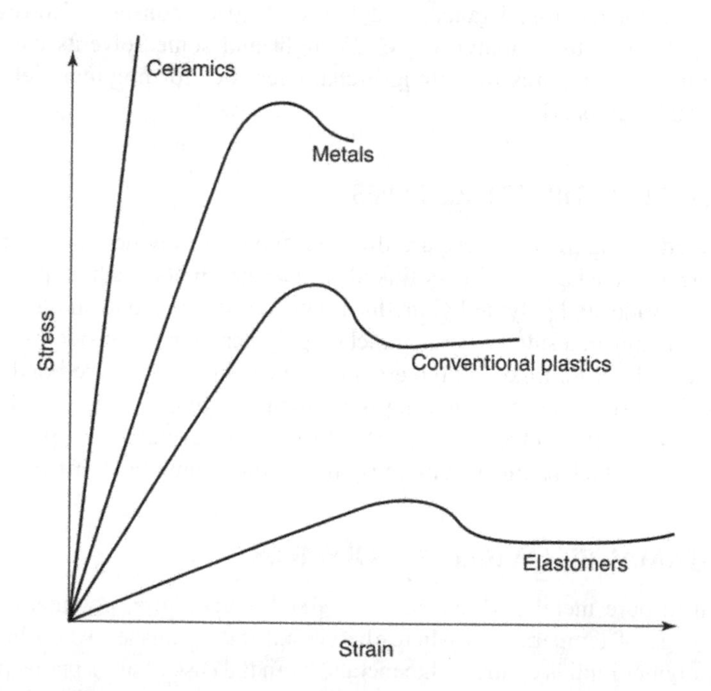

FIGURE 1.2 Comparison of Idealized Stress-Strain Diagrams for Metals, Amorphous Polymers, and Elastomers [1].

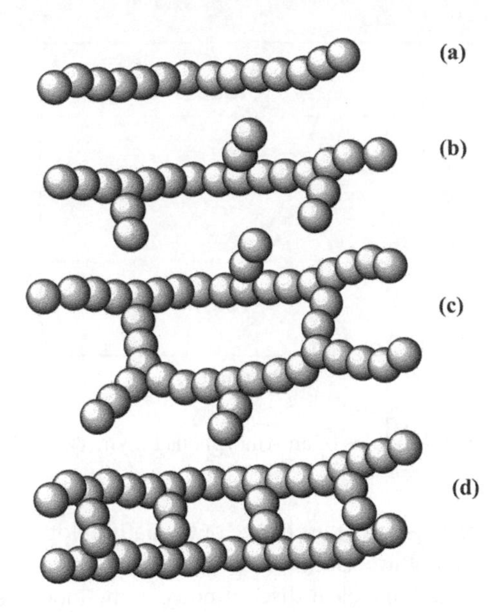

FIGURE 1.3 Molecular Chain Configurations in Polymers: (a) Linear, (b) Branched, (c) Cross-linked, and (d) Ladder.

super-cooled liquid polymer transforms to an extremely rigid form due to extremely high viscosity [2]. Below T_g, the resulting structure of the polymer is essentially disordered. Figure 1.4 displays the different changes in the specific volume as a function of temperature recorded in a polymer. Several physical parameters like the inherent viscosity, heat capacity, modulus, and thermal expansion (measured by coefficient of thermal expansion, CTE) will significantly change at T_g.

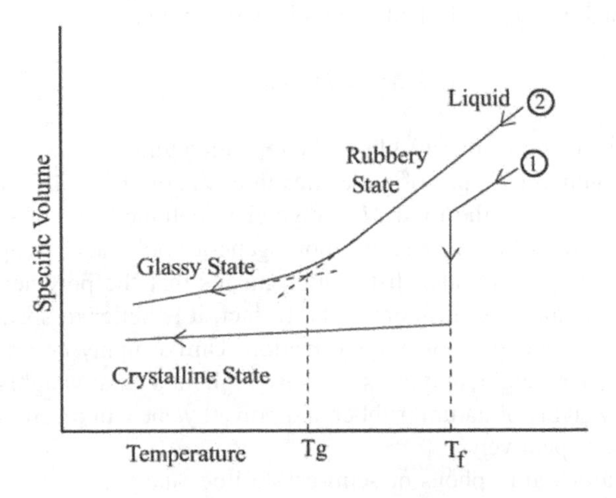

FIGURE 1.4 Specific Volume Versus Temperature Relationship for Amorphous and Semi-crystalline Polymers (T_g, Glass Transition Temperature; T_m, Melting Temperature) [2].

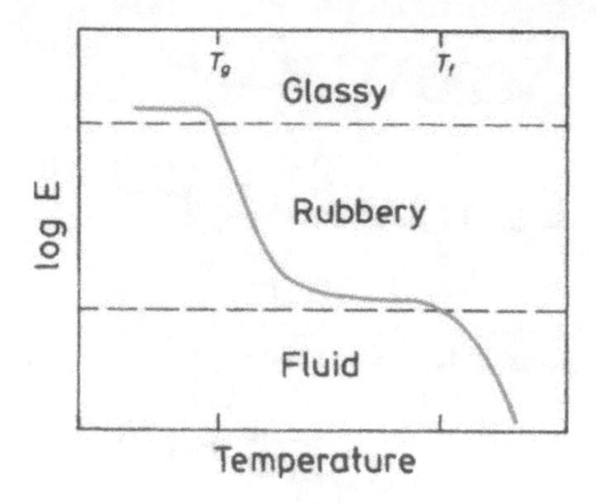

FIGURE 1.5 Variation of Modulus of an Amorphous Polymer with Temperature [2].

For instance, Figure 1.5 shows the change of the natural logarithm of elastic modulus as a function of temperature. Both transitions from the glassy to rubbery and rubbery to fluid states resulted in a clear discontinuity in the modulus. The T_g is directly dependent on the inherent structure of any polymer, hence if a polymer has a rigid backbone structure and/or bulky branch groups, then its T_g value will be relatively high. The T_g phenomenon is also detected for amorphous ceramics, such as glasses.

1.4 THE MOLECULAR WEIGHT OF POLYMER

The molecular weight of polymers is one of the most relevant parameters that determines their properties. Several mechanical properties increase with the increasing of molecular weight, whereas their processing will be very difficult due to the increase of their viscosities. The second key parameter is degree of polymerization (D_P), which determines the number of mers (basic unit) in any polymer. The molecular weight (M_W) and the D_P are linked by relationship below:

$$M_w = D_p x(M_w)u \tag{1.1}$$

where $(M_W)_u$ is the molecular weight of the repeating unit.

Due to the numerous types of molecules in each polymers, different M_w and D_P can be determined, and the total M_W polymer is obtained by a distribution function. A narrower distribution reflects a homogeneous polymer having many repeating identical units, but a wider distribution means that the polymer is constituted from of a large number of different units. In fact, it is better to specify an average molecular weight or degree of polymerization. Unlike many common substances with low molecular weight, polymers can have high molecular weights. For instance, the molecular weights of natural rubber and polyethylene can reach greater than 10^6 and 10^5 dalton, respectively.

Polymers can be amorphous or semi-crystalline, and the degree of crystallinity in any polymer can range between 30% and 90%. The elastic modulus and strength of a polymer increases when its crystallinity increases, although a totally crystalline polymer is unrealistic. Longer and complicated molecular chains in polymers can easily tangle, presenting large segments of molecular chains that may still be trapped between crystalline parts and cannot reorganize themselves into an ordered molecular assembly feature of a fully crystalline state. The molecular arrangement in a polymer can also be affected by the degree of crystallization; linear molecules with small side groups can easily crystallize, whereas branched-chain molecules with bulky side groups never easily crystallize. Therefore, a linear high-density polyethylene can reach up to 90% crystallization, but a branched polyethylene can attain only about 65% crystallization.

Polymerization of polymeric monomers occurs either through condensation or through addition of a catalyst. In condensation polymerization, a stepwise reaction of molecules occurs, and in each step a molecule of a simple compound, generally water, forms as a by-product. In addition to polymerization, monomers can be joined to form a polymer with the help of a catalyst without producing any by-products. For example, the linear addition of ethylene molecules (-CH_2) results in polyethylene, with the final mass of polymer being the sum of monomer masses.

1.5 POLYMER MATRIX-BASED COMPOSITES

Polymer matrix–based composites (PMCs) are used mainly to hold the reinforcing phase and protect it from the environment. The reinforcing in a PMC provides high strength and stiffness compared to a ceramic matrix composite, where the reinforcement is used for improving the fracture toughness. The PMC is tailored so that the mechanical charge to which the structure is applied is primarily supported by the reinforcing agent. Another role of the polymeric matrix is to gather fibers and to transfer loads between them. Polymer matrix–based composites are generally divided into three categories: thermoplastic matrix composites, thermosetting matrix composites, and elastomeric matrix composites. The distinction is based on the processing technique, type of reinforcing phase, cost, and the final use of the resulted composite.

Polymer matrix–based composites offer several advantages over ceramic matrices and metallic matrix–based composites; this includes their light weight coupled with high stiffness and strength along the direction of the reinforcement. This excellent combination makes polymer composites very useful for aeronautics, automobiles, and other moving structures. In addition, the superior corrosion-resistance and fatigue-resistance compared to metallic matrices are other advantages of polymeric composites, which widens their uses in coating and painting applications. However, a polymer matrix can easily decompose at a higher temperature range; this limits its service temperature below about 350°C. Researchers are looking to enhance this temperature limit through the development of new bulky polymer structures or by reinforcing with nanoparticles like carbon nanotubes, graphene, and ceramics.

1.5.1 Thermoplastic Matrices

Thermoplastic polymer matrices can be softened or plasticized repeatedly by applying a thermal energy, without major changes in their properties if treated with certain precautions. Such polymers are polyolefin, nylons, linear polyesters and polyether,

PVC, sealing wax, etc. Thermoplastic resins matrices or engineering plastics, like some polyesters, polyetherimide, polyamide imide, polyphenylene sulfide, poly(ether ketone)(PEEK), and liquid crystal polymers consist of long, discrete molecules that melt to a viscous liquid at the processing temperature range of 260°C to 371°C. However, after forming, they are cooled to an amorphous, semi-crystalline, or crystalline solid [3].

The degree of crystallinity significantly affects the final properties of any thermoplastic matrix. Contrary to the curing process of thermosetting matrices, the processing of thermoplastics is reversible, hence the resin can be reformed into any other shape by a simple reheating to the processing temperature. However, thermoplastic polymers generally exhibit inferior high-temperature strength and chemical stability compared to thermosets, while they are more resistant to both cracking and impact damage. However, recently, synthesized high-performance thermoplastics, like the semi-crystalline PEEK, showed excellent high-temperature strength and better solvent resistance.

From a manufacturing aspect, thermoplastics provide a promising solution for the future as they can be cooled rapidly. This promotes thermoplastic matrices for use in a number of industries, especially automotive. Today, thermoplastics are widely used with discontinuous fiber fillers, such as chopped glass or carbon/graphite. Table 1.1 summarizes the properties and products of some thermoplastic polymers.

TABLE 1.1
Thermoplastic Properties and Products

Plastic Name	Products	Properties
Polyamide	Bearings, gear wheels, casings for power tools, hinges for small cupboards, curtain rail fittings and clothing	Creamy color, *tough*, fairly *hard*, resists wear, *self-lubricating*, good resistance to chemicals and machines
Polymethyl methacrylate	Signs, covers of storage boxes, aircraft canopies and windows, covers for car lights, wash basins and baths	Stiff, hard but scratches easily, durable, brittle in small sections, good electrical insulator, machines and polishes well
Polypropylene	Medical equipment, laboratory equipment, containers with built-in hinges, "plastic" seats, string, rope, kitchen equipment	Light, hard but scratches easily, tough, good resistance to chemicals, resists work fatigue
Polystyrene	Toys, especially model kits, packaging, "plastic" boxes and containers	Light, hard, stiff, transparent, brittle, with good water resistance
Low-density polyethylene (LDPE)	Packaging, especially bottles, toys, packaging film, and bags	Tough, good resistance to chemicals, flexible, fairly soft, good electrical insulator
High-density polyethylene (HDPE)	Plastic bottles, tubing, household equipment	Hard, stiff, able to be sterilized

1.5.2 Thermosetting Polymers

Thermosetting and thermoplastic polymeric matrices are used for producing various high-performance composites materials. However, due to their high modulus, strength, durability, thermal stability, and chemical resistance imparted by high cross-linking density, thermosetting matrices are preferred over thermoplastic ones in the composites industries [4–13]. Various methods are undertaken in the preparation of resins, including polycondensation, polyaddition, and thermal cycling. The polymerization of thermosetting monomers is generally made by heat in the presence of or in absence of a catalyst. The properties and applications of some thermosetting polymers are collected in Table 1.2.

As seen in Table 1.2, thermosetting resins have many outstanding properties which make them suitable to be matrices for a versatile polymer composites. These thermosetting composites make them suitable for a number of industrial, engineering, and aerospace applications such as protective coatings, structural adhesives, electrical laminates, and matrices for advanced composite materials.

TABLE 1.2
Properties and Application of Some Thermosetting Polymers

Thermoset	Properties	Applications
Epoxy	High strength and modulus, good adhesion to various substrates, easy processability, extreme durability, excellent chemical resistance and stability under UV exposure	Coatings, electrical, marine, automotive, aerospace and civil infrastructure
Polyimide	Good processability, solvent resistant, good thermal stability	Microelectronics and aerospace
Phenolic	Good heat and flame resistance, ablative characteristics, and low cost	Electronic ablation, abrasive, adhesive, coatings
Cyanate ester	Low k, low water uptake, excellent mechanical properties, dimensional and thermal stability, flame resistance, good adhesive properties, high T_g	Microelectronics and aerospace
Benzoxazine	Near-zero shrinkage, low water absorption, higher T_g, high char yield, and no release of by-products during curing	Microelectronics, aerospace, automobiles, civil infrastructure
Unsaturated polyester	High strength and modulus, excellent chemical, solvent and salt water resistance, good thermal and electrical properties, outstanding adhesion to substrates	Coatings
Bismaleimide	Excellent thermal and oxidative stability, electrical and mechanical properties, and relatively low propensity to moisture absorption	Aerospace composites
Phthalonitrile	High thermal and oxidative stability, absence of T_g before the thermal decomposition, strength to weight advantages, low water uptake	Marine, aerospace

1.5.3 Elastomers

Elastomers are high molar-weight materials able to undertake both physical and chemical transformations as heat is used, which generally causes undesirable changes in the final properties of these polymers. The most important characteristic of elastomeric materials is elastic recovery when they are exposed to a mechanical deformation in compression or tension mode [14–16]. In addition, they show a good toughness under both static and dynamic stresses, high abrasion-resistance, and improved swelling-resistance to water and chemical solvents. Similar to other polymers, elastomers have viscoelastic properties, which can be tailored for many relevant applications, such as tyres, vibration and shock isolation, and damping. These outstanding properties are prevented over a wide range of temperatures and are stable under climatic change conditions and even in ozone-rich atmospheres.

Rubbers are elastomeric compounds made of various monomer units consisting of polymers that are thermally cured to enhance their toughness and resilience. Figure 1.6 illustrates the various steps for processing rubbers. Rubbers can also easily adhere to many kinds of other materials, resulting in different hybrid structures. When combined with engineering fibers, like rayon, polyamide, polyester, glass, or steel-cord, their tensile strengths are significantly enhanced with a decrease in extendibility. By linking elastomers to metals, their resulting hybrid materials showed a good elasticity from elastomers and improved rigidity from metals. The desired property of any elastomer-based product depends on the choice of

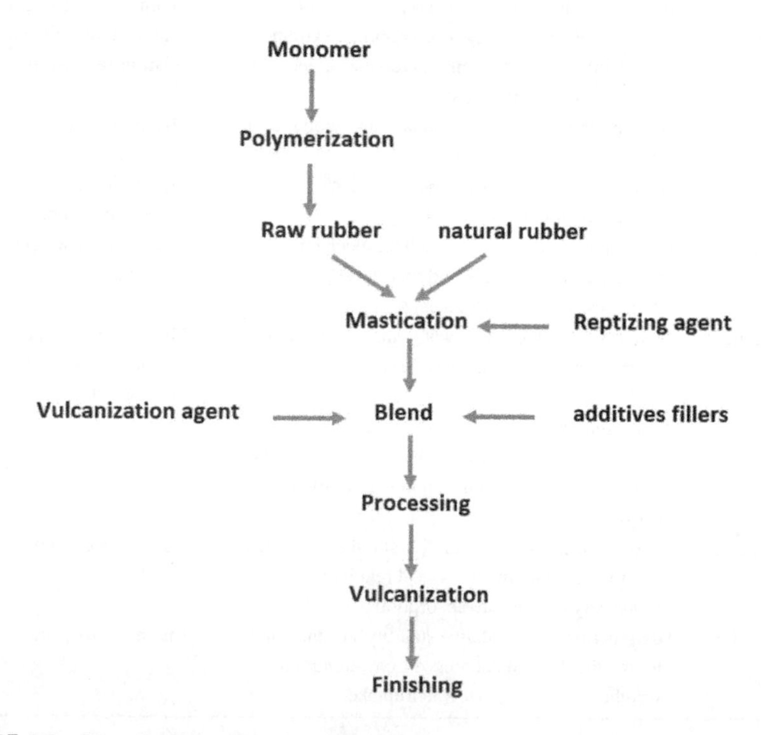

FIGURE 1.6 Manufacturing Process of Rubber.

the suitable rubber, the composition, the elaboration technique, and its final shape. According to the type and fraction of rubbers and additives in a compound, it can be vulcanized with many different mechanical features, including hardness, elasticity, and strength.

In practice, the viscoelasticity of elastomers is easily identified. By stretching a cross-linked elastomeric band, temperature increases in the band from friction of viscous deformation. This energy, which is responsible for the recovery of deformed rubber, is related to the entropy of the elastomeric material [17, 18].

In elastomers, the temperature domain of the elastic behavior is limited by the glass transition temperature (T_g). Below the glass transition temperature, the motion of rubbery molecular chains is significantly restricted, and only low elastic deformations can be observed. Thus, elastomeric polymers are very rigid and fragile materials below the T_g [19]. The thermal expansion coefficient (CTE) of rubbers varies considerably according to the elastomer structure, the type of reinforcing fillers used, and the inherent properties of the rubber compound. In the most cases, the linear CTE of elastomers is five- to twenty-fold compared to that of steels [20]. By consequence, the heat shrinkage of molded elastomeric products can reach several percent [21, 22].

1.5.4 Biopolymers

Due to environmental awareness of synthetic polymers in terms of environmental pollution, depletion of fossil resources, and greenhouse gas emissions, alternative polymers issued from renewable resources are in great demand to replace these antimicrobial materials [23, 24], especially for use in packaging and disposable domains. These renewable bioresource polymers are known as biopolymers, which are extensively utilized as additives to ameliorate the antimicrobials, antioxidants, antifungal agents, color, and other nutrients [25]. Under some conditions of temperature, moisture, and oxygen availability, biodegradation can provoke fragmentation or disintegration of these polymers without generating toxic or hazardous residues [26]. Biopolymers can be subdivided into several categories based on the type of production process and the origin of raw materials. These classes include natural biopolymers, synthetic biodegradable polymers, and biopolymers prepared by microbial fermentation [27].

Biopolymers, such as aliphatic polyesters, which have repeating units that are linked via ester bound and can be degraded by enzymes, are actually fabricated on a semi-commercial scale by several companies that produce biodegradable plastics [28, 29]. Polylactic acid (PLA), polyglycolate (PGA), poly(3-hydroxybutyrate), polycaprolactone (PCL), polyhydroxyvalerate (PHV), and their copolymers poly(butylene succinate) (PBSu), poly(ethylene succinate) (PESu), poly(propylene adipate) (PPAd), etc., are the most commonly used aliphatic polyesters in packaging applications, mulch films, tissue engineering, implants, drug delivery, etc. [30]. Compared to synthetic polymers, biopolymers are produced in relatively small quantities but with increasing use.

The most useful biopolymers in our daily life are cellulose, chitin, and chitosan [31]. Figure 1.7 shows the structures of these biopolymers. Cellulose is the most abundant

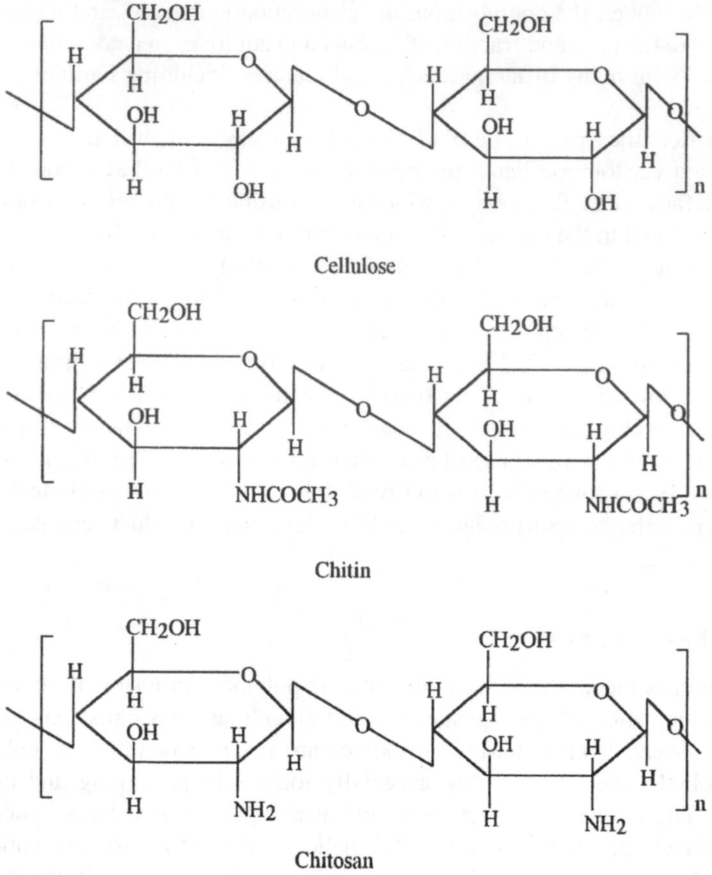

Cellulose

Chitin

Chitosan

FIGURE 1.7 Structures of Cellulose, Chitin, and Chitosan [31].

renewable biopolymer representing about 40% of all organic matter. It is a skeletal polysaccharide ubiquitous in the plant kingdom and is considered the commonest naturally-occurring crystalline polymer. It is usually in a fibrous structure, and when it is used to reinforce amorphous lignin and hemicelluloses, it provides a woody composite structure. The primary structure of cellulose is mainly a regular unbranched linear sequence of 1→4-linked β-D-glucose. Microfibrils can be formed by the neighboring chains through hydrogen-bonding phenomena.

Chitin is a natural amino polysaccharide which is fabricated yearly, almost like cellulose, while chitosan is produced through the deacetylation reaction of chitin. Both chitin and chitosan biopolymers show different structures according to the various chemical and mechanical modifications to produce novel properties, functions, and applications [31]. They are useful materials for biomedical applications due to their biocompatibility, biodegradability, and non-toxicity, apart from their antimicrobial activity and low immunogenicity. These biopolymers can be processed into different forms such as gels, sponges, membranes, beads, and scaffolds [32, 33].

1.6 PROPERTIES OF POLYMERIC MATRICES

In this section, the various mechanical, thermal, barrier, and electrical properties of polymeric matrices will be described along with some examples.

1.6.1 Mechanical Properties of Polymers

Amorphous polymers follow Hooke's law and exhibit a linear elastic response to applied stress. The elastic strain in glassy polymers never exceeds 1%. However, elastomeric polymers exhibit a nonlinear elastic behavior with a huge elastic range (a few percent strain). The large elastic strain in elastomers is due to a redistribution of the tangled molecular chains under the applied stress.

Highly cross-linked thermosetting resins such as epoxies, polyimides, and polyesters, show a high modulus and strength, but they are extremely brittle. The fracture energy of thermosetting resins (100 to 200 $J.m^{-2}$) is only about 10 times that of inorganic glasses (10 to 30 $J.m^{-2}$). However, thermoplastic resins like polymethylemethacrylate (PMMA) exhibit fracture energies on the order of 1,000 $J.m^{-2}$ due to their large free volume, which facilitates the absorption of the energy associated with crack propagation. The fracture toughness of hard and brittle thermosets can be improved by adding small amounts of soft rubbery particles into the brittle matrix through a simple mechanical blending technique or the copolymerization of their mixture.

Polymers are viscoelastic materials exhibiting combined properties of solids and viscous liquids. These characteristics are both time- and temperature-dependent. In addition, several other factors can influence the mechanical properties of polymers, such as the molecular weight, processing technique, extent and distribution of crystallinity, constitution of polymer, and maximum use temperature [34–37]. For example, it is reported that mechanical properties of amorphous polymers, like thermosetting polymers, can be increased rapidly by increasing their molecular weight as depicted in Figure 1.8 [38]. However, for thermoplastic materials having

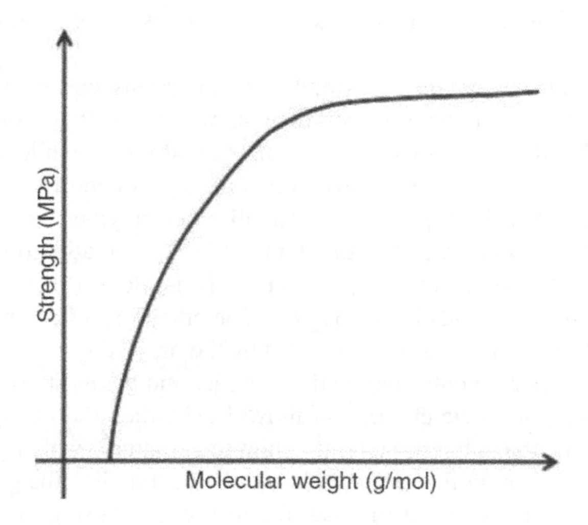

FIGURE 1.8 Variation of Tensile Strength with Molecular Weight of the Polymer [38].

high molecular weights, exceeding threshold region, moderate changes in molecular weight do not significantly affect their mechanical properties, such as yield stress or modulus.

In addition, the modulus of polymers also increases as the cross-link density of polymers rises. Crystalline and filled polymers resemble polymers with low cross-link density at temperatures below the melting temperature, T_m. On the other hand, the crystallinity of the polymer improved the strength because in the crystalline phase the intermolecular bonding is more abundant. Thus, the deformation of a polymer resulted at higher strength provoking oriented chains. The mechanical properties of the polymer are significantly dependent on the temperature. Generally, by increasing the temperature, the elastic modulus and tensile strength are reduced, while the ductility is enhanced. In addition, at a fixed temperature, when the polymer is strained to a given value, then the stress needed to keep this strain decreases with time. This phenomenon is known as stress relaxation and occurs because the molecules of polymers relax with time, and to maintain the level of strain, somewhat smaller values of stress are required.

Polymers reveal a strong temperature dependence of their elastic moduli. Under the T_g, the polymer behaves as a hard and rigid solid, with an elastic modulus as high as 5 to 7 GPa. Above T_g, this modulus decreases sharply, and the polymer exhibits a rubbery behavior. Upon heating the polymer to above its fusing temperature, T_m, at which the polymer transforms to a fluid, the modulus abruptly drops.

1.6.2 Thermal Properties of Polymers

The glass transition temperature (T_g) is characterized as the amorphous region of any polymer, but the crystalline region is defined by the melting point. In addition, the T_g is the second-order transition while the melting point is the first-order transition. The value of T_g is not unique because the glassy state is not in equilibrium, and it is related to many factors, including molecular weight, measurement method, as well as the rate of heating or cooling. Approximate T_g values of some polymers are collected in Table 1.3.

The semi-crystalline polymers exhibit two transitions corresponding to their crystalline and amorphous regions. So, they show true melting temperatures (T_m) at which the ordered phase transforms to a disordered phase, while the amorphous regions soften over a temperature range called the T_g. In contrast, amorphous polymers do not have a melting point, whereas all other polymer types possess a T_g. The T_m of a polymer is enhanced as the double bonds, aromatic groups, and bulky or large side groups existed on the polymeric chains, due to their high ability to restrict the flexibility of these chains [39–47]. The branching of chains provokes the decrease of T_m, because defects are generated by the branching.

The T_g of a polymer is controlled by the mobility and flexibility of the polymeric chains [39]. If the polymeric chains can move freely, then the glassy state can be easily converted to the rubbery state at a low temperature, which gives a low T_g value. However, if the mobility of these chains is reduced, then the glassy state will be more stable, and it is difficult to break the restriction causing the immobility of

TABLE 1.3
Glass Transition Temperatures of Some Polymers [38]

Polymer	T_g (°C)
Polytetrafluoroethylene	−97
Polypropylene (isotactic)	+100
Polystyrene	+100
Poly(methylmethacrylate) (atactic)	+105
Nylon 6,6	+57
Polyethylene (LDPE)	−120
Polyethylene (HDPE)	−90
Polypropylene (atactic)	−18
Polycarbonate	+150
Poly(vinyl acetate) (PVAc)	+28
Polyester(PET)	+69
Poly(vinyl alcohol) (PVA)	+85
Poly(vinyl chloride) (PVC)	+87

the polymer chains at the lower temperature, thus high energy is needed to make the chains free, which increases the T_g.

- Intermolecular forces: Strong intermolecular forces lead to increased T_g.
- Chain stiffness: The presence of the stiffening groups in the polymer chain decreases the flexibility of the chain, causing high T_g value.
- Cross-linking: The cross-links between chains inhibit rotational motion, which increases the T_g.
- Bulky pendant groups: the presence of bulky pendant groups, such as a benzene ring, can restrict rotational freedom, leading to higher T_g [40].
- Flexible pendant groups: the presence of flexible pendant groups limits the packing of the chains by increasing the rotational motion, giving a lower T_g value [41].
- Plasticizers: they increase their chain flexibility by reducing the intermolecular cohesive forces between the polymer chains, which results in a reduce T_g.
- Molecular weight: T_g is enhanced with the molecular weight.

The thermal decomposition of polymers can proceed by heat alone, or by the combined effect of both heat and oxygen [49–53]. In the majority of polymers, the thermal decomposition processes are boosted by oxygen, decreasing the minimum decomposition temperatures [38]. Bubbles of volatile fuel are detected during the thermal decomposing of polymers, around the time of ignition or even prior to ignition; many thermoplastics demonstrate that most volatile formation issues from the bulk of the polymer, not its surface, and thus the critical decomposition condition is still anaerobic. The thermal decomposition of polymers can be subdivided into several general chemical mechanisms. The first three mainly characterize the conversion of an involatile polymer macromolecule into several fragments sufficiently small to be volatile.

1. Random-chain scission, in which chain scissions occur at apparently random locations in the polymer chain.
2. End-chain scission, in which individual monomer units are successively removed at the chain end.
3. Chain-stripping, in which atoms or groups not part of the polymer chain (or backbone) are cleaved.
4. Cross-linking, in which bonds are generated between polymer chains.

The combustion trend of a polymer can be explained in terms of the properties of the volatiles, including their composition, reactivity, and rate of formation. Thermal stability can be quantitatively evaluated from the temperature dependence of decomposition. Detailed studies by Madorsky [48] in the 1960s of the effects of chemical structure on the thermal stability of polymeric materials summarizes the factors controlling the thermal decomposition of polymers. Table 1.4 lists the effects of chemical structure on the thermal stability of polymers, and gives several examples of that behavior.

TABLE 1.4
Factors Affecting the Thermal Stability of Polymers

Factor	Effect on Thermal Stability	Examples	T_h/K
Chain branching	Weakens	Polymethylene	688
		Polyethylene	679
		Polypropylene	660
		Polyisobutylene	621
Double bonds in polymer backbone	Weakens	Polypropylene	660
		Polyisoprene	596
Aromatic ring in polymer backbone	Strengthens	Poly-1,4-phenylene methylene	703
		Polystyrene	637
High molecular weight	Strengthens	PMMA B ($M_W = 5.1 * 10^6$)	600
		PMMA A ($M_W = 1.5 * 10^5$)	556
Cross-linking	Strengthens	Polydivinyl benzene	672
		Polystyrene	637
Oxygen in the polymer backbone	Weakens	Polymethylene	688
		Polyethylene oxide	618
		Polyoxymethylene	<473

Note:
T_h: Temperature degradation, at which 50% of a small polymer sample will volatilize in 30 min in an inert atmosphere.

1.6.3 Electrical Properties of Polymers

Compared to metals, polymers are good insulating materials, however some new types of them are conductors. These polymers are called conjugated polymers or conductive polymers. They exhibit an electrical conductivity in the range of 1 to 10^5 S/cm. Figure 1.9 contains some structure of conjugated polymers with their electrical conductivities [54].

Polymers can exhibit both a low or high dielectric constant, and it directly depends on their inherent permittivity, which represents the ability of a material to polarize under the electrical field. Generally, when the dielectric constant is high, polarization will be huge in the electric field [55–58]. Aromatic polymers and polyvinyl fluoride have very low dielectric materials and are widely used for capacitors. However, the presence of highly polarizable groups like aromatic rings, bromine, and iodine will enhance the dielectric constant of polymers [59]. Table 1.5 summarizes the dielectric permittivity of various polymers.

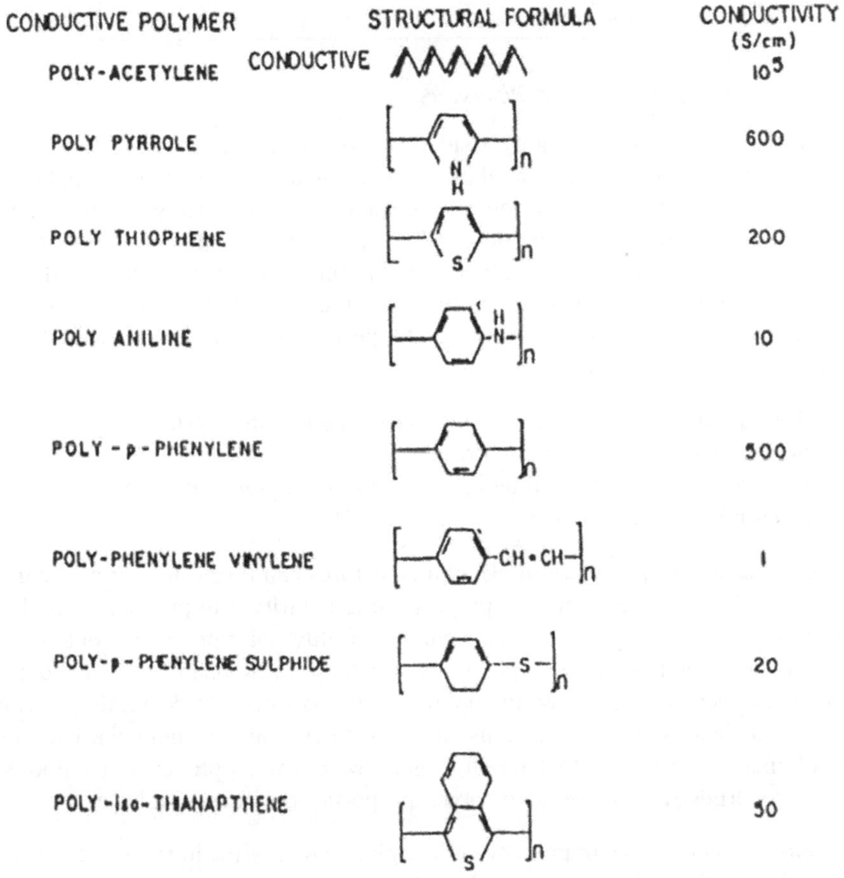

CONDUCTIVE POLYMER	STRUCTURAL FORMULA	CONDUCTIVITY (S/cm)
POLY-ACETYLENE	CONDUCTIVE	10^5
POLY PYRROLE		600
POLY THIOPHENE		200
POLY ANILINE		10
POLY-p-PHENYLENE		500
POLY-PHENYLENE VINYLENE	-CH•CH-	1
POLY-p-PHENYLENE SULPHIDE	-S-	20
POLY-iso-THIANAPTHENE		50

FIGURE 1.9 Conductivity of Some Conjugated Polymers [54].

TABLE 1.5
Low Dielectric Polymers

Materials	Dielectric Constant
Non-fluorinated aromatic polyimides	3.2–3.6
Fluorinated polyimide	2.6–2.8
Poly(phenyl quinoxaline)	2.8
Poly(arylene ether oxazole)	2.6–2.8
Polyquinoline	2.8
Silsesquioxane	2.8–3.0
Poly (norborene)	2.4
Perfluorocyclobutane polyether	2.4
Fluorinated poly (arylene ether)	2.7
Polynaphthalene	2.2
Poly (tetrafluoroethylene)	1.9
Polystyrene	2.6
Poly(vinylidene fluoride-co hexafluoropropylene)	12
Poly(ether ketone ketone)	3.5

1.6.4 Barrier Properties of Polymers

Barrier properties of polymers are considered a critical physical property to be understood and optimized for the potential uses of their materials in numerous application areas. The barrier properties of polymers evaluate the resistance to the permeation of gases, liquids, and vapor compounds. The general theory of permeation of a gas or liquid through a polymer [60] indicates that permeation is the result of a diffusion term and a solubility constant of the permeant in the polymer, both of which are usually independent of each other. The process of permeation through a polymer barrier constitutes four main steps:

- Absorption of the permeating species into the polymer surface;
- Solubility into the polymer matrix;
- Diffusion through the wall along a concentration gradient;
- Desorption from the outer surface.

A good design of polymeric molecular structures can result in improved barrier properties. However, generally the property, like polarity, can produce a good gas barrier, while provoking a poor water barrier. Highly polar polymers containing a huge amount of hydroxyl groups (poly(vinyl alcohol) or cellophane) are excellent gas barriers, but they are considered the poorest water barriers. Moreover, they become poor gas barriers as they plasticize using water. In contrast, the nonpolar hydrocarbon polymers, such as polyethylene, show good water barrier properties but poor gas barrier properties. To achieve good barrier properties polymer must have:

- Appropriate degree of polarity comparable to that of the nitrile radical, the chlorine, fluorine, acrylic, or ester group;
- High chain stiffness;

- Inertness;
- Close chain-to-chain packing by symmetry, order, crystallinity, or orientation;
- Good bonding or attraction between chains;
- High T_g.

For a given polymer, the permeability rate of gas or liquid is related to several parameters, such as chemical composition, temperature, molecular structure, molecular weight, humidity, PH, additives, etc. The permeation of gas changes significantly depending upon the chemical composition of the polymer [61–69]. This is the key single factor in controlling the barrier properties of polymers. The effect of packing is also important. For example, linear polyethylene, which has a simple molecular structure with good packing, has an improved O_2 permeation compared to polymethyl pentene, which exhibits poor packing features [70, 71]. Crystallinity plays an equally crucial role in the barrier performance of polymers, hence if a polymer can undergo more than one state of crystallinity, like polyamide, it will show enhanced barrier in the more crystalline state because the crystallites are impermeable. For crystalline polymers, molecular orientation can markedly affect permeation performance. In amorphous polymers, the orientation reduces the permeation by only 10% to 15%, while for crystalline ones, a decrease of more than 50% has been reported.

The molecular structure of the permeating gas or liquid is another determinant factor in permeation rates of polymer. Concerning liquids, three parameters control their permeation through a polymer [72–76], which are:

- Molecular size of the liquid: small molecules permeate more easily than large molecules.
- Molecular shape: streamlined shapes (*p*-xylene) penetrate more rapidly than bulky shapes (o-xylene).
- Polarity of the liquid: nonpolar molecules (toluene) permeate more rapidly than highly polar molecules (aniline) in nonpolar polymers. This latter effect is reversed in the case of polar polymers.

1.7 CONCLUSIONS

Polymers can be divided into thermoplastics, thermosets, and elastomers and have been widely used as engineering materials, not only for making chairs and tables but also in aerospace, high-voltage insulation, and automotive parts. This is because they have numerous advantages over metallic and ceramic materials such as light weight, a low cost, ease of processing, etc. However, due to some inferior characteristics like thermal and mechanical performances for high-performance applications, they are generally reinforced with inorganic fillers or engineering fibers to create polymeric composites. The introduction of high-performance ceramics micro- and nanofillers resulted in significant new and improved properties, including good mechanical properties, low CTE values, high thermal conductivity, high electrical conductivity, good anticorrosion properties, and improved barrier properties. In addition, the introduction of combined systems of ceramics and engineering fibers like carbon, basalt, glass, and aramid fibers into various polymeric matrices resulted in a huge improvements in the final properties of their composites.

REFERENCES

1. Mitchell, B. S. 2004. *An Introduction to Materials Engineering and Science for Chemical and Materials Engineers.* Hoboken: Wiley-Interscience.
2. Chawla, K. K. 2012. *Composite Material—Science & Engineering.* New York: Springer-Verlag.
3. Lee, W. I., Springer, G. S. 1987. A model of the manufacturing process of thermoplastic matrix composites. *Journal of Composite Materials* 21: 1017–1055.
4. Lin, S. C., Pearce, E. M. 1994. *High Performance Thermosets: Chemistry, Property and Applications.* Munich: Hanser.
5. Kopf, P. W., Little, A. D. 1988. Phenolic resins In: Mark, H. F., Bikales, N. M., Overberger, C. G. *Encyclopedia of Polymer Science and Engineering*, 2nd ed. Wiley, New York, vol 11, p 45.
6. Knop, A., Pilato, L. A. 1985. *Phenolic Resins: Chemistry, Applications and Performance, Future Directions.* Berlin Heidelberg: Springer.
7. Nair, C. P. R., Bindu, R. L., Ninan, K. N. 1997. Recent advances in phenolic resins, In: *Metals, Materials and Processes.* Mumbai, India: Meshap Science Publishers, vol 9(2), p 179.
8. Knop, A., Schieb, W. 1979. *Chemistry and Applications of Phenolic Resins.* Berlin Heidelberg: Springer.
9. Hamerton, I. 1994. *Chemistry and Technology of Cyanate Ester Resins.* Glasgow: Blackie Academic and Professional.
10. Bauer, M., Bauer, J. 1994. Aspects of the kinetics, modelling and simulation of network build-up during cyanate ester cure. In: Hamerton, I. (ed.) *Chemistry and Technology of Cyanate Ester Resins.* Glasgow: Blackie Academic and Professional, Glasgow, p 58.
11. Galy, J., Gerard, J. F., Pascault, J. P. 1991. In: Abadie M. J. M. Sillion B (Eds.). *Polyimides and Other High Temperature Polymers.* Amsterdam: Elsevier Science, p 245.
12. Santhosh Kumar, K. S., Nair, C. P. R. 2010. *Polybenzoxazines: Chemistry and Properties. Blends and Composites of Polybenzoxazines.* England, Shawbury, Shrewsbury, Shropshire: iSmithers Rapra Publishing.
13. Rumdist, S., Jubsilp, C., Tiptipakorn, S. 2013. *Alloys and Composites of Polybenzoxazine Properties and Applications.* Singapore: Springer.
14. Martinoty, P., Stein, P., Finkelmann, H. et al., 2004. Mechanical properties of mono-domain side chain nematic elastomers. *The European Physical Journal E* 14: 311–321.
15. Termonia, Y. 1989. Molecular model for the mechanical properties of elastomers. 1. Network formation and role of entanglements. *Macromolecules* 22(9): 3633–3638.
16. Ferry, J. D., Fitzgerald, E. R., Grand, L. D., Williams, M. L. 1952. Temperature dependence of dynamic mechanical properties of elastomers, relaxation distributions. *Industrial and Engineering Chemistry* 44(4): 703–706.
17. Barbin, W. W., Rodgers, M. B. 1994. The science of rubber compounding. In: Mark, J. E., Erman, B., Eirich, F. R. (Eds.). *Science and Technology of Rubber.* San Diego, CA: Academic Press: 419–469.
18. White, J. L. 1969. Elastomer rheology and processing. *Rubber Chemistry and Technology* 42: 257–338.
19. Zlatkevich, L.Y., Nikolskii, V. G. 1973. Dependence of the glass transition temperature on the composition of elastomer mixtures. *Rubber Chemistry and Technology* 46: 1210–1217.

20. Shubin, S. N., Freidin, A. B., Akulichev, A. G. 2016. Elastomer composites based on filler with negative thermal expansion coefficient in sealing application. *Archive of Applied Mechanics* 86: 351.
21. Bhowmick, A. K. 2008. *Current Topics in Elastomers Research.* Boca Raton, FL: CRC Press.
22. Petrović, Z. S., and Ferguson, J. Polyurethane elastomers. *Progress in Polymer Science* 16: 695–836.
23. Jenck, J. F., Agterberg, F., Droescher, M. J. 2004. Products and processes for a sustainable chemical industry: A review of achievements and prospects. *Green Chemistry* 6: 544–556.
24. Kümmerer, K. 2007. Sustainable from the very beginning: Rational design of molecules by life cycle engineering as an important approach for green pharmacy and green chemistry. *Green Chemistry* 9: 899–907.
25. Clarinval, A. M., Halleux, J. 2005. *Classification of Biodegradable Polymers*, pp. 3–31. Boca Raton, FL: CRC Press.
26. Chandra, R., Rustgi, R. 1998. Biodegradable polymers. *Progress in Polymer Science* 23: 1273–1335.
27. Rhim, J. W., Park, H. M., Ha, C. S. 2013. Bio-nanocomposites for food packaging applications. *Progress in Polymer Science* 38: 1629–1652.
28. Bikiaris, D. N. 2013. Nanocomposites of aliphatic polyesters: an overview of the effect of different nanofillers on enzymatic hydrolysis and biodegradation of polyesters. *Polymer Degradation and Stability* 98: 1908–1928.
29. Shimao, M. 2001. Biodegradation of plastics. *Current Opinion in Biotechnology* 12: 242–247.
30. Gandini, A. 2008. Polymers from renewable resources: A challenge for the future of macromolecular materials. *Macromolecules* 41: 9491–9504.
31. Ravi Kumar, M. N. V. 2000. A review of chitin and chitosan applications. *Reactive & Functional Polymers* 46: 1–27.
32. Koh, H. C., Park, J. S., Jeong, M. A. et al. 2008. Preparation and gas permeation properties of biodegradable polymer/layered silicate nanocomposite membranes. *Desalination* 233: 201–209.
33. Trznadel, M. 1995. Biodegradable polymer materials. *International Polymer Science & Technology* 22: 58–65.
34. Nunes, R. W., Martin, J. R., Johnson, J. F. 1986. Influence of molecular weight and molecular weight distribution on mechanical properties of polymers. *Polymer Engineering Science* 22: 205–228.
35. Seitz, J. T. 1993. The estimation of mechanical properties of polymers from molecular structure. *Journal of Applied Polymer Science* 9: 1331–1351.
36. Perego, G., Cella, G. D., Bastioli, C. 1996. Effect of molecular weight and crystallinity on poly(lactic acid) mechanical properties. *Journal of Applied Polymer Science* 59: 37–43.
37. El-Hadi, A., Schnabel, R., Straube, E., Müller, G., Henning, S. 2002. Correlation between degree of crystallinity, morphology, glass temperature, mechanical properties and biodegradation of poly (3-hydroxyalkanoate) PHAs and their blends. *Polymer Testing* 21: 665–674.
38. Balani, K., Verma, V., Agarwal, A., Narayan, R. 2015. *Biosurfaces: A Materials Science and Engineering Perspective* (1st ed.). Hoboken, New Jersey: John Wiley & Sons, Inc.
39. Chow, T. S. 1980. Molecular interpretation of the glass transition temperature of polymer-diluent systems. *Macromolecules* 13(2): 362–364.
40. Fryer, D. S., Peters, R.D., Kim, E. J., et al. 2001. Dependence of the glass transition temperature of polymer films on interfacial energy and thickness. *Macromolecules* 34(16): 5627–5634.

41. Forrest, J. A., Dalnoki-Veress, K., Dutcher, J. R. 1997. Interface and chain confinement effects on the glass transition temperature of thin polymer films. *Physical Review E* 56: 5705.
42. Forrest, J. A., Dalnoki-Veress, K., Stevens, J. R., Dutcher, J. R. 1996. Effect of free surfaces on the glass transition temperature of thin polymer films. *Physical Review Letters* 77: 2002.
43. Keddie, J. L., Jones, R. A. L., Cory, R.A. 1994. Interface and surface effects on the glass-transition temperature in thin polymer films. *Faraday Discussion* 98: 219–230.
44. Dalnoki-Veress, K., Forrest, J. A., Murray, C., Gigault, C., Dutcher, J. R. 2001. Molecular weight dependence of reductions in the glass transition temperature of thin, freely standing polymer films. *Physical Review E* 63: 031801.
45. Kwei, T. K. 1984. The effect of hydrogen bonding on the glass transition temperatures of polymer mixtures. *Journal of Polymer Science: Polymer Letters Edition Banner* 22(6): 307–313.
46. Nies, C. W., Messing, G. L. 1984. Effect of glass-transition temperature of polyethylene glycol-plasticized polyvinyl alcohol on granule compaction. *Journal of the American Ceramic Society* 67(4): 301–304.
47. Fox, T. G., Loshaek, S. 1955. Influence of molecular weight and degree of crosslinking on the specific volume and glass temperature of polymers. *Journal of Polymer Science Banner* 15(80): 371–390.
48. Madorsky, S. L. 1964. *Thermal Degradation of Organic Polymers.* New York: Interscience, John Wiley.
49. Friedman, H. L. 1963. Kinetics and gaseous products of thermal decomposition of polymers. *Journal of Macromolecular Science: Part A—Chemistry* 1: 57–59.
50. Jones, J. I. 1968. The synthesis of thermally stable polymers: A progress report. *Journal of Macromolecular Science, Part C: Polymer Reviews* 2: 303–386.
51. Bandyopadhyay, P. K., Shaw, M. T., Weiss, R. A. 1985. Detection and analysis of aging and degradation of polyolefins: A review of methodologies. *Polymer-Plastics Technology and Engineering* 24: 187–241.
52. Bockhorn, H., Hornung, A., Hornung, U. 1999. Mechanisms and kinetics of thermal decomposition of plastics from isothermal and dynamic measurements. *Journal of Analytical and Applied Pyrolysis* 50: 77–101.
53. Patel, P., Hull, T. R., McCabe, R. W., Flath, D., Grasmeder, J., Percy M. 2012. Mechanism of thermal decomposition of poly(ether ketone) (PEEK) from a review of decomposition studies. *Polymer Degradation and Stability* 95: 709–718.
54. Kumar, D., Sharma, R. C. 1998. Advances in conductive polymers. *European Polymer Journal* 34: 1053–1060.
55. Havriliak, S., Negami, S. 1967. A complex plane representation of dielectric and mechanical relaxation processes in some polymers. *Polymer* 8: 161–210.
56. Bur, A. J. 1985. Dielectric properties of polymers at microwave frequencies: a review. *Polymer* 26: 963–977.
57. Baker, W. O., and Yager, W. A. 1942. The relation of dielectric properties to structure of crystalline polymers. II. Linear polyamides. *Journal of the American Chemical Society* 64 (9): 2171–2177.
58. Haase, W., Pranoto, H., Bormuth, F. J. 1985. Dielectric properties of some side chain liquid crystalline polymers. *Berichte der Bunsengesellschaft für physikalische Chemie* 89: 1229–1234.
59. Nalwa, H. 1999. *Handbook of Low and High Dielectric Constant Materials and Their Applications.* London: Academic Press.
60. Miller, K. S., Krochta, J. M. 1997. Oxygen and aroma barrier properties of edible films: A review. *Trends in Food Science & Technology* 8: 228–237.
61. Zhang, T., Litt, M. H., Rogers, C. E. 1994. Sulfone-containing polymers as high barrier materials. *Journal of Polymer Science Part B Polymer Physics* 32: 1671–1676.

62. Lee, W. M. 1980. Selection of barrier materials from molecular structure. *Polymer Engineering & Science* 20: 65–69.
63. Lagaron, J. M., Catalá, R., Gavara, R. 2004. Structural characteristics defining high barrier properties in polymeric materials. *Materials Science and Technology* 20: 1–7.
64. Jabarin, S. A., Lofgren, E. A. 1986. Effect of water absorption on physical properties of high nitrile barrier polymers. *Polymer Engineering & Science* 26: 405–409.
65. Ashley, R. J. 1985. Permeability and plastics packaging. In: Comyn J. (ed.) *Polymer Permeability*. Dordrecht: Springer.
66. Wang, Z. F., Wang, B., Qi, N., Zhang, H.F., Zhang, L.Q. 2005. Influence of fillers on free volume and gas barrier properties in styrene-butadiene rubber studied by positrons. *Polymer* 46: 719–724.
67. Mrkić, S., Galić, K., and Ivanković, M. 2007. Effect of temperature and mechanical stress on barrier properties of polymeric films used for food packaging. *Journal of Plastic Film & Sheeting* 23: 239–256.
68. Ghosal, K., Freeman, B. D. 1994. Gas separation using polymer membranes: an overview. *Polymers for Advanced Technologies* 5: 673–697.
69. Polyakova, A., Connor, D. M., Collard, D. M., Schiraldi, D. A., Hiltner, A., Baer, E. 2001. Oxygen-barrier properties of polyethylene terephthalate modified with a small amount of aromatic comonomer. *Journal of Polymer Science Part B Polymer Physics* 39: 1900–1910.
70. Zekriardehani, S., Jabarin, S. A., Gidley, D. R., Coleman, M. R. 2017. Effect of chain dynamics, crystallinity, and free volume on the barrier properties of poly(ethylene terephthalate) biaxially oriented films. *Macromolecules* 50: 2845–2855.
71. Hu, Y. S., Hiltner, A., Baer, E. 2006. Solid state structure and oxygen transport properties of copolyesters based on smectic poly(hexamethylene 4,4′-bibenzoate). *Polymer* 47: 2423–2433.
72. Hu, Y. S., Mehta, S., Schiraldi, D. A., Hiltner, A., Baer, E. 2005. Effect of water sorption on oxygen-barrier properties of aromatic polyamides. *Journal of Polymer Science Part B: Polymer Physics* 43: 1365–1381.
73. Hu, Y. S., Hiltner, A., Baer, E. 2005. Improving oxygen barrier properties of poly(ethylene terephthalate) by incorporating isophthalate. II. Effect of crystallization. *Journal of Applied Polymer Science* 98: 1629–1642.
74. Hiltner, A., Liu, R. Y. F., Hu, Y. S., Baer, E. 2005. Oxygen transport as a solid-state structure probe for polymeric materials: A review. *Journal of Polymer Science Part B: Polymer Physics* 43: 1047–1063.
75. Stannett, V., Yasuda, H. 1963. Liquid versus vapor permeation through polymer films. *Journal of Polymer Science Part B: Polymer Physics* 1: 289–293.
76. Long, R. B. 1965. Liquid permeation through plastic films. *Industrial & Engineering Chemistry Fundamentals* 4: 445–451.

2 Ceramics: Processing, Properties, and Applications

2.1 INTRODUCTION

Ceramics are inorganic compounds constituted from both metallic and nonmetallic elements. Examples include all ceramics, such as alumina ($A1_2O_3$) (A1, metal; O, nonmetal), titanium carbide (TiC) (Ti, metal; C, nonmetal), and titanium dioxide (TiO_2) (Ti, metal; O, nonmetal). These kinds of materials are widely used in high-tech applications, and they are classified into traditional ceramics, such as clay, tile, porcelain, and glass, or modern engineering ceramics, which includes the carbides, borides, oxides, and nitrides of various elements [1]. Till now, many kinds of modern ceramics have been industrially produced. including aluminum nitride (AlN), boron carbide (B_4C), boron nitride (BN), silicon carbide (SiC), titanium diboride (TiB_2), silicon nitride (Si_3N_4), silicon oxide (SiO_2), zirconium dioxide (ZrO_2), barium titanate ($BaTiO_3$), and ceramic superconductors. The development of these technical ceramics started around the World War II era, and several innovations and equipment based on these ceramic materials have been developed to satisfy defense requirements. Compared to metals, ceramics are very hard and brittle, and their mechanical properties are more likely to degrade [2].

In this chapter, we will review the different crystalline structures of ceramics and will also provide an overview about their processing techniques, their general properties, and their most potential applications. As the aim of this book is to evaluate and study the properties of polymer/ceramic composite materials and due to the wide range of properties demonstrated by ceramics, we will only give some examples about the processing technique, the physical and mechanical properties, and the applications of ceramics, while we will cite many potential and up-to-date references from literature for those needing more details.

2.2 THE STRUCTURE OF CERAMICS

Ceramics can have either a crystalline or amorphous structure. Crystalline ceramics can be divided into several groups according to their mono-crystal structures [3]. For example, when the anions (negatively-charged nonmetallic ions) and cations (positively-charged metallic ions) form two interpenetrating FCC lattices, a coordination number of 6 (six nearest neighbors) is obtained, this structure is similar to that of the rock salt structures, including NaCl, MgO, and LiF (Figure 2.1a). However, in the cesium chloride (CsCl) structure (Figure 2.1b), the coordination number for

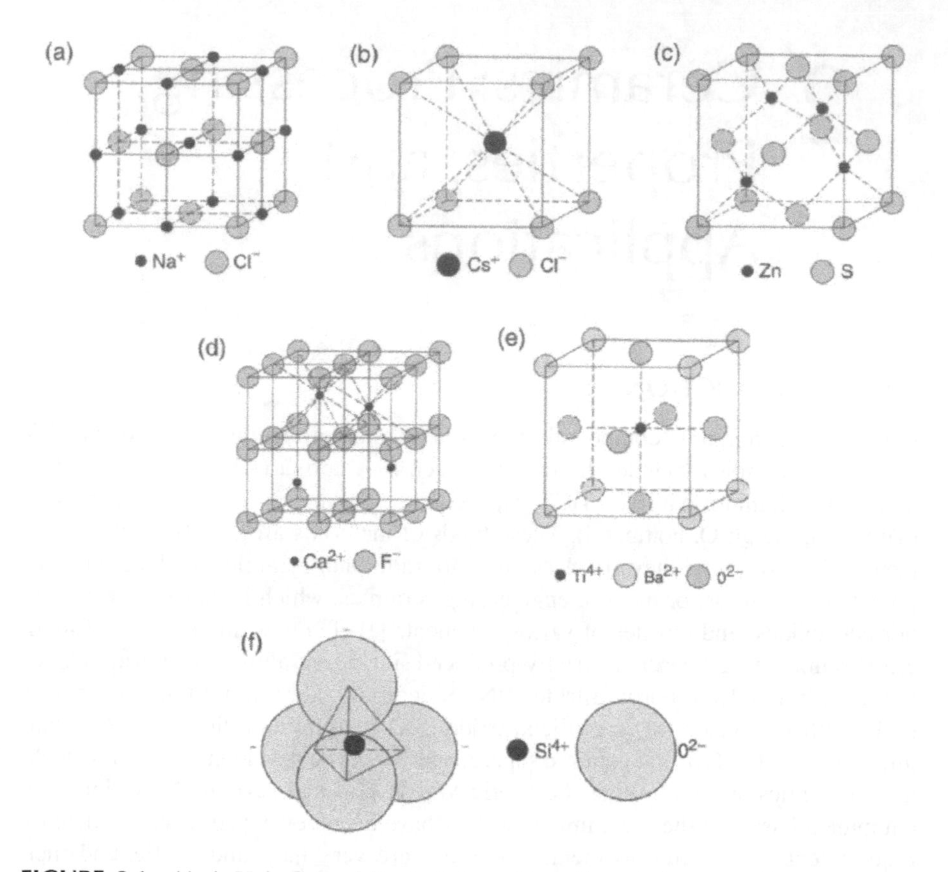

FIGURE 2.1 (a) A Unit Cell of Rock Salt, or Nacl Crystal Structure, Formed as Two Interpenetrating FCC Lattices, One Composed of The Na + Ions and the Other Cl^- Ions; (b) A Unit Cell of Cesium Chloride (CsCl) Crystal Structure with the Cs + Ion at the Center of the cell and Cl^- Ions at the Corners of the Cube; (c) A Unit Cell of Zinc Blende (ZnS) Crystal Structure. Each Zn Atom Is Bonded to Four S Atoms and Vice Versa. All Corner and Face Center Positions Are Occupied by S Atoms, and the Tetrahedral Positions Are Occupied by Zn Atoms; (d) A Unit Cell of Caf_2 Crystal Structure. Calcium Ions Are at the Centers of Cubes, and Fluoride Ions Are at the Corners; (e) A Unit Cell of Perovskite Crystal Structure Displayed by Barium Titanate ($Batio_3$). Ba^{2+} Ions Are Positioned at All Eight Corners of the Cell, and a Single Ti^{4+} Ion Is at the Center of the Cube. The O^{2-} Ions Are Located at the Center of Each of the Six Faces of the Cube. (f) A Silicon–Oxygen Tetrahedron in Which Each Si Atom Is Bonded to Four 0 Atoms, Which Are Located at the Four Corners of a Tetrahedron, with Si Atom at the Center of the Tetrahedron. The SiO Tetrahedron Is the Repeating Unit in Silicate Minerals [6].

both anions and cations is 8, with eight anions located at the corners of a cube and a single cation residing at the cube center. In the ZnS crystal structure (Figure 2.1c), all corner and face centers are taken by sulfur atoms, whereas the interior tetrahedral positions are occupied by zinc atoms. Other ceramics that have a zinc blende structure are zinc telluride (ZnTe) and silicon carbide (SiC). In the crystal structure of calcium fluoride (CaF_2) (Figure 2.1d), Ca^{2+} ions are situated at the center of cubes while the fluorine ions occupy the corner positions.

Due to the presence of unequal charges on the anion and cation (F^- and Ca^{2+}), there are twice as many F^{-1} ions as there are Ca^{2+} ions. When more than one kind of cation are presented in a ceramic compound such as barium titanate ($BaTiO_3$), a perovskite structure (Figure 2.1e) is constructed. In such a structure, Ba^{2+} ions are located at cube corners, Ti^{4+} ions at cube center, and O^{2-} ions at the center of cube faces. Another crystalline structure similar to perovskites is detected for spinel compounds like magnesium aluminate ($MgAl_2O_4$). In this case, Mg^{2+} and Al^{3+} ions take tetrahedral and octahedral positions, respectively, but the O^{2-} ions occupy an FCC lattice. Ferrite crystals are ceramic magnets with a spinel-like structure. A variety of ceramic materials have a siliconoxygen tetrahedron (SiO_4^-) structure as a repeating unit, with each unit having a $^-4$ charge on it because each Si^{4+} is bonded to four O^{2-} ions. In each tetrahedron, a silicon (Si) atom is covalently joined to four oxygen (O) atoms positioned at the comers of a tetrahedron, with the center of the tetrahedron took by the silicon atom (Figure 2.1f).

On the other hand, silica (SiO_2) represents a simple silicate mineral that has, as a repeating unit, the SiO^- tetrahedron. In either vitreous or fused SiO_2, the main crystal unit is SiO_4^- tetrahedron, but there is no long-range order in the arrangement of this unit. Traditional glasses are silica-based ceramics to which oxides of Al, Ca, Mg, B, Na, and K, etc. have been attached; these oxides decrease the melting point and reduce the viscosity of glass, thus markedly easing the shaping of glass into very complex objects [4, 5]. Glass is an amorphous solid solution of various oxides, since SiO_2 content exceeds 50% and is considered the main constituent. Glasses show a disordered or liquid-like atomic structure. As a solid-solution alloy of ceramic oxides, glasses do not exhibit a single melting temperature and they are characterized by a melting range. Several other naturally-existing ceramics also are constituted from SiO_2; this includes kaolinite clay, talc, and mica [6]. In addition to the SiO_4, tetrahedral structures can be detected in other chemical compositions, such as AlO_4, SiN_4, and AlN_4. These tetrahedrons have a similar size as SiO_4, thus, they can replace each other in silicate structures. It has been reported that two-thirds of the Si element in α-Si_3N_4 can be substituted by Al atom without modifying the Si_3N_4 crystalline structure, where the amount of nitrogen is replaced by oxygen. This type of ceramic mixture is called a sialon, which is mainly composed from a solid solution of both Si_3N_4 and Al_2O_3. Similar types of sialons have also been produced, in particular those based on θ-Si_3N_4 and Si_2N_2O structures.

2.3 PROCESSING TECHNIQUES OF CERAMICS

Many ceramic fillers, such as metal oxides or carbides, can be fabricated simply by mixing powdered reactants and heating them to afford the targeted product. Even though the reaction conditions are relatively easy to estimate, there are some shortcomings, like the inhomogeneity of the starting raw materials and the difficulty controlling particle size and morphology. To overcome these drawbacks, several new processing techniques are used to produce ceramic fillers. Some of them are applied also to powdered metals. For instance, either crystalline ceramics or powdered metals can be prepared using solid-state powder metallurgy techniques, which are based on pressing and sintering (heating) of very fine raw powders. In addition, other special slurry-based techniques like extrusion, tape-casting, slip-casting, and injection-molding can similarly be used to shape ceramics. In contrast to powdered ceramics, amorphous ones such as glasses are elaborated on by other techniques that are also used to molten polymers, including blowing, pressing, and rolling techniques.

In the next sections, we will describe these synthetic techniques, which are mainly utilized to prepare oxide and non-oxide ceramics (Figure 2.2) [7].

2.3.1 Synthesis from the Solid Phase

High-pressure/high-temperature (HP-HT) technique can form new compounds/phases under solid-state conditions that are not achievable under ambient temperature and/or pressure. The utilization of high pressures combined with high temperatures and compositional variables gives a chance to prepare totally new types of ceramic materials and/or to tailor their various electrical, structural, and magnetic properties suitable for several high-performance applications. But, because of the high processing costs and relatively low production yield, industrial HP-HT synthesis is usually limited to the elaboration of diamond and c-BN [8].

Nevertheless, in the last few years, numerous HP ceramic materials have been discovered and studied. Some of these HP-phases ceramics have attracted potential interest for use in industrial applications because of their high mechanical properties, such as elastic moduli and hardness. For example, cubic spinel-type silicon nitride (c-Si_3N_4) is one the most promising ceramic to be industrially produced in the near future. It can be prepared using both amorphous and crystalline (alpha and

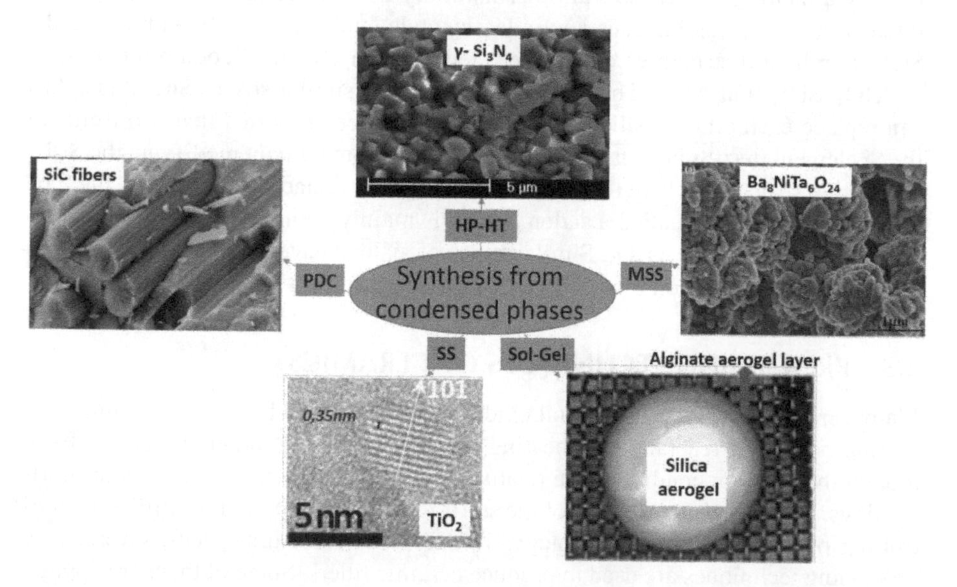

FIGURE 2.2 Overview of the Ceramic Synthetic Methods from Condensed Phases [7].

beta) Si_3N_4 at high pressures of 15 to 30 GPa and high temperatures in the range of 2,200 to 2,800 K using LH-DAC [9]. In addition, a novel crystalline boron oxynitride (BON) ceramic has been produced under static pressures of 15 GPa and temperatures exceeding 1,900°C, using various molar ratios of B_2O_3 and hexagonal BN as raw materials [10].

TiB_2 was synthesized under a high-pressure sintering (HPS) by mixing the elemental raw material powders using a high-pressure self-combustion synthesis technique (HPCS). The sintering and preparation conditions were taken as a pressure of 3 GPa in the temperature and time ranges 2,250 to 2,750 K and 5 to 300 s, respectively. A high sintering temperature (2,750 K) was required to provide dense monolithic TiB_2 compacts, where a remarkable grain growth was detected in the compacts sintered at higher temperatures (Figure 2.3) [11]. Dense and entirely nano-crystalline diamond–SiC ceramics were also prepared from a liquid Si and nanodiamond at a high-pressure (77 kbar) and high-temperature (1,400 to 2,000°C) conditions [12].

2.3.2 Molten Salt (MS) Synthesis

Molten salt (MS) synthesis efficiently produces crystalline, chemically-purified, single-phase ceramic powders due to its simplicity, versatility, and cost-effectiveness. In fact, molten inorganic salts are used as high-temperature solvents, mimicking the crystal nucleation and growth conditions in conventional solvents. In this synthesis technique, a huge amount of salt having a low melting point is heated higher than

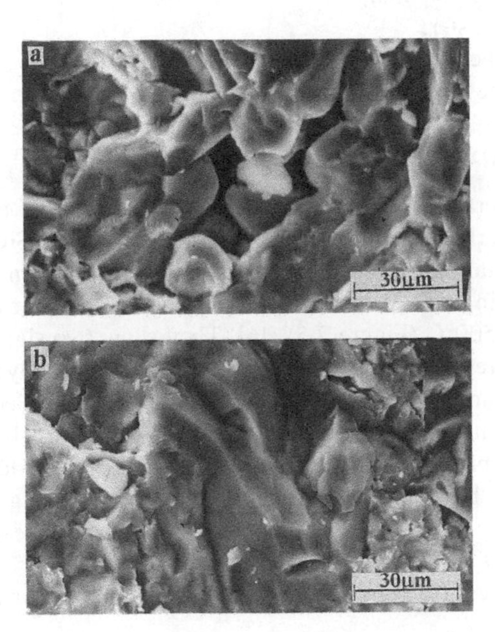

FIGURE 2.3 SEM Fractographs of HPS TiB_2 Compacts Sintered at 3 GPa, and (a) 2,500 K and (b) 2,750 K for 300 s. The High-temperature Sintering Resulted in Exaggerated Grain Growth [11].

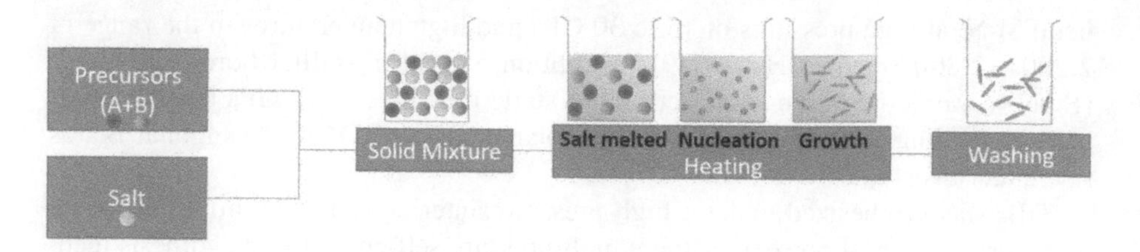

FIGURE 2.4 Scheme of the General Molten Salt Synthesis Procedure [7].

its melting temperature, thus playing the role of the solvent (Figure 2.4). This characteristic gives it many advantages over solid-state reactions or mild synthetic techniques using conventional solvents. For instance, using mild synthetic conditions, the creation of strong covalently bonded ceramic solids, including borides, carbide, and nitrides, produced poorly crystalline or even amorphous ceramic materials [13–15]. To surpass the energy barrier of ceramic crystal nucleation, a higher temperature is generally needed, which, in turn, exceeds the processing window for the majority of organic solvents under atmospheric pressure. This case stimulated the synthesis in inorganic MS media, which provides several benefits as compared to the other synthetic routes manipulating conventional solvents. Benefits include:

- An operational temperature window from 100°C to over 1,000°C without developing dangerous pressures;
- High oxidizing potential;
- High mass transfer;
- High thermal conductivity;
- Low viscosities and densities.

2.3.3 Polymer-to-ceramic Transformation Synthesis

Polymer-derived ceramics (PDCs) present a new category of technical ceramics, especially in the ternary and multinary systems like SiCO, SiCN, Si(M)CO, and Si(M)CN (M = metal), prepared generally by the thermal treatment of suitable preceramic polymers in an inert or reactive atmosphere and under temperatures ranging from 800 to 1,500°C (Figure 2.5) [16]. They show two significant advantages. First, the ceramics resulting from polymeric precursors generally provide a chemical composition that cannot be affordable by other techniques. Second, the possibility of joining the shaping and synthesis of ceramics means that constituents can be designed at the precursor phase using conventional plastic-forming techniques, including spinning, blowing, injection-molding, warm pressing, and resin transfer molding, before they are converted into ceramics under heat treatments at temperatures exceeding 800°C.

The typical process of the production of PDCs contains four main stages as follows [17–20]:

- The synthesis of preceramic polymers from their basic monomers;
- Cross-linking of these polymers at relatively low temperatures (100 to 400°C);

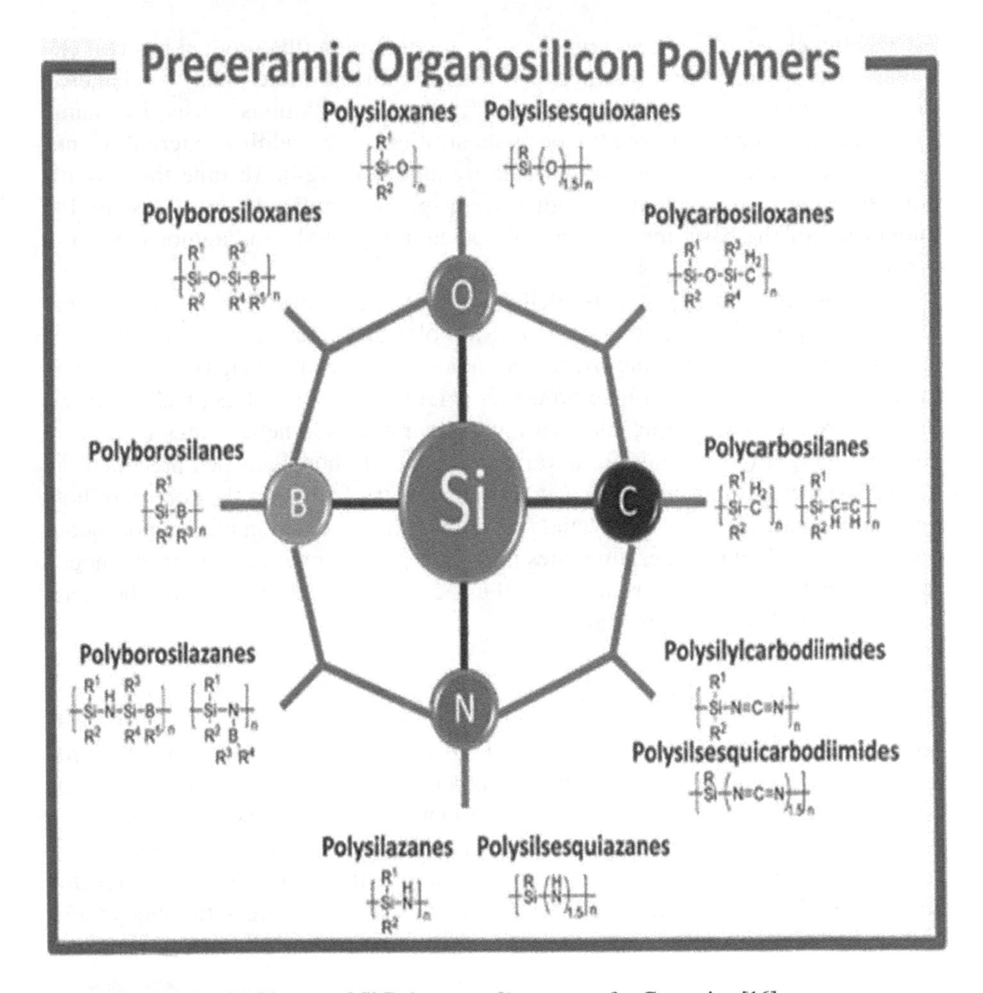

FIGURE 2.5 Main Classes of Si Polymer as Precursors for Ceramics [16].

- A ceramization process via pyrolysis at high temperatures in the range of 800 and 1,500°C to provide amorphous ceramics;
- Subsequent annealing at much higher temperatures to afford a (poly)crystalline ceramic materials.

2.3.4 Sol–Gel

Synthesis of ceramic by wet chemical methods is a special synthetic technique that produces advanced ceramics having controllable dimensions ranging from micro- to nanoscale; shapes such as powder, fibers, films, or monoliths; and exhibiting high chemico-physical reactivity as well as high-purity control. This method can be utilized to prepare oxidic and non-oxidic ceramics; binary, ternary, and multicomponent ceramic, pure and doped ceramics, and stable or metastable ceramic materials. In addition, the liquid-phase synthesis process allows the producing of fine powders, thin fibers, films, and aerogels [21–23].

In many cases, this method involves the precipitation (the product has two elements) or co-precipitation (the product constituted from more than two elements) of solid particles from soluble precursors in aqueous solutions caused by changing pH, temperature, and precursor concentration, or by adding external agents, such as oxidizing, reducing, and/or stabilizing agents [24]. To tune the ceramic powder properties, which are designed for a specific application, it is essential to understand of the basic mechanisms of nucleation, growth, and agglomeration of ceramics.

The sol–gel technique can be defined as the formation of an oxidic network through the polycondensation reaction of a molecular precursor in a liquid. It has been widely used to fabricate oxidic and non-oxidic ceramic materials using both the hydrolytic and nonhydrolytic processes (Figure 2.6). The sol–gel technique was mainly developed searching for new low-temperature synthetic routes to prepare complex inorganic materials from various chemically homogeneous precursors in shorter synthesis times. In addition, sol–gel chemistry facilitates the control of both the ceramic particle morphology and size. However, the creation of a homogeneous precursor at ambient temperature does not guarantee the homogeneity throughout a reaction, and many sol–gel routes have thus been developed to deal with the phase segregation during synthesis [25].

2.3.5 Solvothermal Synthesis

Solvothermal research was first developed during laboratory simulations of natural hydrothermal phenomena. Later, the hydrothermal synthesis was discovered as an important process for synthetizing ceramic material. This method used single- or heterogeneous-phase reactions in the presence of either aqueous or nonaqueous solvent above room temperature and at pressures higher than 1 atm in a closed reactor. Solvothermal synthesis is mainly controlled by two key parameters: thermodynamic

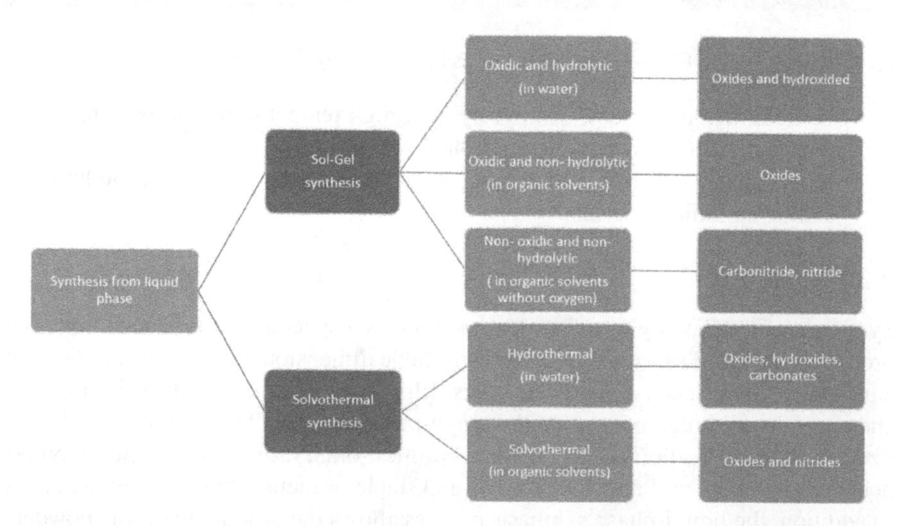

FIGURE 2.6 Methods to Synthesize Ceramic Materials from Liquid Phases [7].

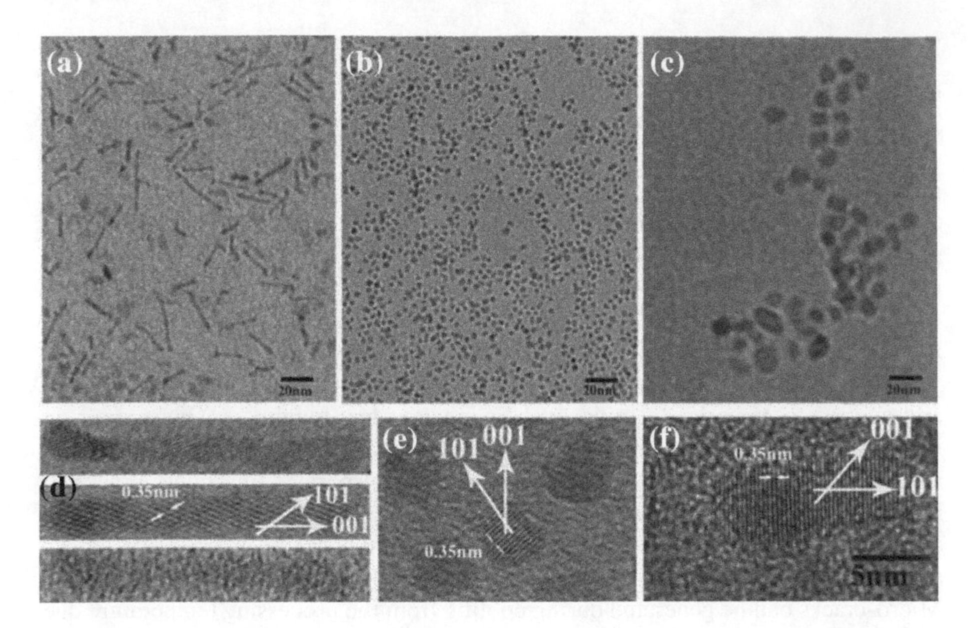

FIGURE 2.7 An Example of TiO_2 Nanoparticles Solvothermally Synthesized in Hexane [26].

parameters and kinetic parameters. This process facilitates the production of very complex ceramics having the desired physicochemical properties in a temperature range of 100 and 400°C without any annealing step. Three main synthetic routes can be distinguished in solvothermal processes, including the precipitation of the solid phase when the reactants are soluble in the solvent, the decomposition of the insoluble precursor in the solvent, and crystallization of amorphous starting ceramic materials (Figure 2.7) [26].

The solvothermal approach provides many advantages over other conventional and nonconventional ceramic synthesis techniques. Compared to solid-state processes, the solvothermal method favors processes of wet synthesis procedure, such as diffusion, adsorption, dispersion, reaction rate, and crystallization. Compared to solvothermal process, ceramic-powder processing includes several processing steps, which need high-energy consumption like mixing, milling, and calcination. In addition, this method can directly precipitate the crystallized ceramic powders from their solutions. This allows good control of the rate and homogeneity of the nucleation, growth, and aging stages, which results in enhanced control of size and morphology of the crystallites and a markedly decreased aggregation state [27].

2.4 PROPERTIES OF CERAMICS

Ceramics exhibit various outstanding properties such as good chemical and corrosion resistance, good wear properties, improved thermal and electrical conductivities and low-cost production. In the following sections we will summarize the different properties demonstrated by these materials.

2.4.1 Mechanical Properties

Ceramics are characterized by their high brittleness and hardness, which represents a serious limitation to the application of ceramics under conditions of high load or shock. Compared to metallic compounds, the number of slip systems for dislocation movement through polycrystalline ceramics are limited [28, 29]. However, for both ceramics and metals, the slip direction is the closest packed direction, and the slip plane is the closest packed crystal plane. The brittle behavior of crystalline ceramics is ascribed to the presence of charged ions, which impose certain restrictions on the number of slip systems. For amorphous ceramics, slip by dislocation motion is not found as there is no crystalline phase in these materials [30].

In the last few decades, several new techniques in materials design were invented and applied to ceramics to mitigate their inherent low toughness. One promising method is to enhance the disorientation between grains to hinder crack propagation across the grain boundaries. Another technique incorporates tiny micro-cracks in noncubic crystalline ceramics ($A1_2O_3$, TiO_2, etc.) to increase the fracture energy, and thus the toughness of these ceramics. These subcritical microscopic cracks afford an extra mechanism to dissipate the energy although they reduce the strength. Micro-cracks can be generated during cooling from the processing temperature due to anisotropic thermal expansion of noncubic ceramic crystals. The third method to toughen ceramics is the sintering of ceramics [31–34]. Table 2.1 summarizes some of the mechanical properties of various types of ceramics [35].

2.4.2 Thermal Properties

Thermal properties, such as thermal expansion coefficients, thermal conductivity, and thermal shock resistance are crucial features of ceramics. Many ceramics are either ionic or strong covalently-bonded solids, which have a low linear coefficient of thermal expansion (CTE) in the range of 0.5×10^{-6} to 12×10^{-6} °C^{-1} [36, 37]. It is reported that crystalline ceramics, other than those having a cubic structure, show

TABLE 2.1
Some Mechanical Properties of Ceramics [35]

Ceramic Code	Elastic Modulus (GPa)	Hardness (GPa)	Density (g/cm^3)
SiC	410	28.2	3.10
AlN	330	11.1	3.26
Al_2O_3	300	11.75	3.97
SiO_2	73	6.02	2.20
TiO_2	230	9.33	3.97
B_4C	432	38.0	2.52
TiC	316	29.5	4.85
Cr_2O_3	—	29.5	5.21
BN	46.9	---	2.27
ZrO_2	200	11	5.5
Si_3N_4	303	16.9	3.2

an anisotropy in the thermal expansion, where the expansion is greater along certain crystallographic directions than others [38]. The CTE of oxide ceramic (Al_2O_3) and nitride ceramic (Si_3N_4) is very low. Silicate ceramics such as steatite, superpyrostat, and cordierite also demonstrate low thermal expansion. Moreover, some ceramics showed even negative CTE values, such as the case for the ZrW_2O_8 and HfW_2O_8 due to the transverse thermal vibrations of bridging oxygen atoms. These are responsible for coupled rotations of the essentially rigid polyhedral building blocks of their structures [39].

Ceramic materials are utilized in many demanding applications because of their excellent thermal insulation. They are widely used as thermal insulation material in microelectronic packaging applications due to their low thermal conductivity. The thermal conductivity of ceramic is usually inferior to that of metals but higher than that of polymers. Some oxide ceramics, like zirconia (ZrO_2) and silica (SiO_2) exhibit very low thermal conductivity compared to other types of ceramic materials, however, alumina (Al_2O_3), silicon nitride (Si_3N_4), and aluminium nitride (AlN) have a relatively high thermal conductivity [40–42]. Figure 2.8 shows the thermal conductivity of beryllium oxide (BeO)-doped silicon carbide (SiC) ceramic at a temperature range of 5 to 1,300 K. From this figure, it appears that SiC ceramic can exhibit a high thermal conductivity of 270 $W.m^{-1}.K^{-1}$ by adding only a small amount of BeO [42].

Thermal stresses are related to geometrical, thermal, and physical boundary conditions that govern a property called thermal shock resistance [43–45]. The silicate ceramic and the compound Si_3N_4 show average values, whereas oxide ceramics such as Al_2O_3 and ZrO_2, as well as silicate ceramic materials, exhibit relatively low thermal shock-resistance values [46, 47].

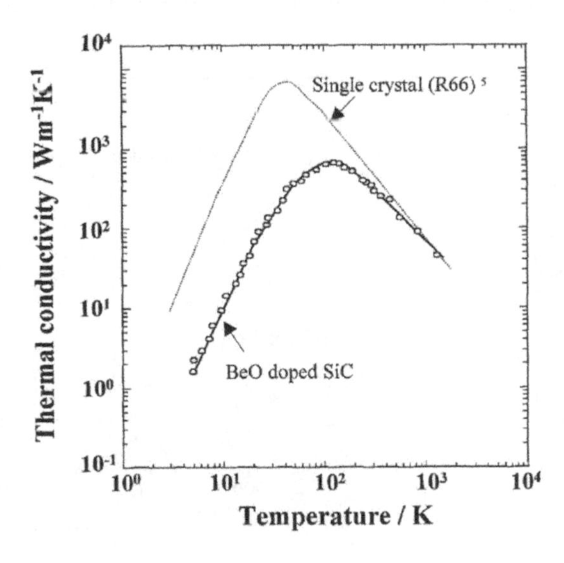

FIGURE 2.8 Temperature Dependence of Thermal Conductivity of Single-crystal SiC (R66) and Polycrystalline SiC Doped with BeO [42].

The maximum temperature of use is also a critical criterion when choosing ceramic materials for high-temperature applications [48]. Generally, ceramics show a high thermal stability, which is better than some pure metals like aluminum and magnesium. For example, pure Al_2O_3 can resist temperatures up to 1,800°C in air atmosphere. The thermal-load capacity of non-oxide silicate ceramics is markedly higher as they are exposed to inert atmosphere rather than air. ZrO_2 exhibits an extremely high thermal resistance of nearly 2,000°C [49]. Compared to the pure SiC ceramic particles, the weight-loss rate of the ZrB_2–SiC-coated samples decelerates, which infers a better oxidation protection ability of this binary ceramic powder system (Figure 2.9).

2.4.3 Electrical and Electronic Properties

Many crystalline ceramics exhibit interesting electrical and electronic properties and they are usually applied in several types of modem devices. Crystalline ceramics like Rochelle salt [$NaK(C_4H_4O_6).{}_4H_2O$], barium titanate ($BaTiO_3$), and strontium titanate ($SrTiO_3$) also show ferroelectric behavior. These crystals exhibit a dipole moment even in the absence of an electric field, which means that the centers of positive and negative charges in their crystalline structure cannot coincide [51]. Under an electric field, a mechanical moment is generated on the crystal due to spatial separation of positive and negative charges (assuming that the electric field is not exactly aligned with charge centers). Ferroelectric ceramics are utilized in transducers to convert one type of excitation into another (e.g., optical to electrical). This ferroelectric behavior disappears above a Curie temperature. Many types of ceramics, such as quartz, tourmaline (an aluminosilicate that contains boron), Rochelle salt, $BaTiO_3$, and lead zirconate titanate (PZT) have the ability to react to an imposed electrical excitation by varying their dimensions, and conversely, electrically respond to

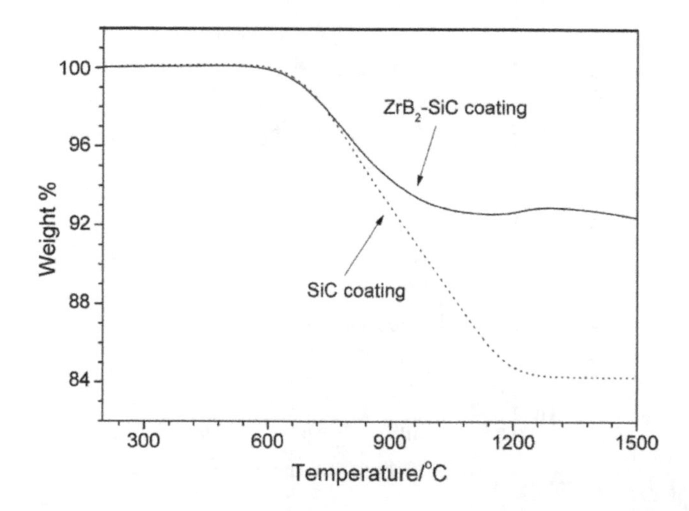

FIGURE 2.9 TGA Curves of the ZrB_2–SiC-coated Sample and the SiC-coated Sample in Simulated Air from 200 to 1,500°C [50].

mechanical excitation [52]. Several parameter can control the electrical and dielectric properties of ceramic, such as sintering temperature, processing technique, size, and shape of ceramics. For example, the grain size and the relative density of the $BaTiO_3$ ceramics synthesized by the sol–gel technique is enhanced with the increasing of sintering temperature, which in turns increased the dielectric permittivity of this ceramic filler (Figure 2.10) [53].

2.4.4 Oxidation and Corrosion Resistance

Ceramics resist harsh corrosive environments even at high temperatures compared to other materials. Particularly, oxides, carbides, silicides, and borides of refractory metals Zr-, Ti-, Hf-, Mo-, and Ta-based ceramics showed strong resistance to oxidation and corrosion at higher temperatures. Many oxide, carbide, and boride ceramics can be grown to form small-diameter fibers for decreasing the number of flaws in their cross section and enhance their strength and modulus while keeping their inherent corrosion resistance. SiO_2 or its derivatives was utilized to improve the oxidation resistance of thermal protection materials [54]. In addition, Y_2O_3-stabilized ZrO_2 fibers have superior resistance to corrosive environments rich in alkali-metal chlorides and carbonates up to nearly 700°C and can withstand short-term exposure to mineral acids even at their boiling point [55]. These fibers demonstrate excellent resistance to oxidizing atmospheres at

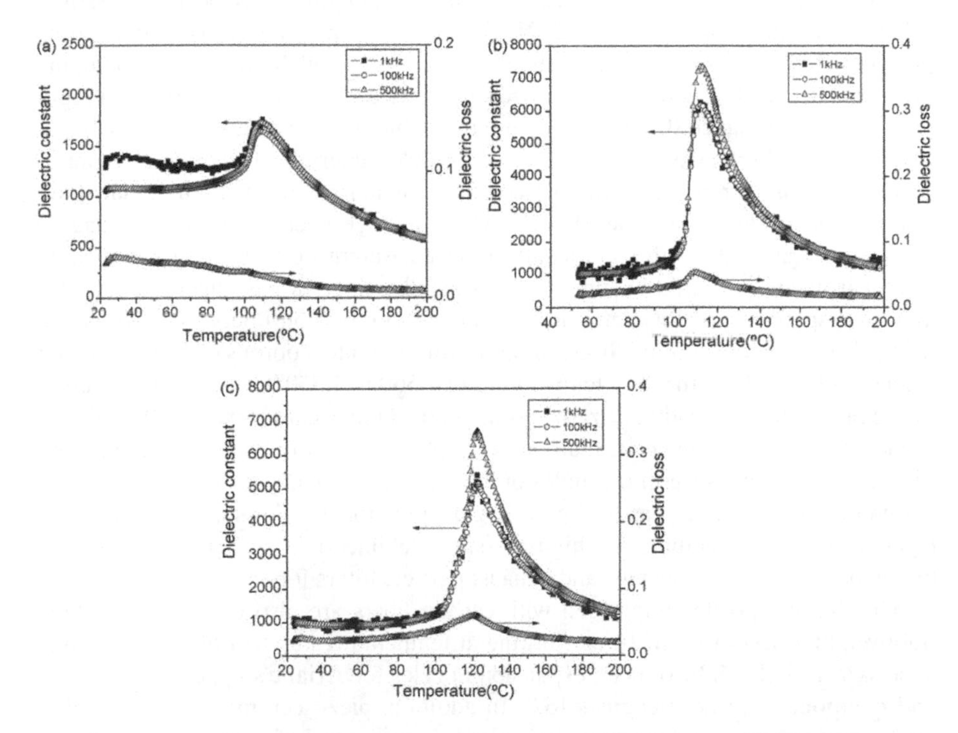

FIGURE 2.10 Temperature Dependence of the Dielectric Constant and Dielectric Loss for the $BaTiO_3$ Ceramics Using the Measurement Frequencies of 1 kHz, 100 kHz and 500 kHz, (a) T = 1,200°C; (b) T = 1,300°C; (c) T = 1,400°C [53].

higher temperatures. Similarly, Y_2O_3-stabilized ZrO_2 and Y_2O_3-stabilized CeO_2 showed excellent improved corrosion resistance at elevated temperatures [56, 57].

Borides and boride of refractory metals Ti-, Zr-, Hf-, and Ta-based ceramics exhibit high melting temperatures, good hardness, low volatility, and higher thermal shock resistance and thermal conductivity. These materials demonstrate good oxidation resistance, and further enhancements can be achieved by adding other ceramics like SiC, which can effectively increase the oxidation resistance of both ZrB_2 and H_fB_2.

2.5 APPLICATIONS OF CERAMICS

Advanced ceramic materials have been widely used in many fields of modern technologies, including communication and information technology, energy and environmental technology, automotive and aerospace technology, as well as in our daily life and medicine [58–64]. Due to their outstanding dielectric, ferroelectric, piezoelectric, pyroelectric, ferromagnetic, magnetoresistive, ionical, electronical, superconducting, and electro-optical characteristics, functional ceramics have been suitable for versatile applications.

2.5.1 Transportation Industry

The commercial uses of advanced ceramics in diesel engines have been successfully achieved for many parts, such as Si_3N_4–fuel-injector links, check balls and riming plungers, fuel pump rollers, ZrO_2 injector metering, and high-pressure pumping plungers [65]. Today, ceramics are increasingly used to meet future emission regulations and to enhance the fuel economy. In the automotive industry, there is great demand for lighter, stronger, and more affordable materials. Thus, the transportation industry is increasingly utilizing composite materials and foam materials [66]. For example, ceramic composite brakes exhibiting high-friction coefficients ranging between 0.4 and 0.6, with significantly reduced weight up to 60% less compared to cast iron, are particularly suited to many applications in passenger cars, trucks, and high-speed trains. SiC ceramics reinforced with carbon fibers were developed using either a vapor or liquid Si-precursor infiltration into a porous carbon form with oriented fiber architecture at a temperature of 1,500 to 1,550°C. Such ceramic composite brakes were already utilized by some in the luxury car market in 2001, and by reducing the production costs, many other applications should appear in the future. Figure 2.11 collects several examples of ceramic components currently used by the automobile industry, such as the ceramic glow plug for the diesel engine, the multilayer piezo-ceramic actuator for high-pressure fuel injection, the ceramic composite brake, ball and roller bearings, and exhaust particle filters [66].

The SiC composites reinforced with carbon fibers are also produced for many lightweight spacecraft structures working at temperatures up to 1,600°C, including nose skirts for the X38 vehicle, expansion nozzles for Ariane's upper-stage engine, and components in aero-engines [67]. In addition, piezo-ceramics are very suitable for low-frequency vibration damping and shape control of lightweight aerospace structures. Piezo-driven control flaps of helicopter rotor blades showed decreased noise and reduced rotor-induced vibrations inside the cabin by almost 50% [67].

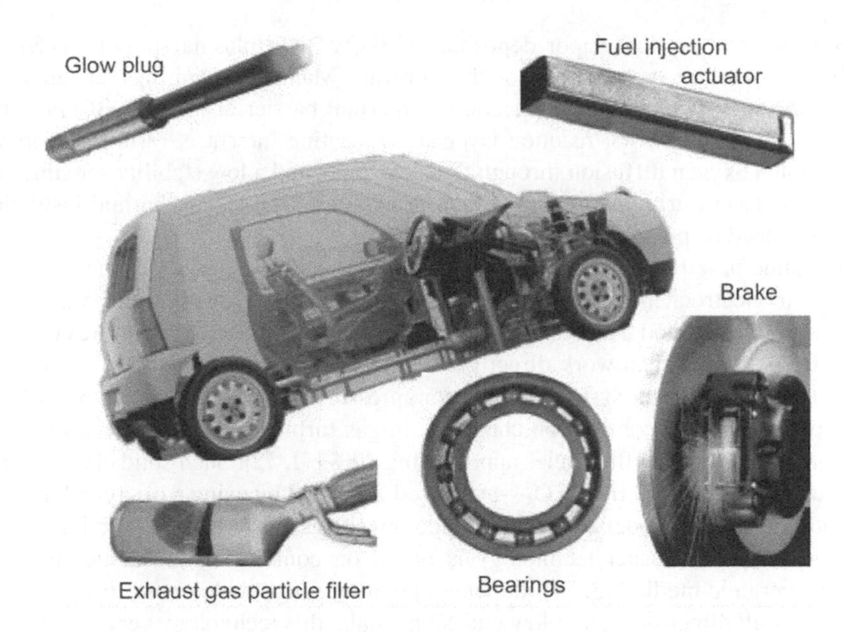

FIGURE 2.11 Examples of Advanced Ceramic Components Currently Being Developed for Application in Automobiles [66].

Piezoelectric multilayer actuators designed for high-pressure common rail fuel injection in diesel engines provide an improved ignition operation, showing reduced noise and low pollutant emissions.

Porous ceramic structures *diesel particle filter* devices (DPFs) are used in both trucks and the passenger car industry. For instance, honeycomb substrates from cordierite and SiC fiber structures coated with catalysts are extensively applied as diesel particulate filters [68]. Regeneration of the carbon particle–loaded ceramic filler can be achieved either by heating or under a catalytic combustion. Reduction of CO_2 is ensured by rising the thermal efficiency of an engine when energy-recovery systems are utilized. Compared to a conventional engine exhibiting a thermal efficiency of approximately 40% to 42%, a turbo compound-diesel engine equipped with heat-insulating ceramic components (combustion chamber, exhaust manifold) and additional exhaust gas energy-recovery systems was predicted to attain a thermal efficiency of nearly 65% [68].

2.5.2 Energy

Ceramics offer good thermal barrier coatings and are used to protect thermally-loaded metallic components in gas turbines and diesel engines so that the inlet temperature can be enhanced, or the cooling system size can be reduced, and the lifetime can be significantly increased [69]. The main advantage of ceramic-barrier layers is to decrease the fuel consumption due to their higher efficiencies and prolonged service durations. Under stationary conditions, thermal protection coatings prepared from Y_2O_3-stabilized ZrO_2 having a thickness in the range of 100 to 300 μm were found to keep a temperature gradient of 60 to 70°C from the hot gas to the metallic substrate.

Electron-beam physical vapor deposition (EB-PVD) or plasma spraying (PS) are utilized to produce such coatings on the substrate. Major shortcomings related to the long-term high-temperature application of thermal barrier coatings (TBCs) are the construction of an Al_2O_3 reaction layer at the coating interface, resulting from an accelerated oxygen diffusion through the ZrO_2 layer and a low stability for sintering of the top layer during utilization, which provokes a decrease in thermal insulation and enhanced in-plane elasticity modulus [70].

Ceramic-based solid oxide fuel cells (SOFCs) provide a clean, ecofriendly technology to electrochemically produced electricity and heat, at high yields and very low ratios of NO_X and SO_X emissions. Due to their high usage temperatures exceeding 800°C, SOFCs can work directly on natural gas, without the need for a high-cost, external reformer system. In addition, pressurized SOFCs also can be utilized as replacements for combustion chambers in gas turbines; such hybrid gas-turbine power systems attain efficiencies approaching 70% [71, 72]. Such solid electrolyte is fabricated from conductive Y_2O_3—stabilized ZrO_2 and by using a p—type $LaMnO_3$ as cathode (air electrode) and a Ni/ZrO_2 composite as anode (fuel electrode).

Porous media–burner technology is based on combustion in cavities of inert porous ceramic media [73, 74]. Its wide operating range with regard to its thermal power, small dimensions, and low emissions make this technology very suitable for power-plant and automobile engineering, as well as for local heating equipment and many other industrial applications. Because of the high conduction of heat, combustion temperatures can be decreased markedly relative to the free–flame-burning system. The most used ceramic materials are ZrO_2 foams, SiC foams and wavy structures of $A1_2O_3$ fibers, and C/SiC composite ceramics, which exhibit typical cavity dimensions of about 5,710 mm. Figure 2.12 reveals a fibrous SiC structure utilized in a porous media burner at temperatures up to 1,600°C.

2.5.3 Environment

Today, the elimination of NO_X gas is one of the major environmental problems. This poisonous gas is produced during the high-temperature combustion of liquid or gaseous fuels, especially by automobile exhaust. An effective strategy of removing NO_X is the usage of catalytic combustion, a technique that has been explored

FIGURE 2.12 Application of Fibrous Ceramic Structures in Porous Burner at Temperatures up to 1,600°C [66].

in the past few decades [75, 76]. The catalyst support is characterized by its high surface area, its superior mechanical properties, its good thermal shock resistance, and its ability to work in environments where a flame temperature exceeds 1,300°C. Porous $A1_2O_3$ supports with high surfacc areas and strong grain bonding, produced by low-temperature sintering of $Al(OH)_3$-rich $A1_2O_3$ powder, were produced to meet the above-mentioned requirements [77].

Porous $CaZrO_3$/MgO ceramic system having a homogenous 3D network structure, narrow pore size (~1 μm), and a high porosity of 3,050 vol.% were developed and revealed improved structural stability at high temperatures [78]. Hot gas filtration products with enhanced temperature stability up to 900°C were elaborated from rigid granular ceramic structures consisting of SiC bonded with siliceous material or of Al_2O_3 modified with low amounts of SiO_2 particles [79].

Multilayer ceramic membranes that afford enhanced stability compared to unfilled polymer membranes, when a high fluid pressure or temperature is used, were recently developed via a tape-casting method for both micro- and ultrafiltration of colloidal fluids [80]. Figure 2.13 reveals a high-performance, multi-segment ceramic filter module utilized for cleaning waste fluid in the paper industry. Moreover, such ceramic membranes are widely used in many other industries, including the food and beverage industry, pharmaceutical industry, and biotechnology.

2.5.4 Biomedical Applications

Biocompatible structural ceramics for orthopedic (hip and knee joints) and dental implants (all-ceramic dental bridges) have been produced with ameliorated mechanical, chemical, and wear properties, showing long lifetimes and improved safety of the implants under complex loading conditions [81]. Table 2.2 lists some clinical uses and examples of bioceramics. Microporous bioceramics are mainly developed for repairing or changing of damaged body parts. They can be found as crystalline like hydroxyapatite, semi-crystalline such as bioactive glass-ceramic, or even amorphous such as bioactive glass.

FIGURE 2.13 High-performance Turbulent Flow Filtration Module Used for Cleaning Waste Fluids in the Paper Industry (Courtesy KERAFOL, Eschenbach, Germany) [80].

TABLE 2.2
Some Clinical Uses and Examples of Bioceramics

Ceramic	Applications
Al_2O_3, ZrO_2, hydroxyapatite, bioactive glass	Orthopedic
Hydroxyapatite, bioactive glass ceramic	Coatings for bioactive bonding
Calcium phosphate salts, tricalcium phosphate	Bone space fillers
Al_2O_3, hydroxyapatite, bioactive glasses	Dental implants
Hydroxyapatite, bioactive glass ceramic	Spinal surgery

ZrO_2 ceramic material has good mechanical properties for producing medical devices, while ZrO_2-stabilized Y_2O_3 has adequate properties for such use [82–84]. Orthopedic research found that these material can be used to fabricate hip head prostheses due to the biocompatibility of ZrO_2 materials with bone or muscle. The zirconia implant abutments can also be utilized to ameliorate the aesthetic outcome of implant-supported rehabilitations. The use of ceramic/ceramic as bearing surfaces for hip joint prostheses was developed to further reduce wear rate compared with the other combinations such as metal-on-polyethylene and ceramic-on polyethylene. Two new types of mixed-oxide ceramics (Al_2O_3 and Y_2O_3-stabilized ZrO_2) femoral heads and acetabular cups containing different proportions of Al_2O_3 and ZrO_2 were compared with commercial ZrO_2 in terms of wear behavior in a hip joint simulator (Figure 2.14) [85].

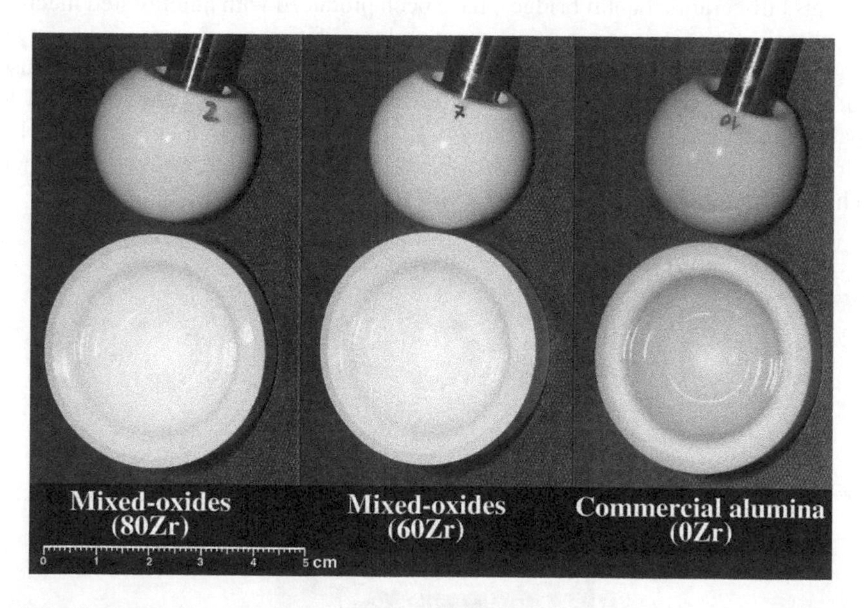

FIGURE 2.14 ZrO_2 Samples of All Three Types of Femoral Heads and Acetabular Cups Biomaterials: 0Zr (100% Al_2O_3); 60Zr (60%ZrO_2 + 40%Al_2O_3); 80Zr (80%ZrO_2 + 20%Al_2O_3) [85].

TABLE 2.3
Some Electronic and Electrical Application Areas Requiring Ceramic Powders

Application Area	Used Ceramic
Capacitors	$BaTiO_3$, other titanates and zirconates, Pb-based "relaxers"
Piezoelectric transducers	$Pb(Zr, Ti)O_3$, $BaTiO_3$, Al_2O_3, AIN, SiC, cordierite
Magnets	Ferrites, high-Tc superconductors
Electro-optics	$(Pb, La)(Zr, Ti)O_3$
Thermistors	$BaTiO_3$
Varistors	TiO_2
Solid oxide fuel cells	ZrO_2
Spark plugs/insulators	Al_2O_3

2.5.5 Electronic and Electrical Applications

The lead zirconate titanate (PZT) ceramic crystals are applied to transform voltage to motion and vice versa. This phenomenon, known as piezoelectric behavior, allows the use of these ceramics in radio transmitters and receivers to control specific broadcast frequencies. Piezoelectric ceramics are also utilized in traditional electrical watches, ultrasound generators, phonograph pickups, microphones, and in SONAR (Sound Navigation and Ranging) for detecting underwater sound [86–89]. Till now, PZT ceramics are the basic material for producing piezoelectric ceramics used in these equipment. Ceramics having ferromagnetic characteristics are introduced in powder form to stealth coatings that absorb radar and other electromagnetic radiation and avoid detection. The electrically conductive ceramic are added to carpet and floor tile to reduce the effect of static electricity. Ceramics are also utilized to fabricate electrical conductors, capacitors, resistors, and dielectrics for applying them in microelectronic devices, and as substrates for packaging these corposants. Table 2.3 offers some electrical applications areas based on ceramics.

2.6 CONCLUSIONS

Ceramics are constituted from both metallic and nonmetallic elements and exhibit a high hardness, good compressive strength, high elastic modulus, low thermal expansion, low density, and improved dielectric properties, etc. Ceramic materials can be designed with improved mechanical, electrical, chemical, or magnetic properties using suitable processing techniques and by composite approaches to meet a wide range of applications. They have many advantages such as light-weight, and higher temperature resistance under aggressive environments compared to metals which are mainly used in coatings and sensors applications.

REFERENCES

1. Richerson, D. W. 1992. *Modern Ceramic Engineering: Properties, Processing, and Use in Design* (2nd ed.). New York: Marcel Dekker.
2. Wachtman, J. B. Jr., 1989. *Structural Ceramics, Treatise on Materials Science and Technology,* Vol. 29. San Diego: Academic.
3. Teufer, G. 1962. The crystal structure of tetragonal ZrO_2. *Acta crystallographica* 15: 1187–1187.
4. Uchino K., Sadanaga, E., Hirose, T. 1989. Dependence of the crystal structure on particle size in barium titanate. *Journal of the American Ceramic Society* 72: 1555–1558.
5. Budworth, D. W. 1970. *Introduction to Ceramic Science*. Oxford: Pergamon Press.
6. Kumar, A., Dahotre, N. B., Asthana, R. 2006. *Materials Processing and Manufacturing Science.* Oxford: Butterworth-Heinemann.
7. Gonzalo-Juan, I., Riedel, R. 2016. Ceramic synthesis from condensed phases. *ChemTexts* 2:6.
8. Demazeau, G. 1993. Growth of cubic boron nitride by chemical vapor deposition and high-pressure high-temperature synthesis. *Diamond and Related Materials* 2: 197–200.
9. Zerr, A., Miehe, G., Serghiou, G., et al. 1999. Synthesis of cubic silicon nitride. *Nature* 400: 340–342.
10. Bhat, S., Wiehl, L., Molina-Luna, L, et al. 2015. High-pressure synthesis of novel boron oxynitride $B_6N_4O_3$ with sphalerite type structure. *Chemistry of Materials* 27: 5907–5914.
11. Bhaumik, S. K., Divakara, C., Singh, A. K., et al. 2000. Synthesis and sintering of TiB_2 and TiB_2–TiC composite under high pressure. *Materials Science and Engineering: A* 279: 275–281.
12. Ekimov, E. A., Gavriliuk, A. G. 2000. High-pressure, high-temperature synthesis of SiC–diamond nanocrystalline ceramics. *Applied Physics Letters* 77: 954.
13. Lei, F., Yan, B., Chen, H. H. et al. 2009. Molten salt synthesis, characterization, and luminescence properties of Gd_2MO_6:Eu_3? (M = W, Mo) phosphors. *Journal of American Ceramic Society* 92: 1262–1267.
14. Gong, M., Lu, F., Kuang, X. et al. 2015. Molten salt synthesis, polymorphism, and microwave dielectric properties of $Ba_8NiTa_6O_{24}$ perovskite. *Journal of American Ceramic Society* 98: 2451–2458.
15. Liu, X., Antonietti, M., Giordano, C. 2013. Manipulation of phase and microstructure at nanoscale for SiC in molten salt synthesis. *Chemistry of Materials* 25: 2021–2027.
16. Colombo, P., Mera, G., Riedel, R., et al. 2010. Polymer derived ceramics: 40 years of research and innovation in advanced ceramics. *Journal of American Ceramic Society* 93: 1805–1837.
17. Mera, G., Riedel, R., Poli, F., et al. 2009. Carbon-rich SiCN ceramics derived from phenyl-containing poly(silylcarbodiimides). *Journal of European Ceramic Society* 29: 2873–2883
18. Michelle Morcos, R., Mera, G., Navrotsky, A. 2008. Enthalpy of formation of carbon-rich polymer-derived amorphous SiCN ceramics. *Journal of American Ceramic Society* 91: 3349–3354.
19. Mera, G., Tamayo, A., Nguyen, H. et al. 2010. Nanodomain structure of carbon-rich silicon carbonitride polymer derived ceramics. *Journal of American Ceramic Society* 93: 1169–1175.
20. Gao, Y., Mera, G., Nguyen, H. et al. 2012. Processing route dramatically influencing the nanostructure of carbon-rich SiCN and SiBCN polymer-derived ceramics. Part I: Low temperature thermal transformation. *Journal of European Ceramic Society* 32: 1857–1866.

21. Nassar, E. J., Ciuffi, K. J., Calefi, P. S., et al. 2011. Biomaterials and sol–gel process: a methodology for the preparation of functional materials. In: Pignatello PR (Ed.), *Biomaterials Science and Engineering*. Hampshire, U.K.: InTech.
22. Xu, Y., Su, D., Feng, H. et al. 2015. Continuous sol–gel derived SiOC/HfO_2 fibers with high strength. *RSC Advances* 5: 35026–35032.
23. Hao, D., Yang, Z., Jiang, C. et al. 2014. Synergistic photocatalytic effect of TiO_2 coatings and p-type semiconductive SiC foam supports for degradation of organic contaminant. *Applied Catalysis Part B* 144: 196–202.
24. Osman, N., Jani, A. M., Talib, I. A. 2006. Synthesis of Y_b-doped Ba(Ce, Zr)O_3 ceramic powders by sol–gel method. *Ionics* 12: 379–384.
25. Danks, A. E., Hall, S. R., Schnepp, Z. 2016. The evolution of "sol–gel" chemistry as a technique for materials synthesis. *Materials Horizon* 3: 91–112.
26. Gonzalo-Juan, I., McBride, J. R., Dickerson, J. H. 2011. Ligand-mediated shape control in the solvothermal synthesis of titanium dioxide nanospheres, nanorods and nanowires. *Nanoscale* 3: 3799–3804.
27. Riman, R. E., Suchanek, W. L., Lencka, M. M. 2002. Hydrothermal crystallization of ceramics. *Annales de Chimie Science des Matériaux* 27: 15–36.
28. Rice, R. W. 2000. *Mechanical Properties of Ceramics and Composites: Grain and Particle Effects*. New York: Marcel Dekker.
29. Cannon, W. R., Langdon, T. G. 1983. Review: Creep of ceramics: Part 1, mechanical characteristics. *Journal of Materials Science* 18: 1–50.
30. Claussen, N., Steeb, J., Pabst, R. F. 1977. Effect of induced microcracking on the fracture toughness of ceramics. *American Ceramic Society Bulletin* 56: 559–562.
31. Davidge, R. W. 1979. *Mechanical Behavior of Ceramics*. New York: Cambridge University Press.
32. Evans, A. G., McMeeking, R. M. 1986. On toughening of ceramics by strong reinforcements. *Acta Metallurgica* 34: 2435–2441.
33. Ostrowski, T., Rodel, J. 1999. Evolution of mechanical properties of porous alumina during sintering and hot pressing. *Journal of the American Ceramic Society* 82: 3080–3086.
34. Takahashi, M., Suzuki, S. 1990. Compaction behaviour and mechanical characteristics of ceramic powders. In: Cheremisinoff, N. P. (Ed.). *Handbook of Ceramics and Composites*, Vol. 1, pp. 65–97. New York: Marcel Dekker.
35. Ramdani, N., Liu, W. B., Wang, J., Derradji, M. 2017. *Ceramic Based Polybenzoxazine Micro- and Nanocomposites*. In: H Ishida (Ed.). *Advanced and Emerging Polybenzoxazine Science and Technology*, Amsterdam: Elsevier. (pp. 861–919).
36. Adler, S. B. 2011. Chemical expansivity of electrochemical ceramics. *Journal of the American Ceramic Society* 84: 2117–2119.
37. Hirose, Y., Doi, H., Kamigaito, O. 1984. Thermal expansion of hot-pressed cordierite glass ceramics. *Journal of Materials Science Letters* 3: 153–155.
38. Tvergaard, V., Hutchinson, J. W. Microcracking in ceramics induced by thermal expansion or elastic anisotropy. *Journal of the American Ceramic Society* 71: 157–166.
39. Evans, J. S. O., Mary, T. A., Vogt, T., Subramanian, M. A., Sleight, A. W. 1996. Negative thermal expansion in ZrW_2O_8 and HfW_2O_8. *Chemistry of Materials* 8: 2809–2823.
40. Kurokawa, Y., Utsumi K., Takamizawa, H. 1988. Development and microstructural characterization of high-thermal-conductivity aluminum nitride ceramics. *Journal of the American Ceramic Society* 71: 588–594.
41. Molisani, A. L., Goldenstein, H., Yoshimura, H. N. 2017. The role of CaO additive on sintering of aluminum nitride ceramics. *Ceramics International* 43, 16972–16979.
42. Nakano, H., Watari, K., Kinemuchi, Y., Ishizaki, K., Urabe, K. 2004. Microstructural characterization of high-thermal-conductivity SiC ceramics. *Journal of the European Ceramic Society* 24: 3685–3690.

43. Hasselman, D. P. H. 1969. Unified theory of thermal shock fracture initiation and crack propagation in brittle ceramics. *Journal of the American Ceramic Society* 52: 600–604.
44. Hasselman, D. P. H., Youngblood, G. E. 1978. Enhanced thermal stress resistance of structural ceramics with thermal conductivity gradient. *Journal of the American Ceramic Society* 61: 49–52.
45. She, J., Ohji, T., Deng, Z. Y. 2002. Thermal Shock Behavior of Porous Silicon Carbide Ceramics. *Journal of the American Ceramic Society* 85: 2125–2127.
46. Becker, P. F. 1981. Transient thermal stress behavior in ZrO_2-Toughened Al_2O_3. *Journal of the American Ceramic Society* 64: 37–39.
47. Hasselman, D. P. H. 1985. Thermal stress resistance of engineering ceramics. *Materials Science and Engineering* 71: 251–264.
48. Ivanov, D. A., Balabanov, A. S., Fomina, G. A., Alad'ev, N. A., Val'yano, G. E. 1998. The problem of thermal stability of ceramic materials. *Refractories and Industrial Ceramics* 39: 80–85.
49. Chang, C. H., Gopalan, R., Lin, Y. S. 1994. A comparative study on thermal and hydrothermal stability of alumina, titania and zirconia membranes. *Journal of Membrane Science* 91: 27–45.
50. Li, L., Li, H., Yin, X., Chu, Y., Chen, X., Fu, Q. 2015. Oxidation protection and behavior of in-situ zirconium diboride–silicon carbide coating for carbon/carbon composites. *Journal of Alloys and Compounds* 645: 164–170.
51. Ching, W. Y. 1990. Theoretical studies of the electronic properties of ceramic materials. *Journal of the American Ceramic Society* 73: 3135–3160.
52. Colombo, P., Gambaryan-Roisman, T., Scheffler, M., et al. 2001. Conductive ceramic foams from preceramic polymers. *Journal of the American Ceramic Society* 84: 2265–2268.
53. Li, W., Xu, Z., Chu, R., Fu, P., Hao, J. 2009. Structure and electrical properties of $BaTiO_3$ prepared by sol–gel process. *Journal of Alloys and Compounds* 482: 137–140.
54. Tsai, T., Barnett, S. A. 1995. Bias sputter deposition of dense yttria-stabilized zirconia films on porous substrates. *Journal of the Electrochemical Society* 142: 3084–3087.
55. Ruddell, D. E., Stoner, B. R., Thompson, J. Y. 2003. The effect of deposition parameters on the properties of yttria-stabilized zirconia thin films. *Thin Solid Films* 445: 14–19.
56. Park, S. Y., Kim, J. H., Kim, M. C. et al. 2005. Microscopic observation of degradation behavior in yttria and ceria stabilized zirconia thermal barrier coatings under hot corrosion. *Surface and Coatings Technology* 190: 357–365.
57. Ahmadi-Pidani, R., Shoja-Razavi, R., Mozafarinia, R. et al. 2012. Evaluation of hot corrosion behavior of plasma sprayed ceria and yttria stabilized zirconia thermal barrier coatings in the presence of Na_2SO_4+V_2O_5 molten salt. *Ceramics International* 38: 6613–6620.
58. Asanabe, S. 1987. Applications of ceramics for tribological components. *Tribology International* 20: 355–364.
59. Clinard Jr., F. W. 1979. Ceramics for applications in fusion systems. *Journal of Nuclear Materials* 85/86: 393–404.
60. Heimke, G. 1989. Advanced ceramics for biomedical applications. *Angewandte Chemie International Edition in English* 28: 111–116.
61. Jo, W., Dittmer, R., Acosta, M. et al. 2012. Giant electric-field-induced strains in lead-free ceramics for actuator applications—Status and perspective. *Journal of Electroceramics* 29: 71–93.
62. Bowen, H. K. 1980. Basic research needs on high temperature ceramics for energy applications. *Materials Science and Engineering* 44: 1–56.

63. Wang, S. F., Zhang, J., Luo, D. W. 2013. Transparent ceramics: Processing, materials and applications. *Progress in Solid State Chemistry* 41: 20–54.
64. Medvedovski, E. 2010. Ballistic performance of armour ceramics: Influence of design and structure. Part 1. *Ceramics International* 36: 2103–2115.
65. Mandler, W. F., Yonushonis, T. M. 2001. In 25th Annual Conference on Composites for Advanced Ceramic Material Structuring A (M. Singh, T. Jessen, Eds.). The American Ceramic Society, Westerville, OH, p. 3 (*Ceram. Eng. Sci Proa*, 22/3).
66. P. Greil. 2002. Advanced engineering ceramics. *Advanced Engineering Materials* 4: 247–254.
67. Flegel, H. A. 2001. New Technologies in the Light of Materials. In Heinrich, J. G., Aldinger, F. (Eds.) *Ceramic Materials and Components for Engines*, (p. 2–6). Weinheim, Germany: Wiley-VCH.
68. Kawamura, H. 2001. Practical use of Ceramic Components and Ceramic Engines. In Heinrich, J. G., Aldinger, F. (Eds.), *Ceramic Materials and Components for Engines* (E, (p. 27–32) Weinheim, Germany: Wiley-VCH.
69. Brindley, W. J. 1997. Thermal barrier coatings of the future. *Journal of Thermal Spray Technology* 6: 3–4.
70. Cao, X.Q., Vassen, R., Stoever, D. 2004. Ceramic materials for thermal barrier coatings. *Journal of the European Ceramic Society* 24: 1–10.
71. Singhal, S. C. 1999. In 9th *CIMTEC World Fomm on New Materials*, Symp. V117 Innovative Materials in Advanced Energy Technologies (P. Vincencini, Ed.), *Techna Srl.,* Italy, p. 3.
72. Antolini, E., Gonzalez, E. R. 2009. Ceramic materials as supports for low-temperature fuel cell catalysts. *Solid State Ionics* 180: 746–763.
73. Trimis, D., Durst, F. 1996. Combustion in a Porous Medium-Advances and Applications. *Combustion Science and Technology* 121: 153–168.
74. AbdulMujeebu, M., Abdullah, M. Z., AbuBakar, M. Z. et al. 2009. A review of investigations on liquid fuel combustion in porous inert media. *Progress in Energy and Combustion Science* 35: 216–230.
75. Frank, B., Emig, G., Renken, A. et al. 1998. Kinetics and mechanism of the reduction of nitric oxides by H_2 under lean-burn conditions on a Pt–Mo–Co/α-Al_2O_3 catalyst. *Applied Catalysis B: Environmental* 19: 45–57.
76. Zarur, A. J., Ying, J. Y. 2000. Reverse microemulsion synthesis of nanostructured complex oxides for catalytic combustion. *Nature* 403: 65–97.
77. Deng, Z. Y., Fukasawa, T., Ando, M. et al. 2001. High-Surface-Area Alumina Ceramics Fabricated by the Decomposition of $Al(OH)_3$. *Journal of the American Ceramic Society* 84: 485–491.
78. Suzuki,Y., Awano, M., Kondo, N. et al. 2002. Effect of In-doping on the microstructure and CH_4-sensing ability of porous $CaZrO_3$/MgO composites. *Journal of the European Ceramic Society* 22: 1177–1182.
79. Suzuki, Y., Morgan, P. E. D., Ohji, T. 2000. New Uniformly Porous $CaZrO_3$/MgO Composites with Three-Dimensional Network Structure from Natural Dolomite. *Journal of the American Ceramic Society* 83: 2091–2093.
80. Westerheide, R., Adler, J. 2001. In J. G. Heinrich, F. Aldinger (Eds.), *Ceramic Materials and Components for Engines* (p. 73). Weinheim, Germany: Wiley-VCH.
81. Rahaman, M. N., Yao, A. 2007. Ceramics for prosthetic hip and knee joint replacement. *Journal of the American Ceramic Society* 90: 1965–1988.
82. Manicone, P. F., Iommetti, P. R., Raffaelli, L. 2007. An overview of zirconia ceramics: Basic properties and clinical applications. *Journal of Dentistry* 35: 819–826.
83. Liang, X., Qiu, Y., Zhou, S. et al. 2008. Preparation and properties of dental zirconia ceramics. *Journal of University of Science and Technology Beijing, Mineral, Metallurgy, Material* 15: 764–768.

84. Abd El-Ghany, O. S., Sherief, A. H. 2016. Zirconia based ceramics, some clinical and biological aspects: Review. *Future Dental Journal* 2: 55–64.
85. Affatato, S., Goldoni, M., Testoni, M., et al. 2001. Mixed oxides prosthetic ceramic ball heads. Part 3: effect of the ZrO_2 fraction on the wear of ceramic on ceramic hip joint prostheses. A long-term in vitro wear study. *Biomaterials* 22: 717–723.
86. Harris, J. H. 1998. Sintered aluminum nitride ceramics for high-power electronic applications. *Journal of Metals* 50: 56–60.
87. Rödel, J., Webber, K. G., Dittmer, R. et al. 2015. Transferring lead-free piezoelectric ceramics into application. *Journal of the European Ceramic Society* 35: 1659–1681.
88. Haertling, G. H., Land, C. E. 1971. Hot-pressed (Pb, La)(Zr, Ti)O_3 ferroelectric ceramics for electrooptic applications. *Journal of the American Ceramic Society* 54: 1–11.
89. Haertling, G. H. 1991. Ferroelectric thin films for electronic applications. *Journal of Vacuum Science & Technology A* 9: 414–420.

3 Functionalization Methods of Ceramic Particles

3.1 INTRODUCTION

Ceramic fillers have been widely investigated by many researchers due to their versatile, advantageous properties. Their unique physical properties result from their crystalline structures, which at atomic scale can be thought of as hexagonal cubic, or even orthorhombic, structures. Besides their strong hardness characteristics, their excellent electrical and thermal conducting properties, combined with their good toughness and low coefficient of thermal expansion (CTE) value make them promising materials to be used in a different applications, such as microelectronic components, chemical sensors and actuators, chemical and genetic probes, and cutting tools. Ceramics are thus expected to improve their unique properties in multidisciplinary fields by reinforcing polymers.

Unfortunately, these kinds of reinforcing agents suffer from high brittleness. Another problem is a strong tendency of ceramic nanofillers for aggregation in the polymeric matrices, especially at higher ratios where they form large bundles thermodynamically stabilized by physical entanglements between these inorganic particles; this can occur during the synthesis of their polymer/ceramics composites. The presence of these aggregates in addition to the insolubility of ceramics in water and organic solvents appear to be another drawback for the engineering of ceramics in polymer nanocomposites. It has been reported that the uniform dispersion of ceramics is relatively difficult to achieve in the large majority of polymers, especially for thermoplastic ones.

The chemical functionalization and silane treatment of the ceramic outer surface (surface-anchoring of reactive functional groups), represents a prominent solution for the tuning of interactions between ceramics gust and the host polymer matrix, improving the ability of their dispersion. These surface-modification methods of ceramics can be categorized into three main approaches. The first one is noncovalent functionalization by using silane treatment, used in different techniques such as the addition of surfactants, and polymer wrapping or polymerization-filling techniques, which allow the unaltered particles to keep their physical properties. However, the interaction between the wrapping molecules and the ceramics are still usually weak, affecting the efficiency of the property transfer between the ceramic particles and the host polymeric matrix; in particular, there may be an inferior load transfer for mechanical properties.

The other approaches depend upon the covalent grafting of functional groups on the ceramic outer surface. These chemical functions can be utilize as anchoring sites for macromolecule chains and to ameliorate the interaction with the polymer matrix. The covalent bonding between ceramics and the polymer chains ensures an optimal interfacial strength and thus a better load transfer. Nevertheless, the covalent functionalization can influence the ceramic properties, depending on the nature and density of the functional sites.

To increase the interfacial adhesion and dispersion of ceramic fillers into a polymeric matrix for composites, small amount of silane molecules is used to treat these fillers to improve their mechanical performance, thermal conductivity, electrical insulation, and to decrease their thermal expansion coefficients, and so forth. However, excess of the coupling agent must be avoided as it can negatively influence the final composite's properties.

3.2 SILANE TREATMENT METHODS

The dispersion of ceramic particles in the polymer matrix significantly impacts the final properties of composites. Improved dispersion may be attained by surface chemical modification of the ceramic particles or physical methods like a high-energy ball-milling process or ultrasonic treatment. The large differences in the properties of polymer and ceramic fillers can generally provoke phase separation. Thus, the interfacial interaction between two phases of nanocomposites is the most critical factor affecting the performance of the resulting materials. Table 3.1 summarizes the various types of silane molecules use for the treatment of silica nanoparticles.

TABLE 3.1
Typical Silane Coupling Agents Used for Surface Modification of Silica Nanoparticles

Code	Name	Chemical Structure
APMDES	Aminopropylmethydiethoxysilane	$H_2N(CH_2)_3(CH_3)Si(OC_2H_5)_2$
APMDMOS	(3acryloxypropyl)methydimethoxysilane	$CH_2{=}CHCOO(CH_2)_3(CH_3)Si(OCH_3)_2$
APTES	3-aminopropyltriethoxysilane	$H_2N(CH_2)_3Si(OC_2H_5)_3$
APTMS	3-aminopropyltrimethoxysilane	$H_2N(CH_2)_3Si(OCH_3)_3$
APTMS	(3-acryloxypropyl)trimethoxysilane	$CH_2{=}CHCOO(CH_2)_3Si(OCH_3)_3$
APTMS	Aminophenyltrimethoxysilane	$H_2NPhSi(OCH_3)_3$
TESPT	Bis(triethoxysilylpropyl)tetrasulfane	$(C_2H_50)_3Si(CH_2)_3S_4(CH_2)_3Si(OC_2H_5)_3$
DDS	Dimethyldichlorosilane	$(CH_3)_2SiCl_2$
GPS	3-glycidoxypropyltrimethoxysilane, 3-glycidyloxypropyltrimethoxysilane	$CH_2(O)CHCH_2O(CH_2)_3Si(OCH_3)_3$
ICPTES	3-isocyanatopropyltriethoxysilane	$OCN(CH_2)_3Si(OC_2H_5)_3$
MMS	Methacryloxymethyltriethoxysilane	$CH_2{=}C(CH_3)COOCH_2Si(OC_2H_5)_3$
MPTES	Methacryloxypropyltriethoxysilane	$CH_2{=}C(CH_3)COO(CH_2)_3Si(OC_2H_5)_3$
MPTS	Mercaptopropyl triethoxysilane	$SH(CH_2)_3Si(OC_2H_5)_3$
MTES	Methyltriethoxysilane	$CH_3Si(OC_2H_5)_3$
PTMS	Phenyltrimethoxysilane	$PhSi(OCH_3)_3$
VTES	Vinyltriethoxysilane	$CH_2{=}CHSi(OC_2H_5)_3$
VTS	Vinyltrimethoxysilane	$CH_2{=}CHSi(OCH_3)_3$

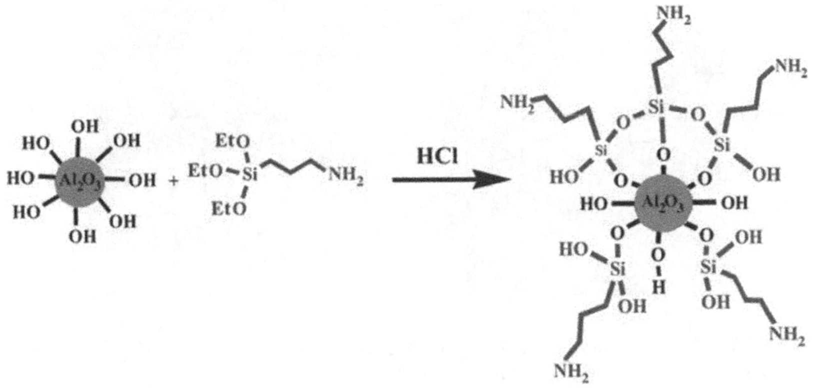

FIGURE 3.1 Modification of Al_2O_3 Nanoparticles with KH550 [1].

Silane is the most extensively used coupling agent for the treatment of a variety of ceramic particles. The polar surfaces of the ceramic fillers are treated by grafting silane coupling agents to enhance dispersion stability in many organic media. In the Ghezelbash et al. study, the surface of nano-alumina (Al_2O_3) was treated with a γ-aminopropyltriethoxysilane silane-coupling agent (KH550) in the presence of HCl acid (Figure 3.1), which introduces organic functional groups on the surface of Al_2O_3 nanoparticles [1]. The effect of silsesquioxane coating of Al_2O_3 nanoparticles on their dispersion and on the interfacial strength between nanoparticles and low-density polyethylene composites was investigated. The surface chemistry of the nanoparticles was tailored from hydroxyl groups to alkyl groups with different lengths by reacting methyltrimethoxysilane (C_1), octyltriethoxysilane (C_8) or octadecyltrimethoxysilane (C_{18}) with Al_2O_3 nanoparticles. The inter-particle distance of the nanocomposite based on C_8-coated nanoparticles revealed only a small deviation from the ideal value, showing a very improved particle dispersion in the polymer. The interfacial adhesion between nanoparticles and matrix was also evaluated. The composite based on C_{18}-coated Al_2O_3 nanoparticles had the highest strain at cavitation/necking suggesting a high interfacial adhesion between nanoparticles and polymer [2].

Kumar et al., [3] treated SiO_2 nanofillers with aminopropyl triethoxysilane (APTS) silane in a solution of 70/30 wt.% of acetone/water, and pH in the range 4.5 to 5.5 controlled with acetic acid. As shown in Figure 3.2(A), the surface of nano-SiO_2 filler after treatment with silane molecules indicated that the nanoparticles were well-covered by silane molecules and were also homogenously dispersed compared to those raw nanofillers [Figure 3.2(B)]. This was ascribed to the presence of siloxane bond and to the organic phase mobility and flexibility features of this nanofiller in the stress support of the sample as stated by Gharazi et al., [4].

Pan et al. reported the use of silane coupling agent KH550-modified hexagonal boron nitride (hBN) platelets to fill the PTFE matrix using a cold pressing and sintering method. This study revealed that after surface treatment, the interfacial adhesion between hBN platelets and PTFE matrix was improved and the in-plane orientation degree of hBN platelets in the polymeric matrix decreased, which effectively

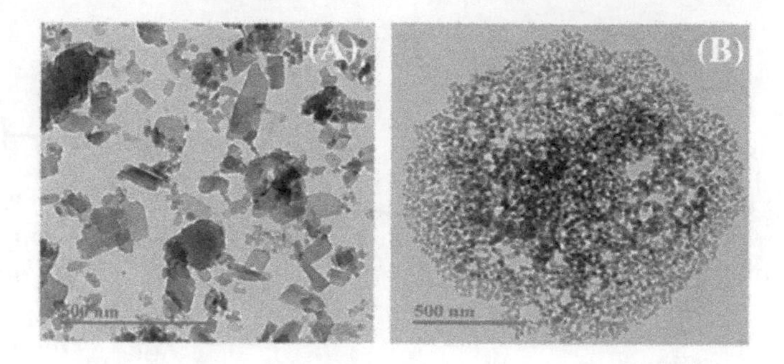

FIGURE 3.2 Transmission Electron Microscopy Images of Nano-SiO_2 Filler A, After Silane Treatment B, Before Silane Treatment [3].

enhanced the thermal conductivity of the resulted composites [5]. ZrO_2 with an enhanced optical transmittance of 65% is obtained by grafting these nanofillers using silane coupling agent, 3-isocyanatopropyltriethoxysilane (IPTES) at a 1:1 ligand-to-nanofillers molar ratio [6]. As demonstrated in Figure 3.3, the hBN nanoparticles were treated using three types of silane molecules including γ-methacryloxypropyltrimethoxysilane, (MPTS): A methacryloxy-terminated silane-coupling agent, carbodiimide treatment (CDI), and DMAP (4 dimethylaminopyridine) [7].

To promote good interfacial adhesion, the hexadecyltrimethoxysilane silane-coupling agent was used for best $CaCO_3$-reinforced polypropylene and propylene-ethylene copolymer-based nanocomposites [8]. Plueddemann et al. [9] first reported the surface treatment of nanoparticles by means of silane coupling agents. They discovered that the surface-treatment ameliorates the compatibility of ceramic particle and polymer chains, which directly increased the properties of their resulting composite materials.

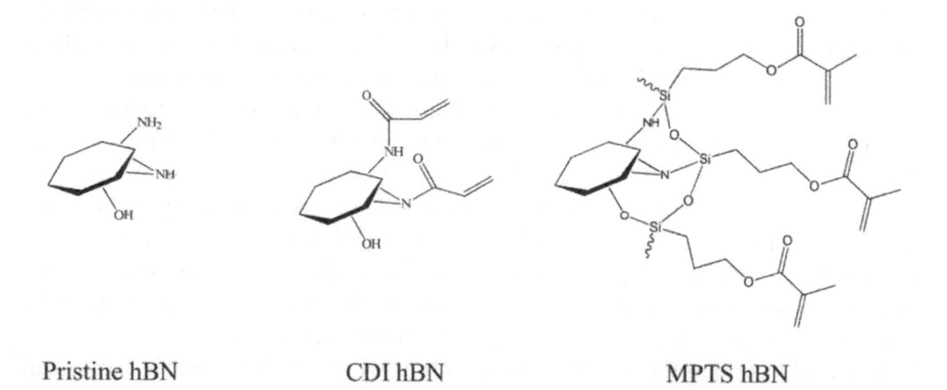

FIGURE 3.3 Schematic Representation of the Chemical Modifications Made to Pristine hBN. CDI: Carbodiimide and MPTS: γ-Methacryloxypropyltrimethoxysilane [7].

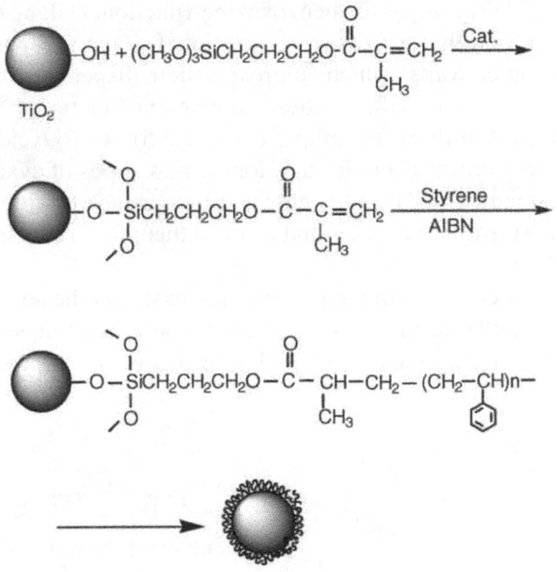

FIGURE 3.4 Reaction Scheme for the Deposition of MPS on Nano TiO_2 Particles and the Formation of the Polystyrene-TiO_2 Particles via Free Radical Polymerization; MPS: 3-(trimethoxysilyl)propylmethacrylate [10].

Rong et al. [10] modified the outer surface of nano-TiO_2 particles using 3-(trimethoxysilyl)propylmethacrylate (MPS) (Figure 3.4) and then in situ polymerized them with styrene. Tuan et al. [11] treated the surface of TiO_2 nanofillers via the 3-glycidoxypropyltrimethoxysilane (GPS) and utilized them for reinforcing the high-density polyethylene (HDPE) composites. Prabakaran et al. [12] prepared amine-functionalized TiO_2 using 3-aminopropyltrimethoxysilane (APTMS) and TiO_2 nanofillers and studied their influences on the dielectric properties of polyvinylidene fluoride-co-hexaflouropropylene (PVDF) matrix. Duraibabu et al. [13] changed the surface of Al_2O_3 nanoparticles using APTMS and the treated-Al_2O_3 nanoparticles were used to fill epoxy resin. Mandhakini et al. [14] investigated the tribological properties on the effect of adding GPS-modified Al_2O_3 in a bismalimide/epoxy-blend-based matrix.

Chena et al. [15] produced a colloidal SiO_2 through a sol–gel process and modified their external surfaces using various kinds of silane-coupling agents, including methyltri-ethoxysilane (MTES), octyltriethoxysilane (OTES), vinyltriethoxysilane (VTES), and methacryloxypropyltrimethoxysilane (MATMS). They also evaluated the effect of ano-SiO_2 surface-treatment on the properties of acrylic-based polyurethane matrix. Ariraman et al. [16] prepared ZrO_2 nanoparticles through a sol–gel technique and treated them by GPS. These nano-ZrO_2 were introduced into cyanate ester/azomethine blends to evaluate their effects on the dielectric properties of their resulted nanocomposites.

Vengatesan et al. [17] grafted the benzoxazine-functional silane onto SiO_2 nanoparticles. They reported that benzoxazine-functional silane was perfectly attached to the nano-SiO_2 outer walls, which improved their dispersion into the benzoxazine matrix. Devaraju et al. [18] synthesized glycidyl functional SiO_2 using GPS through a simple post-grafting technique (Figure 3.5). The glycidyl-treated SiO_2 has been used as reinforcing agent for developing new types of cyanate ester nanocomposites. Ariraman et al. [19] also changed the surface of silica nanoparticles using GPS via a post-grafting process and utilized them for producing cyanate ester nanocomposites.

The surface treatments of ceramic particles are based on the adsorption of silane molecules on their outer surfaces. It is the most useful technique to improve the dispersion stability of ceramic nanoparticles in the various polymeric matrices,

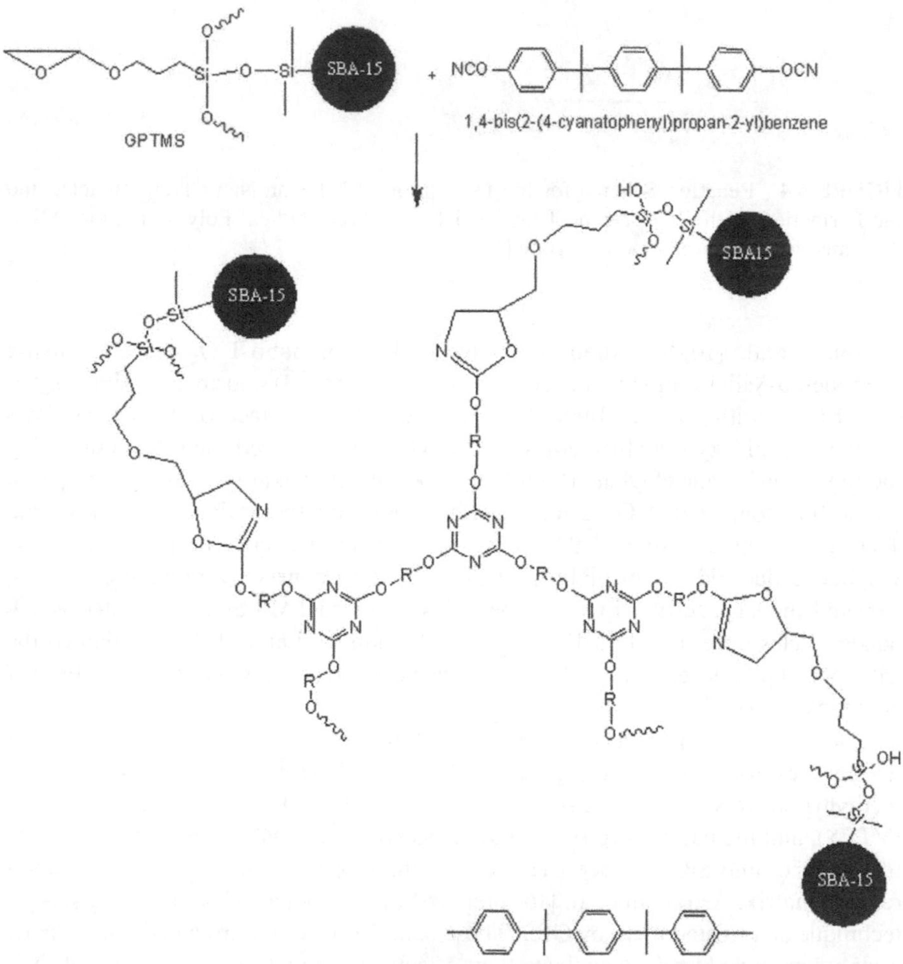

FIGURE 3.5 Schematic Representation of Glycidyl Functionalized Mesoporous Silica (GSBA-15) Reinforced Cyanate Ester (GSBA-15/BCE) Nanocomposites [18].

whether they are thermoplastics, thermosets, or even elastomers. The concept of silane-coupling agents was first reported by Plueddemann et al. [20]. Silane-coupling agents were also utilized to improve the compatibility between particles and polymeric matrix surfaces aiming to ameliorate the properties of composite materials [21, 22]. Whereas the outer surface of the unmodified ceramic nanoparticles is covered only with –OH groups, the surface of the silane-modified ones is covered with 3-methacryloxypropyl-trimethoxysilane molecules. The surface-treated nanoparticles perform differently within polymer matrices or organic solvents compared to the unmodified ones, hence the modified nanoparticles show comparatively good dispersion in both media [23].

The outer surface of ceramic particles may also be treated through reactions with metal alkoxides, epoxides, such as propylene oxide, and alkyl or aryl isocyanates [24]. Guo et al. [25] treated silica nanoparticles with 3-(trimethoxysilyl) propyl methacrylate (MPS) silane-coupling agents, and found that the grafting ratio of MPS on the surface of nano-SiO_2 increased with increasing MPS content. Kim and White [26] modified nano-SiO_2 with silane-coupling agents bearing different aliphatic chain lengths. The surface treatment of TiO_2 nanoparticles has been reported using several silane-coupling agents, including 3-aminopropyltriethoxysilane, n-propyltriethoxysilane and 3-methacryloxypropyltrimethoxysilane [27, 28].

Recently, Sabzi et al. [29] studied the surface modification of nano-TiO_2 using aminopropyltrimethoxysilane (APS) and scrutinized its effect on the final properties of a polyurethane composite coating, revealing an increase in both mechanical and UV-protective properties of the pure urethane. Fourier transform infrared spectroscopy (FTIR) analysis showed that the hydroxyl groups on the surface of the nano-TiO_2 particles (Ti-OH) formed reactive sites for the reaction with alkoxy groups of APS silane molecules. In another study, the dispersion stability of nano-TiO_2 into organic solvents was much improved by modifying the particle outer surface by a silane-coupling agent [30]. Zhao et al. [31] studied the surface modification of nano-TiO_2 using two types of silane-coupling agents, which are 3-aminopropyltrimethoxysilane (APTMS) and 3-isocyanatopropyltrimethoxysilane (IPTMS). The process of nanoparticle surface modification by silane-coupling agents is shown in Figure 3.6.

Majedi et al. [32] developed a novel type of adsorbent based on citric acid/ 3-aminopropyltriethoxysilane-modified ZrO_2 nanoparticles that exhibits a thermal stability at about 180°C. They reported that particle size and pore size distribution were influenced by this surface modification with silane molecules. Truong et al. [33] modified the Al_2O_3 nanoparticle surface with two different silane coupling agents, (3-chloropropyl)triethoxysilane and (octyl)triethoxysilane to increase the hydrophobic interactions within the syndiotactic polypropylene matrix. Guo et al. [34] successfully modified Al_2O_3 nanoparticles with a bi-functional coupling agent, (3-methacryloxypropyl)trimethoxysilane, through a facile neutral solvent method. Mallakpour and Barati [35] investigated the surface treatment of TiO_2 nanoparticles by reaction with a γ-aminopropyltriethoxy silane-coupling agent. The silane molecules were adsorbed on the outer surface of the ceramic nanoparticles at their hydrophilic end and interacted with hydroxyl groups that are preexisting on the nanoparticle's surface.

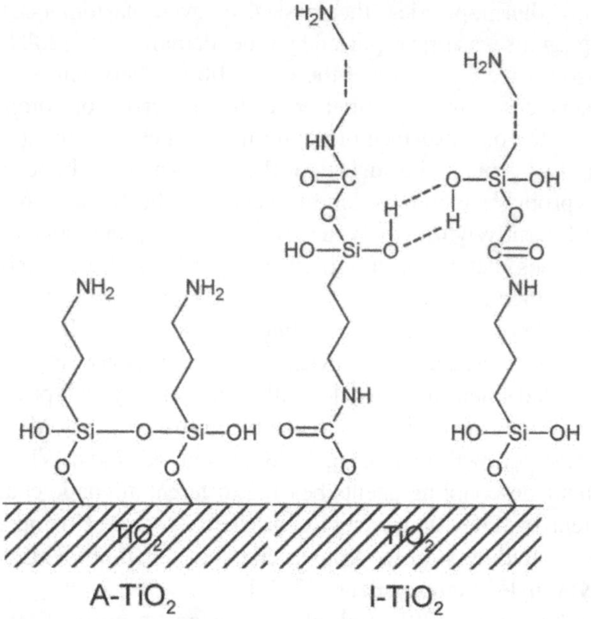

FIGURE 3.6 Crosslinked Net Structure Created on TiO_2 Nanoparticles by Condensation Reaction of APTMS and IPTMS Silane Molecules [31].

3.3 GRAFTING WITH SYNTHETIC POLYMERS

Another approach to treat the surfaces of ceramic particles is based on grafting synthetic polymers directly to the substrate surface, which increases the chemical functionality and alters the surface topology of the native ceramics. Such polymer-grafted ceramic particles are considered to be organic/inorganic hybrid-composite particles. The grafting of polymers for the surface treatment of ceramic filler can be achieved in two manners; the first one is based on grafting of the end-functionalized polymers, which react with the desired surface, while the second one is based on the growing polymer chains from an initiator-terminated, self-assembled monolayer. A high yield of grafting ratio of polymer-grafted nanoceramics has been recorded by starting the graft polymerization from the surface of the nanofiller with polymer-initiating groups [36]. The polymerization technique varied from radical, anionic, or cationic polymerization methods, including propagation of the grafted polymers from the surface of the ceramic nanoparticle.

Arrachart et al. [37] functionalized the surface of TiO_2 nanoparticles using undecenylphosphonic acid to elaborate TiO_2/PMMA nanocomposite, since the treated-TiO_2 and MMA were mixed via in situ bulk copolymerization. Hojjati et al. [38] coated the PMMA chains on TiO_2 spherical surfaces via the RAFT polymerization under supercritical carbon dioxide ($scCO_2$) as the green solvent. Hu et al. [39] modified the surface of ZrO_2 using MMA and produced transparent PMMA/ZrO_2 nanocomposites from MMA-grafted ZrO_2 and MMA via the in situ bulk polymerization. Ou et al. [40] functionalized the surface of TiO_2 nanoparticles using

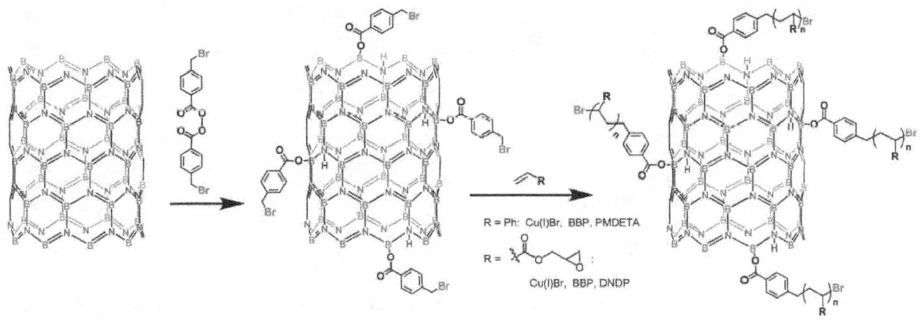

FIGURE 3.7 Grafting of Polymer Brushes in BNNTs via a Surface-initiated ATRP Process [42].

toluene-2,4-diisocyanate (TDI) and used this as a nanofiller for the polypropylene/polyamide blend. Wu et al. [41] changed the surface of TiO_2 nanotubes using phenyl dichloro phosphate and introduced them into the polystyrene (PS) matrix through in situ bulk polymerization.

Ejaz et al. [42] grafted polyglycidyl methacrylate (PGMA) and PS on the surface of boron nitride nanotubes (BNNTs) by the surface-initiated ATRP technique (Figure 3.7). Pasetto et al. [43] grafted the polymeric chains on the surface of crystalline mesoporous SiO_2 (OMS) particles via surface-initiated atom-transfer radical polymerization (SI-ATRP) using MMA or styrene. They investigated the effect of the polymerization conditions on the OMS particle structure.

The methacrylate groups easily penetrate the aggregated ceramic particles and react with the activated sites on their outer surface. The interstitial volume inside the ceramic particle aggregates was partly filled with grafted polymeric chains, while aggregated particles become further separated. In addition, the surfaces of the particles become more hydrophobic, which is critical for increasing the miscibility of both the filler and matrix [44].

Two techniques have been found in the literature to covalently graft polymer chains on the surface of ceramic particles. The first method is the "grafting to" method, in which the end-functionalized polymers react with an appropriate surface. The second method is the "grafting from" method, in which polymer chains are grown from an initiator-terminated self-assembled monolayer [45–47].

By initiating the graft polymerization from initiating groups placed on the particles' surfaces, an increased percentage of successful grafts usually resulted for the many polymer-grafted ceramic particles. The polymerization behavior, which may include radical, anionic, and cationic polymerization methods, includes propagating the grafted polymers from the surface of the particle [48]. Tsubokawa et al. [49] carried out the grafting of hyperbranched polymers having pendant azo groups on SiO_2 nanoparticle surfaces, and they subsequently initiated a radical post-graft polymerization of vinyl monomers from the azo groups of those polymer chains. The method of producing controlled/living radical polymerization from the surfaces of nano-SiO_2 particles using atom-transfer radical polymerization systems was given by von Werne and Patten [50]. Rong et al. [51] studied surface modification of nano-sized

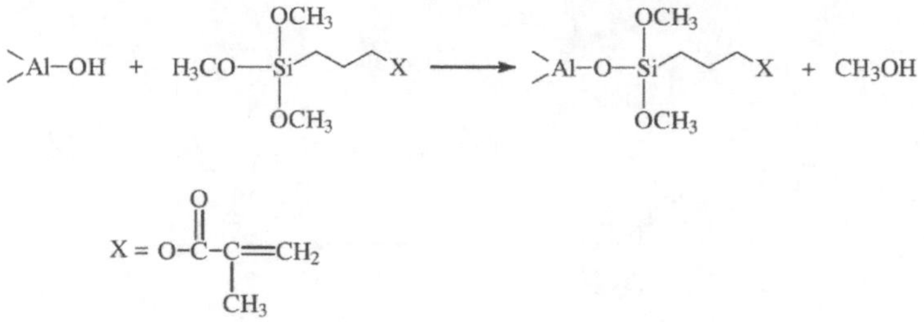

$$\text{>Al–OH} + \text{H}_3\text{CO–Si(OCH}_3)_2\text{–(CH}_2)_3\text{–X} \longrightarrow \text{>Al–O–Si(OCH}_3)_2\text{–(CH}_2)_3\text{–X} + \text{CH}_3\text{OH}$$

$$\text{X} = \text{O–C(=O)–C(CH}_3\text{)=CH}_2$$

FIGURE 3.8 Schematic Drawing of the Blocking Effect of the Growing Polymeric Radicals and/or Grafted Polymer Chains on the Diffusion of Radicals to Alumina Surface [51].

Al_2O_3 particles by grafting polystyrene and polyacrylamide (PAAM) on their surfaces as depicted in Figure 3.8. Sidorenko et al. [52] investigated the radical polymerization of styrene and methyl methacrylate (MMA) on the outer surface of TiO_2 particles, which adsorbed hydroperoxide macro-initiators. Wang et al. [52] reported the synthesis of poly(methyl methacrylate)-grafted TiO_2 nanoparticles through a photocatalytic polymerization process. Under sunlight illumination PMMA chains were grafted directly from the surfaces of the TiO_2 nanoparticles in water.

Shirai et al. [54] studied a radical-graft polymerization of vinyl monomers on the surface of polymethylsiloxane-coated TiO_2 nanoparticles that were further treated with alcoholic hydroxyl groups initiated by the azo groups introduced on Ti/Si-R-OH (Figure 3.9). The graft polymerization of vinyl monomers initiated by a system consisting of trichloroacetyl groups on the TiO_2 surface interacting with $Mo(CO)_6$ was also carried out (Figure 3.9).

Fan et al. [55] reported the surface-initiated graft polymerization of methyl methacrylate from TiO_2 nanoparticle surfaces using a biomimetic initiator. Yokoyama et al. [56] carried out the radical grafting of a biocompatible polymer,

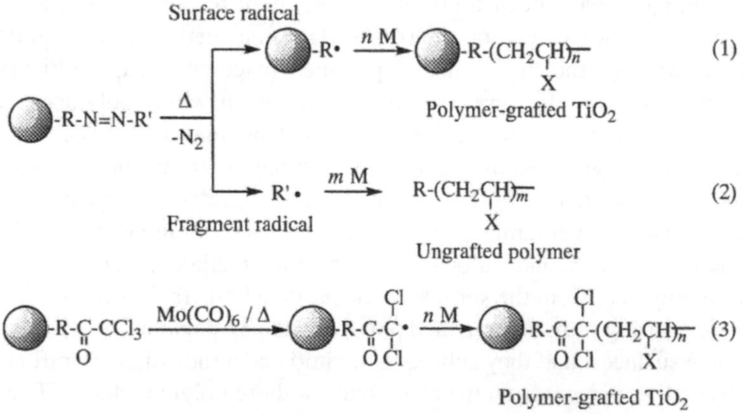

Surface radical: ●-R• $\xrightarrow{n\text{M}}$ ●-R-$(CH_2CH(X))_n$– (1) Polymer-grafted TiO_2

●-R-N=N-R' $\xrightarrow[-N_2]{\Delta}$

Fragment radical: R'• $\xrightarrow{m\text{M}}$ R-$(CH_2CH(X))_m$– (2) Ungrafted polymer

●-R-C(=O)-CCl_3 $\xrightarrow{Mo(CO)_6 / \Delta}$ ●-R-C(=O)-CCl_2• $\xrightarrow{n\text{M}}$ ●-R-C(=O)-CCl_2-$(CH_2CH(X))_n$– (3) Polymer-grafted TiO_2

FIGURE 3.9 The Radical Graft Polymerization of Vinyl Monomers Initiated by Azo Groups Introduced onto Ti/Si-R-OH [54].

2-methacryloyloxyethyl phosphorylcholine (MPC) on the surfaces of nano-SiO_2 particles, which was started either by azo groups previously introduced on the silica surface or by a system containing $Mo(CO)_6$ and trichloroacetyl groups on the ceramic nanoparticle surface.

The graft polymerization of vinyl monomers on SiO_2 and TiO_2 nanoparticles was studied by Shirai and Tsubokawa [57] by initiating a system consisting of trichloroacetyl groups on the nanoparticle surfaces along with molybdenum hexacarbonyl. Liu and Wang [58] reported on the grafting of poly(hydroethyl acrylate) (PHEA) from the surface of ZnO nanoparticles using copper-mediated surface-initiated atom-transfer radical polymerization (SI-ATRP). The surfaces of nano-SiO_2 particles were treated with poly(ethylene glycol) methacrylate (PEGMA) or poly(propylene glycol) methacrylate (PPGMA) for improving dispersibility in the polymeric matrix. SiO_2 nanoparticles were first treated with triethoxyvinylsilane (VTES) molecules to facilitate the incorporation of reactive groups, and PEG or PPG molecules were then grafted on the particle surface through UV-photopolymerization [59]. The mechanism involved in the grafting of PEG or PPG on SiO_2 nanoparticles using UV-photopolymerization is depicted in Figure 3.10.

In order to incorporate the functional groups on the surface of the AlN particles, a silane coupling agent was utilized as it can be grafted easily with the functional groups on the covalent functionalized MWCNTs. The salinization reaction of the AlN was qualitatively and quantitatively confirmed by mean of FTIR and X-ray photoelectron spectroscopy (Fourier transform infrared spectrocopy) [60].

Mallik et al. incorporates a poly(octadecyl acrylate) (pODA)-based organic phase on SiO_2 particles, which is controlled by 2-vinyl-4,6-diamino-1,3,5-triazine (AT), for a chromatography stationary phase [61]. Zhang et al. [62] fabricated a novel support from zwitterionic polymer-grafted SiO_2 nanoparticles (p-SNPs), and Candida rugosa

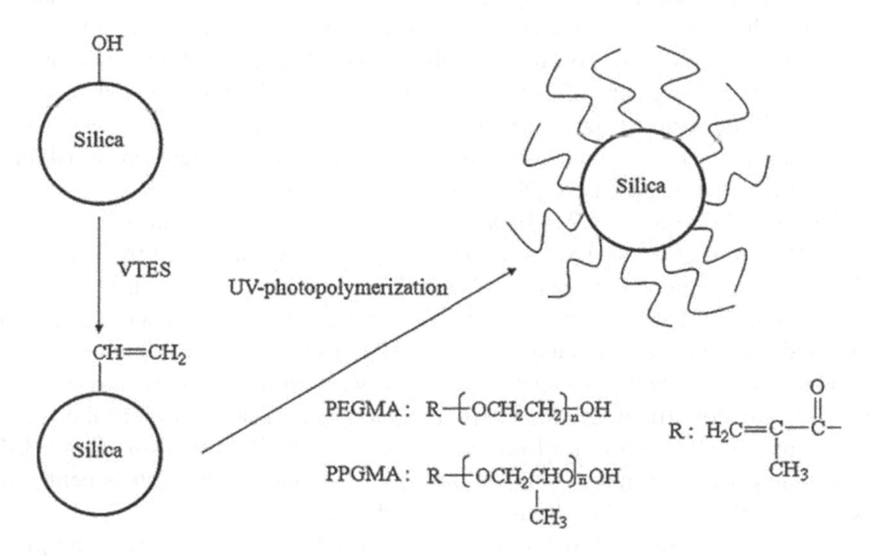

FIGURE 3.10 Grafting of PEG or PPG onto Silica Nanoparticles by UV-photopolymerization [59].

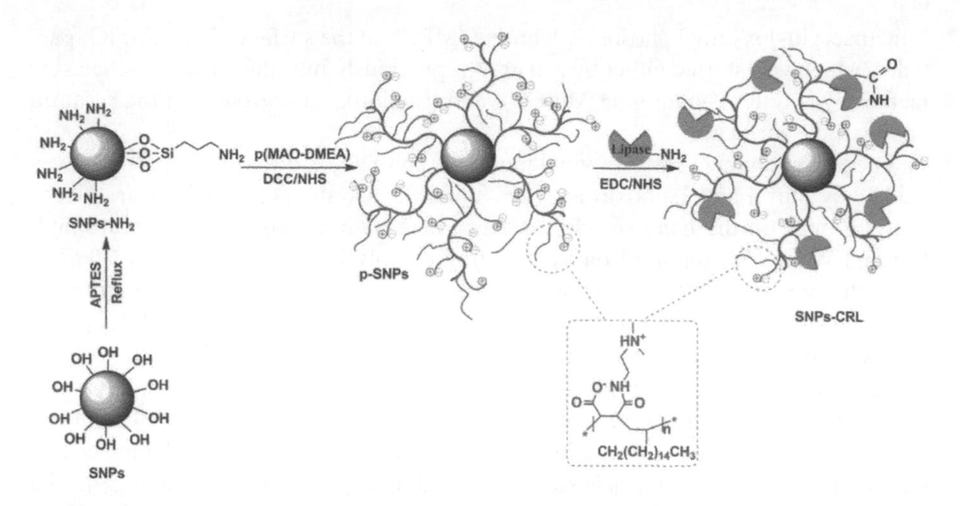

FIGURE 3.11 Schematic for the Preparation of Zwitterionic Polymer-grafting onto the Support and Lipase Immobilization [62].

lipase (CRL) was covalently linked to these p-SNPs (Figure 3.11). This zwitterionic polymer was produced via the reaction between poly(maleic anhydride-alt-1-octadecene) and N,N-dimethylenediamine and bored cetane-side chain.

Reversible addition-fragmentation chain transfer (RAFT) polymerization was used for producing polyacrylamide brushes grafted onto SiO_2 nanoparticles (SiO_2-*g*-PAAm) *via* "grafting to" and "grafting from" technique in a tailored manner. These methods can ensure the control both the polymer's molecular weight and the density of grafted chains [63]. Kutcherlapati et al. reported on the synthesis of poly(*N*-vinyl imidazole) (PNVI)-grafted SiO_2 nanoparticles (SiNP) by using RAFT polymerization through a *grafting-from* approach to show that the self-assembled structure of SiNP was responsible for the improvement of physical properties of SiNP-based nanocomposites. During multistep synthetic processes, well-defined PNVI chains with tailored molecular weights and surface-chain densities resulted from the RAFT agent–anchored SiNP surface using *N*-vinyl imidazole (NVI) as a monomer [64].

Mesocellular silica foam (MCF) was prepared and treated by grafting polyacrylamide (PAAm) onto its outer surface. Grafting process was achieved by free radical polymerization using vinyltrimethoxysilane (VTMS) as coupling agent. The dependency of the final PAAm ratio with respect to monomer and initiator contents, reaction time, and temperature was studied [65]. Meléndez-Ortiz et al. explored new antimicrobial metallopolymer nanoparticles, which were produced via surface-initiated reversible addition–fragmentation chain transfer polymerization of a cobaltocenium having methacrylate monomer from SiO_2 nanofillers as illustrated in Figure 3.12. These nanoparticles formed a complex with β-lactam antibiotics, such as penicillin, rejuvenating the bactericidal activity of the antibiotic [66].

Zhang and Sun developed a novel nanoparticle-based carrier treated by grafting a highly hydrophilic zwitterionic polymer, poly(carboxybetaine methacrylate) (pCBMA), on nano-SiO_2 particles (SNPs). For comparison, an uncharged polymer,

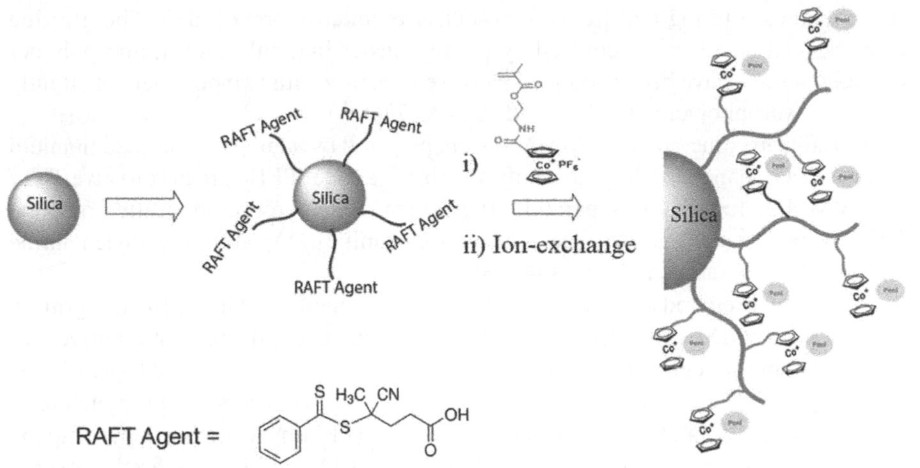

FIGURE 3.12 Synthesis of Cobaltocenium-Containing Silica Nanoparticles by Surface-Initiated RAFT Polymerization [66].

poly(glycidyl methacrylate) (pGMA), was also grafted onto SNPs to fabricate SNPs-pGMA. The two types of polymer-grafted SNPs were characterized and successfully applied for the covalent bonding of two enzymes, catalase and lipase [67]. Three different types of polypropylene-grafted SiO_2 having different grafting chain lengths were synthesized as shown in Figure 3.13 [68].

Polyaniline-grafted SiO_2 nanocomposites (PANI-SiO_2) were prepared by in situ surface chemical oxidative polymerization of aniline with the NH_2-modified silica nanoparticles [69]. Hyper-branched poly(styrene) (HPS) prepared by the copolymerization of styrene (St) and a branching monomer divinylbenzene (DVB) in the presence of a chain transfer agent *n*-dodecanethiol ($C_{12}SH$) was grafted onto the surface of nano-SiO_2 via emulsion polymerization (donated as HPS-SiO_2) [70]. In their research, Li et al. produced polymer dots (PDs) by a moderate hydrothermal treatment of polyvinyl alcohol, and PD-grafted TiO_2 (PDs-TiO_2) nanohybrids with Ti-O-C bonds were obtained by a facile in situ hydrothermal treatment of PDs and $Ti(SO_4)_2$ [71].

The effects of surface treatment of TiO_2 nanoparticles using phosphonic acid molecules on the structure of polymer nanocomposites has been investigated by

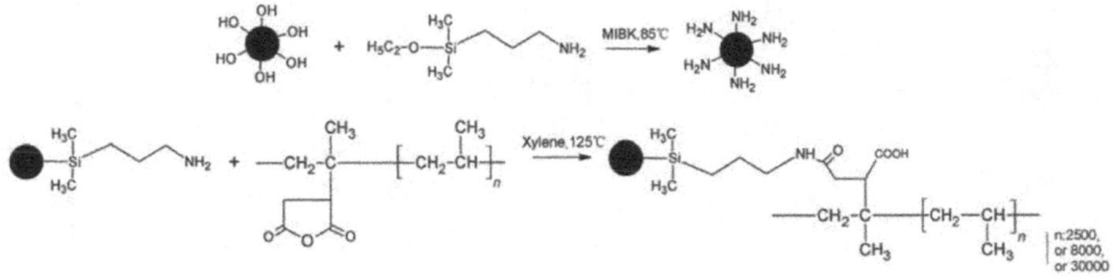

FIGURE 3.13 Schematic of Surface Modification of the Silica Nanoparticles [68].

small-angle scattering and transmission electron microscopy (TEM). The grafting of phosphonic acids was achieved by phase transfer into chloroform, and polymer nanocomposites have been produced by solvent casting with two polymers of slightly different hydrophobicity, PMMA and PEMA [72].

Mixed-matrix membranes (MMMs) were prepared by adding synthesized titanium dioxide (TiO_2) nanoparticles into a polyether block amide (PEBA) matrix to give TiO_2/PBEA MMMs for CO_2/N_2 separation. In addition, amine groups are grafted onto the TiO_2 nanofillers via one-step reaction with dopamine (DA) and polyethyleneimine (PEI) as demonstrated in Figure 3.14 [73].

Yang et al. prepared a series of novel poly(methyl methacrylate) (PMMA) grafted TiO_2 (TiO_2-g-PMMA) nanoparticles and utilized them as pro-oxidant additives for elaborating of biodegradable low-density polyethylene composites (TiO_2-g-PMMA/LDPE) [74]. TiO_2 nanoparticles were treated with 3-methacryloxypropyltrimethoxysilane (γ-MPS) and grafted to polyetherimide (PEI) in *N*-methyl-2-pyrrolidone (NMP) using an anchoring grafting method [75]. TiO_2 surface was firstly covered with polydopamine (PDA) followed by immobilization of an ATRP initiator, α-bromoisobutyryl bromide (BiBB). PMMA and poly(butyl acrylate) (PBA) were then grafted from the PDA-modified TiO_2 of different size (25 and 300 nm) in DMF at room temperature via supplemental activator reducing agent (SARA) ATRP using only 100 ppm of the copper catalyst [76].

Polyacrylamide grafted onto silica nanoparticles (nano-SiO_2-*g*-PAAm) was elaborated via a "grafting-to" approach via controlled reversible addition-fragmentation chain transfer (RAFT) polymerization system with fixed molecular weight and graft density [77]. In another work, the preparation, characterization, and adsorption properties of a novel type of biocompatible cellulose grafted TiO_2 made by the

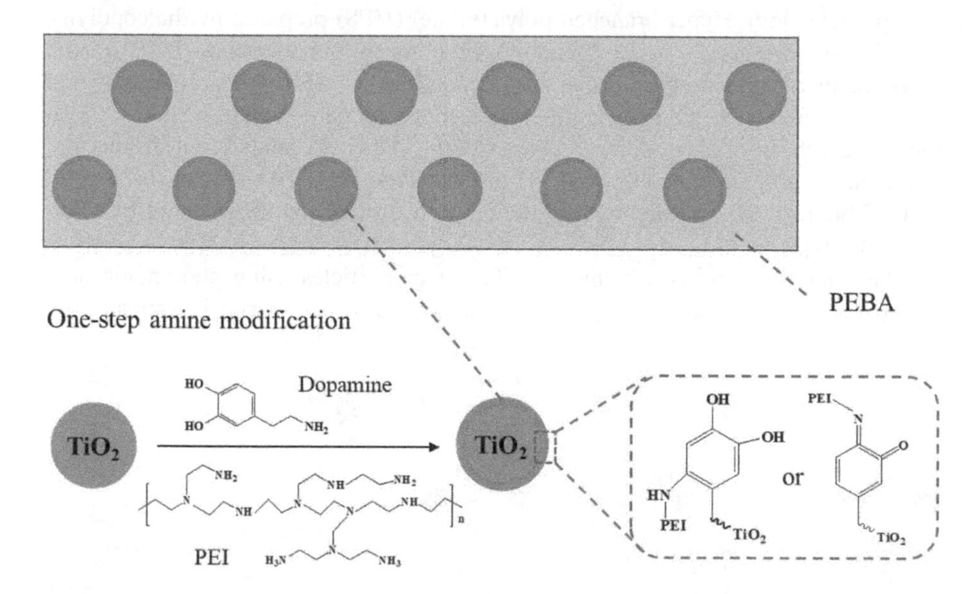

FIGURE 3.14 Schematic Structure of the DP- or PEI-grafted TiO_2/PEBA Mixed-matrix Membranes [73].

click reaction were studied, and the synthetic procedure of this bio-nanocomposite is drawn in Figure 3.15 [78].

Ensafi et al. grafted TiO_2 nanoparticles on the diethyl aminoethyl methacrylate (DEAEMA), which was previously polymerized and cross-linked simultaneously using direct atom transfer radical polymerization (ATRP) [79]. A polylactic acid (PLA) grafting of TiO_2 nanoparticles was also achieved using in-situ polycondensation of lactic acid [80]. A simple method to graft covalently poly(maleic anhydride-alt-1-alkenes) to γ-Al_2O_3 nanofillers was described by Amirilargani et al. The 1-alkenes changed from 1-hexene, 1-decene, 1-hexadecane to 1-octadecene. The grafting reaction occurred between the reactive anhydride moieties of the polymer and surface-free hydroxyl groups, providing very stable bonds [81]. Al_2O_3 were also grafted to citric acid and ascorbic acid before they added to the blend of poly(vinyl alcohol)@poly(vinyl pyrrolidone) (50/50) blend [82].

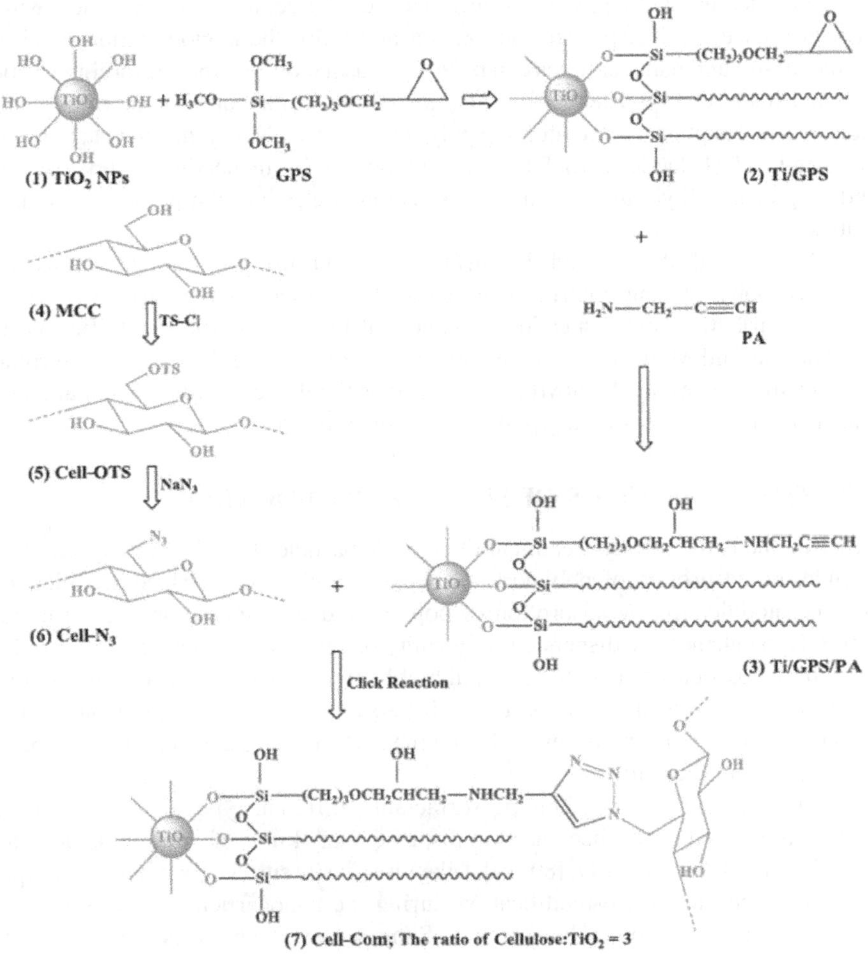

FIGURE 3.15 Synthetic Procedure for Preparation of Cellulose Nanocomposites [78].

Surface-initiated reversible addition-fragmentation chain transfer (SI-RAFT) polymerization has been widely applied to synthesize several polymers grafted from nanoparticles (NPs) for including them into polymer matrices. It was stated that this technique can ameliorate the NPs dispersion by decreasing unfavorable interactions between the ceramic NPs and polymeric matrices. The SI-RAFT polymerization was utilized to graft poly(hexyl, lauryl, and stearyl methacrylate) on SiO_2 nanoparticles, providing good control of the polymerizations [83].

3.4 SURFACE MODIFICATION OF NANOMATERIALS USING SURFACTANTS

Physical treatment of ceramics is generally achieved using surfactants or macromolecules. The polar groups in the surfactants are selectively adsorbed on the ceramic nanoparticle surface due to the presence of electrostatic interactions. The surfactant decreases the physical forces between the ceramic nanoparticles, which in turn reduce the interparticle interaction and tailor the agglomeration, thus the surfactant-treated nanoceramics can be utilized as nanofillers for reinforcing various polymer matrices [84]. Zhu et al. [85] changed the surface of SiO_2 nanofiller using oleic acid and applied it for reinforcing polylactide matrix. Nakayama and Hayashi [86] produced poly(L-lactic acid)/TiO_2 nanocomposites by dispersing carboxylic acid and long-chain alkyl amine-treated TiO_2 nanoparticles into the poly(L-lactic acid) matrix.

Rahmani et al. [87] treated the surface of $CaCO_3$ using steric acid and used to reinforce polypropylene matrix. They studied the effect of the surface treatment on the distribution and dispersion of nano-$CaCO_3$ into the thermoplastic matrix. Mallakpour and Mani [88] changed surface of ZrO_2 using 2, 3, 4, 5-tetrabromo-6-[(4-hydroxyphenyl)carbamoyl] benzoic acid as the flame-retardant agent and used it as a nanofiller for reinforcing poly(amide-imide) matrix.

3.5 OTHER METHODS OF SURFACE MODIFICATION

Other methods for surface treatment of ceramic particles have been reported; these include the adsorption of polymeric dispersants and in situ surface-modification. Surface modification by adsorption of polymeric dispersants is one of the easiest methods to enhance the dispersion uniformity of ceramics in aqueous systems. The hydrophilic particles can be dispersed in highly polar organic solvents using anionic or cationic polymeric dispersants. These dispersants engender steric repulsive forces among the polymeric chains and enhance the surface charge, which provides better dispersibility of these fillers.

As an example of using anionic surfactants, different types of polycarboxylic acids and their salts have been used to disperse many kinds of ceramic oxide nanoparticles, like TiO_2 and Al_2O_3 [89–91]. Likewise, *in situ* surface treatment techniques that are based on surface modification during the nanoparticle synthesis process, have also been described in the literature. Some of these techniques used the reverse micelle method, thermal decomposition of organometallic compounds and polyol methods [92–94]. The capping agents or surfactants, such as tri-octylphosphine oxides (TOPO), oleic acid, and amines, are dissolved in the synthesis solution to avoid the agglomeration of ceramics nanofillers. Surfactant-capped nanoparticles that have been prepared using *in situ* surface-modification technique can be further treated to control their surface features.

3.6 CONCLUSIONS

The surface-modification of ceramic fillers are potentially useful in the development of hybrid polymer/ceramic composites. The surface-modification of ceramic fillers can significantly improve both the miscibility and interfacial interaction of the ceramic filler with polymeric matrices, demonstrating unique superior properties, such as improved mechanical, optical, electronic, gas-barrier, and flame-retardancy properties. Different approaches have been conducted to modify the outer surface of ceramic fillers including treatment with silane-coupling agents, treatment by surfactants like acids and bases, and grafting of functional polymerizable groups. The choice of surface-modification technique depends on the type of ceramic as well as the type of matrix.

REFERENCES

1. Ghezelbash, Z., Ashouri, D., Mousavian, S. et al. 2012. Surface modified Al_2O_3 in fluorinated polyimide/Al_2O_3 nanocomposites: Synthesis and characterization. *Bulletin of Materials Science* 35: 925–931.
2. Liu, D., Pourrahimi, A. M., Olsson, R. T. et al., 2015. Influence of nanoparticle surface treatment on particle dispersion and interfacial adhesion in low-density polyethylene/aluminium oxide nanocomposites. *European Polymer Journal* 66: 67–77.
3. Kumar, S. R., Patnaik, A., Bhat, I. K. 2018. Wear behavior of light-cured dental composite reinforced with silane-treated nanosilica filler. *Polymers for Advanced Technologies* 29: 1394–1403.
4. Gharazi, S., Ershad-Langroudi, A., Rahimi, A. 2011. The influence of silica synthesis on the morphology of hydrophilic nanocomposite coating. *Scientia Iranica* 18: 785–789.
5. Pan, C., Kou, K., Jia, Q. et al. Improved thermal conductivity and dielectric properties of hBN/PTFE composites via surface treatment by silane coupling agent. *Composites Part B: Engineering* 111: 83–90.
6. Puguan, J. M. C., Kim, H. 2017. ZrO_2-silane-graft-PVdFHFP hybrid polymer electrolyte: Synthesis, properties and its application on electrochromic devices. *Electrochimica Acta* 230: 39–48.
7. Goldin, N., Dodiuk, H., Lewitus, D. 2017. Enhanced thermal conductivity of photopolymerizable composites using surface modified hexagonal boron nitride fillers. *Composites Science and Technology*. 10.1016/j.compscitech.2017.09.001.
8. Joshi, M., Brahma, S., Roy, A. et al. 2017. Nano-calcium carbonate reinforced polypropylene and propylene-ethylene copolymer nanocomposites: Tensile vs. Impact behavior. *Fibers and Polymers* 18: 2161–2169.
9. Plueddemann, E. P., Clark, H. A., Nelson, L. E. et al. 1962. New silane coupling agents for reinforced plastics. *Modern Plastic* 39: 135–187.
10. Rong, Y., Chen, H. Z., Wu, G., et al. 2005. Preparation and characterization of titanium dioxide nanoparticle/polystyrene composites via radical polymerization. *Materials Chemistry and Physics* 9: 370–374.

11. Tuan, V. M., Jeong, D. W., Yoon, H. J. et al. 2014. Using rutile TiO_2 nanoparticles reinforcing high density polyethylene resin. *International Journal of Polymer Science* 2014: 1–7.
12. Prabakaran, K., Mohanty, S., Nayak, S. K. 2014. Influence of surface modified TiO_2 nanoparticles on dielectric properties of PVdF–HFP nanocomposites. *Journal of Materials Science: Materials in Electronics* 25: 4590–4602.
13. Duraibabu, D., Alagar, M., Ananda Kumar, S. Studies on mechanical, thermal and dynamic mechanical properties of functionalized nanoalumina reinforced sulphone ether linked tetraglycidyl epoxy nanocomposites. 2014. *RSC Advances* 4: 40132–40140.
14. Mandhakini, M., Lakshmikanthan, T., Chandramohan, A. 2014. Effect of nanoalumina on the tribology performance of C4-ether-linked bismaleimide-toughened epoxy nanocomposites. *Tribological Letters* 54: 67–79.
15. Chena, G., Zhoua, S., Gua, G. et al. 2005. Effects of surface properties of colloidal silica particles on redispersibility and properties of acrylic-based polyurethane/silica composites. *Journal of Colloid and Interface Science* 281: 339–350.
16. Ariraman, M., Sasi Kumar, R., Alagar, M. 2014. Design of cyanate ester/azomethine/ ZrO_2 nanocomposites high-k dielectric materials by single step sol–gel approach. *Journal of Applied Polymer Science* 131: 41097–41102.
17. Yang, H., Hu, Y. 2012. Covalent functionalization of graphene with organosilane and its use as a reinforcement in epoxy composites. *Composites Science and Technology* 72: 737–743.
18. Devaraju, S., Vengatesan, M. R., Selvi, M. et al. 2013. Mesoporous silica reinforced cyanate ester nanocomposites for low k dielectric applications. *Microporous and Mesoporous Materials* 157: 157–164.
19. Ariraman, M., Sasi Kumar, R., Alagar, M. 2014. Studies on FMCM-41 reinforced cyanate ester nanocomposites for low k applications. *RSC Advances* 4: 57759–57767.
20. Plueddemann, E. P., Clark, H. A., Nelson, L. E. et al. 1962. Silane coupling agents for reinforced plastics. *Modern Plastics* 39: 135.
21. Owen, M. J. 2002. Coupling agents: Chemical bonding at interfaces. *Adhesion Science and Engineering* 2: 403–431.
22. Plueddemann, E. P. 1970. Adhesion through silane coupling agents. *Journal of Adhesion* 2: 184–201.
23. Uyanik, M. 2008. Synthesis and characterization of TiO_2 nanostars. PhD Thesis. Saarbrucken, Germany: Saarland University.
24. Lin, F. 2006. Preparation and characterization of polymer TiO_2 nanocomposites via In-situ polymerization. Master's Thesis. Ontario, Canada: University of Waterloo.
25. Guo, Y. K., Wang, M. Y., Zhang, H. Q. et al. 2008. The surface modification of nanosilica, preparation of nanosilica/acrylic core- shell composite latex, and its application in toughening PVC matrix. *Journal of Applied Polymer Science* 107: 2671–2680.
26. Kim, K. J., White, J. L. 2002. Silica surface modification using different aliphatic chain length silane coupling agents and their effects on silica agglomerate size and processability. *Composite Interfaces* 9: 541–556.
27. Ukaji, E., Furusawa, T., Sato, M. et al. 2007. The effect of surface modification with silane coupling agent on suppressing the photo-catalytic activity of fine TiO_2 particles as inorganic UV filter. *Applied Surface Science* 254: 563–569.
28. Tang, E., Liu, H., Sun, L. et al. 2007. Fabrication of zinc oxide/poly(styrene) grafted nanocomposite latex and its dispersion. *European Polymer Journal* 43: 4210–4218.
29. Sabzi, M., Mirabedini, S. M., Zohuriaan-Mehr, J. et al. 2009. Surface modification of TiO_2 nano-particles with silane coupling agent and investigation of its effect on the properties of polyurethane composite coating. *Progress in Organic Coatings* 65: 222–2288.

30. Wang, C., Mao, H., Wang, C. et al. 2011. Dispersibility and hydrophobicity analysis of titanium dioxide nanoparticles grafted with silane coupling agent. *Industrial and Engineering Chemistry Research* 50: 11930–11934.
31. Zhao, J., Milanova, M., Warmoeskerken, M. M. C. G. et al. 2012. Surface modification of TiO_2 nanoparticles with silane coupling agents. *Colloids and Surfaces A* 413: 273–279.
32. Ma, S. R., Shi, L. Y., Feng, X. et al. 2005. Graft modification of ZnO nanoparticles with silane coupling agent KH570 in mixed solvent. *Journal of Shanghai University* 12: 278–282.
33. Majedi, A., Davar, F., Abbasi, A. 2018 Citric acid-silane modified zirconia nanoparticles: Preparation, characterization and adsorbent efficiency. *Journal of Environmental Chemical Engineering* 6: 701–709.
34. Guo, Z., Pereira, T., Choi, O. et al. 2006. Surface functionalized alumina nanoparticle filled polymeric nanocomposites with enhanced mechanical properties. *Journal of Materials Chemistry* 16: 2800–2808.
35. Mallakpour, S., Barati, A. 2011. Efficient preparation of hybrid nanocomposite coatings based on poly(vinyl alcohol) and silane coupling agent modified TiO_2 nanoparticles. *Progress in Organic Coatings* 71: 391–398.
36. Prucker, O., Ruhe, J. 1998. Synthesis of poly(styrene) monolayers attached to high surface area silica gels through self-assembled monolayers of azo initiators. *Macromolecules* 31: 592–601.
37. Arrachart, G., Karatchevtseva, I., Heinemann, A. et al. 2011. Synthesis and characterisation of nanocomposite materials prepared by dispersion of functional TiO_2 nanoparticles in PMMA matrix. *Journal of Materials Chemistry* 21, 13040–13046.
38. Hojjati, B., Charpentier, P. A. 2010. Synthesis of TiO_2-polymer nanocomposite in supercritical CO_2 via RAFT polymerization. *Polymer* 51: 5345.
39. Hu, Y., Gu, G., Zhou, S. 2011. Preparation and properties of transparent PMMA/ZrO_2 nanocomposites using 2-hydroxyethyl methacrylate as a coupling agent. *Polymer,* 52: 122–129.
40. Ou, B., Li, D., and Liu, Y. 2009. Effect of multiwalled carbon nanotubes on the crystallization and hydrolytic degradation of biodegradable poly(l lactide), *Composites Science and Technology* 69: 627–632.
41. Wu, Y., Song, L., Hu, Y. 2012. Thermal properties and combustion Behaviors of polystyrene/surface-modified TiO_2 nanotubes nanocomposites. *Polymer-Plastics Technology and Engineering* 51: 647–653.
42. Ejaz, M., Rai, S.C., Wang, K. et al. 2014. Surface-initiated atom transfer radical polymerization of glycidyl methacrylate and styrene from boron nitride nanotubes. *Journal of Materials Chemistry C* 2: 4073–4088.
43. Pasetto, P., Blas, H., Audouin, F. et al. 2009. Mechanistic insight into surface-initiated polymerization of methyl methacrylate and styrene via ATRP from ordered mesoporous silica particles. *Macromolecules* 42: 5983–5995.
44. Rong, M. Z., Zhang, M. Q., Zheng, Y. X. et al. 2001. Structure-property relationships of irradiation grafted nano inorganic particle filled polypropylene composites. *Polymer* 42: 167–183.
45. Tran, Y., Auroy, P. 2001. Synthesis of polystyrene sulfonate brushes. *Journal of the American Chemical Society* 123: 3644–3654.
46. Mansky, P., Liu, Y., Huang, E. et al. 1997. Controlling polymer-surface interactions with random copolymer brushes. *Science* 275: 1458–1460.
47. Prucker, O., Ruhe, J. 1998. Synthesis of poly(styrene) monolayers attached to high surface area silica gels through self-assembled monolayers of azo initiators. *Macromolecules* 31: 592–601.
48. Kickelbick, G. 2003. Concepts for the incorporation of inorganic building blocks into organic polymers on a nanoscale. *Progress in Polymer Science* 28: 83–114.

49. Tsubokawa, N., Kogure, A., Sone, Y. 1995. Grafting of polyesters from ultra- fine inorganic particles: copolymerization of epoxides with cyclic acid anhydrides initiated by COOK groups introduced onto the surface. *Journal of Polymer Science Part A Polymer Chemistry* 28: 1923–1933.
50. von Werne, T., Patten, T. E. 2001. Atom transfer radical polymerization from nanoparticles: A tool for the preparation of well-defined hybrid nanostructures and for understanding the chemistry of con- trolled/"living" radical polymerizations from surfaces. *Journal of the American Chemical Society* 123: 7497–7505.
51. Rong, M. Z., Ji, Q. L., Zhang, M. Q. et al. 2002. Graft polymerization of vinyl monomers onto nanosized alumina particles. *European Polymer Journal* 38: 1573–1582.
52. Sidorenko, A., Minko, S., Gafijchuk, G. et al. 1999. Radical polymerization initiated from a solid substrate: grafting from the surface of an ultrafine powder. *Macromolecules* 32: 4539–4543.
53. Wang, X., Song, X., Lin, M. et al. 2007. Surface initiated graft polymerization from carbon-doped TiO_2 nanoparticles under sunlight illumination. *Polymer* 48: 5834–5838.
54. Shirai, Y., Kawatsura, K., Tsubokawa, N. 1999. Graft polymerization of vinyl monomers from initiating groups introduced onto polymethylsiloxane-coated titanium ioxide modified with alcoholic hydroxyl groups. *Progress in Organic Coatings* 36: 217–224.
55. Fan, X., Lin, L., Messersmith, P. B. 2006. Surface-initiated polymerization from TiO_2 nanoparticle surfaces through a biomimetic initiator: A new route toward polymer–matrix nanocomposites. *Composites Science and Technology* 66: 1195–1201.
56. Yokoyama, R., Suzuki, S., Shirai, K. et al. 2006. Preparation and properties of biocompatible polymer-grafted silica nanoparticles. *European Polymer Journal* 42: 3221–3229.
57. Shirai, Y., Tsubokawa, N. 1997. Grafting of polymers onto ultrafine inorganic particle surface: graft polymerization of vinyl monomers initiated by the system consisting of trichloroacetyl groups on the surface and molybdenum hexacarbonyl. *Reactive and Functional Polymers* 32: 153–160.
58. Liu, P., Wang, T. 2008. Poly(hydroethyl acrylate) grafted from ZnO nanoparticles via surface-initiated atom transfer radical polymerization. *Current Applied Physics* 8: 66–70.
59. Shin, Y., Lee, D., Lee, K. et al. 2008. Surface properties of silica nanoparticles modified with polymers for polymer nanocomposite applications. *Journal of Industrial and Engineering Chemistry* 14: 515–519.
60. Kim, M., Park, S., Park, J. 2017. Effect of the grafting reaction of aluminum nitride on the multi-walled carbon nanotubes on the thermal properties of the poly(phenylene sulfide) composites. *Polymers* 9: 452–467.
61. Mallik, A. K., Noguchi, H., Han, Y., 2018. Enhancement of thermal stability and selectivity by tntroducing aminotriazine comonomer to poly(octadecyl acrylate)-grafted silica as chromatography matrix. *Separations* 5: 15–21.
62. Zhang, C., Dong, X., Guo, Z., et al. 2018. Remarkably enhanced activity and substrate affinity of lipase covalently bonded on zwitterionic polymer-grafted silica nanoparticles. *Journal of Colloid and Interface Science*. doi: https://doi.org/10.1016/j.jcis.2018.02.039.
63. Ghasemi, S., Karim, S. 2018. Organic/inorganic hybrid composed of modified polyacrylamide grafted silica supported Pd nanoparticles using RAFT polymerization process: Controlled synthesis, characterization and catalytic activity. *Materials Chemistry and Physics* 205: 347–358.
64. Kutcherlapati, S. R., Koyilapu, R., Jana, T. 2018. Poly(N-vinyl imidazole) grafted silica nanofillers: Synthesis by RAFT polymerization and nanocomposites with polybenzimidazole. *Journal of Polymer Science Part A: Polymer Chemistry* 56: 365–375.

65. Meléndez-Ortiz, H. V., Saucedo-Zuñig, N., Puente-Urbina, B. et al. 2018. Polymer-grafted mesocellular silica foams: Influence of reaction conditions on the mesostructure and polymer content. *Materials Chemistry and Physics* 203: 333–339.
66. Pageni, P., Yang, P., Chen, Y. P. et al. 2018. Charged Metallopolymer-Grafted Silica Nanoparticles for Antimicrobial Applications. *Biomacromolecules* 19: 417–425.
67. Zhang, L., Sun, Y. 2018. Poly(carboxybetaine methacrylate)-grafted silica nanoparticle: A novel carrier for enzyme immobilization. *Biochemical Engineering Journal* 132: 122–129.
68. Wang, W., Wu, J. 2018. Interfacial influence on mechanical properties of polypropylene/polypropylene-grafted silica nanocomposites. *Journal of Applied Polymer Science* 135: 45887.
69. Ma, P., Tan, J., Cheng, H., et al. 2018. Polyaniline-grafted silica nanocomposites-based gel electrolytes for quasi-solid-state dye-sensitized solar cells. *Applied Surface Science* 427: 458–464.
70. Xu, X., Zeng, Y., Zhang, F. A. 2018. Enhancement of polystyrene composites by hyper-branched polymer-grafted nano-SiO_2. *Journal Plastics, Rubber and Composites Macromolecular Engineering* 47: 266–272.
71. Li, G., Wang, F., Liu, P., et al. 2018. Polymer dots grafted TiO_2 nanohybrids as high performance visible light photocatalysts. *Chemosphere* 197: 526–534.
72. Genix, A.C., Schmitt-Pauly, C., Alauzun, J.G., et al. 2017. Tuning local nanoparticle arrangements in TiO_2–polymer nanocomposites by grafting of phosphonic acids. *Macromolecules* 50: 7721–7729.
73. Zhu, H., Yuan, J., Zhao, J. et al. 2018. Enhanced CO_2/N_2 separation performance by using dopamine/polyethyleneimine-grafted TiO_2 nanoparticles filled PEBA mixed-matrix membranes. *Separation and Purification Technology*. doi.org/10.1016/j.seppur.2018.02.020.
74. Yang, W., Song, S., Zhang, C. et al. 2018. Enhanced photocatalytic oxidation and biodegradation of polyethylene films with PMMA grafted TiO_2 as pro-oxidant additives for plastic mulch application. *Polymer Composites* 39(10): 3409–3417.
75. Zhu, P., Liu, B., Bao, L. 2018. Preparation of double-coated TiO_2 nanoparticles using an anchoring grafting method and investigation of the UV resistance of its reinforced PEI film. *Progress in Organic Coatings* 104: 81–90.
76. Kopeć, M., Spanjers, J., Scavo, E. et al. 2018. Surface-initiated ATRP from polydopamine-modified TiO_2 nanoparticles. *European Polymer Journal* 106: 291–296.
77. Ghasemi, S., Karim, S. 2018. Controlled synthesis of modified polyacrylamide grafted nano-sized silica supported Pd nanoparticles via RAFT polymerization through "grafting to" approach: Application to the Heck reaction. *Colloid and Polymer Science* 296: 1323–1332.
78. Fallah, Z., Isfahani, H. N., Tajbakhsh, M. et al. 2018. TiO_2-grafted cellulose via click reaction: an efficient heavy metal ions bioadsorbent from aqueous solutions. *Cellulose* 25: 639–660.
79. Ensafi, A. A., Khoddami, E., Nabiyan, A. et al. 2017. Study the role of poly(diethyl aminoethyl methacrylate) as a modified and grafted shell for TiO_2 and ZnO nanoparticles, application in flutamide delivery. *Reactive and Functional Polymers* 116: 1–8.
80. Shaikh, T., Rathore, A., Kaur, D. H. 2017. Poly (lactic acid) grafting of TiO_2 nanoparticles: A Shift in Dye Degradation Performance of TiO_2 from UV to Solar Light. *Chemistry Select* 2: 6901–6908.
81. Amirilargani, M., Merlet, R. B., Nijmeijer, A., et al. 2018. Poly (maleic anhydride-alt-1-alkenes) directly grafted to γ-alumina for high-performance organic solvent nanofiltration membranes. *Journal of Membrane Science* 564: 259–266.
82. Mallakpour, S., Sadeghzadeh, R. 2017. Facile and green methodology for surface-grafted Al_2O_3 nanoparticles with biocompatible molecules: preparation of the

poly(vinyl alcohol)@poly(vinyl pyrrolidone) nanocomposites. *Polymers for Advanced Technologies* 28: 1719–1729.
83. Khani, M. M., Woo, D., Mumpower, E. L., et al. 2017. Poly(alkyl methacrylate)-grafted silica nanoparticles in polyethylene nanocomposites. *Polymer.* doi: 10.1016/j.polymer.2016.12.046.
84. Zou, H., Wu, S., Shen, J. 2008. Polymer/silica nanocomposites: Preparation, characterization. *Chemical Reviews* 108: 3893–3957.
85. Zhu, A., Diao, H., Rong, Q. et al. 2010. Preparation and proper-ties of polylactide–silica nanocomposites. *Journal of Applied Polymer Science* 116: 2866–2873.
86. Nakayama, N., Hayashi, T. 2007. Preparation and characterization of poly(L-lactic acid)/TiO_2 nanoparticle nanocomposite films with high transparency and efficient photodegradability. *Polymer Degradation and Stability* 92: 1255–1264.
87. Rahmani, M., Ghasemia, F. A., Payganeh, G. 2014. Effect of surface modification of calcium carbonate nanoparticles on their dispersion in the polypropylene matrix using stearic acid. *Mechanics & Industry* 15: 63–67.
88. Mallakpour, S., Mani, L. 2014. Improvement of the interactions between modified ZrO_2 and poly(amide-imide) matrix by using unique biosafe diacid as a monomer and coupling agent. *Journal Polymer-Plastics Technology and Engineering* 53: 1574–1582.
89. Sato, K., Kondo, S., Tsukada, M. et al. 2007. Influence of solid fraction on the optimum molecular weight of polymer dispersants in aqueous TiO_2 nanoparticle suspensions. *Journal of the American Ceramic Society* 90: 3401–3406.
90. Palmqvist, L., Holmberg, K. 2008. Dispersant adsorption and viscoelasticity of alumina suspensions measured by quartz crystal microbalance with dissipation monitoring and in situ dynamic rheology. *Langmuir* 24: 9989–9996.
91. Nsib, F., Ayed, N., Chevalier, Y. 2006. Dispersion of hematite suspensions with sodium polymethacrylate dispersants in alkaline medium. *Colloids and Surfaces A* 286: 17–26.
92. Pileni, M. P. 2003. The role of soft colloidal templates in controlling the size and shape of inorganic nanocrystals. *Nature Materials* 2: 145–150.
93. Murray, C. B., Kagan, C. R., Bawendi, M. G. 2000. Synthesis and characterization of monodisperse nanocrystals and close-packed nano-crystal assemblies. *Annual Review of Materials Science* 30: 545–610.
94. Feldmann, C. 2003. Polyol-mediated synthesis of nanoscale functional materials. *Advanced Functional Materials* 13: 101–107.

4 Processing Methods of Polymer/Ceramic Composites

4.1 INTRODUCTION

The use of ceramic fillers as reinforcing agent in polymers is highly attractive for producing new kinds of composites suitable for high-performance applications. Most of the literature about polymer composites is based on the use of ceramics and clays as fillers [1–15]. It has been shown that addition of ceramics in synthetic thermoplastic and thermosetting polymers, like epoxy, phenolic, benzoxazine, rubber, polyethylene, silicon rubber, conjugated polymers, etc. Improved mechanical, thermal, electrical, anti-aging and barrier properties have resulted from the combination of ceramic and polymeric matrices. There are several methods of producing polymer/ceramic micro- and nanocomposites, such as the direct-melt processing, solution-blending, and sol–gel techniques. All these processing approaches have their own advantages and shortcomings.

Most of the published literature on polymer/ceramic composites is based on the solution-blending technique, where a polymeric monomer is dissolved in a desired solvent along with ceramic filler followed by eliminating the solvent to give the targeted composite. The solution-blending technique cannot be used for large production of composites as dissolution of polymer in the solvent and subsequent evaporation of the solvent pose several engineering difficulties and environmental issues. However, this method can afford good uniformity in the dispersion of ceramic, especially nanoparticles into the polymeric matrices, which can result in a huge improvement in the mechanical properties of the composites. Polymer/ceramic composites have been prepared like most common synthesis polymer matrix composites, using different synthesis techniques [16–19] including:

- Sol–gel technique
- In situ polymerization
- Solution blending
- Melt processing

The selected technique of preparing any polymer/ceramic has an important effect on the final performance of the resulting composite material. Several methods have been used to achieve this purpose, such as solution-blending technique, sol–gel technique, direct-melt processing, and in situ polymerization, etc. This chapter is dedicated to the various synthesis methods used for producing polymer/ceramic micro- and nanocomposite. According to the nature and reactivity of the ceramic fillers, as well as the processing parameters of polymeric matrices, a selection of synthetic technique can be made, and this can control the final properties of the prepared composite materials.

4.2 SOL–GEL TECHNIQUE

The ceramic nanofillers can be processed using a sol–gel technique to reinforce the various polymeric matrices, where these nanofillers can be dissolved in either aqueous or organic media in the presence of their matrices, forming interpenetrated networks between ceramics and organic moieties at mild temperatures; this structure improves the compatibility between the hybrids constituents and constructs a strong interfacial adhesion between them. This technique has been extensively used to elaborate many kinds of polymer nanocomposites reinforced by ceramic fillers including ZrO_2, SiO_2, Al_2O_3, $CaCO_3$, and TiO_2. In this technique, metal alkoxides, silane coupling agents, and many polymer precursors have been utilized for the elaboration of polymer/ceramic nanocomposites.

Sun et al. [20] reported the synthesis of polypropylene/SiO_2 nanocomposites using a two-step sol–gel technique: diffusion of TEOS through the PP matrix by a supercritical carbon dioxide ($SCCO_2$) as a carrier and a swelling agent, followed by hydrolysis/condensation reactions among the precursor molecules confined in the polymer network. The results affirmed the construction of nano-sized silica networks, which were found to be uniformly dispersed in the PP matrix (Figure 4.1). Jain et al. [21–23]

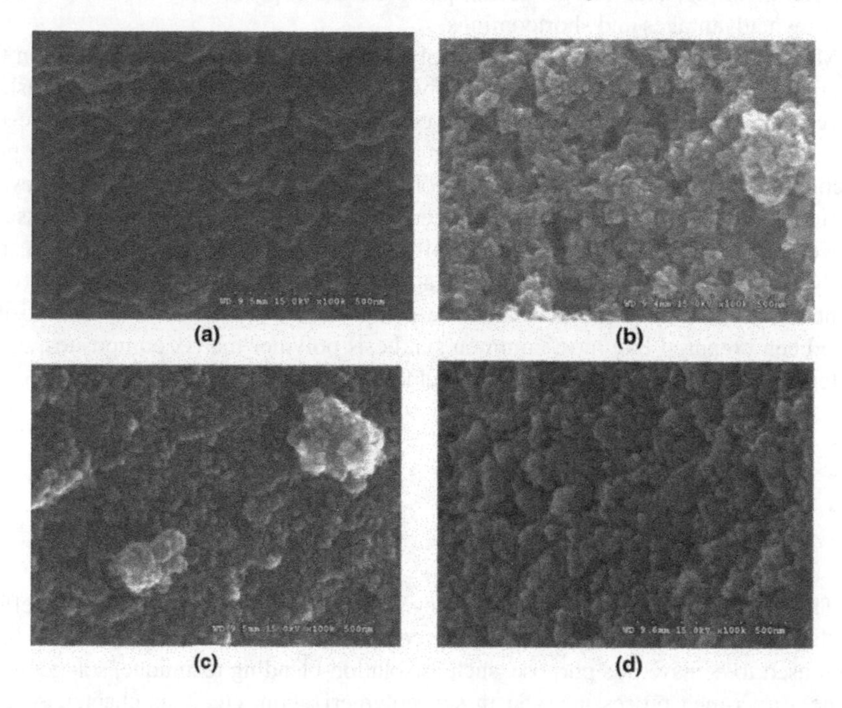

FIGURE 4.1 SEM Images of the Silica Network from the TGA Residues of the Hybrids Obtained at 600°C. Silica Contents in Hybrid are (a) 1.9%, (b) 4.7%, (c) 8.5%, and (d) 11.0% [20].

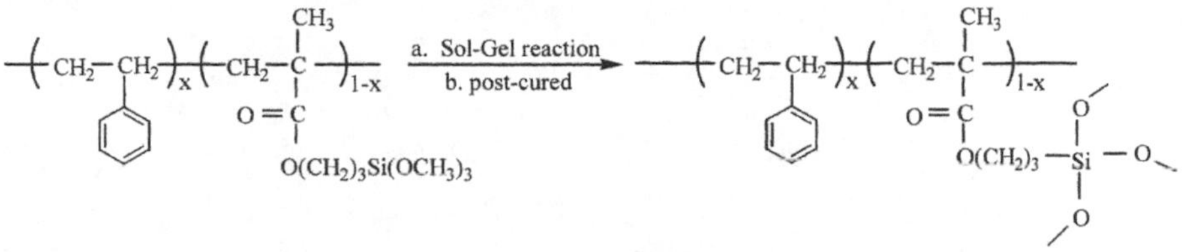

FIGURE 4.2 Preparation of Organic–inorganic Hybrid Film by Sol–gel Process [24].

utilized also the diffusion of TEOS into PP powder in the presence or absence of a grafting agent, followed by the hydrolysis/condensation step. The morphology of these composites kept stable even after a further processing step in a mini-extruder.

Another well-known sol–gel process includes the polymerization of organic functional groups like vinyl or epoxy groups and free radical or cationic, from a preformed sol–gel network [24, 25]. Hsiue et al. [24] studied the synthesis of new polystyrene/SiO_2 nanocomposites using a sol–gel technique as illustrated in Figure 4.2. The miscibility of the polystyrene (PS)/SiO_2 systems was improved by covalent introduction of SiO_2 nanoparticles into the polymeric matrix. On the other hand, sol–gel hydrolysis and condensation of a precursor, like tetraethyloxysilane (TEOS), tetrabutyl titanate, or aluminum iso-propoxide can be carried out starting from a preformed organic polymer, such as epoxy, polybenzoxazine [26], polyetherimide [27], polyvinyl alcohol [28], polyamides [29], polyimide (PI), and several other polymers [30].

Wu [31] investigated the synthesis of a new series of polycaprolactone (PCL)/TiO_2 and PCL-g-AA/TiO_2 nanocomposites from tetra isopropyl ortho titanate (TTIP) and PCL as the ceramic precursor and the continuous phase, respectively. Table 4.1 contains compositions of various sol–gel liquid solutions used to produce these nanocomposites. Hu and Marand [32] explored the in situ synthesis of TiO_2 nanoparticles within a poly(amide–imide) matrix via a sol–gel method. The elaborated nanocomposite films showed an excellent optical transparency. Garcia et al. [33] also synthesized a new series of nylon-6/SiO_2 nanocomposites by a sol–gel method.

TABLE 4.1
Compositions of Various Sol–gel Liquid Solutions for Preparation of Hybrid Materials [31]

TiO_2 (wt. %)	3	7	10	13	16
PCL or PCL-*g*-AA (g)	38.80	37.20	36.00	34.80	33.60
TTIP (g)	4.26	9.93	14.18	18.44	22.70
Isopropanol/[TTIP][a]	17	17	17	17	17
Sol A					
[acetic acid]/[TTIP][a]	0.01	0.01	0.01	0.01	0.01
[HCl]/[TTIP][a]	0.08	0.08	0.08	0.08	0.08
[H_2O]/[TTIP][a]	4.0	4.0	4.0	4.0	4.0

[a] The mole ratio of isopropanol, acetic acid, HCl, and H_2O to TTIP.

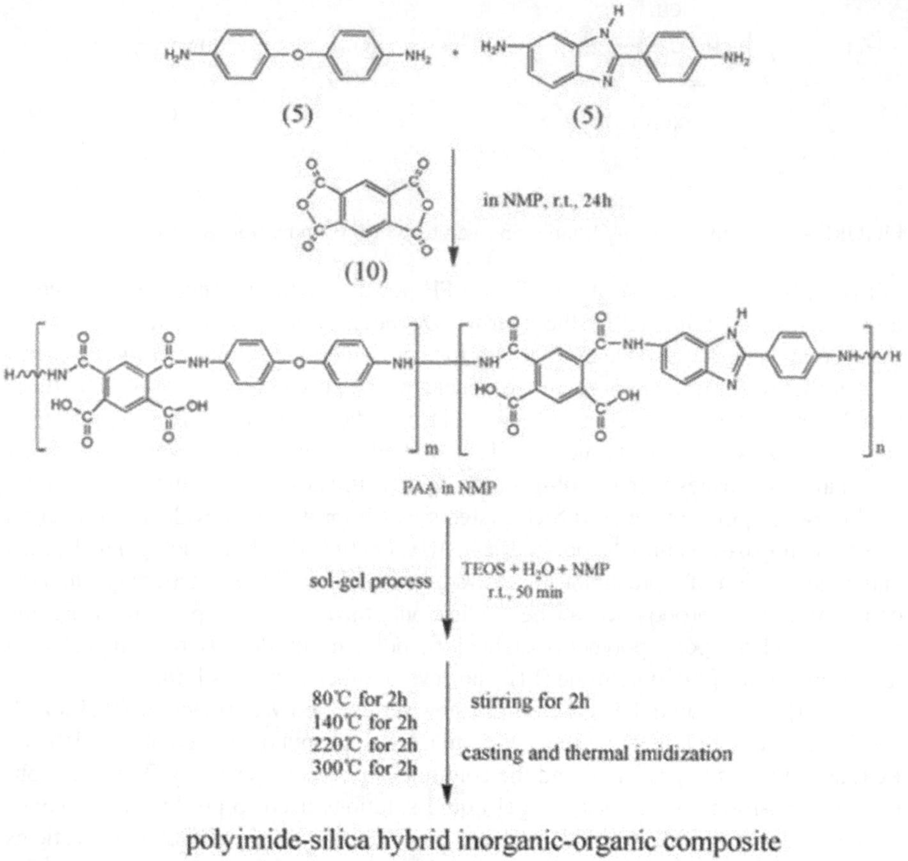

in NMP, r.t., 24h

PAA in NMP

sol-gel process

TEOS + H_2O + NMP
r.t., 50 min

80℃ for 2h
140℃ for 2h
220℃ for 2h
300℃ for 2h

stirring for 2h

casting and thermal imidization

polyimide-silica hybrid inorganic-organic composite

FIGURE 4.3 Sol–gel assisted synthesis of Polyimide/SiO_2 Hybrid Films [34].

Polyimide/SiO_2 nanocomposites were elaborated by sol–gel method without adding any coupling agent [34]. As depicted in Figure 4.3, a diamine containing a benzimidazole group, 2-(4-aminophenyl)-5-aminobenzimidazole (PABZ), was incorporated to copolymerize with 4,4'-oxydianiline (ODA) and pyromellitic dianhydride (PMDA) for producing the polyimide matrix. The compatibility between matrix and SiO_2 was enhanced via the huge content of hydrogen bonds formed between ceramic filler and the –NH– group on benzimidazole from the novel diamine.

In another work, Cui et al. [35] prepared UV-cured polymer/ZrO_2 hybrid nanocomposites using an auto-hydrolysis sol–gel process from zirconium oxychloride octahydrate coordinated with an organic amine. From the experimental point of view, as shown in Figure 4.4, the zirconium oxychloride octahydrate ($ZrOCl_2{\cdot}8H_2O$) precursor was first used to produce clear and transparent ZrO_2 nanoparticles via an auto-hydrolysis sol–gel technique in the absence of water. Triethylamine (TEA) and dimethylaminoethyl methacrylate (DMAM) were used as acid-binding agents to manage the pH value. On the other hand, silane coupling agents, acetylacetone (acac) and isopropyl tri(dioctyl)pyrophosphate titanate (titanate coupling agent, NDZ-201),

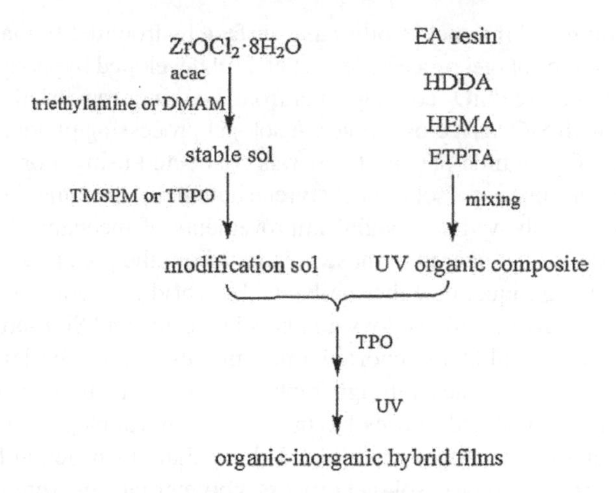

FIGURE 4.4 Synthetic Routes of the ZrO_2/polymer Hybrid [35].

were used to prevent the ZrO_2 nanoparticles and enhance the interaction between the matrix chains and the ceramic component during the UV-curing of these coatings.

Jothibasu et al. [36] prepared a transparent PS/SiO_2 hybrid using maleimide-grafted PS, tetraethoxysilane (TEOS), and APTES via in situ sol–gel technique and using the Michael-addition reaction. Using an in situ sol–gel technique followed by a thermal-curing process, Selvi et al. [37] prepared a new kind of polybenzoxazine/SiO_2–TiO_2 nanocomposites from dimethylol-functional benzoxazine monomers, TEOS, 3-(isocyanatopropyl) triethoxysilane (ICPTS), and titaniumisopropoxide (TIPO) (see Figure 4.5). They discovered that the newly developed nanocomposites exhibited higher surface energy than that of an unfilled matrix. Devaraju et al. [38] also produced polybenzoxazine/SiO_2 nanocomposite films using a sol–gel technique followed by heat-treatment process.

Ivanković et al. [39] conducted a simultaneous polymerization and sol–gel reaction of GPS, MMA, and poly(oxypropylene)diamine. The produced nanocomposites

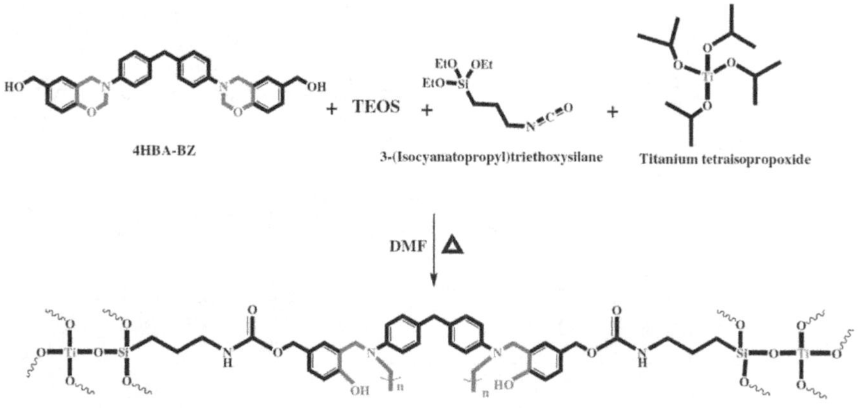

FIGURE 4.5 Synthesis of Polybenzoxazine–silica–titania Hybrid Nanomaterials [37].

exhibit much improved thermal stability and surface hydrophilicity than the PMMA matrix. Based on the sol–gel process, Jena et al. [40] developed hyperbranched waterborne polyurethane-urea/SiO_2 coating films from 3-aminopropyltriethoxysilane as a coupling agent with SiO_2 as a cross-linker. A sol–gel processing of aqueous SiO_2 precursor to solid SiO_2 ceramic nanoparticles was conducted using a one-pot synthesis technique in the presence of a solution of styrene butadiene elastomer [41]. The resulting nanocomposites showed substantial improvements of mechanical and dynamic mechanical properties compared to those prepared from the precipitated nano-SiO_2.

Using sol–gel technique, new thermally-stable hybrid nanocomposites were prepared from chemically modified epoxy and novolac resins and SiO_2-attached carbon nanotubes [42]. Zhou and Qiao, reported a new process for synthesizing raspberry-like hybrid colloidal particles through biphasic sol–gel technique. Poly(ethylene oxide)-(PEO-) functionalized latexes having various morphologies such as sphere, vesicle, and fiber were produced via a nitroxide mediated surfactant-free emulsion polymerization. In the biphasic sol–gel process, polymer latexes were first dispersed into an aqueous solution of L-arginine and a basic amino acid utilized as catalyst. Tetraethyl orthosilicate (TEOS), as the SiO_2 source, was carefully poured on top of the reactor to provide a two-phase solution. TEOS is slowly dissolved into the aqueous solution to produce SiO_2 particles, which can be adsorbed on the latex surface in the presence of hydrogen bonding generated between PEO and Si–OH. The size of SiO_2 shell particles can be easily tailored by managing the TEOS loading [43].

Recently, Abdollahi et al. [44] developed a facile sol–gel method for producing three hybrid composites constituted from tetraethyl orthosilicate oligomer-modified epoxy resin (MER), (3-glycidyloxypropyl) trimethoxysilane-modified novolac resin (MNR), epoxidized novolac resin (ENR), and SiO_2 nanoparticles as illustrated in Figure 4.6. By using tetraethyl silicate (TEOS) and polyethylene glycol (PEG),

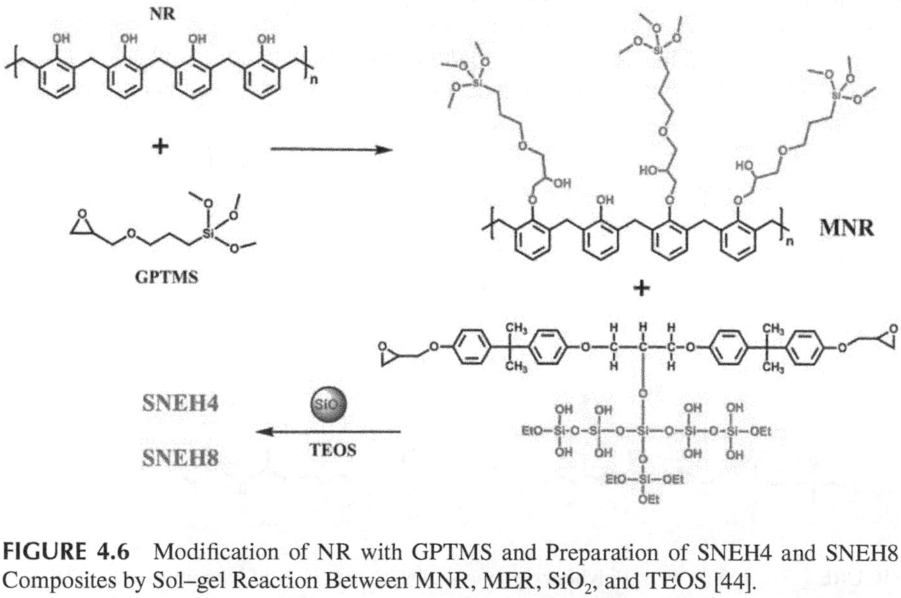

FIGURE 4.6 Modification of NR with GPTMS and Preparation of SNEH4 and SNEH8 Composites by Sol–gel Reaction Between MNR, MER, SiO_2, and TEOS [44].

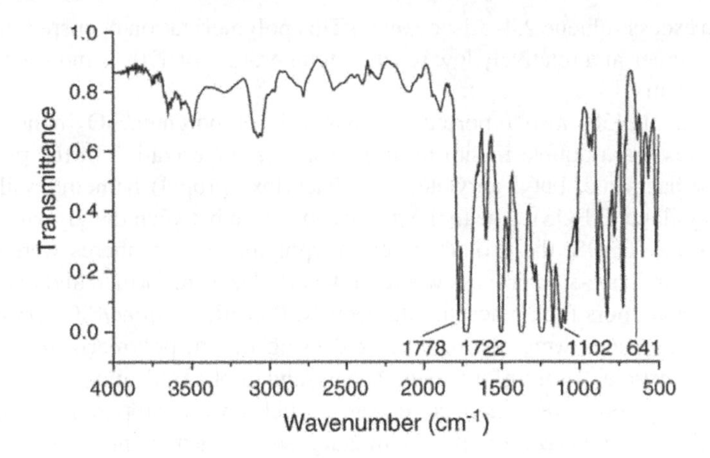

FIGURE 4.7 FTIR Spectrum of Silica–alumina Co-doped Polyimide Film [47].

Weng et al. also fabricated new polyethylene glycol (PEG)/SiO_2 nanocomposites through a controllable sol–gel method [45]. Phenol-formaldehyde (PF)/SiO_2 hybrid ceramers were prepared via a sol–gel technique at different SiO_2-filler ratios to prepare silica having carbon/carbon (C/C) hybrids, and results revealed that the density of the C/C composites increasing the SiO_2 loading [46]. In addition, the SiO_2 fillers were homogenously dispersed on the surfaces of the PF matrix with some particles located at the pores due to the decomposition of the PF matrix.

According to the work of Jing et al. [47], a new class of polyimide/silica–alumina (PI/SiO_2–Al_2O_3) composite films can be prepared by synthesizing ceramic oligomers through sol–gel reactions and coupling with polymerization. The FTIR spectrum of a PI/SiO_2–Al_2O_3 sample is drawn in Figure 4.7 and reveals that the imide stretching of matrix resin appeared at 1,720 and 1,780 cm^{-1}, whereas the characteristic absorption peaks of Si–O–Si and Al–O–Al were located at 1,100 cm^{-1} and 640 cm^{-1}, respectively. Another hybrid based on PF reinforced by SiO_2 nano-rods was also prepared by the modified sol–gel technique, as reported by Streckova et al. [48]. The reinforcing of PF with nano-SiO_2 produced a high-performance hybrids suitable for insulating coating deposited on FeSi particles.

4.3 IN SITU POLYMERIZATION TECHNIQUE

Relative to other methods, this processing technique offers many advantages including better compatibility of the compositions, and their resulting composites generally demonstrate improved dispersion and exhibit better mechanical properties. In situ polymerization is another method in which ceramic nanoparticles are first dispersed in an organic monomer, and the resulting mixture is polymerized using a technique similar to bulk polymerization.

For example, using sodium caprolactamate as a catalyst and caprolactam-treated SiO_2 as an initiator, polyamide 6/SiO_2 nanocomposites were prepared via the in-situ ring-opening anionic polymerization of *ε*-caprolactam [49]. In addition, the initiator precursor, isocyanate-functionalized silica, was produced by reacting commercial

SiO_2 with excess toluene 2,4-diisocyanate. This polymerization occurred in a highly efficient manner at a relatively low reaction temperature of 170°C and short reaction times of 6 hours.

Zhang et al. [50] also reported the synthesis of polymer/SiO_2 nanocomposite microspheres via a double in situ mini-emulsion polymerization in the presence of methyl methacrylate, butyl acrylate, γ-methacryloxy(propyl) trimethoxysilane, and tetraethoxysilane (TEOS). Due to the phase separation between the growing polymer molecules and TEOS, the produced ceramic/polymer microspheres were prepared successfully in a one-step process where SiO_2 particles were formed and the polymerization of monomers took place simultaneously. Polyethersulfone/SiO_2 crystal structure nanocomposites were also synthesized using in situ polymerization of mixed nano-SiO_2 particles as reinforcement agents and diphenolic monomers at 160°C [51]. The SiO_2/polystyrene nanocomposite particles were prepared via an in situ mini-emulsion polymerization of sodium lauryl sulfate surfactant (SLS), hexadecane co-stabilizer, and methacryloxy(propyl)trimethoxysilane-treated SiO_2 fillers [52].

A nanocomposite based on poly(ε -caprolactone)-castable polyurethane elastomer reinforced with SiO_2 nanoparticles of various surface properties was produced via in situ polymerization as reported by Chen et al. [53]. Polypropylene nanocomposites filled with SiO_2 nanospheres produced by the sol–gel reaction were prepared using in situ polymerization of a racemic $Et(Ind)_2ZrCl_2$/methylaluminoxane (MAO) system [54]. Two different routes were utilized depending on the interaction between the nano-SiO_2 particles with the selected catalytic system. Morphological analysis revealed a good dispersion of SiO_2 nanospheres into the polypropylene matrix. In addition, novel electrically-conductive nanocomposite aerogels were successfully elaborated using in situ oxidative polymerization of pyrrole (Py) using ammonium persulfate (APS) as an oxidizing agent in SiO_2 gels [55].

A formation mechanism of guava-like polymer/SiO_2 nanocomposite particles by in situ emulsion co-polymerization systems based on the increased homogenous nucleation was proposed by Qi et al. [56]. Low nano-SiO_2 filled nylon 6 (PA6) nanocomposites were prepared using in situ polymerization. The influence of nano-SiO_2 treated with two surfactants containing amino groups and alkyl chains, respectively, on the interfacial structure and properties of nylon 6/SiO_2 nanocomposites was evaluated [57]. LLDPE/SiO_2 nanocomposites were also synthesized by in situ polymerization of ethylene/1-hexene with a methylaluminoxane/metallocene catalyst [58]. Lin et al. [59] described an in situ synthesis process of novel ceramic particles/polymer electrolyte as shown in Figure 4.8. Improved chemical/mechanical interactions between monodispersed 12 nm-SiO_2 nanospheres and poly(ethylene oxide) (PEO) chains were elaborated by in situ hydrolysis, which markedly deceased the crystallization of PEO and eased the polymer segmental motion for ionic conduction.

The synthesis of LLDPE/TiO_2 nanocomposites via in situ polymerization of ethylene/1-hexene with a zirconocene/MMAO catalyst was described by Ekrachan et al. [60]. The size of the nano-TiO_2 particles can affect the catalytic activity in the polymerization system, whereas using larger nano-TiO_2 resulted in better catalytic activity in the polymerization system as they provided more space for monomer attack. In Wu's study [61], a silicic acid and tetra isopropyl ortho titanate ceramic precursor and a metallocene polyethylene-octene elastomer (POE) or acrylic acid–grafted

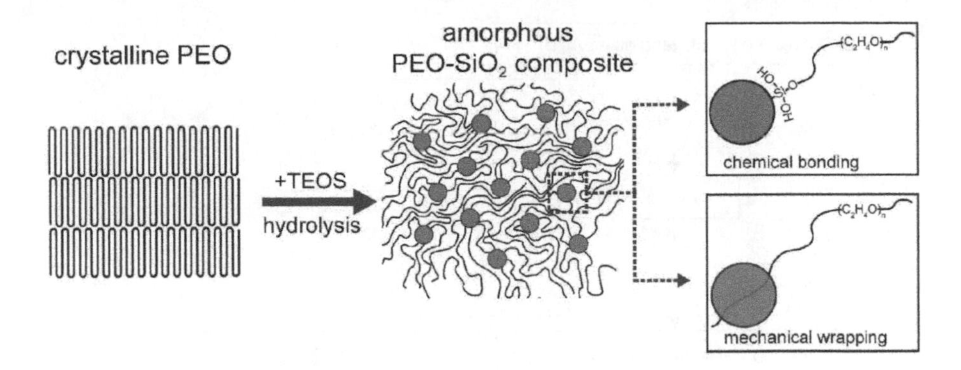

FIGURE 4.8 Schematic Figures Showing the Procedure of In Situ Hydrolysis and Interaction Mechanisms Among PEO Chains and $MUSiO_2$ [59].

metallocene polyethylene-octene elastomer (POE-g-AA) were utilized for producing new hybrids (POE/SiO_2-TiO_2 and POE-g-AA/SiO_2-TiO_2) by in situ sol–gel process. Well-organized polystyrene-TiO_2 nanocomposites having core-shell structures were produced by two steps, including surface-initiated atom transfer radical polymerization (ATRP) of styrene and in situ chemical oxidative polymerization of aniline monomers from the surfaces of the TiO_2 nanoparticles [62].

Aiming to improve the dispersion state, superfine Al_2O_3 was encapsulated with poly(methyl methacrylate) (PMMA) using in situ emulsion polymerization as reported by Liu et al. [63]. Polyimide/Al_2O_3 nano-hybrids were also produced via simple in situ polymerization and thermal-curing processes [64]. Polyaniline/γ-Al_2O_3 composites were fabricated by in situ polymerization in the presence of HCl as dopant by incorporating γ-Al_2O_3 nanoparticles into an aniline solution [65].

The spectral results confirmed that γ-Al_2O_3 nanoparticles were linked to the organic chains and influenced the absorption characteristics of the prepared composite through the interaction between the matrix and nano-sized γ-Al_2O_3. Using the in situ polymerization, the LLDPE/Al_2O_3 nanocomposites were synthesized from a dried-modified methylaluminoxane (d-MMAO)/zirconocene catalyst [66]. Novel Al_2O_3/polyimide nanocomposite films based on 4,4'-oxydianiline (ODA) and pyromellitic dianhydride (PMDA) were produced by including various ratios of nano-sized Al_2O_3 via in situ polymerization as illustrated in Figure 4.9 [67].

The crystallization behavior of polypropylene (PP) copolymer prepared by in situ reactor copolymerization in the presence or absence of a nucleating agent and/or nano-$CaCO_3$ particles was investigated by thermal analysis and by polarized light microscopy [68]. A water-based nanocomposite was fabricated via in situ polymerization of methyl methacrylate in the presence of nano-$CaCO_3$ in the form of aqueous solution of surfactant [69]. The PMMA/$CaCO_3$ composite can be also prepared by in situ soapless emulsion polymerization technique in the aqueous suspension of nano calcium carbonate [70].

To produce PMMA/$CaCO_3$ nanocomposite, the methyl methacrylate (MMA) emulsion polymerization was achieved in the presence of nanometer nano-$CaCO_3$ previously treated with γ-methacryloxypropyltrimethoxysilane (MPTMS). The reaction

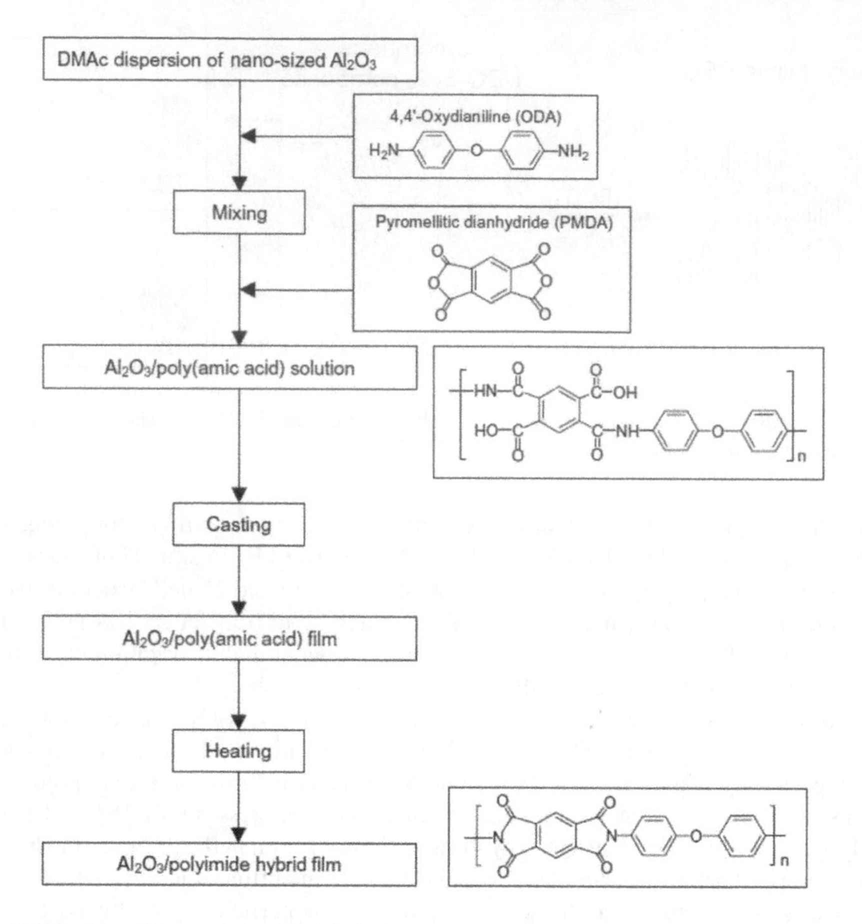

FIGURE 4.9 Preparation of Al_2O_3/polyimide Hybrid Films by In Situ Polymerization Method [67].

between nano-$CaCO_3$ and MPTMS, and the grafting of PMMA onto nano-$CaCO_3$ were monitored by FTIR technique [71]. Novel abrasion-resistant PMMA/$CaCO_3$ nanocomposites were similarly produced using in situ polymerization process [72]. Poly(vinyl chloride) (PVC)/$CaCO_3$ nanocomposites were elaborated by in situ polymerization of vinyl chloride (VC) in the presence of $CaCO_3$ nano-filler [73].

Zhuang et al. [74] used an in situ polymerization method to develop a series of TiO_2/polylactide (PLA) nanocomposites at various TiO_2 contents. The size of the organically-treated TiO_2 particles was studied by the X-ray diffraction (XRD) technique. Scanning electron microscope (SEM) showed that nano-TiO_2 particles were well-dispersed into the PLA matrix evenly at lower TiO_2 contents as depicted in Figure 4.10.

Using tetrabutyl titanate as a precursor, hydrochloric acid as catalyst, and γ-methacryloxpropyltrimethoxysilane (KH-570) as modifier, TiO_2/acrylate nanocomposite emulsions were prepared by in situ emulsion polymerization reaction [75]. PMMA/TiO_2 composite particles were produced via in situ emulsion polymerization of MMA in the presence of TiO_2 particles. Before polymerization, the TiO_2 particles

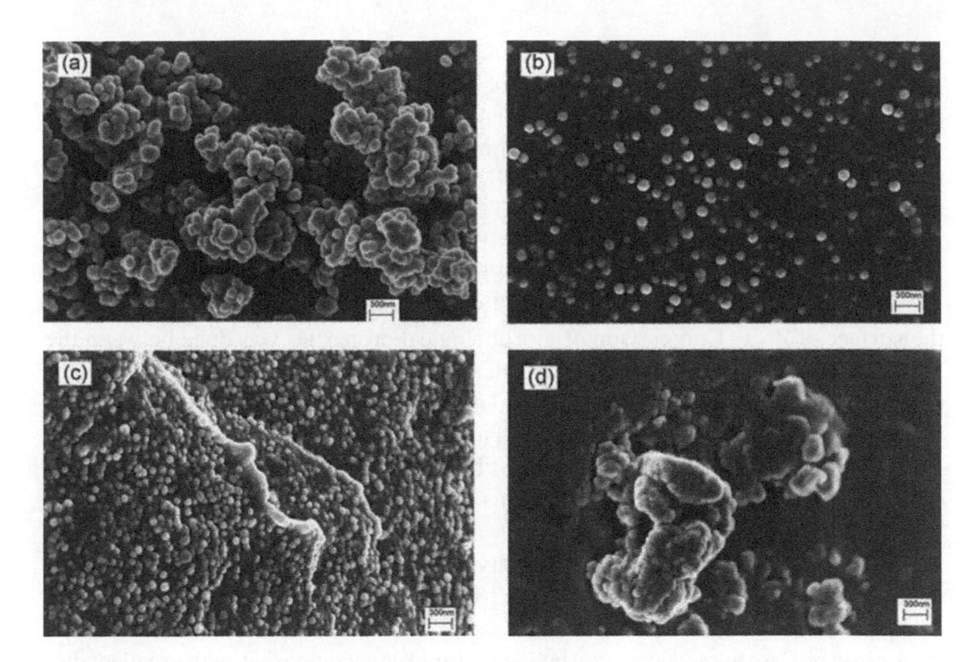

FIGURE 4.10 SEM Photographs of TiO_2 and TiO_2-x/PLA Composites (a) TiO_2, (b) TiO_2-1/PLA (1.0 wt.%), (c) TiO_2-3/PLA (3.0 wt. %), and (d) TiO_2-10/PLA (10.0 wt.%) [74].

were treated by a silane coupling agent to ensure that PMMA reattached to nano-TiO_2 with covalent bond bindings [76]. The 6-palmitate ascorbic acid–modified TiO_2 nanoparticles were incorporated in methyl methacrylate monomer (MMA) and mixed to give nanocomposite sheets after a bulk polymerization in a glass sandwich cell in the presence of 2,2'-azobisisobutyronitrile as initiator [77]. Cellulose/polypyrrole-TiO_2 composites were fabricated using in situ oxidative chemical polymerization of pyrrole using $FeCl_3$ as oxidant. The effect of a pyrrole ratio on the structure and properties of prepared composites was scrutinized [78].

PMMA/TiO_2 nanocomposites were also developed using the in situ radical polymerization of MMA in a toluene solution of surface-modified TiO_2 nanoparticles [79]. The in situ polymerization method was similarly used for synthesizing polyester/TiO_2 nanocomposites as reported by Evora and Shukla [80]. Low-cost PU/TiO_2 nanocomposites were synthesized by an in situ polymerization technique with improved thermal stability and mechanical properties exceeding those recorded for the neat PU thermoset [81].

Novel nano-Ag-TiO_2/PU coatings were prepared by solution combustion and grafting from a polymerization method, where nAg-TiO_2 was chemically grafted to the skeleton of the PU matrix with a DMPA [82]. In their work, the differences between dispersing dry and colloidal TiO_2 nanoparticles into epoxy resin was investigated. This approach revealed an improved colloidal nanoparticle dispersion with a minimum aggregation compared to the dispersion of dry nanoparticles [83]. In this method, ceramic nanofiller dispersion and polymerization of the matrix take place simultaneously. Abdul Kaleel et al. [84] produced polyethylene (PE)/TiO_2 nanocomposites from ethylene, metallocene catalysts, and TiO_2 nanoparticles via the in situ polymerization process. Rong et al. [85] prepared a series of PS/TiO_2 nanocomposites by in situ radical polymerization technique of styrene monomer and MPS-modified TiO_2.

Hu et al. [86] developed a new type of PMMA/ZrO_2 nanocomposites from MMA-functionalized ZrO_2 nanoparticles and MMA by means of in situ bulk polymerization. Transmission electron microscopy (TEM) analysis revealed that the nano-ZrO_2 are homogenously distributed into the PMMA matrix (Figure 4.11). Ultra-high molecular-weight polyethylene (UHMWPE)/ZrO_2 composites were produced by in situ polymerization of ethylene using a Ti-based Ziegler-Natta catalyst located on the surface of ZrO_2. Comparison of morphological properties was done between the in situ polymerized and mechanically-blended composites, where the composites prepared by in situ polymerization showed more homogenous dispersion of ZrO_2 and improved interfacial properties compared to the mechanically prepared ones [87].

Similarly, UHMWPE/ZrO_2 composites having a good dispersion state were elaborated using in situ polymerization to be used as bearing material for artificial joints [88]. In another study, both the nano-SiO_2 and nano-ZrO_2 were utilized as reinforcing agents for producing linear low-density polyethylene (LLDPE) nanocomposites via the in situ polymerization of ethylene/1-octene with a zirconocene/MAO catalyst in the presence of these nanofillers [89].

A novel process to produce core-shell–structured nanocomposites with excellent dielectric performance was reported by Yang et al. [90]. This approach is based on the grafting of polystyrene (PS) from the surface of $BaTiO_3$ using in situ RAFT polymerization.

Core-shell–structured $BaTiO_3$/PMMA nanocomposites were successfully fabricated by in situ atom transfer radical polymerization (ATRP) of MMA monomer from the surface of $BaTiO_3$nanoparticles [91]. Yang et al. [92] prepared thiol-epoxy elastomers by a thiol-epoxide nucleophilic ring-opening reaction in the presence of micro-sized BN via in situ polymerization. In another work, new aryl boron-containing PF resin composites showing a significant higher thermal stability were formed by reacting phenyl boronic acid (PBA) with phenolic resin. Boron oxide was constructed during the char formation from the cleavage of B-C and B-O-C bonds

FIGURE 4.11 TEM Analysis of 15 wt.% of ZrO_2 Dispersed in PMMA/ZrO_2 Nanocomposite [86].

during the pyrolysis process, which effectively avoided the release of volatile carbon oxides and retained the carbon. In addition, the presence of boron into the carbon lattice enhanced the crystallite height and lowered the interlayer spacing [93].

4.4 SOLUTION-BLENDING TECHNIQUE

The addition of inorganic ceramic filler into the various polymer matrix can be achieved by melt blending or solution blending method. This technique is more suitable and practicable for producing polymer/ceramic micro- and nanocomposites in large scale. The ceramic fillers bear reactive and nonreactive sites, and they can react or interact with the polymeric matrices, offering perfect polymer composite materials. In addition, in this technique, the nanoceramic cannot form agglomeration within the polymer nanocomposites, which helps improve their mechanical and thermal performance.

In this technique, the ceramic particles are added after the base polymeric matrix (generally thermoset) is dissolved in an organic solvent or melted by heat. The resulting mixture is agitated to produce a uniform suspension. The composites are fabricated by eliminating the solvent or thermal polymerizing of the mixture. The most convenient technique for producing thermosetting polymer/ceramic micro- and nanocomposites is the melt blending of matrices in the presence of their reinforcing phases. However, in this technique, the high tendency of the ceramic fillers to form agglomerates resulted in a low degree of dispersion in the various thermosets.

The solution-blending method involved mixing ceramic fillers and polymeric matrices in a liquid state generated from the addition of a solvent or from the melted monomers. Improved mixing quality can generally obtained using this technique, which promotes its application in producing such composite materials. In a study, Yan et al. [94] used the solution-blending method to elaborate a series of BMI/ZrO_2 nanocomposites. In their work, the N-(2-aminoethyl)-γ-aminopropylmethyldimethoxy silane (JH-53)–treated and –untreated nano-ZrO_2 of 20 to 40 nm were ultrasonically dispersed into an acetone solution. After that, BMI prepolymer solution was added to the resulting mixture and heated to 120°C while stirring for 30 min to evaporate the solvent. The resulting dried composite was then thermally cured at different temperatures and post-cured 250°C for 6 h.

Ramdani et al. [95] produced polybenzoxazine-based networks reinforced with γ-aminopropyltrioxysilane (KH-550)–treated Si_3N_4 nanoparticles having a diameter of 50 nm, using a solution-blending method in dichloromethane (DCM). They also fabricated a series of polybenzoxazine/TiC nanocomposites by a solution casting in THF (tetrahydrofuran) [96]. Choi and Kim [97] studied the synthesis of thermally-conductive epoxy nanocomposite by adding hybrids of Al_2O_3/AlN fillers at different sizes to an ETDS/DDM epoxy matrix using a solution-blending technique. In the same context, boron-phenolic resin/APTES-modified silicon nitride (B-BPF/Si_3N_4) micro-composites binders for grinding wheels were prepared by mixing 10 g of resins with 0.9 g of m-Si_3N_4 in ethyl alcohol (EA)(50 ml) at 50°C for 5 h and evaporating the solvent from the mixture at 130°C for 3 h [98].

New kinds of polyimide/modified-alumina (PI/KH-550-treated Al_2O_3) nanocomposites were recently developed by Ghezelbash et al. [99] via a solution-blending method. In their work, Lin et al. [100] used a nonaqueous sol method to produce 3-methacryloxypropyltrimethoxysilane modified nano-SiO_2 (MPS-SiO_2) in N,N-dimethylformamide (DMF) substituting alcoholic solvents, and epoxy acrylate resins (EA) based on novolac epoxy resin (EP). As shown in Figure 4.12, epoxy acrylate copolymers (EPAc/SiO_2) having core/shell structure were produced

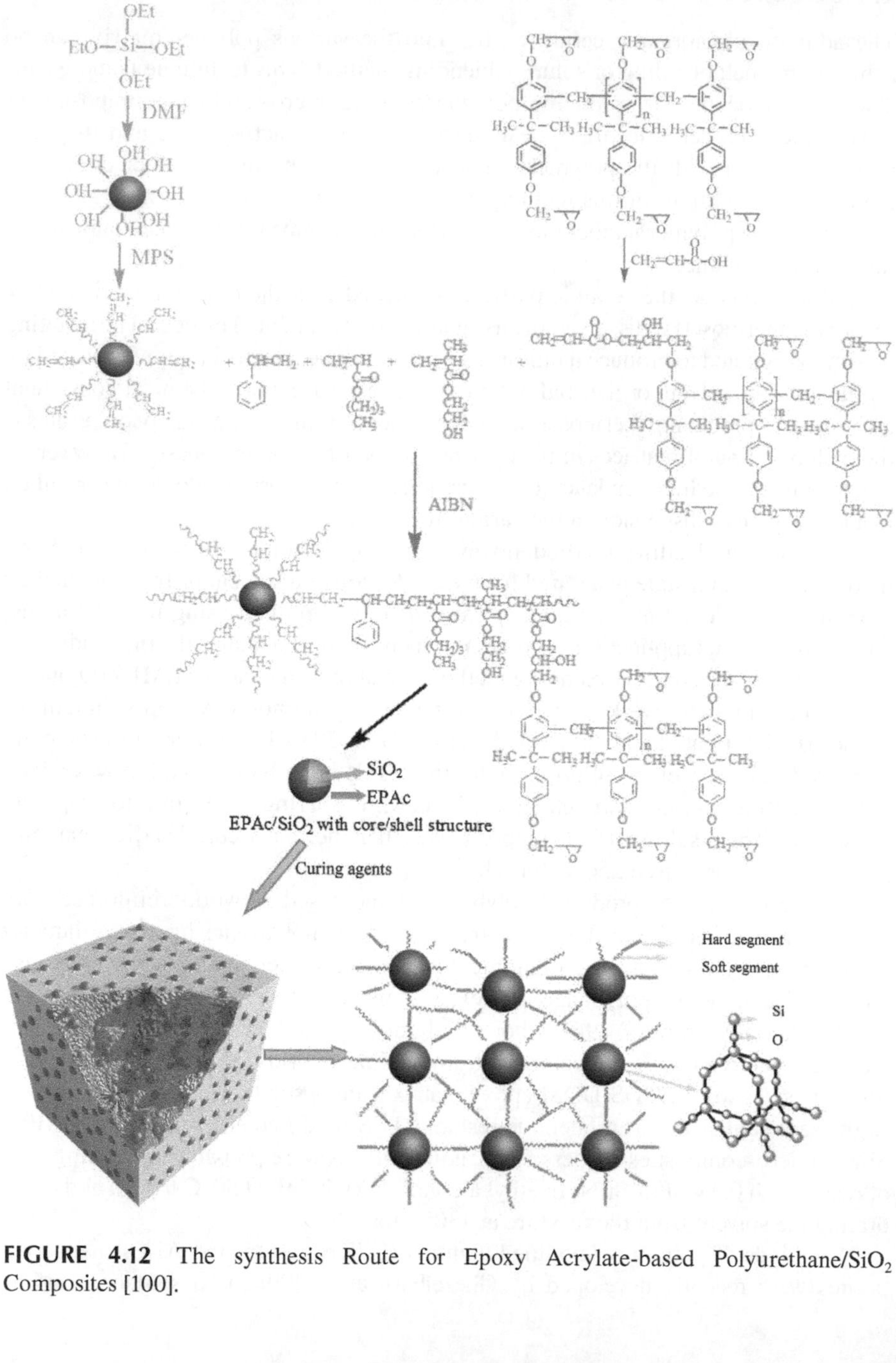

FIGURE 4.12 The synthesis Route for Epoxy Acrylate-based Polyurethane/SiO_2 Composites [100].

using one-step in-situ solution blending of EA, acrylic monomers and a selected amount of treated SiO_2 sol as core.

In direct-melt processing, the thermoset/ceramic hybrid is obtained by direct dispersion of the selected ceramic micro- and nanoparticle into the melted thermoset. A compression-molding technique is generally used for curing. In a study, bisphenol E cyanate ester/alumina (BECy/Al_2O_3) nanocomposites were also prepared using this technique [101]. Surface-treated SiC nanowire-filled epoxy resin also synthesized via the melt-process technique [102]. Kajohnchaiyagual et al. developed new polybenzoxazine/Al_2O_3 micro-composites by direct mixing of different Al_2O_3 amounts and benzoxazine precursor at 100°C to enhance the micro-particles wet-out by the thermoset [103]. Many polybenzoxazines' micro- and nanocomposites filled with ceramic likes BN, $BaTiO_3$, $CaCO_3$ were also produced by the melt-processing technique. Recently a wetting process was used to fabricate epoxy/BN composites [104] as illustrated in Figure 4.13.

PF/Al_2O_3 nanocomposites were synthesized by a solution-mixing technique, with good nano-Al_2O_3 dispersion at low concentrations (5 wt.% nano-Al_2O_3), while above this concentration, the stress concentrations were developed due to the formation of large aggregates. Theoretical analyses revealed that strong interfacial interaction and a thick interphase region around the Al_2O_3 nanoparticles resulted [105]. Charpentier et al. [106] developed new kind of polyurethane/TiO_2 hybrids using the solution-blending technique and investigated their self-healing features. Vengatesan et al. [107] used the solution-blending process for fabricating a series of SBA-15/PBZ nanocomposites from benzoxazine functional silane and benzoxazine monomer. The nanocomposites were first produced by eliminating the solvent followed by thermal curing process as depicted in Figure 4.14.

Sasikala et al. [108] prepared a PS/SiO_2 spherical composite from vinyl and amine-functionalized SiO_2 spheres and PS matrix via the solution-blending

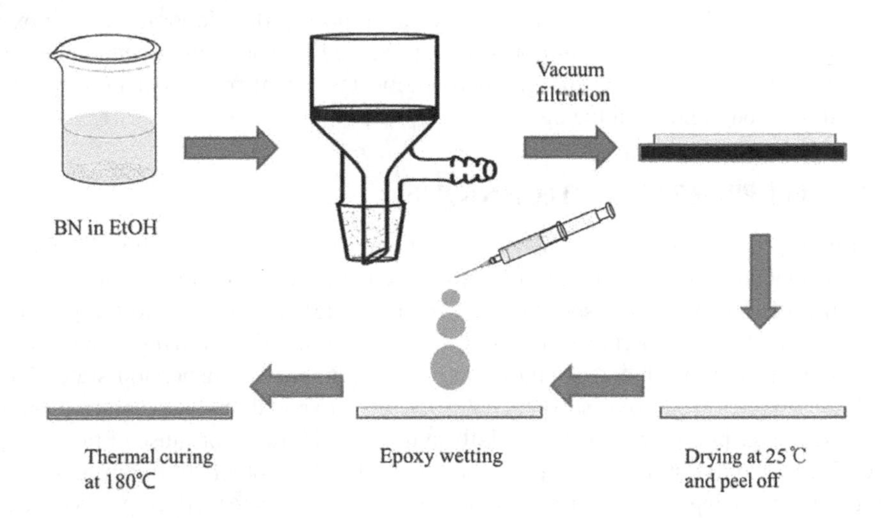

FIGURE 4.13 Schematic Diagram of the Fabrication Procedure of the Epoxy/BN Composite. Reprinted from [104].

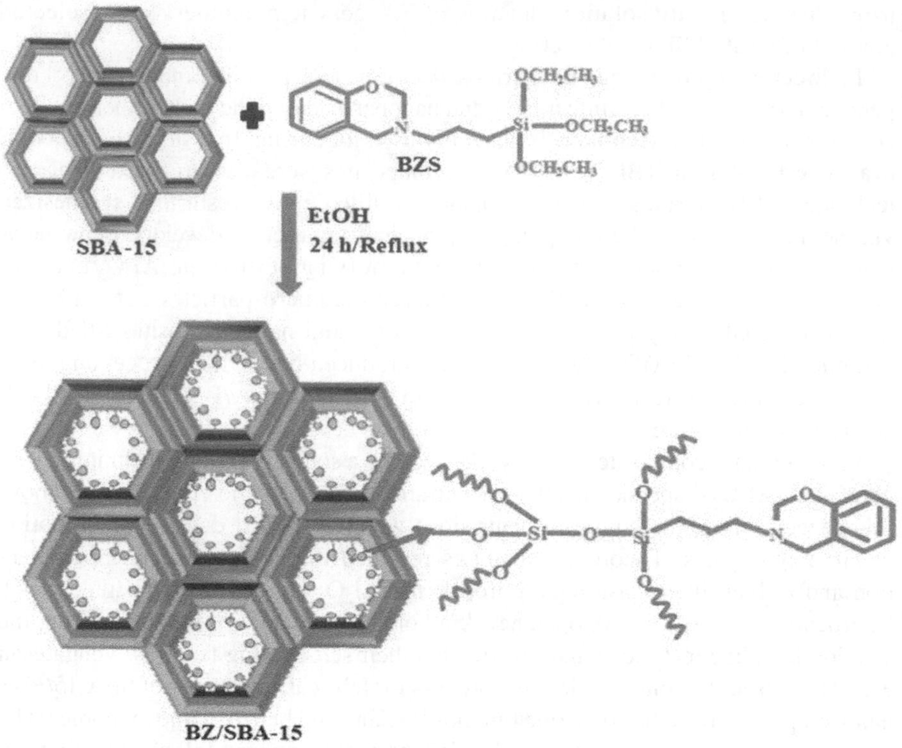

FIGURE 4.14 Schematic Representation of Polybenzoxazine-silica Hybrid (PBZ/SiO_2) [107].

technique. Devaraju et al. [109] also synthesized cyanate ester/ceramic composites via the solution-blending process. Isobe et al. reported the elaboration of epoxy/porous SiO_2 containing 20-nm pores were produced by solution-casting technique in a 10% ethanol solvent. The gas permeation tests confirmed that porous SiO_2 appeared through holes in the nano-hybrids [110].

4.5 MELT-PROCESSING TECHNIQUE

The melt-processing technique is the easiest and most practical method especially for the elaboration of thermoplastic/ceramic composites. This technique is based on a high temperature and shear force to disperse ceramic fillers into the polymer matrix. The high temperature softens the polymeric matrix allowing easier incorporation of the reinforcement phase and offering improved dispersion state. This technique is free from toxic solvents but is less effective in dispersing ceramic nanoparticles in the polymer matrix especially at higher filler ratios because of increasing the viscosity of the polymer/ceramic composites [111]. Another shortcoming of this method is buckling, rolling, or even shortening of ceramic sheets during mixing due to the stronger resulted shear forces that can significantly reduce the aspect ratios which in turns can limit the dispersion of ceramic nanofillers [112].

In this method, the ceramics are uniformly dispersed into the polymer matrix by extrusion, internal mixing, and two-roll milling at high temperatures. This technique is more convenient and preferred process for the developing of polymer hybrid nanocomposites on a large scale. Ou and Li [113] elaborated nanocomposites by adding the TDI-functionalized TiO_2 nanoparticles into the blended matrix of PP/PA6 via the melt blending technique. Xu et al. [114] prepared a series of polyamide 6/SiO_2 nanocomposites by melt blending of polyamide 6 with surface-treated nano-SiO_2, where their crystallization temperature and crystallization rate were inferior to those of unfilled matrices. Reddy and Das [115] also synthesized high-pressure low-density polyethylene (HP-LDPE)/organic functionalized SiO_2 nano-hybrids by the melt-blending method. They found that the organic surface-treatment of nano-SiO_2 imparted an enhancement in thermal stability and mechanical performance of the prepared nanocomposite. Chanmal and Jog [116] elaborated PVDF/$BaTiO_3$ nanocomposites containing 10 to 30 wt.% nanofillers loadings via a simple melt-blending technique to evaluate their dielectric relaxations.

Melt-mixing and hot-pressing processing techniques are useful and are more reproducible, economic, and environmentally benign methods for producing polymer/ceramic hybrid composites. Compared to the solution-casting composites, this process is used to avoid porosity and to enhance the homogeneity of the composites. Thus, a dense polymer/ceramic composite could result from changing the ceramic reinforcing agent. However, during the operating process, the problem related to scalding is retained. Ceramic powder–reinforced polymers were widely produced using the melt-blending technique. Bikiaris et al. [117] produced SiO_2-reinforced isotactic polypropylene (iPP) thermoplastic matrix by incorporating poly(propylene-graft-maleic anhydride) copolymer (PP-g-MA) modified with 0.6 wt.% maleic anhydride as compatibilizer. This phenomenon could contribute to the reaction between the maleic anhydride and the surface hydroxyl of SiO_2, thus the agglomeration of nano-SiO_2 was decreased as confirmed by morphological analysis results.

Using the melt-mixing method, another thermoplastic matrix, polyphenylene sulfide, filled with barium strontium titanate (PPS)/BST composites was prepared at various BST concentrations. By using a twin-screw extruder, the BST nanofillers were homogenously dispersed into the PPS polymer, and the composite specimen containing 70 wt.% BST showed improved dielectric properties [118]. Li et al. [119] included silver-coated $Ba_{0.6}Sr_{0.4}TiO_3$ core–shell nanoparticles into the PVDF matrix using the melt-blending technique. Chao and Liang [120] prepared bisphenol A di-isocyanate composites filled with KH550 silane molecules covered-$BaTiO_3$ ceramic filler at various loadings via a melt-mixing method and investigated their dielectric properties at different frequencies.

4.6 CONCLUSIONS

In this chapter, we have described and exemplified four different synthetic strategies for producing polymer/ceramic particles composites on both the micrometric and nanometric scales, according to whether the components are constructed in situ or ex situ and according to whether the joining takes place through covalent bonds or by physical bonds. Noncovalent approaches, such as heterocoagulation or layer-by-layer deposition, are well-established and are very useful in fabricating highly controlled morphologies. However, the conditions needed for some applications can generate a destabilization of noncovalent attachments. The covalent binding of polymers and ceramic fillers is a potential alternative. For biomedical applications, significant attempts have been made to attach polymeric chains to ceramic nanoparticles, especially in the context of the so-called bioconjugation. However, few studies have been reported on the hierarchical linking of nanoceramics of different features. This is certainly an area to be further studied in the next few years and offers increased possibilities for the production of new types of materials by assembling the treated-ceramic nanoparticles. In the case of polymerization in the presence of surface-modified ceramic particles, one of the key aspects still not understood is the relationship between the functionalization of the particles and the final structure of the composites. A good understanding can provide new structures, like Janus and patchy particles, which have recently become a model. Other techniques based on in situ precipitation offer simplicity and many other advantages, but the control of the nanoparticle morphologies is too difficult and still represents a big challenge. The preparation of hybrid nanoparticles by means of "all in situ" techniques (i.e., by simultaneous polymerization and formation of the ceramic component) is still an unexplored field to be considered, due to the convenience of the "one-pot" synthesis method and because it permits an interpenetration of the matrix chains and ceramic fillers at a size dimension that is not allowed by the other techniques.

REFERENCES

1. Sengupta, R., Chakraborty, S., Bandyopadhyay, S. et al. 2007. A short review on rubber/clay nanocomposites with emphasis on mechanical properties. *Polymer Engineering & Science* 47: 1956–1974.
2. Ganter, M., Gronski, W., Semke, H. et al. 2001. Surface compatibilized layered silicates—a novel class of nanofillers for rubbers with improved mechanical properties. *Kautschuk, Gummi, Kunststoffe* 54: 166–171.
3. Pramanik, M., Srivastava, S. K., Samantaray, B. K. et al. 2003. Rubber-clay nanocomposite by solution blending. *Journal of Applied Polymer Science* 87: 2216–2220.
4. Lim, S. K., Lim, S. T., Kim, H. B. et al. 2003. Preparation and physical characterization of polyepichlorohydrin elastomer/clay nanocomposites. *Journal of Macromolecular Science, Part B* 42: 1197–1199.
5. Jeon, H. S., Rameshwaram, J. K., Kim, G. 2004. Structure-property relationships in exfoliated polyisoprene/clay nanocomposites. *Journal of Polymer Science Part B: Polymer Physics* 42: 1000–1009.
6. Varghese, S., Karger-Kocsis, J. 2003. Natural rubber-based nanocomposites by latex compounding with layered silicates. *Polymer* 44: 4921–4927.
7. Chaichana, E., Jongsomjit, B., Praserthdam, P. 2007. Effect of nano-SiO_2 particle size on the formation of LLDPE/SiO_2 nanocomposite synthesized via the in situ polymerization with metallocene catalyst. *Chemical Engineering Science* 62:899–905
8. Yu, C. B., Wei, C., Lv, J. et al. 2012. Preparation and thermal properties of mesoporous silica/phenolic resin nanocomposites via in situ polymerization. *eXPRESS Polymer Letters* 6:783–793.
9. Wu, Y. P., Zhang, L. Q., Wang, Y. Q. 2001. Structure of carboxylated acrylonitrile-butadiene rubber (CNBR)–clay nanocomposites by co-coagulating rubber latex and clay aqueous suspension. *Journal of Applied Polymer Science* 82: 2842–2848.

10. Balachandran, M., Bhagawan, S. S. 2011. Studies on acrylonitrile—butadiene copolymer (NBR) layered silicate composites: mechanical and viscoelastic properties. *Journal of Composite Materials* 45:2011–2022.
11. Nandi, M., Conklin, J. A., Jr, Salvati, L. Jr. et al. 1991. Molecular level ceramic/polymer composites. 2. Synthesis of polymer-trapped silica and titania nanoclusters. *Chemistry Materials* 3: 201–206.
12. Rezac, M. E., Koros, W. J. 1992. Preparation of polymer–ceramic composite membranes with thin defect-free separating layers. *Journal of Applied Polymer Science* 46: 1927–1938
13. Vollath, D., Szabo, D.V., Fuchs, J. 1999. Synthesis and properties of ceramic-polymer composites. *Nanostructured Materials* 12: 433–438.
14. Wolff, M.F.H., Salikov, V., Antonyuk, S. et al. 2014. Novel, highly-filled ceramic–polymer composites synthesized by a spouted bed spray granulation process. *Composites Science and Technology* 90: 154–159
15. Sobczak-Kupiec, A., Pluta, K., Drabczyk, A. et al. 2018. Synthesis and characterization of ceramic - polymer composites containing bioactive synthetic hydroxyapatite for biomedical applications. *Ceramics International* 44: 13630–13638.
16. Brinker, C., Scherer, G. 1990. *Sol–gel science: physics and chemistry of sol–gel science processing*. Toronto: *Academic Press*, 2–10.
17. Pomogailo, A. D. 2005. Polymer sol–gel synthesis of hybrid nanocomposites. *Colloid Journal* 67: 658–677.
18. Bounor-Legaré, V., Cassagnau, P. 2014. In situ synthesis of organic–inorganic hybrids or nanocomposites from sol–gel chemistry in molten polymers. *Progress in Polymer Science* 39: 1473–1497.
19. Yen, M. S., Kuo, M. C. 2012. Sol–gel synthesis of organic–inorganic hybrid materials comprising boehmite, silica, and thiazole dye. *Dyes and Pigments* 94: 349–354.
20. Sun, D., Zhang, R., Liu, Z. et al. 2005. Polypropylene/silica nanocomposites prepared by in-situ sol–gel reaction with the aid of CO_2. *Macromolecules* 38:5617–5624.
21. Jain, S., Goossens, J. G. P., Van Duin, M. 2006. Synthesis, characterization and properties of (vinyl triethoxy silane-grafted PP)/silica nanocomposites. *Macromolecular Symposia* 233: 225–234.
22. Jain, S., Goossens, H., Picchioni, F. et al. 2005. Synthetic aspects and characterization of polypropylene–silica nanocomposites prepared via solid-state modification and sol–gel reactions. *Polymer* 46: 6666–6681.
23. Jain, S., Goossens, J. G. P., Van Duin, M. et al. 2005. Effect of in situ prepared silica nanoparticles on non-isothermal crystallization of polypropylene. *Polymer* 46: 8805–8818.
24. Hsiue, G. H., Kuo, W. J., Jeng, R. J. 2000. Microstructural and morphological characteristics of PS-SiO_2 nanocomposites. *Polymer* 41: 2813–2825.
25. Song, K. Y., Crivello, J. V., Ghoshal, R. 2001. Synthesis and photoinitiated cationic polymerization of organic inorganic hybrid. *Chemistry of Materials* 13: 1932–1942.
26. Silveira, K. F., Yoshida, I. V. P., Nunes, S.P. 1995. Phase separation in PMMA/silica sol–gel systems. *Polymer* 36: 1425–1434.
27. Nunes, S. P., Peinemann, K. V., Ohlrogge, K. et al. 1999. Membranes of poly(ether imide) and nanodispersed silica. *Journal of Membrane Science* 157: 219–226.
28. Suzuki, F., Onozato, K., Kurokawa, Y. 1990. A formation of compatible poly (vinyl alcohol)/alumina gel composite and its properties. *Journal of Applied Polymer Science* 39: 371–381.
29. Sengupta, R., Bandyopadhyay, A., Sabharwal, S., et al. 2005. Polyamide 6,6/in situ silica hybrid nanocomposites by sol–gel technique: synthesis, characterization and properties. *Polymer* 46: 3343–3354.
30. Hsiue, G. H., Chen, J. K., Liu, Y. L. 2000. Synthesis and characterization of nanocomposite of polyimide-silica hybrid from nonaqueous sol–gel process. *Journal of Applied Polymer Science* 76: 1608–1618.

31. Wu, C. S. 2004. In situ polymerization of titanium isopropoxide in polycaprolactone: properties and characterization of the hybrid nanocomposites. *Journal of Applied Polymer Science* 92: 1749–1757.
32. Hu, Q., Marand, E. 1999. In situ formation of nanosized TiO_2 domains within poly(amide–imide) by a sol–gel process. *Polymer* 40: 4833–4843.
33. Garcia, M., van Vliet, G., ten Cate, M. G. J., et al. 2004. Large-scale extrusion processing and characterization of hybrid nylon-6/SiO_2 nanocomposites. *Polymers for Advanced Technologies* 15:164–172
34. Zhang, P., Chen, Y., Li, G. et al. 2011. Enhancement of properties of polyimide/silica hybrid nanocomposites by benzimidazole formed hydrogen bond. *Polymers for Advanced Technologies* 23:1362–1368.
35. Cui, Y., Chen, Z., Liu, X. 2016. Preparation of UV-curing polymer-ZrO_2 hybrid nanocomposites via auto-hydrolysis sol–gel process using zirconium oxychloride octahydrate coordinated with organic amine. *Progress in Organic Coatings* 100: 178–187.
36. Jothibasu, S., Ashok Kumar, A., Alagar, M., 2007. Synthesis of maleimide substituted polystyrene-silica hybrid utilizing michael addition reaction. *Journal of Sol–Gel Science and Technology* 43: 337–345.
37. Selvi, M., Devaraju, S., Vengatesan, M.R. et al. 2014. Development of polybenzoxazine–silica–titania (PBZ–SiO_2–TiO_2) hybrid nanomaterials with high surface free energy. *Journal of Sol–Gel Science and Technology* 72: 518–526.
38. Devaraju, S., Vengatesan, M.R., Ashok Kumar, A., et al. 2011. Polybenzoxazine–silica (PBZ–SiO_2) hybrid nanocomposites through in situ sol–gel method. *Journal of Sol–Gel Science and Technology* 60: 33.
39. Ivanković, M., Brnardić, I., Ivanković et al. 2009. Preparation and properties of organic–inorganic hybrids based on poly(methyl methacrylate) and sol–gel polymerized 3-glycidyloxypropyltrimethoxysilane. *Polymer* 50: 2544–2550.
40. Jena, K. K., Sahoo, S., Narayan, R. et al. 2011. Novel hyperbranched waterborne polyurethane-urea/silica hybrid coatings and their characterizations. *Polymer International* 60: 1504–1513.
41. Vaikuntam, S. R., Stöckelhuber, K. S., Eshwaran Subramani Bhagavatheswaran, E. S. et al. 2018. Entrapped styrene butadiene polymer chains by sol–gel-derived silica nanoparticles with hierarchical raspberry structures. *The Journal of Physical Chemistry B* 122: 2010–2022.
42. Amin Abdollahi, A., Roghani-Mamaqani, H., Salami-Kalajahi, M. et al. 2018. Preparation of organic-inorganic hybrid nanocomposites from chemically modified epoxy and novolac resins and silica-attached carbon nanotubes by sol–gel process: Investigation of thermal degradation and stability. *Progress in Organic Coatings* 117: 154–165.
43. Zhou, S.Z., Qiao, X.G. 2018. Synthesis of raspberry-like polymer@silica hybrid colloidal particles through biphasic sol–gel process. *Colloids and Surfaces A: Physicochemical and Engineering Aspects* 553: 230–236.
44. Abdollahi, A., Roghani-Mamaqani, H., Salami-Kalajahi, M. et al. 2018. Preparation of hybrid composites based on epoxy, novolac, and epoxidized novolac resins and silica nanoparticles with high char residue by sol–gel method. *Polymer Composites*. doi.org/10.1002/pc.24631
45. Weng, Z., Wu, K., Luo, F. et al. 2018. Fabrication of high thermal conductive shape-stabilized polyethylene glycol/silica phase change composite by two-step sol gel method. *Composites Part A: Applied Science and Manufacturing* 110: 106–112.
46. Ma, C. C. M., Lin, J. M., Chang, W. C. et al. 2002. Carbon/carbon nanocomposites derived from phenolic resin–silica hybrid ceramers: microstructure, physical and morphological properties. *Carbon* 40: 977–984.
47. Zhong, J., Zhang, M., Jiang, Q. et al. 2006. Synthesis and characterization of silica–alumina co-doped polyimide film. *Materials Letters* 60: 585–588.

48. Streckova, M., Fuzer, J., Kobera, L. et al. 2014. A comprehensive study of soft magnetic materials based on FeSi spheres and polymeric resin modified by silica nanorods. *Materials Chemistry and Physics* 147: 649–660.
49. Yang, M., Gao, Y., He, J. P., et al. 2007. Preparation of polyamide 6/silica nanocomposites from silica surface initiated ring-opening anionic polymerization. *eXPRESS Polymer Letters* 1: 433–442.
50. Zhang, J., Liu, N., Wang, M. et al. 2010. Preparation and characterization of polymer/silica nanocomposites via double in situ miniemulsion polymerization. *Journal of Polymer Science Part A: Polymer Chemistry* 48: 3128–3134.
51. Shariatmadar, F. S., Mohsen-Nia, M. 2012. PES/SiO_2 nanocomposite by in situ polymerization: Synthesis, structure, properties, and new applications. *Polymer Composites* 33: 1188–1196.
52. Zhang, S. W., Zhou, S. X., Weng, Y. M. et al., Synthesis of SiO_2/polystyrene nanocomposite particles via miniemulsion polymerization. *Langmuir* 21: 2124–2128.
53. Chen, X. D., Zhou, N., and Zhang, H. 2009. In-situ polymerization and characterization of poly (ε -caprolactone) urethane/SiO_2 nanocomposites. *Journal of Physics: Conference Series* 188: 012025.
54. Zapata, P., Quijada, R. 2012. Polypropylene nanocomposites obtained by in situ polymerization using metallocene catalyst: Influence of the nanoparticles on the final polymer morphology. *Journal of Nanomaterials* 2012:1–6.
55. Muller, D., Pinheiro, G. K., Bendo, T. et al. 2015. Synthesis of conductive PPy/SiO_2 aerogels nanocomposites by in situ polymerization of pyrrole. *Journal of Nanomaterials* 2015: 1–6.
56. Qi, D., Liu, C., Chen, Z. et al. 2016. Formation mechanism of guava-like polymer/SiO_2 nanocomposite particles in in situ emulsion polymerization systems. *Colloids and Surfaces A: Physicochemical and Engineering Aspects* 489: 265–274
57. Xu, Q. J., Wang, S. B., Chen, F. F. et al. 2016. Studies on the interfacial effect between Nano-SiO_2 and nylon 6 in nylon 6/SiO_2 nanocomposites. *Nanomaterials and Nanotechnology* 6:31.
58. Jongsomjit, B., Chaichana, E., Praserthdam, P. 2005. LLDPE/nano-silica composites synthesized via in situ polymerization of ethylene/1-hexene with MAO/metallocene catalyst. *Journal of Materials Science* 40: 2043–2045.
59. Lin, D., Liu, W., Liu, Y. et al., 2016. High ionic conductivity of composite solid polymer electrolyte via in situ synthesis of monodispersed SiO_2 nanospheres in poly(ethylene oxide). *Nano Letters* 16: 459–465.
60. Ekrachan, C., Somsakun, P., Okornn M. et al. 2012. LLDPE/TiO_2 nanocomposites produced from different crystallite sizes of TiO_2 via in situ polymerization. *Chinese Science Bulletin* 57: 2177–2184.
61. Wu, C. S. 2005. Synthesis of polyethylene-octene elastomer/SiO_2-TiO_2 nanocomposites via in situ polymerization: Properties and characterization of the hybrid. *Journal of Polymer Science Part A: Polymer Chemistry* 43: 1690–1701.
62. Abbasian, M., Ahmadkhani, L. 2016. Synthesis of conductive PSt-g-PANi/TiO_2 nanocomposites by metal catalyzed and chemical oxidative polymerization. *Journal Designed Monomers and Polymers* 19: 7585–7595.
63. Liu, H., Ye, H., Lin, T. et al. 2008. Synthesis and characterization of PMMA/Al_2O_3 composite particles by in situ emulsion polymerization. *Particuology* 6: 207–213.
64. Li, H., Liu, G., Liu, B. et al. 2007. Dielectric properties of polyimide/Al_2O_3 hybrids synthesized by in-situ polymerization. *Materials Letters* 61: 1507–1511.
65. Qi, Y. -N., Xu, F., Ma, H. -J. et al. 2008. Thermal stability and glass transition behavior of PANI/γ-Al_2O_3 composites. *Journal of Thermal Analysis and Calorimetry* 91: 219–223.
66. Desharun, C., Jongsomjit, B., Praserthdam, P. 2008. Study of LLDPE/alumina nanocomposites synthesized by in situ polymerization with zirconocene/d-MMAO catalyst. *Catalysis Communications* 9: 522–528.

67. Wu, J., Yang, S., Gao, S. et al. 2005. Preparation, morphology and properties of nano-sized Al_2O_3/polyimide hybrid films. *European Polymer Journal* 41: 73–81.
68. Zhu, W., Zhang, G., Yu, J. et al. 2004. Crystallization behavior and mechanical properties of polypropylene copolymer by in situ copolymerization with a nucleating agent and/or nano-calcium carbonate. *Journal of Applied Polymer Science* 91: 431–438.
69. Bhanvase, B. A., Gumfekar, S. P., Sonawane, S. H. 2009. Water-based PMMA-nano-$CaCO_3$ nanocomposites by in situ polymerization technique: Synthesis, characterization and mechanical properties. *Journal Polymer–Plastics Technology and Engineering* 48: 939–944.
70. Wu, W., He, T., Chen, J., et al. 2006. Study on in situ preparation of nano calcium carbonate/PMMA composite particles. *Materials Letters* 60: 2410–2415.
71. Jian-ming, S., Yong-zhong, B., Zhi-ming, H. et al. 2004. Preparation of poly (methyl methacrylate)/nanometer calcium carbonate composite by in-situ emulsion polymerization. *Journal of Zhejiang University–SCIENCE A* 5: 709–713.
72. Avella, M., Errico, M. E., Martuscelli, E. 2001. Novel PMMA/$CaCO_3$ nanocomposites abrasion resistant prepared by an in situ polymerization process. *Nano Letters* 1:213–217.
73. Xie, X. L., Liu, Q. L., Li, R. K. Y. et al. 2004. Rheological and mechanical properties of PVC/$CaCO_3$ nanocomposites prepared by in situ polymerization. *Polymer* 45: 6665–6673.
74. Zhuang, W., Liu, J., Zhang, J. H., et al. 2008. Preparation, characterization, and properties of TiO_2/PLA nanocomposites by in situ polymerization. *Polymer Composites* 30: 1074–1080.
75. Ye, C., Li, H., Cai, A. et al. 2011. Preparation and characterization of organic nano-titanium dioxide/acrylate composite emulsions by in-situ emulsion polymerization. *Journal of Macromolecular Science, Part A Pure and Applied Chemistry* 48: 309–314.
76. Yang, M., Dan, Y. 2006. Preparation of poly(methyl methacrylate)/titanium oxide composite particles via in-situ emulsion polymerization. *Journal of Applied Polymer Science* 101: 4056–4063.
77. Džunuzović, E., Marinović-Cincović, M., Vuković, J. et al. 2009. Thermal properties of PMMA/TiO_2 nanocomposites prepared by in-situ bulk polymerization. *Polymer Composites* 30: 737–742.
78. El Nahrawy, A. M., Ahmed A. Haroun, A. A., Hamadneh, I. et al. 2017. Conducting cellulose/TiO_2 composites by in situ polymerization of pyrrole. *Carbohydrate Polymers* 168: 182–190.
79. Dzunuzovic, E., Jeremic, K., Nedeljkovic, J. M. 2007. In situ radical polymerization of methyl methacrylate in a solution of surface modified TiO_2 and nanoparticles. *European Polymer Journal* 43: 3719–3726.
80. Evora, V. M. F., Shukla, A. 2003. Fabrication, characterization, and dynamic behavior of polyester/TiO_2 nanocomposites. *Materials Science and Engineering A* 361: 358–366.
81. da Silva, V. D., dos Santos, L. M., Subda, S. M. et al. 2013. Synthesis and characterization of polyurethane/titanium dioxide nanocomposites obtained by in situ polymerization. *Polymer Bulletin* 70: 1819–1833.
82. Sadu, R. B., Chen, D. H., Kucknoor, A. S. et al. 2014. Silver-doped TiO_2/polyurethane nanocomposites for antibacterial textile coating. *BioNanoScience* 4: 136–148.
83. Elbasuney, S. 2014. Dispersion characteristics of dry and colloidal nano-titania into epoxy resin. *Powder Technology* 268: 158–164.
84. Abdul Kaleel, S.H., Bahuleyan, B.K., Masihullah, J., et al. 2011. Thermal and mechanical properties of polyethylene/doped-TiO_2 nanocomposites synthesized using in situ polymerization. *Journal of Nanomaterials* 2011: 1–6.
85. Rong, Y., Chen, H. Z., Wu, G. et al. 2005. Preparation and characterization of titanium dioxide nanoparticle/polystyrene composites via radical polymerization. *Materials Chemistry and Physics* 91: 370–374.

86. Hu, Y., Zhou, S., Wu, L. 2009. Surface mechanical properties of transparent poly(methyl methacrylate)/zirconia nanocomposites prepared by in situ bulk polymerization. *Polymer* 50: 3609–3616.
87. Park, H. J., Kwak, S.Y., Kwak, S. 2005. Wear-resistant ultra high molecular weight polyethylene/zirconia composites prepared by in situ ziegler-natta polymerization. *Macromolecular Chemistry and Physics* 206: 945–950.
88. Kwak, S., Noh, D. I., Chun, H. J. et al. 2009. Effect of γ-ray irradiation on surface oxidation of ultra high molecular weight polyethylene/zirconia composite prepared by in situ ziegler-natta polymerization. *Macromolecular Research* 17: 603–608.
89. Jongsomjit, B., Panpranot, J., Praserthdam, P. 2007. Effect of nanoscale SiO_2 and ZrO_2 as the fillers on the microstructure of LLDPE nanocomposites synthesized via in situ polymerization with zirconocene. *Materials Letters* 61: 1376–1379.
90. Yang, K., Huang, X., Xie, L. et al. 2012. Core–shell structured polystyrene/$BaTiO_3$ hybrid nanodielectrics prepared by in situ RAFT polymerization: A route to high dielectric constant and low loss materials with weak frequency dependence. *Macromolecular Rapid Communications* 33: 1921–1926.
91. Xie, L., Huang, X., Wu, C. 2011. Core-shell structured poly(methyl methacrylate)/$BaTiO_3$ nanocomposites prepared by in situ atom transfer radical polymerization: a route to high dielectric constant materials with the inherent low loss of the base polymer. *Journal of Materials Chemistry* 21: 5897–5906.
92. Yang, X., Guo, Y., Luo, X. et al. 2018. Self-healing, recoverable epoxy elastomers and their composites with desirable thermal conductivities by incorporating BN fillers via in-situ polymerization. *Composites Science and Technology* 164: 59–64.
93. Kakiage, M., Tahara, N., Yanagidani, S., Yanase, I., Kobayashi, H. 2011. Effect of boron oxide/carbon arrangement of precursor derived from condensed polymer-boric acid product on low-temperature synthesis of boron carbide powder. *Journal of the Ceramic Society of Japan*, 119 (6): 422–425.
94. Yan, H., Ning, R. Liang, G. et al. 2007. The effect of silane coupling agent on the sliding wear behavior of nanometer ZrO_2/bismaleimide composites. *Journal of Materials Science* 42: 958–965.
95. Ramdani, N., Derradji, M., Feng, T. T. et al. 2015. Preparation and characterization of thermally-conductive silane-treated silicon nitride filled polybenzoxazine nanocomposites. *Material Letters* 155: 34–37.
96. Ramdani, N., Wang, J., Liu, W. B. 2014. Preparation and thermal properties of polybenzoxazine/TiC hybrids. *Advanced Materials Research* 887–888: 49–52
97. Choi, S., Kim, J. 2013. Thermal conductivity of epoxy composites with a binary-particle system of aluminum oxide and aluminum nitride fillers. *Composites: Part B* 51: 140–147
98. Lin, C. T., Lee, H. T., Chen, J. K. 2015. Preparation and properties of bisphenol-F based boron-phenolicresin/modified silicon nitride composites and their usage as binders for grinding wheels. *Applied Surface Science* 330: 1–9.
99. Ghezelbash, Z., Ashouri, D., Mousavian, S. et al. 2012. Surface modified Al_2O_3 in fluorinated polyimide/Al_2O_3 nanocomposites: Synthesis and characterization. *Bulletin of Materials Science* 35: 925–931.
100. Lin, J., Wu, X., Zheng, C. et al. 2014. Synthesis and properties of epoxy-polyurethane/silica nanocomposites by a novel sol method and in-situ solution polymerization route. *Applied Surface Science* 303: 67–75.
101. Thunga, M., Akinc, M., Kessler, M. R. 2014. Tailoring the toughness and CTE of high temperature bisphenol E cyanate ester (BECy) resin. *Express Polymer Letters* 8: 336–344.
102. Nhuapeng, W., Thamjaree, W., Kumfu, S. et al. 2008. Fabrication and mechanical properties of silicon carbide nanowires/epoxy resin composites. *Current Applied Physics* 8: 295–299.

103. Kajohnchaiyagual, J., Jubsilp, C., Dueramae, I. et al. Thermal and mechanical properties enhancement obtained in highly filled alumina-polybenzoxazine composites. *Polymer Composites* 35: 2269–2279.
104. Kim, K., Kim, J. H. 2014. Fabrication of thermally conductive composite with surface modified boron nitride by epoxy wetting method. *Ceramics International* 40: 5181–5189.
105. Etemadi, H., Shojaei, A. 2014. Characterization of reinforcing effect of alumina nanoparticles on the novolac phenolic resin. *Polymer Composites* 35:1285–1293.
106. Charpentier, P. A., Burgess, K., Wang, L., et al. 2012. Nano-TiO_2/polyurethane composites for antibacterial and self-cleaning coatings. *Nanotechnology* 23: 425606.
107. Vengatesan, M. R., Devaraju, S., Dinakaran, K. et al. 2012. SBA-15 filled polybenzoxazine nanocomposites for low-k dielectric applications. *Journal of Materials Chemistry* 22: 7559.
108. Sasikala, T. S., Nair, B. P., Pavithran, C., Sebastian, M. T. 2012. Improved dielectric and mechanical properties of polystyrene-hybrid silica sphere composite induced through bifunctionalization at the interface. *Langmuir* 28: 9742–9747.
109. Devaraju, S., Prabunathan, P., Selvi, M., et al. 2013. Low dielectric and low surface free energy flexible linear aliphatic alkoxy core bridged bisphenol cyanate ester based POSS nanocomposites. *Frontiers in Chemistry* 1: 19.
110. Isobe, T., Nishimura, M., Matsushita, S. et al. 2014. Gas separation using Knudsen and surface diffusion I: Preparation of epoxy/porous SiO_2 composite. *Microporous and Mesoporous Materials* 183: 201–206.
111. Mathieu, L. M., Bourban, P. E., Månson, J. A. E. 2006. Processing of homogeneous ceramic/polymer blends for bioresorbable composites. *Composites Science and Technology* 66: 1606–1614
112. Zou, H., Wu, S., Shen, J. 2008. polymer/silica nanocomposites: preparation, characterization, properties, and applications. *Chemical Reviews* 108: 3893–3957.
113. Ou, B., Li, D. 2009. The effect of functionalized-TiO_2 on the mechanical properties of PP/PA6/functionalized-TiO_2 nanocomposites prepared by reactive compatibilization technology. *Journal of Composite Materials* 43: 1361–1372.
114. Xu, Q., Chen, F., Li, X., et al. 2013. The effect of surface functional groups of nanosilica on the properties of polyamide 6/SiO_2 nanocomposite. *Polish Journal of Chemical Technology* 3: 20–24.
115. Reddy, C. S., Das, C. K. 2005. HLDPE/organic functionalized SiO_2 nanocomposites with improved thermal stability and mechanical properties. *Composite Interfaces* 11: 687–699.
116. Chanmal, C. V., Jog, J. P. 2008. Dielectric relaxations in PVDF/$BaTiO_3$ nanocomposites. *eXPRESS Polymer Letters* 2: 294–301.
117. Bikiaris, D. N., Vassiliou, A., Pavlidou, E. et al. 2005. Compatibilisation effect of PP-g-MA copolymer on iPP/SiO_2 nanocomposites prepared by melt mixing. *European Polymer Journal* 41:1965–1978.
118. Hu, T., Juuti, J., Jantunen, H. 2009. RF-properties of BST-PPS composites. *Journal of the European Ceramic Society* 27: 2923–2926.
119. Li, K., Wang, H., Xiang, F., et al. 2009. Surface functionalized $Ba_{0.6}Sr_{0.4}TiO_3$/poly(vinylidene fluoride) nanocomposites with significantly enhanced dielectric properties. *Applied Physics Letters* 95: 202904.
120. Chao, F., Liang, G. 2009. Effects of coupling agent on the preparation and dielectric properties of novel polymer–ceramic composites for embedded passive applications. *Journal of Materials Science* 20: 560–564.

5 Mechanical Properties of Polymer/Ceramic Composites

5.1 INTRODUCTION

The first aim of introducing ceramic fillers to polymers is to enhance their inferior mechanical performance, thus the mechanical properties of polymer/ceramic micro- and composites are significantly studied. In fact, one of the critical requirements of polymer composites is to tailor their strength/stiffness and their toughness values as much as possible. Therefore, it is necessary to evaluate the mechanical performance of such composites from different aspects. Many features, including tensile strength, impact strength, flexural strength, hardness, fracture toughness, etc., have been utilized to characterize the various kinds of polymer/ceramic nanocomposites.

The ceramic fillers have many outstanding mechanical characteristics, such as good wear-resistance, higher tensile strength and modulus, good fatigue resistance, and high microhardness [1]. Similar to other composites, the extent of the improvement in the mechanical properties depends on many key factors, such as the reinforcement-phase particle size and fraction and its degree of dispersion in the thermosetting matrix, interface bonding, and the filler aspect ratio. For example, a crucial aspect for thermoset/ceramic nanocomposites is that all the property enhancements can be obtained at lower filler loading compared to micro-composite once. The addition of ceramic fillers into the various thermosetting matrices can significantly affect their mechanical properties in different ways. Ma et al. developed TDE-85 and bismalimide (BMI) composites reinforced with $AlBO_3$ whiskers pre treated with KH-550 silane by a casting method. The mechanical properties of these composites was controlled by changing the size, the amount, and the mixing process [2].

A ternary nanocomposite composed from bismaleimide/polyetherimide/silica (BMI/PEI/SiO_2) showing a uniform filler dispersion was prepared through a sol–gel reaction. The silane coupling agent KH-550 was used to enhance the interaction between the matrix and SiO_2 nano-filler. The mechanical performance of this hybrid were markedly improved with increasing silica loading [3]. Pinto et al. reviewed the effect of TiO_2 on epoxy resins and found that even at low filler contents, the filling with TiO_2 nanoparticles can significantly ameliorate the crucial mechanical properties, such as tensile modulus, tensile strength, toughness and fracture toughness, fracture energy, flexural modulus, flexural strength, elongation at break, fatigue crack propagation resistance, abrasion, pull-off strength, and fracture surface properties [4].

The evolution of mechanical and fracture properties of α-Al_2O_3/epoxy resins composites as a function of average filler size, size distribution, particle shape, loading, and epoxy cross-link density was studied by McGrath et al. Results show that only small changes in particle size, shape, and size distribution had a slight impact on the final properties, while the resin cross-link density and filler loading had a significant influence, provoking major variations in all properties [5]. Novel polypropylene (PP)/SiO_2 nanocomposites were developed using PP and silica nanospheres (SNSs) joined to poly(5-hexen-1-ol-*co*-propylene) (PPOH), which is a functionalized PP having hydroxyl groups that was found to play a key role as not only an elastomer but also an interface modifier [6]. The toughness of the nanocomposite was noticeably improved by blending PPOH6.4 without a significant loss of stiffness. In contrast, the toughness was hardly improved using PPOH1.3.

5.2 TENSILE PROPERTIES

Tensile properties are the most investigated mechanical properties of polymer/ceramic composites, hence Young's modulus, tensile strength, and the elongation at break are three main parameters studied. These vary with the ceramic filler loading, but the variation trends are different. In a study, Hasan et al. [7] investigated the influence of the size and shape of the Al_2O_3 fillers, and their ratios on the mechanical properties of a PMMA/Al_2O_3 composite were studied using nanoindentation measurements. It was observed that both Al_2O_3 whiskers and spherical Al_2O_3 nanoparticles filled PMMA matrix improved the tensile properties of the composites, but the improvement was significantly higher with reinforcement by Al_2O_3 whiskers. The best performance was obtained by the addition of 3 wt.% of Al_2O_3 whiskers in the PMMA matrix in terms of the tensile properties of the resulted composite.

Three kinds of polypropylene-grafted silica (PGS-2 K, PGS-8 K and PGS-30 K) with various grafting chain lengths were elaborated by Wang et al. [8]. After melt-blending PGS with polypropylene (PP), the PP/PGS interface properties and the influence of PP/PGS interfaces on tensile properties of nanocomposites were evaluated. The resulting nanocomposites benefited from an improved dispersion, and a strong matrix/particle interface not only showed increased Young's modulus and yield stress, but also the strain at break remains in line with the unfilled PP, which is in contrast to the conventional wisdom that the gain in modulus and strength must be at the expense of the decreased break strain. The tensile testing showed that the stiffness of the epoxy polymer could be significantly improved using only small amounts of Al_2O_3, even if some these fillers were in an agglomerated state [9].

The presence of carbon black (CB) and SiO_2 in the formulation of acrylonitrile butadiene/chlorosulphonated polyethylene rubber (NBR/CSM) blends at maximum loading of 30 wt.% provoke satisfactory changes in the polymer properties: 152% increase in tensile strength, 116%, in elongation at break, and 142% modulus at 100% elongation according to synergistic effect between the combined fillers [10]. Table 5.1 collects the mechanical properties of PP/SiO_2 nanocomposites grafted with different polymers at a fixed SiO_2 fraction [11].

As Table 5.1 shows, except for PEA, all the monomers of the grafting polymers having different degrees of miscibility with PP showed a reinforcement effect on the tensile strength of their nanocomposites. Thus, a further understanding of the modified nanoparticles and their role in the composites have to be scrutinized. That is,

TABLE 5.1
Mechanical Properties of PP (Melting Flow Index = 8.5 g/10 min)-based Nanocomposites (Content of SiOz = 3.31 vol.%) Filled with Different Polymer-grafted SiO_2) [11]

Grafting Polymers	Nanocomposite						
	[a]PS	[b]PBA	[c]PVA	[d]PEA	[e]PMMA	[f]PMA	neat PP
Tensile strength (MPa)	34.1	33.3	33.0	26.8	35.2	33.9	32.0
Young's modulus (GPa)	0.92	0.86	0.81	0.88	0.89	0.85	0.75
Nm Elongation-to-break (%)	9.3	12.6	11.0	4.6	12.0	11.9	11.7
Area under tensile stress-strain curve (MPa)	2.4	3.3	2.3	0.8	3.2	2.9	2.2

Grafting conditions: Irradiation dose =10 Mrad; weight ratio of monomer/SiO_2 = 20/100; all the systems used acetone as solvent when they were irradiated, except for methyl acrylic acid/SiO_2 system with ethanol as solvent.

[a]PS: polystyrene, [b] PBA: polybutyl acrylate, [c] PVA: polyvinyl acetate, [d] PEA: polyethyl acrylate, [e] PMMA: polymethyl methacrylate, [f] PMA: polymethyl acrylate.

interdiffusion and entanglement of the grafting polymer segments with the PP molecules, instead of a miscibility between the grafting polymer and the matrix, can determine the interfacial in these PP nanocomposites. This result reflects that for a PP matrix having a higher molecular weight should be entangled more effectively with the agglomerated nano-SiO_2, leading to a higher tensile strength increment.

The influence of adding 0.5 to 5 wt.% of nano-alumina particles to tensile properties of LDPE matrix was studied. The results revealed that the incorporation of 1 wt.% of nano-Al_2O_3 successfully increased the tensile and elongation at the break of the LDPE/Al_2O_3 nanocomposites. The addition of >1 wt.% of Al_2O_3 nanoparticles generated agglomeration and uneven distribution of the particles throughout the LDPE matrix [12].

Nano-silica-filled polypropylene composites were produced using a conventional compounding technique in which the nano-SiO_2 are grafted by polystyrene molecules using irradiation beforehand. A strong interfacial stress-transfer efficiency is shown by both strengthening and toughening effects perceived in tensile tests. The role of the modified nano-SiO_2 in improving the tensile properties of these nanocomposites was related to the percolation concept. A double percolation of yielded regions is presented to reveal the special effect generated by the addition of nano-SiO_2 particles at low-filler loading [13].

According to Lee et al. [14] the incorporation of SiO_2 nanoparticles somehow increased the mechanical properties of the resulting sulfonated polyimide membranes, irrespective of the type of SiO_2. For the IXSPI membranes, the tensile strength increased more than 6%, and the elongation almost tripled owing to an extensible interpenetrating polymer network (IPN) structure containing flexible urethane (–NH–C O–O–) groups. The improvement of tensile properties was markedly observed in sulfonated polyimide/SiO_2 nanocomposite membranes having both

SiO_2 and UAN. This is due to the homogeneous dispersion of SiO_2 nanoparticles, as well as a high extension of IPN based on UAN. New phenolic/silica composites were produced and studied by Rajulu et al. The effect of orientation of SiO_2 fibers and the coupling agent on the tensile properties of the composites was evaluated [15].

Yao et al. [16] used silane-containing epoxy groups (EP)-treated SiO_2 composite to reinforce unsaturated polyester (UPE). The in situ polymerized UPR/SiO_2 composite exhibited better strengthening and toughening effects than those prepared by a mechanical-blending technique due to the formation of a heterogeneous network structure in such composites via the chemical reaction among the epoxy group and the carboxyl group of UPR. The Young's modulus of nanocomposite systems composed of UPE and EP resins reinforced with α-Al_2O_3 nanoparticles having 30 nm to 40 nm and 200 nm in diameter had been evaluated. Results indicated that the Young's modulus value increased with increasing alumina nano-filler contents, while the estimated Young's moduli values were compared with some classical theoretical models, revealing that the experimental results agree with literature data [17].

A casting technique was used to prepare a new series of UPE/Al_2O_3 nanocomposites with good dispersion below a 5 wt.% ratio. The nanocomposites exhibited higher tensile properties than those of pristine UPR due to the ductile fracture nature of UPR, which was converted to brittle fracture by adding nano-Al_2O_3 [18]. The influence of Al_2O_3 particles' shapes and dimensions on the tensile properties of epoxy/Al_2O_3 composites were explored by Lim and co-workers. Results revealed that the shape of the Al_2O_3 nanoparticles did not only affect their dispersion in the epoxy matrix while it affected the tensile modulus and tensile strength of the nanocomposites. The fracture-surface analysis showed that several toughening mechanisms, such as particles' pull-out, crack pinning, plastic yielding, and deformation were the main factors responsible for the increments of the fracture toughness of the epoxy/Al_2O_3 nanocomposites [19].

Thermoplastic polyurethane (TPU)-modified BN nanotubes' epoxy nanocomposites showed promising improvements, including significant increases in yield strength, modulus and strain at failure compared to untreated BN/epoxy, and tailorable properties that become more TPU-like, with high strain at failure, as the BN: TPU weight ratio decreased [20].

As illustrated in Figure 5.1, the fracture pattern of nano-SiO_2/epoxy composite cured with dimethylbenzanthracene (DMBA), which revealed a brittle fracture composed of radial striations; while nano-SiO_2/epoxy composite cured with methyltetrahydrophthalic anhydride (MeTHPA) fractures in a stick–slip manner, revealing organized spaced "rib" markings. Whether the content of nano-SiO_2 was, the elastic moduli of the epoxy/nano-SiO_2 composites were higher than those of the pure epoxy resins, while the elongation at break decreased with the enhancing of nano-SiO_2 loadings. The composites cured by MeTHPA had improved elastic modulus, strength and larger elongation compared to those cured by DMBA [21].

The 3-methacryloxypropyltrimethoxysilane-modified silica (MPS-SiO_2) nanoparticles were synthesized in *N,N*-dimethylformamide (DMF) to overcome the problem that SiO_2 sol cannot be directly added in PU systems if ethanol is used as solvent. Epoxy acrylate copolymers/silica EPAc/SiO_2 was prepared by one-step in-situ solution polymerization of EA and acrylic monomers directly adding a

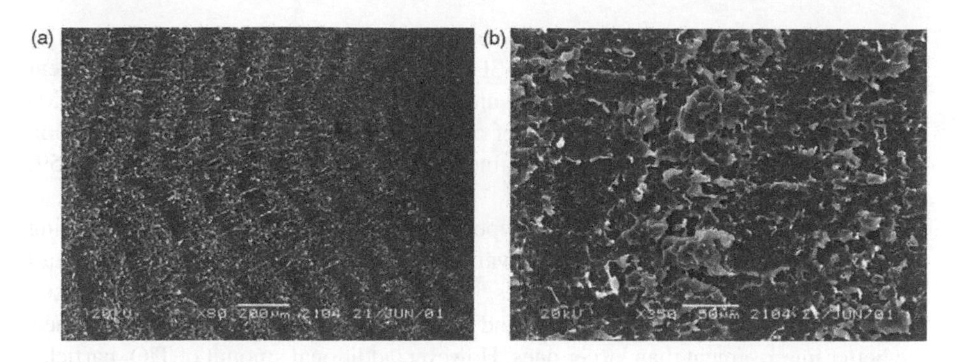

FIGURE 5.1 Fracture Morphology of 1 wt.% Nano-SiO_2/epoxy Composite Cured by MeTHPA: (a) Overall View of the Fracture Surface; (b) Magnification of the Periodic Stick–slip Motion Zone [21].

certain amount of MPS/SiO_2 sol as core. In addition, incorporation of MPS/SiO_2 into EPUAs/SiO_2 greatly enhanced other physicochemical properties of EPUAs/SiO_2; it also greatly enhanced both the tensile strength and elongation at break [22].

The tensile properties of four different epoxy polymers reinforced with well-dispersed SiO_2 nanoparticles were evaluated by Hsieh et al. [23]. They found that the Young's modulus steadily increased as the volume ratio of the nano-SiO_2 was enhanced. The presence of SiO_2 nanoparticles led to significant improvements in the toughness of these epoxy thermosets, while the extent of improvement was related to the epoxy structure/property relationships and the adhesion of SiO_2 to each epoxy matrix interface. The two toughening mechanisms were (a) localized shear bands initiated by the stress concentrations around the periphery of the nano-SiO_2 particles, and (b) debonding of the SiO_2 nano-filler followed by subsequent plastic void growth of the epoxy polymer.

Modeling studies have confirmed the key roles of the stress versus strain behavior of the epoxy polymer and the nano-SiO_2/epoxy-polymer interfacial adhesion in influencing the degree of the two toughening mechanisms of the nanoparticle-filled polymers [23].

Silica/poly-dicyclopentadiene (SiO_2/poly-DCPD) nanocomposites with enhanced tensile properties were successfully elaborated by in situ ring-opening metathesis polymerization (ROMP) by means of reaction-injecting molding (RIM) technique. To improve the interfacial compatibility between SiO_2 nanoparticles and poly-DCPD matrix, polystyrene (PS) was chemically grafted onto the surface of SiO_2 nanoparticles to obtain oil-soluble polystyrene-coated silica spheres with core-shell structure (PS@SiO_2). The improved tensile properties of SiO_2/poly-DCPD nanocomposites were attributed to the improved interfacial compatibility between the PS shell layer of PS@SiO_2 spheres and poly-DCPD matrix, which resulted from the swelling of DCPD monomer in the PS shell before polymerization.

Hybrid nanocomposites prepared from a hybrid filler system composed of BN and MWCNT filled an EP resin were investigated by Ulus et al. The tensile strengths of BN/EP composites increased for BN fractions up to 0.5 wt.% but decreased with rising BN loading. The MWCNT/BN-hybrid reinforcement improved the tensile properties of the unfilled epoxy [25]. Trivedi et al. successfully evaluated the Young's modulus of BNNT/EP composites by a 3D nanoscale representative volume element (RVE) based on continuum mechanics and using the finite element method (FEM) and compared with an extended rule of mixtures. They observed that the addition of 5 vol.% BNNTs in any epoxy matrix increased its tensile strength by 20% to 55% according to the resin composition [26].

Al-Turaif investigated the tensile properties of toughened epoxy composite using two micro- and nano-TiO_2 particles at various filler fractions. By adding only small amounts of nano and submicron particles, improved tensile properties included tensile stress, elongation at break, toughness, and Young's modulus. Smaller particles generate better improvement than larger ones. However, additional amount of TiO_2 particles decreases the tensile properties. It was concluded that the better tensile properties can be achieved at an optimal of TiO_2 content that is related to the particle size. A correlation between tensile and flexural stress as function of particle size was reported [27].

The stress-strain curves of Al_2O_3/natural rubber (NR) nanocomposites at different ceramic filler contents are depicted in Figure 5.2. In this figure, the strain-induced crystallization parameters of NR is clearly detected. The modulus of NR increases with the increase in strain at the small strain region and then is sharply boosted due to the growth of induced crystallization. For prepared nanocomposites, the tensile strength and modulus values are higher than that of unfilled NR, but the strain-induced crystallization is reduced. This was attributed to improved dispersion of nano-Al_2O_3 particles, and to the formation of a strong force between the nanofiller layers and the NR organic chains.

The monotonical enhancement in the tensile strength and strain is found in the nanocomposite of NBR/nano-$CaCO_3$ [29], which represents all synthetic elastomers.

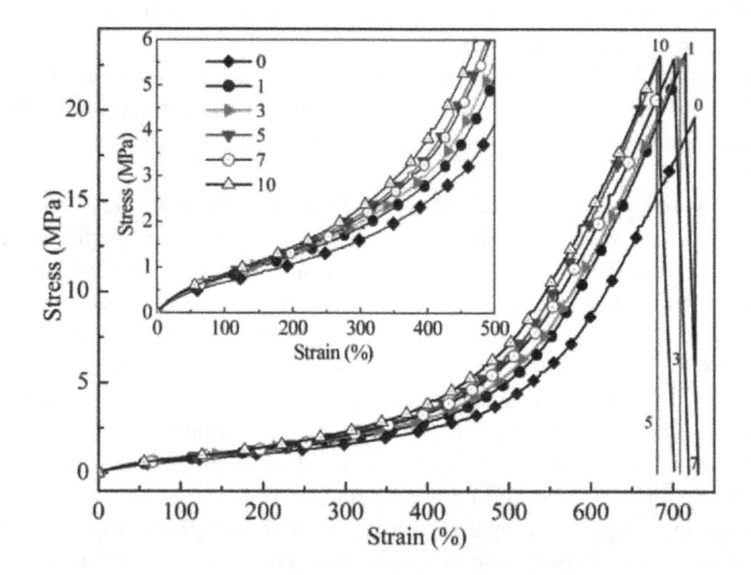

FIGURE 5.2 Stress-strain Curves of a KH570-Al_2O_3/NR Nanocomposites with Different Filler Content [28].

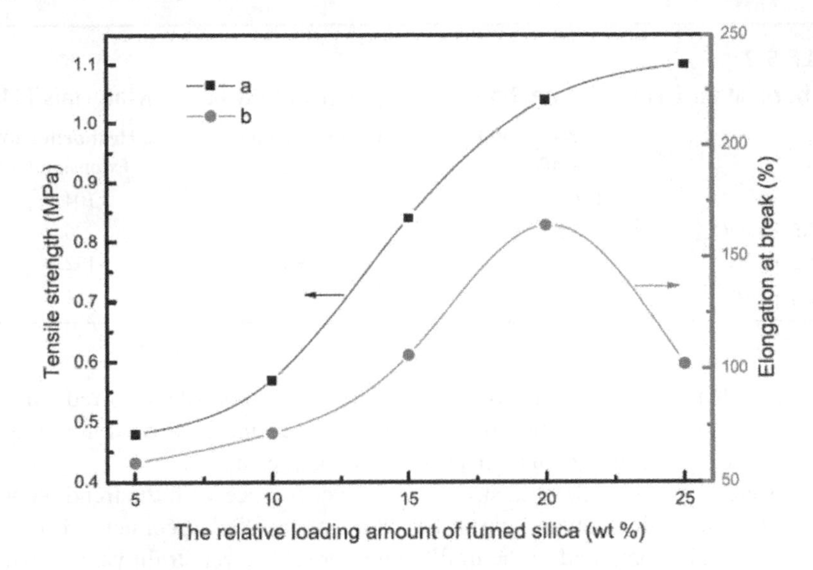

FIGURE 5.3 Effects of the Relative Loading Amount of Fumed Silica on Mechanical Properties of the Novel RTV Silicone Rubbers: (a) Tensile Strength (b) Elongation at Break [30].

The incorporation of $CaCO_3$ nanoparticles increases the stress of their resulting NBR nanocomposite for the similar degree of strain and the strength rises with rising the nano-filler loadings. For example, the tensile strength increases by 24% and 42%, respectively, for 5 wt.% and 10 wt.% of nano-$CaCO_3$ filler contents. The synergistic effect of fumed SiO_2 nanoparticles on the tensile properties of silicone rubbers are studied in [30] as illustrated in Figure 5.3. The tensile strength of the unfilled elastomer enhances from 0.48 to 1.10 MPa as the filler loading increase from 0 to 25 wt.% (Figure 5.3a), whereas the elongation at break increase from 58.6 to 164.1% at 20 wt.% nano-filler ratio and thereafter reduces with further increase of nano-SiO_2 (Figure 5.3b).

5.3 FLEXURAL PROPERTIES

Comparing flexural strength and modulus data of phenolic resin/n-SiC nanocomposites (Table 5.2), the positive effect produced by adding n-SiC into the polymeric matrix clearly appeared [31]. The most significant enhancement in the strength and modulus is recorded for the samples filled with only 1 wt.% of n-SiC nanofillers. In terms of flexural strength, the filling with 0.5 and 2 wt.% of n-SiC gives an increase of 13% compared to an unfilled matrix, however, the incorporation of 1 wt.% n-SiC produced an enhancement in the flexural strength as high as 33%. Concerning flexural modulus, the addition of 0.5, 2 wt.%, 1 wt.% of n-SiC to phenolic resin resulted in an average value of nearly 10%, 15%, and 30% higher than that of the pure resin, respectively. These prominent results can be justified by the fact that the SiC nanoparticles are embedded in the phenolic resin and homogenously covered, generating an improved mechanical interface between these phases. Also this can be addressed

TABLE 5.2
Mechanical and Thermo-mechanical Properties of PR-based Materials [31]

Sample	Flexural strength (MPa)	Flexural modulus (GPa)	Heat deflection Extension (%)
PR	111.1 ± 8.76	2.91 ± 0.12	104.2
PR+0.5% n-SiC	125.5 ± 13.8	2.34 ± 0.35	126.7
PR+1% n-SiC	148 ± 5.6	2.28 ± 0.15	132.1
PR+2% n-SiC	126.2 ± 7.8	2.21 ± 0.06	129.7

by the fact that the included nanofillers can effectively contribute to reducing the crack propagation effect in the polymeric matrix, serving as a reinforcing agent, increasing both the strength and stiffness of phenolic resin.

The average values of flexural strain were in accordance with the trend recorded for the strength and modulus values, hence all nano-filled specimens revealed a decrease of strain compared to the unfilled thermoset. Lower strain values provoke a higher strength and modulus values.

It is clearly appeared from the Figure 5.4a that the flexural modulus of the epoxy nanocomposites increased with the increasing in the of Al_2O_3 nanorods content up to 1 wt.% while the modulus decreased on further addition of nanorods [32]. On the other hand, the modulus of nanocomposite sample at 1.5 wt.% is much higher compared to that of unfilled epoxy resin. The Al_2O_3 nanorods restrict the motion of epoxy chains which generate this increment in the flexural modulus. At higher Al_2O_3 ratios, the flexural modulus was further decreased due to agglomeration of Al_2O_3 nanorods. The increment in the flexural modulus of these nanocomposites as a result of adding Al_2O_3 nanorods also reduced with increasing the post-cure temperature. A similar trend was recorded for the flexural strength of nanocomposites as shown Figure 5.4b. This resulted increment by the reinforcement with Al_2O_3 alumina nanorods was decreased because of the damage in the cross-linked epoxy chain after a post curing at elevated temperature.

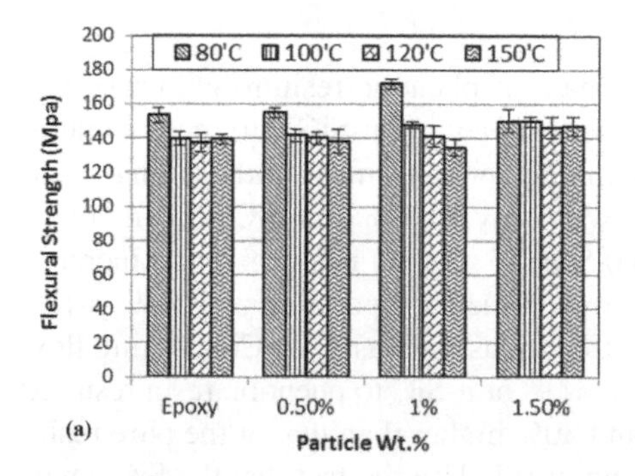

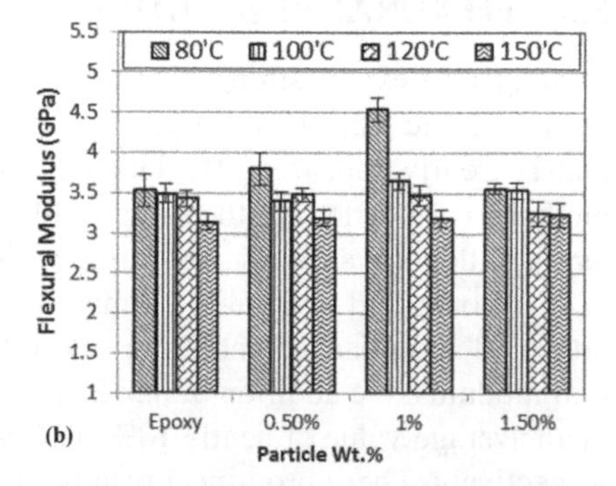

FIGURE 5.4 Flexural Properties of Epoxy and Composites with Post-curing Temperature: (a) Flexural Modulus and (b) Flexural Strength [32].

Yan et al. studied the mechanical properties of low-viscosity bismalimide-based nanocomposites reinforced with surface-treated Si_3N_4 nanoparticles. Compared to the pure matrix, the flexural strength of the prepared nanocomposites enhanced by almost 21%, at only 3 wt.% of nano-Si_3N_4 [33]. New type epoxy/bismaleimide-triazine nanocomposites were prepared using a nano-SiO_2, TDE-85 epoxy resin, a BMI and a BADCy. TDE-85/BMI/BADCy/SiO_2 nanocomposites exhibited improved the flexural strength compared with the pure copolymers' matrix as an appropriate amount of nano-SiO_2 was added [34]. The changes in the flexural properties of phthalonitrile/micro-SiC composites were evaluated. The addition of micro-SiC increased both the flexural strength and modulus as reported by Derradji et al. [35]. The addition of a binary hybrid filler of SiO_2/GO into the phenolic-resin foams enhanced both the mechanical strengths and flame-retardancy characteristics. The flexural of the PF matrix foam reinforced with 1.5 wt.% SiO_2/GO increased by almost 32% [36].

A new type of cross-linked polybenzoxazine/SiO_2 hybrid particles was used to ameliorate the mechanical properties and the structural homogeneity of PF matrix as reported by Yuan et al. The inclusion of the hybrid particles did not affect the thermal stability and flammability of PF matrix, while the flexural strength of the PF resin was significantly improved with the addition of SiO_2 nano-fillers to reach a maximum enhancement of 36% [37]. Effect of including different TiO_2 particle sizes and weight fractions on flexural strength and failure mode of composite-containing epoxy resin was studied by Zhou et al. The addition of 1 wt.% nano-TiO_2 to the epoxy matrix resulted in the highest mechanical performance, while above this fraction, the strength of the composite decreased due to poor dispersion. Micro-sized TiO_2 particles have little effect on strength of epoxy at lower loading, but the strength of composite increased as the size of particles was decreased to nano scale. However, degradation in strength was found in the 5-nm TiO_2/epoxy system due to agglomeration [38].

5.4 MICROHARDNESS

Hardness refers to the characteristic of any material that resists the shape change when force is applied to it. It is recommended for many applications, and it is one of the most mechanical parameters of materials. There are three different kinds of hardness: scratch hardness (resistance to fracture or plastic deformation due to friction from a sharp object), indentation hardness (resistance to plastic deformation due to impact from a sharp object), and rebound hardness (height of the bounce of an object dropped on the material). In addition, there are different measurement systems of hardness, including Brinell hardness, Knoop hardness, Vickers hardness, Shore hardness, etc.

The *Shore A hardness* is the relative hardness of elastic materials, such as soft plastics or their composites, and it can be tested using an instrument known as a *Shore A durometer*. Petrović et al. studied the hardness features of polyurethane filled with silica micro- or nanoparticles. As depicted in Figure 5.5, the *Shore A hardness* values of the PU/silica nanocomposites was steadily enhanced with increasing micro-SiO_2 content, but with increasing the nano-SiO_2 content, this trend was changed after an initial increase. The difference in the hardness-improvement

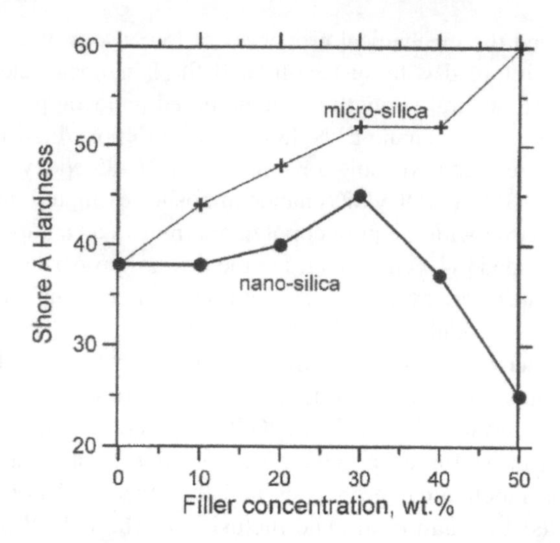

FIGURE 5.5 Shore Hardness of Polyurethane/SiO_2 Nano- and Microcomposites. Reprinted with permission from John Wiley & Sons, from Petrović et al. 2000 [39].

degree were attributed to the fact that micro-SiO_2 is a hard filler, while it appears that nano-SiO_2 is relatively not hard [39].

Rockwell hardness is based on pressing a steel ball (called the indenter) into the flat, smooth surface of a specimen. The change in depth between the initial position at pre-load, and the final position at full load, is considered as the hardness number. The deformation will generally have three constituents:

1. Elastic (Hertzian)
2. Brittle (Griffiths)
3. Ductile (Huber–von Mises).

Li et al. [40] found that the hardness of the composite measured by the Rockwell method decreased with increasing BN nanotubes' content. As illustrated in Figure 5.6, the Rockwell hardness of the unmodified PU reached the value of 80.772.1 HRR, while at 0.5 vol.% and 2 vol.% BN nanotubes filled PU composites attained 79.171.9 HRR and 76.274.2 HRR, respectively. For polymer/ceramic composites, as the fillers are much stiffer and stronger than the matrices, the hardness will increase with increasing the filler content if there is a perfect adhesion and no defects. However, the opposite is recorded for these composites; this is due to the huge number of voids in these composite samples, which are enhanced with increasing the BN nanotube loading. This trend is also recorded for the agglomerated specimen filled by 2 vol.%.

According to Mohandas et al. [41], the addition of ano-Al_2O_3 functionalized with GPS to epoxy resin toughened with bismalimide produced significant improvements in the hardness up to 5 wt.% nano-Al_2O_3 content due to good filler dispersion and interaction with the matrix. SiC/epoxy composites exhibiting promoted mechanical properties were fabricated by using the ultrasonic mixing and casting techniques as

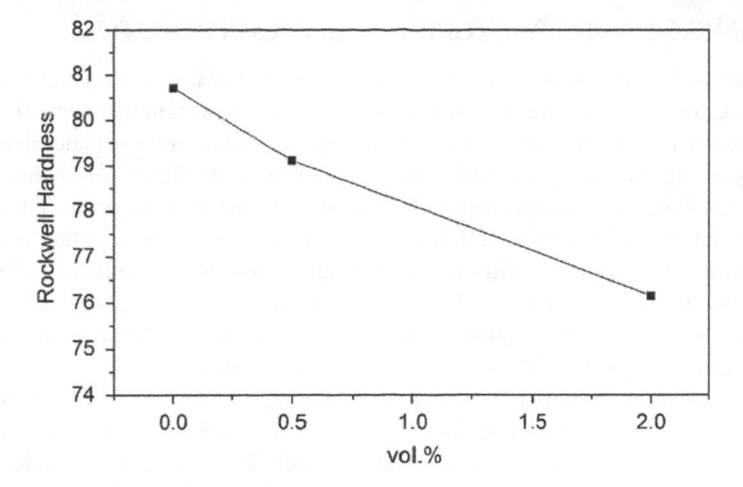

FIGURE 5.6 HRR Hardness Values of Neat PU, 0.5 vol.% and 2.0 vol.% BNNT-reinforced PU Composites [40].

reported by Nhuapeng et al. [42]. The composites with 15 vol.% of SiC nanowires showed the best hardness of composites, which were enhanced by 384% compared to that of the pure epoxy resin.

Demian et al. [43] determined the mechanical behavior of the polyimide (PI) and their composite coatings using microhardness measurements. The Vickers hardness ($HV_{0.01}$) experimental data are shown in Figure 5.7. The PI-coating microhardness attained a value of nearly 27 $HV_{0.01}$, which is much higher than that of the pristine PI-based coatings. If different PTFE weight ratios are included into the PI matrix, a marked decrease in the Vickers microhardness of the composite coatings was obtained. Although, the addition of SiC particles into the PI/PTFE hybrid matrix resulted in a slight increase of microhardness of the composite coatings.

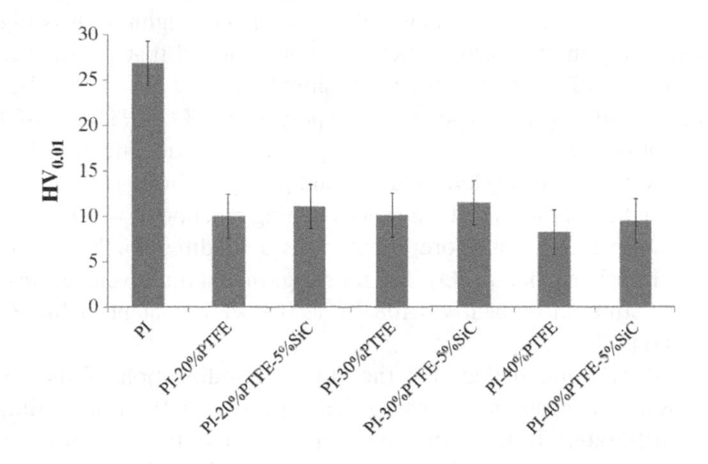

FIGURE 5.7 Vickers Microhardness of PI and PI-based Composite Coatings [43].

5.5 IMPACT AND FRACTURE TOUGHNESS PROPERTIES

Fracture toughness is a property that characterizes the ability of a material containing a crack to resist fracture. It is also one of the most important mechanical parameters of materials. A parameter known as a *stress-intensity factor* is generally used to evaluate the fracture toughness of numerous materials. As the stress-intensity factor of a material attains a critical value (K_{IC}), unstable fracture take place. This critical value of the stress-intensity factor is called the fracture toughness of the material. It can be tested by different technique, including single-edge notch bend (SENB) and indentation fracture toughness (IFT). The addition of ceramic micro- and nanoparticles has been reported as a prominent option to increase the fracture toughness of thermosetting polymers without compromising the stiffness.

The fracture toughness, K_{IC}, of the SiO_2 nanoparticle reinforced epoxy polymers was measured by the SENB method. A K_{IC} value of 0.59 MN.m$^{-3/2}$ was recorded for the unfilled epoxy, while the incorporation of SiO_2 nanoparticles increased the fracture toughness value with a maximum value of 1.42 MN m$^{-3/2}$ was obtained for the epoxy nanocomposite sample filled with 13.4 vol.% nano-SiO_2 [44]. Using the measured modulus, these values were converted to fracture energies, G_{IC}. Thus, the pristine epoxy matrix had a G_{IC} of 103 J/m^2 while the maximum fracture energy of 460 J/m^2 was estimated. This is due to the significant toughening effect of nano-SiO_2 on the epoxy matrix.

It was reported by Rosso et al. [45] that the addition of 5 vol.% SiO_2 nanoparticles could significantly improve the toughness of an epoxy resin. The elastic modulus (E-modulus) from the tensile test was improved by almost 20%, whereas the fracture toughness values, K_{IC}, were increased by nearly 70% and G_{IC} by more than 140%. Moreover, the reinforced nanocomposites were more ductile and revealed a greater yielding than the pure epoxy.

The fracture mechanics tests were recorded on newly developed epoxy/SiO_2 nanocomposites as reported by Ragosta et al. [46]. Results showed that the addition of SiO_2 nanoparticles up to 10 wt.% generate a marked increase in the fracture toughness and an enhancement in the critical crack length for the onset of crack propagation was also observed. This enhancement in toughness was higher than that obtained using micro-sized particles. Xu et al. found that the impact strength of polypropylene (PP) can be enhanced simultaneously when the 3-glycidoxypropyltrimethoxysilane/SiO_2/hyperbranched polyester (SiO_2/GPTS/H20-C19) nanoparticles are introduced [47]. The nano-SiO_2 particles functionalized with poly(butyl acrylate-co-glycidyl methacrylate)-g-diaminodiphenyl sulfone (P(BA-co-GMA)-g-DDS)) were synthesized by ATRP and ring-opening reaction. It was found that SiO_2-P(BA-co-GMA)-g-DDS) was more effective as a modifier for bismalimide resin compared to the pristine nano-SiO_2. The most significant improvement was recorded for the impact strength by nearly 1,108.7% at 0.5 wt.% content of SiO_2/P(BA-co-GMA)-g-DDS) [48].

Liang et al. [49] found that that the surface modification of the ($Al_{18}B_4O_{33}$) whiskers played a key role in improving the impact strength of bismalimide resin, where the amelioration in these properties are better in the case of using BE4 as a coupling agent. The mechanical properties, including the impact strength of polybenzoxazine/$CaCO_3$, were enhanced with the increasing of $CaCO_3$ in the range of 0 to 30 wt.% filler loading [50]. An increase from 195% to 247% of stress at break, and from 109% to 131% of impact strength of unsaturated polyester/silica (UPE/SiO_2) nanocomposites was obtained for varying the ratio of vinyl-modified SiO_2 from 0 to 1 wt.% [51].

The addition of nano-SiO_2 to UPE matrices affects the toughness by enhancing the tortuosity of the crack path locally while micro-scale intercalated tactoids can engender crack deflection. These involved mechanisms depend on localized plasticity for a huge energy dissipation. Since UPE had a very low localized plasticity below a temperature of approximately 90°C, insignificant improvement can be obtained at room temperature toughness by manipulating the micro-/nanostructure of the nanocomposite without introducing another modifier [52]. Kong et al. [53] studied the interfacial structure and related mechanical properties of WD-SC101-SiO_2/UPE composites. It is reported that the interfacial thickness of the sample without coupling agent ranged from 0 to 600 nm; by contrast, the recorded interfacial thickness of a UPE nanocomposite having a 0.3 wt.% coupling agent was about 600 to 1,200 nm. After the coupling agent reacted with the surface hydroxyl of the SiO_2 reinforcement, the thickness of high-modulus interfaces increased while the bending fracture strength reduced with the increase in coupling-agent content.

Wang et al. [54] modified UPR using various concentrations of KH-570-treated SiO_2 nanofillers to improve its lower strength and toughness properties. The addition of 1.5 wt.% KH-570-SiO_2 improved the impact strength of UPE resin by approximately 9.6%. This is related to the uniform dispersion of 3 wt.% KH-570 modified nano-SiO_2 at a temperature of 80°C for 2 hours. Kim et al. [55] studied new epoxy/Al_2O_3 composites cured using γ-rays at 100 kGy under nitrogen atmosphere at room temperature. The incorporation of γ-Al_2O_3 filler revealed an improved fracture toughness by a crack deviation due to the presence of ceramic particles. Macro/microscopic fracture characterizations of epoxy/SiO_2 nanocomposites were identified and some typical fracture parameters like load displacement relationship, fracture load, and fracture toughness were studied by Yao et al. The nanocomposite containing 3 wt.% nanoparticle had the highest fracture toughness and larger deformation-resisting capability [56].

Homogeneously dispersed nano-TiO_2 filled epoxy micro-composites were developed using a mechanical stirring followed by a vertical centrifugal casting and studied by Patnaik et al. The 20 wt.% micro-TiO_2–reinforced graded-epoxy composites exhibited a dramatic enhancement in the impact strength compared to homogeneous composites. Under the same experimental conditions, the morphological analysis revealed that in homogeneous composites with TiO_2, micro-particles are peeled off from the matrix to form holes while in graded composites, the micro-TiO_2 particles remain quite intact to the matrix [57].

Opelt et al. [58] synthesized nano-Al_2O_3 embedded in epoxy resin composites at different ratios. Mechanical tests revealed that as nano-Al_2O_3 content increased in epoxy matrices, fracture toughness of the nanocomposites increased to reach 12% at 1.5 vol.% Al_2O_3 nano-filler loading, without changing the tensile properties. SEM images reflected that cavitation-shear yielding was the predominant mechanism for the epoxy/Al_2O_3 nanocomposites [58].

Li et al. [59] investigated the effect of surface-modification for nano-SiO_2 on the impact properties of epoxy resin, where they showed more effectiveness than that of raw SiO_2 in improving the impact toughness. This is related to large specific surface area and active groups on surface-modified SiO_2 nanoparticles. Similar behavior has been shown for the toughness of epoxy resin-based composite with nano-SiO_2 and standard SiO_2 as shown in the SEM images in Figure 5.8.

The SiC nanofibers at the ratios of 0.1, 0.25, and 0.5 wt.% produced by self-propagating high-temperature synthesis technique (SHS), were used to elaborate the epoxy/SiC composites with good dispersion of SiC nanofiber, as reported by Vijayan et al. Results showed that the maximum increase in fracture toughness was detected for composites consisting of 0.25 wt.% SiC nanofibers [60].

Nanosized-SiO_2 particles ($\approx$ 15 nm) modified with a grafted-block copolymer consisting of a 5- to 20-nm rubbery polyhexylmethacrylate (PHMA) inner block and a 30-nm outer block of matrix-compatible polyglycidylmethacrylate (PGMA) were synthesized to toughen an epoxy. The copolymer-grafted SiO_2 nanoparticles improved the ductility (maximum 60% improvement), fracture toughness (maximum 300% improvement), and fatigue crack growth resistance of the virgin epoxy matrix, while keeping the same modulus at loadings inferior to 2 vol.% of SiO_2 core. On the other hand, at lower graft density and larger molecular weight of the PHMA block, the nanocomposites showed simultaneous improvements in fracture toughness and tensile modulus. The PGMA epoxy compatible block also enhance the fracture energy of the resulted nanocomposites [61].

According to Bittmann et al. [62], the addition of TiO_2-nanoparticles to DGEBA epoxy resin resulted in a simultaneous improvement of bending, impact, and fracture toughness properties, involving several mechanism such as breaking of agglomerates, shear lips, shear yielding, particle debonding and bridging of the crack by the nanoparticles. The influence of SiO_2 bimodal particle size and its content on the fracture toughness was scrutinized by Dittanet et al., [63]. As mixtures of SiO_2 micro- and nanoparticles were used with higher nano-SiO_2 ratio, fracture toughness was increased by approximately 30% due to three main mechanisms: micro-particle-induced matrix void growth, nanoparticle-induced matrix void growth, and nanoparticle-induced matrix shear banding.

The addition of SiO_2 nanoparticles to anhydride-cured epoxy resin evoked a huge increment in the fracture toughness by almost 74%. However, the increase of the critical stress-intensity factor rose according to the filler loading. A plateau-like behavior was detected for the fracture energy release rate at filler loadings higher than 15 wt.%. *Fatigue crack propagation* (FCP) behavior was ameliorated in all the regimes. Particle debonding combined with subsequent plastic void growth and shear yielding of the matrix are identified as the main energy-dissipating mechanisms in these regimes of FCP [64].

Figure 5.9 described the change in the G_{IC} and K_{IC} of epoxy nanocomposites at various nano-SiO_2 loading as reported by Kothmann et al. [64]. At low nano-SiO_2 contents below 10 wt.%, the critical G_{IC} of the nanocomposites showed a linear trend with the filler content. However, a plateau form results when the nano-SiO_2 concentration exceeded 15 wt.%. Above this ratio the interparticles distance of the SiO_2 nanoparticles is smaller than the diameter of the nanoparticles.

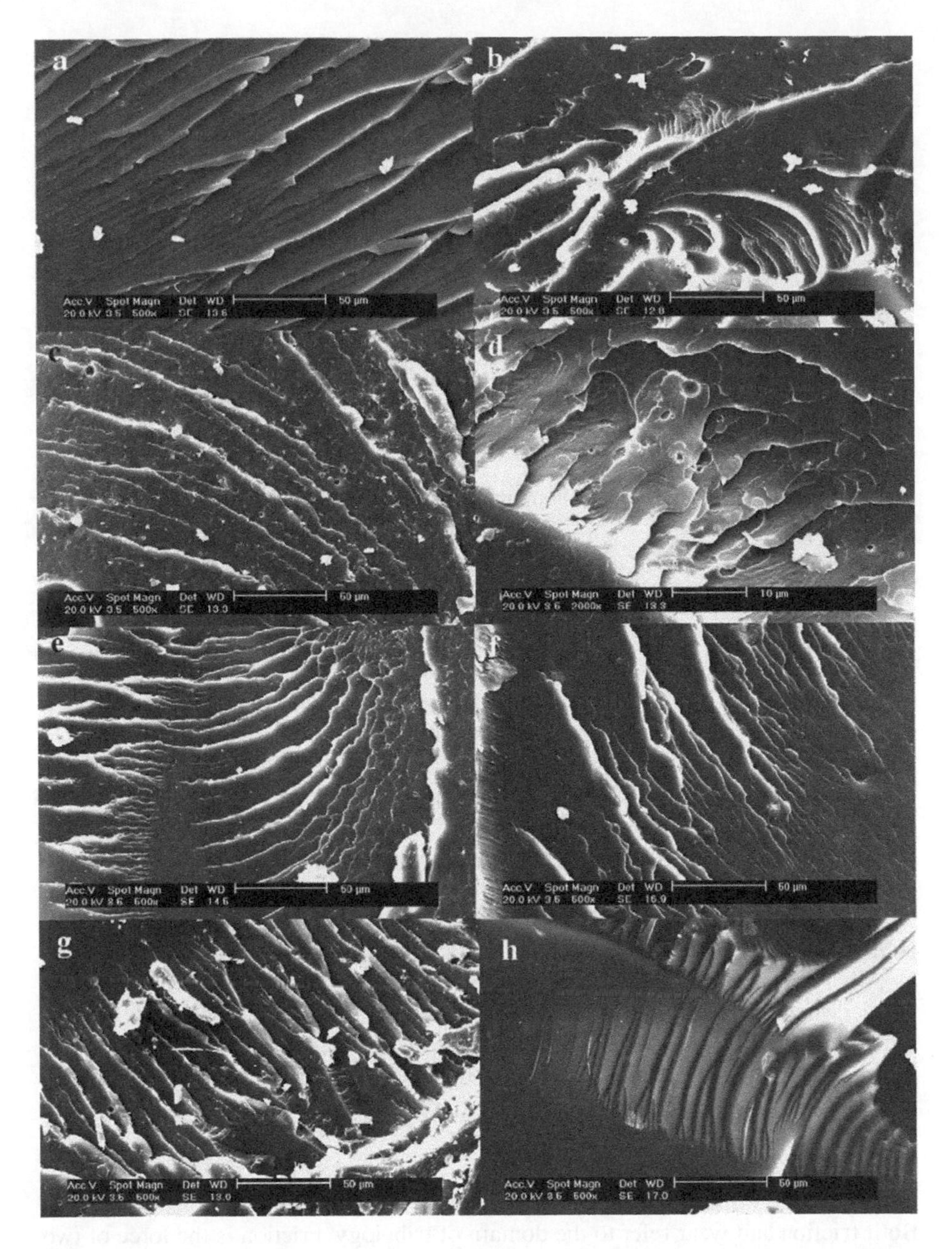

FIGURE 5.8 SEM Images of Fractured Surfaces of Epoxy Resin (a) and Epoxy Resin-based Composites Containing 2 wt.% (b), 4 wt.% (c, d), 6 wt.% (e) Nano-SiO_2 Powder and 2 wt.% (f), 4 wt.% (g), 6 wt.% (h) Standard SiO_2 Formed by Impact; Image (d) Is a Part of Image (c) [59].

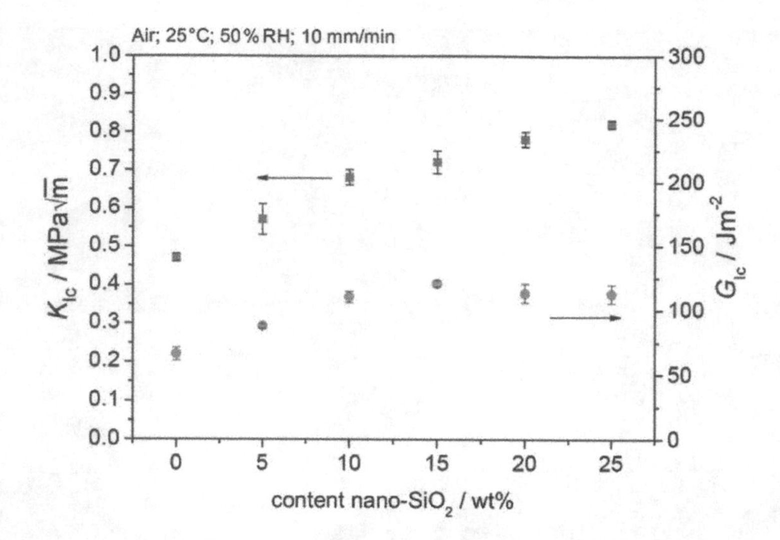

FIGURE 5.9 Influence of Nano-SiO_2 Particles on the Critical Stress Intensity factor, K_{IC}, and the Calculated Critical Energy Release Rate, G_{IC}, of the Anhydride-cured Epoxy/silica Nanocomposites [64].

Rubab et al. [65] investigated the role of crystalline and functional TiO_2 nanoparticles in enhancing mechanical properties of epoxy thermosets. The crystalline TiO_2/epoxy or functional TiO_2/epoxy nanocomposites were reinforced with 2.5 to 12.5 wt.% TiO_2. Functionalized TiO_2/epoxy (FTEN) composites possess greater fracture strength (>64%), modulus (>80%), and toughness (>35%) as compared to pure epoxy, in the presence of TiO_2 nanoparticles. FTEN offered better performance than FTEN composites through appropriate processing of the inorganic reinforcements.

The impact strength (IS) and fracture toughness (K_{IC}) of poly(methyl methacrylate) (PMMA) were significantly improved by adding treated ceramic fillers as reported by Alhare et al. [66]. The IS increased to 56% (8.26 kJ.m^2) and 73% (2.77 MPa.m$^{1/2}$) for K_{IC} when compared with unreinforced PMMA matrix. PMMA denture base reinforced by nitrile butadiene rubber (NBR) particles mixed with treated ceramic fillers are ideally suited for dentistry applications with the ability to withstand high mastication forces.

5.6 TRIBOLOGICAL PROPERTIES

Both friction and wear refer to the domain of tribology. Friction is the force of two surfaces in contact or the force of a medium acting on a moving object, and wear is the erosion of material from a solid surface by the action of another solid. Factors that exert influence on friction and wear characteristics of polymer/ceramic composites are the particle size of filler, morphology of ceramic, and content of the ceramic filler. Table 5.3 lists some polymer/ceramic systems, which are studied for their tribological properties. Experimental results showed high μ at different loads and speeds, while the wear behavior of it was the worst despite its improved thermal

TABLE 5.3
Listing of Matrix, Filler, and the Used Dispersion Technique and Optimized Loading for Some Polymer/ceramic Composites

Matrix type	Ceramic type	Vol.% at lowest wear rate	Dispersing technique
PTFE	Al_2O_3	>50	mechanical
PTFE	Al_2O_3	20	ultrasonication
PTFE	Al_2O_3	0.5	jet mill
PEEK	Si_3N_4	4	ultrasonication
PEEK	SiC	1	ultrasonication
PEEK	SiO_2	4	ultrasonication
PEEK	ZrO_2	3	ultrasonication
PET	Al_2O_3	0.7	melt mixing
PPS	Al_2O_3	2	mechanical

resistance. The worn surface topographies of composites showed that tribological properties were related to the plateaus formed on the contact surfaces [67].

The addition of a reinforcing phase was considered one of the effective methods for improving the tribological features (such as coefficient of friction and wear rate) of the thermoplastic PA6 as stated by García et al. [68]. The friction and wear properties of nylon 6/SiO_2 nanocomposites, prepared using solution-blending technique and compression molding, were tested on a pin over disk tribometer by running a flat pin of steel against a composite disk.

Tanaka and Kawakami [69] conducted a study of various filler particles including ZrO_2, and TiO_2, and other fibers and inorganic particles, hence they found that the TiO_2 filler was the least effective at reducing PTFE wear. In that study, the TiO_2 was the smallest filler at less than 300 nm, while the other fillers had larger particle sizes of several micrometers or more. It was thus hypothesized that such smaller filler particles were transported within the wearing PTFE in its process of transferring to the countersurface, incapable of preventing large-scale reorganization of the PTFE structure. In fact, the TiO_2 filler similarly provided PTFE with 100-fold wear-rate reductions in several instances, but these wear rate reductions were not comparable to those provided by the more effective fillers having particle sizes in excess of a micrometer.

Sawyer et al. [70] reported that the filling with 38-nm Al_2O_3 nanoparticles at 20 wt.% decreased the wear rate of PTFE till 1.2×10^{-6} mm^3/Nm, and upon further modification subsequently reached 1.3×10^{-7} mm^3/Nm when using 80-nm Al_2O_3 filler particles at an optimum 5 wt.% concentration [71]. In the Burris and Sawyer study, a 0.5-μm Al_2O_3 particle size was also studied, but the wear-reducing performance of the 80-nm Al_2O_3 filler particle was more effective.

To improve the wear properties of polymers, extensive researches involved ceramic fillers reinforced polymeric matrices have been reported. Such composites included ZrO_2/polyetheretherketone [72], Al_2O_3/polyphenylene sulphide [73], TiO_2/epoxy [74], Al_2O_3/polyimide [75], and SiC/polyetheretherketone composites, and the results showed great enhancement in their wear-resistance properties. Lai et al. [76] recently stated that the incorporation of SiO_2 nanoparticles ameliorate the frictional coefficient and wear resistance of polyamide composite coatings. According to their recorded results of Su et al. [77], the SiO_2 nanoparticles exhibited their significance in enhancing the wear-resistance and reducing the frictional coefficient of Nomex fabric phenolic composites. The studies published by Luo et al. [78] revealed that the grafting of GMA onto nano-Si_3N_4 augmented the interfacial interaction between ceramic nanoparticles and the epoxy resin through chemical bonding. It showed to be an efficient method to further increase the nano-effect of the nano-Si_3N_4 nanoparticles on the improvements of the tribological properties. The research reports based on the sliding wear performance of nanometer Si_3N_4-filled BMI compositions [79] and nanometer SiC-filled BMI compositions and ZrO_2-filled bismalimide composites [80] also revealed the effectiveness of these ceramics fillers in proving the wear properties of BMI thermosets.

The transient run-in behavior of ultra-low-wearing polytetrafluoroethylene (PTFE) and Al_2O_3 nanocomposites was evaluated for several loading fractions. There was a higher-wearing run-in period followed by ultra-low-wear steady-state performance for these hybrids. The steady-state wear of these composites filled with 2, 5, and 8 wt.% Al_2O_3 attained approximately 6.3×10^{-8}, 5.2×10^{-8}, and 3.1×10^{-8} mm^3/Nm, respectively, which represented more than four orders of magnitude better than that of pure PTFE (5.7×10^{-4} mm^3/Nm). The run-in performance was ameliorated with increasing Al_2O_3 loadings up to 8 wt.%, based on the metric of the best run-in being the least amount of wear before steady-state wear was obtained. The transition from moderate wear to ultralow wear was related to the formation of running films on the wear surface of these composites. These running films showed an improved mechanical characteristics in comparison to the unworn composites and the worn and unworn unfilled PTFE matrix. Similar tribofilms were formed on both the composite wear surface and the surface of the counterface and effectively endeared a new interface upon which sliding appeared [81].

As compared to the virgin PTFE inherent wear rate that reached approximately 0.7×10^{-3} mm^3/Nm, the introduction of 5wt.% of 40- or 80-nm Al_2O_3 ceramic particles was significantly decreased the wear rate to ~10^{-7} mm^3/Nm. At a similar weight ratio of Al_2O_3, composites based on ceramic micro-particle fillers with size ranging from 0.5 to 20 μm reduced the PTFE wear rate to ~10^{-5} mm^3/Nm [82]. PTFE nanocomposites reinforced with Al_2O_3 nanoparticles were less abrasive to the mating steel (304 stainless) counter-surfaces than those with Al_2O_3 micro-fillers, despite the nature of ceramic filler. These results suggested that the PTFE/Al_2O_3 nanocomposites deposits a thinner, well-adhered transfer film that is more stable, as the nano-Al_2O_3 fillers never abrade it. Composites reinforced with both nano- and micro-particles at equal Al_2O_3 ratios behaved like a micro-composite having a higher wear rate of ~10^{-5} mm^3/Nm, which is resulted from the deletion of the transfer film in the presence of more abrasive microscale alumina particles. The friction coefficients (μ) of these PTFE composites were independent of the alumina-filler particle size and were not changed from the 0.18 value recorded for the pure PTFE at the 0.01 m/s sliding speed.

The incorporation of Al_2O_3 particles into a PTFE matrix generated smaller wear debris, thinner transfer films, and reduced wear rates (w_a) as reported by

Burris et al. [83]. The wear rates of the 44-nm nanocomposites were a weaker function of surface roughness and correlated well with transfer film thickness. The μ of the PTFE composites ranged from ~0.12 to reach 0.19 and tended to enhance with rising the ceramic loadings and decreasing surface roughness but was not affected by the thickness of the transfer film. The lowest w_a was detected for every composite on the lapped surface, and no correlation was established between wear rate and *Rq*/*Df*.

5.6.1 FE-SEM of the Worn Surfaces

Images of the worn surfaces recorded by scanning electron microscope (SEM) give some details about both the role of ceramic nanoparticles in reducing wear rate and determining the wear mechanisms. Some wear pattern representatives for every specimen are illustrated in Figure 5.10. It is obvious from the worn surface of neat epoxy that the main wear mechanism involved is fatigue. For unfilled epoxy matrix, the magnified microstructures of the worn surface (Figure 5.10a) reflected that the wear platelets joined with chunks of debris loosely smeared on the worn surface.

These platelets appear due to the production and propagation of surface and subsurface cracks. The small-sized platelets and debris detected on the worn surface referred to the brittle feature of the matrix. The wear surface of the C_4-e-BMI-toughened neat epoxy matrix illustrated in Figure 5.10b showed an adhesive wear, and the reason for the wear behavior is that as indenters slide against the surface of these composites, the actual contact is between asperities of the two surfaces. The physical interactions at these contact regions may generate a junction because of the heating at the interface, and this heat is sufficiently augmented due to the melting of the sliding polymer surface [33]. As the wear sliding started, wear fragments detached from the surface provoking an adhesive wear if the bonding at the interface of the adhering asperities is higher than the strength of the asperities.

Ultimately, this provokes the formation of film or lumps of polymer on the slider contact surface, and this polymer film is clearly seen in the FESEM and directly provided a proof for adhesive wear [34]. The key feature of abrasive wear of a composite is characterized by cutting or ploughing of its surface using rigid ceramic particles or asperities. These cutting points may either be included in the counterface or lie loose within the contact zone. The former case is commonly called two-body abrasion and is detected in the wear of 1, 3, and 5 wt.% Al_2O_3-reinforced nanocomposite shown by the FESEM images (Figure 5.10c, d), which revealed scratches, ploughing generated by the asperities, scoring marks on the worn surface and some debris caused by abrasion frequently appeared as fine cutting chips like those ejected during machining, although at a very small scale, and the latter, three-body abrasion occurred in 10 wt.% Al_2O_3- reinforced nanocomposite.

In that as, the free abrasive readily penetrated the matrix surface and acted as an emery cloth providing an increased wear of counter surface [35]. The FESEM image of worn surface of 5 wt.% Al_2O_3 filled C4 e-BMI-toughened epoxy nanocomposite (Figure 5.10e) is rough and contains voids due to the spherical shape of Al_2O_3 nanoparticles which separated from the BMI matrix during the wear process. Despite

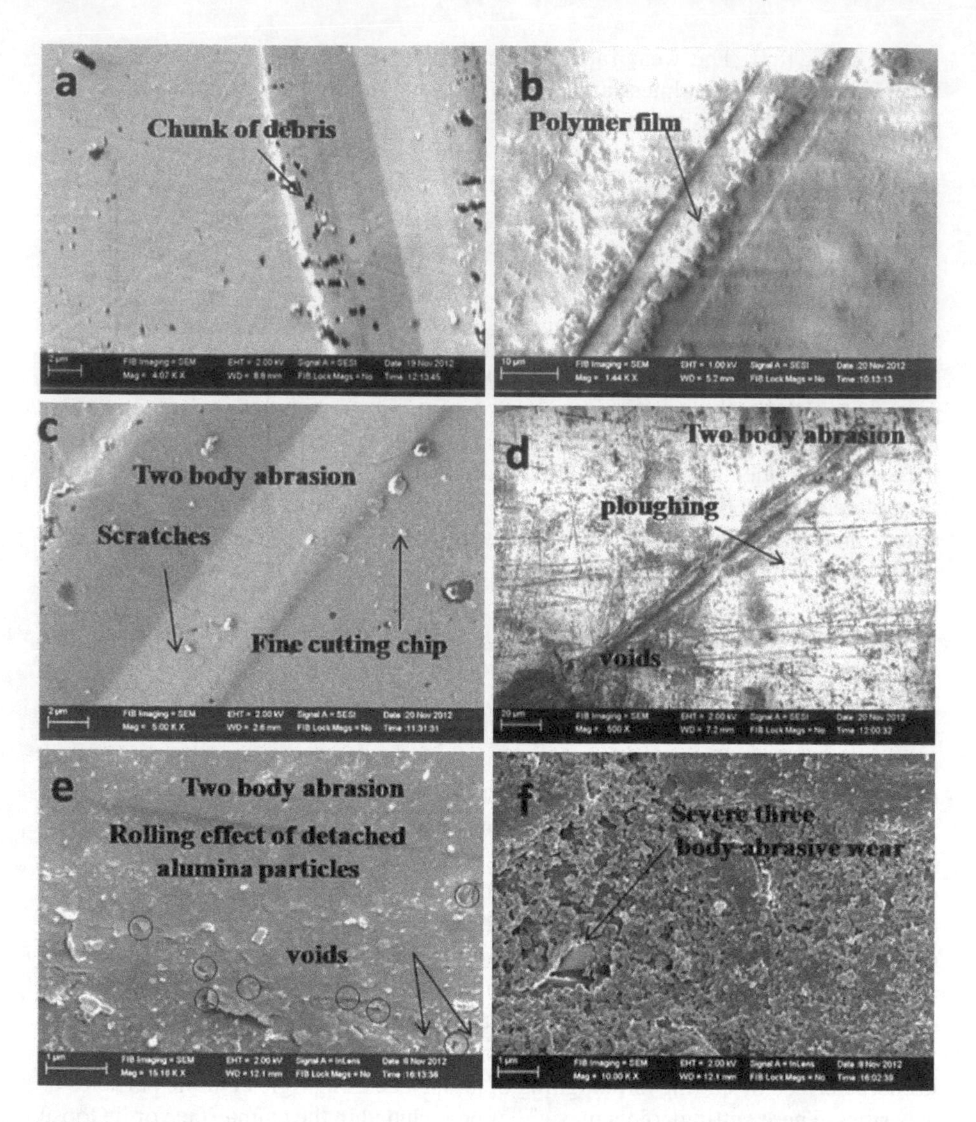

FIGURE 5.10 SEM Images of Worn Surfaces of C_4 e-BMI/Al_2O_3-Toughened Epoxy Nanocomposite Systems [84].

the roughness of worn surface, the wear resistance of this nanocomposite sample is improved according to the wear track deformation. This is traced to the desired rolling influence [36], where the detached particles behaved like rolling balls during sliding; a rolling effect of nano-Al_2O_3 has decreased the shear stress [26]. The matrix cracks in the interfacial area were diminished by this rolling effect, this was done by changing sliding friction into rolling friction between the indenter and the tested specimen. Thus, it is clearly appeared that nano-Al_2O_3 not only played the role of a load carrier in this nanocomposite but also contributed to enhancing the bonding between the transfer film and the steel counter-face.

This promising assumption of rolling particles is justified from the decreasing trend in the wear track deformation. When particle content is high (10 wt.%), the particles started to agglomerate severely leading to a number of loosened clusters of particles in the tribo-system, which could be easily removed during sliding wear conditions. Ultimately, the composites would be easily damaged under repeated frictional stress (Figure 5.10f). The worn-out surface became more rugged as more particles were detached from the matrix under wear load, and thus, it is easier for the nanoparticles to aggregate together. Thus, the agglomerates showing irregular forms provoked a severe three-body abrasive wear between the indenter and the specimen providing poor wear-resistance. The abrasive wear resulting from the hard-wear debris increased the wear of the BMI/Al_2O_3 nanocomposite block.

5.6.2 Mechanisms

New polymer/ceramic micro- and nanocomposites having improved sliding wear resistance were developed to be typical materials for studying the role of the ceramic fillers in the dry sliding-wear behavior of metal/particulate composites. Generally, an epoxy thermoset was selected as a polymeric matrix. The wear of such composites was about 50 times lower than that of unfilled epoxy base materials. Surface cracking and particle pulled-out wear phenomena occurred. During sliding, material waves of a thickness of about 15 pm were constructed on the surface before generating wear debris. A correlation can be established between the filler particle size and the wave thickness. Hence, particles having a size lower than 20 pm cannot be used to reinforce the polymer matrix because as they excluded together with matrix debris.

Due to their embedding ability between the two sliding parts, these fillers can be involved in majority of wear mechanisms by abrading both sliding materials which resulted in high wear coefficients [85]. However, particles with a mean size higher than 20 pm still included and protected the polymer matrix. Composites filled with 20 vol.% of ceramic particles showed excellent wear resistance, while at higher ceramic volume ratios this property was slightly increased. This study also showed that carbide-based ceramics generally increased the wear resistance of composites compared to oxide-based ceramic particles.

The addition of Si_3N_4 nanoparticles to epoxy can significantly reduce frictional coefficients and wear rate of the matrix under dry sliding-wear conditions. Such an improvement was mostly provided at low filler loading (typically less than 1 vol.%). The enhanced interfacial adhesion between nano-Si_3N_4 ceramic particles and the epoxy thermoset, decreased the damping ability, and raised the resistance to thermal distortion of their resulted composites, and tribochemical reactions involving Si_3N_4 nano-filler can also improve tribological performance of these composites. Due to the high frictional surface temperature and contact pressure, some chemical reactions that cannot take place under normal conditions might be completed between the sliding counterfaces.

In general, pure thermosetting systems are not suitable material for sliding wear resistance applications due to their three-dimensional structure as compared with thermoplastics. As depicted in Figure 5.11, the frictional coefficient and specific

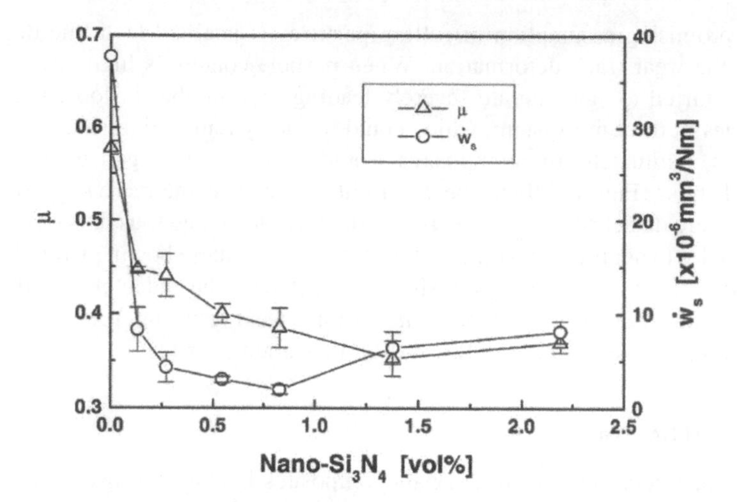

FIGURE 5.11 Frictional Coefficient, μ, and Specific Wear Rate, w_s, of Nano-Si_3N_4/epoxy Composites as a Function of Filler Content [86].

wear rate (w_s) of pure epoxy resin is relatively high. This case is changed significantly by adding nano-Si_3N_4 nanoparticles [86]. It is observed that there was a marked decrease in the values of μ and ˙w_s at a filler loading below 0.27 vol.%. With rising in the fraction of the ceramic nanoparticles, the sharp decrease trend becomes slighter and is changed by a small increase from 1.38 vol.% and go on. This means the nano-Si_3N_4 fillers are more effective in enhancing the tribological properties of epoxy matrix. Micro-SiO_2 is required to provoke a significant reduction in the wear rate of epoxy resin as high as 40 wt.%. This value is higher than the fraction used for the Si_3N_4/epoxy systems.

Burris et al. [87] reported that the incorporation of irregular-shaped Al_2O_3 particles was an effective way to reduce the wear of a PTFE matrix, however it equally led to increased friction. The wear-resistance of PTFE was augmented by almost 3,000 times by adding only 1 wt.% of filler. In addition, steady values of friction and wear were not changed by rising the filler ratio from 1 to 10 wt.%. The μ of the 5 wt.% composite was about 30% lower than those recorded for rest of specimens. The initial transient behavior for these nanomaterials was identified by high wear and low friction, and was found to be more sensitive to filler ratio and longer lasting as the filler loading decreased.

A recent survey on ceramic reinforced polyetheretherketone (PEEK) composites has been published by McCook et al. [88]. A wide range of existent ceramic nanoparticles and micro-particles were utilized to evaluate the wear properties of PEEK. The μ of PEEK composites varied from 0.05 to 0.60; the w_r ranged from 10^{-4} to 10^{-7} mm^3/N.m. Krishna et al. [89] found that for a fixed load and velocity the μ increased with increasing the ceramic loading in the polymer composite. In addition, the specific wear was enhanced with rising both the velocity and load for both the unfilled matrix and their composites while decreased with the increasing of ceramic fractions.

5.6.3 Effect of Particle Size

Polyimide (PI) composites filled with micro SiO_2 were prepared by means of a hot-press molding technique [90]. Experimental results revealed that single incorporation of micro-SiO_2 can affect the friction and wear properties of the PI matrix. However, it is found that a combinative addition of micro SiO_2 with other materials like short carbon fiber and graphite could effectively improve the friction-reducing and anti-wear abilities of the PI polymer. Research results also showed that the reinforced PI composites exhibited better tribological properties under higher PV-product.

The wear behavior of polyphenylene sulfide (PPS) filled with two ceramic nanoparticles was investigated by Bahadur et al. [91]. The ceramic fillers investigated were TiO_2 and SiC, whose sizes varied from 30 to 50 nm. It was found that the wear rate of PPS decreased when TiO_2 was used but increased when SiC fillers was used. The best wear resistance was obtained with 2 vol.% TiO_2. These filled composites had the μ values lower than that of the pure PPS. There was a good trend detected between the transfer film–counterface bond strength and wear resistance.

Kurahatti et al. [92] evaluated the friction and dry sliding wear behavior of nano-ZrO_2 reinforced BMI matrix prepared by a high shear mixer at 0.5, 1, 5, and 10 wt.%. The results indicated that the μ and w_s of BMI can be decreased at low fractions of nano-ZrO_2. The lowest w_s reached 4×10^{-6} mm^3/Nm for 5 wt.% nano-ZrO_2 filled composite which is reduced by almost 78% as compared to that of neat BMI. The incorporation of nano-ZrO_2 particles led to a remarkable enhancement in the wear performance of the composites and provided a good trend with the hardness up to 5 wt.% of filler ratio. The involved wear mechanisms were identified on the SEM photos in the Figure 5.12.

The tribological properties of two kinds of high temperature resistant thermoplastic composites, polyetheretherketone (PEEK) and polyetherimide (PEI) reinforced with sub-micro particles of TiO_2 and ZnS were studied in dry sliding conditions [93]. With the addition of sub-micro ceramic particles, the μ and wear rate of the composites were further decreased, especially at higher temperatures. The γ-Al_2O_3 nanoparticles grafted with two different molecular weights polysulfone (PSU) composites has been prepared by Llorente et al. [94]. The surface-treatment of γ-Al_2O_3 led to an improved nanoparticle dispersion. Abrasive wear analysis revealed that the wear volume loss was reduced as the modified-γ-Al_2O_3 nanoparticles were used. Results also revealed a strong tendency for increased strength and wear resistance.

Research by Yan et al. [95] suggests that the addition of nano-ZrO_2 reduced both the wear coefficient and the wear rate of the pure BMI, and these improvements were more pronounced in the case of using silane-treated nano-ZrO_2. Yan et al. [96] reported that the addition of a 3 wt.% hyperbranched polysilane functionalized SiO_2 nanoparticles (HBPSi-SiO_2) to the benzoxazine/bismaleimide blended matrix significantly reduced the μ and wear rate w_r. The elaborated nanocomposites showed an abrasive wear in the presence of nano-SiO_2 filler.

Tribological tests indicated that the presence of SiO_2 nanoparticles decreased both the μ and the w_r of the BMI resin [97]. These improvements are more significant if surface-modified nano-SiO_2 fillers were used compared to that generated by raw SiO_2 nanoparticles. The difference between the tribological behavior of the pure

FIGURE 5.12 Worn Surface Features of (a) Neat BMI, (b) 5 wt.% Nano-ZrO_2-filled BMI, and (c) 10 wt.% Nano-ZrO_2-filled BMI Nanocomposites [92].

resin and its filled nanocomposites was related to their mechanical properties and wear mechanism. Yan et al. [98] found that at rather low nano-ZrO_2 fractions, the μ and w_s rate of the BMI were decreased.

The specific wear rate of 5 wt.% nano-ZrO_2-filled BMI nanocomposite was decreased by 78% as compared to the pure BMI. The addition of nano-ZrO_2 particles increased the hardness of BMI and wear performance of these nanocomposites. The bismaleimide composite materials filled with graphite and nano-Si_3N_4 were produced by a casting technique. The nano-Si_3N_4 had positive effect on decreasing the friction and wear of bismaleimide resin composites than graphite. For example, the friction coefficient of the pair was lowered from 0.36 to 0.25, and the wear rate was lowered by 72% when the composite contained 1.5 wt.% nano-Si_3N_4.

Yan et al. [99] discovered that the BMI/SiC nanocomposites showed lower μ and wear loss. The lowest w_r was obtained with the nanocomposite containing 6 wt.% nano-SiC, while the highest mechanical properties were recorded with the sample containing 2 wt.% nano-SiC. The wear mechanism of the nanocomposite was mainly adhesion wear, while that of pure BMI resin was essentially fatigue cracking with plastic deformation.

The effect of incorporating nano-Si_3N_4 on the properties of BMI matrix was studied by Yan et al [100]. The BMI thermoset showed high modulus, good heat properties, especially a low viscosity allowing the good dispersion of the nanoceramics and the casting of the composite precursors. These composites exhibited considerably lower μ and w_r than the BMI resin did. The SEM analysis revealed that the wear mechanism of the composites of BMI/Si_3N_4 was essentially adhesion wear, while that of the pure BMI was mainly plastic deformation. The results indicated that a ZrO_2-rich layer was formed on the PI film sourcing from the zirconium n-butoxide after Atomic oxygen (AO) exposure, which decreased the erosion rate and obviously improved the AO resistance of polyimide films [101].

The friction and wear properties of the PF resin reinforced with different fractions of TF-treated nano-TiO_2 were evaluated by Song and Zhang [102] with the good dispersibility of nano-TiO_2 in the phenolic matrix, compared with the cases of untreated nano-TiO_2. Using TF-modified nano-TiO_2 significantly improved the tribological performance, including the wear resistance of the phenolic matrix, especially at extreme wear conditions, i.e., high- contact pressures. The wear properties of three different PF/SiO_2 nanocomposites were evaluated at various filler contents by Taheri-Behrooz et al. It was found that the scratch tests that nano-SiO_2 were inert for the *COF* and w_r of the thermosets [103].

Song et al. [104] prepared new PF/TDI-Al_2O_3 nanocomposites exhibiting higher tribological performance benefiting from the improved interfacial adhesion between the nano-Al_2O_3 and PF matrix by covalent bonds from the hydroxyl groups of Al_2O_3 and –OCN groups of TDI. As compared to the neat PF resin, the addition of only 3 wt.% TDI-Al_2O_3 resulted markedly reduced frictional and enhanced wear-resistance properties. The μ of the composite coating reinforced with different TDI-Al_2O_3 fractions decreased with increasing the sliding speed and the applied load. SEM showed that low content nano-Al_2O_3 or TDI-Al_2O_3 was able to enhance the adhesion of the transfer films of the PF coating to the surface of counterpart ring, which significantly reduced the wear of the PF coating (Figure 5.13) [104].

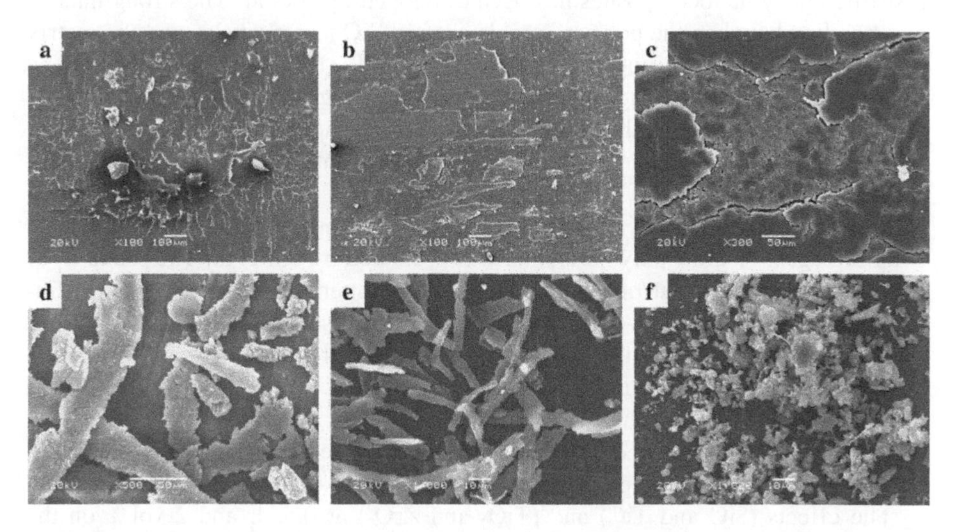

FIGURE 5.13 SEM Pictures of the Worn Surfaces and Wear Debris of the Phenolic Coating Reinforced with 3.0 wt.% TDI-Al_2O_3 Under Different Applied Load (2.56 m/s, 60 min): (a) 420 N; (b) 620 N; (c) 1220 N; (d) Debris of (a); (e) Wear Debris of (b); (f) Wear Debris of (c) [104].

The SiC- and B_4C-based particulate carbon/ceramic composites were studied, and their performance were compared other kinds of composites. It is found that contribution to μ value during sliding can occur through surface adhesion, deformation of asperities, and ploughing by wear debris or hard asperities [105]. The friction and wear behavior of different kinds of SiO_2 particles reinforced PF matrix were discussed by Jin et al. Compared to irregular SiO_2, spherical SiO_2 powders improved the wear resistance, but decreased the μ. Surface-treated spherical SiO_2 was more effective in improving the wear-resistance, but they showed similar *wear friction* compared to other types of irregular SiO_2-filled composite materials. While the irregular SiO_2 is beneficial in decreasing the heat fade and improving recovery, it is harmful to the wear resistance of friction materials [106].

The friction behaviors of polybenzoxazine/ZrO_2 nanocomposites were studied by Wu et al. [107], and the results revealed that the temperature and load sensitivity of nanocomposites increased with the increasing of load and temperature. This behavior was speculated to be due to the effect of the temperature dependence of modulus in the surface topography and strength. But speed sensitivity varies with the temperature, due to the effect of temperature dependence of viscoelastic response in the energy dissipation.

The effects of the glass-to-rubber transition of PBZ matrix on the friction and wear characteristics of PBZ/ZrO_2 nanocomposites were examined. The storage modulus and T_g values of the nanocomposites increased with enhancing ZrO_2 content up to 4 wt.% because of the exceptional mechanical strength of ZrO_2 particles and their interfacial adhesion with PBZ matrix, which restricted the segmental motion of thermoset. The nanocomposites with 4 wt.% ZrO_2 had a relatively higher modulus and T_g, and better capability to stabilize the μ and w_r under the applied conditions [108].

A feasible way of efficiently applying SiC nanoparticles to the preparation of wear resisting epoxy nanocomposites has been developed by Ji et al. The strong interfacial bonding between the polyacrylamide grafted SiC nanoparticles and the matrix accounted for the wear properties improvement [109]. Results indicate that epoxy/SiO_2 nanocomposite coatings had a higher surface roughness and lower μ than the neat EP coating. However, the introduction of surface-capped nano-silica as the filler resulted in inconsistent changes in the friction coefficient and wear rate of the filled composites [110].

Sliding wear behavior of epoxy/Al_2O_3 nanocomposites pre-treated by either silane coupling agent or graft polymerization was evaluated by Shi et al. The μ and w_r of epoxy can be reduced at rather low nano-Al_2O_3 contents, and the surface-modification of the nano-Al_2O_3 was further increased to this positive change. The graft polymerization was more efficient compared to silane-treatment in ameliorating the tribological parameters of matrix. Also, any increased in the impact strength correlated with a decrease in w_s [111]. The wear behavior of PTFE-filled SiO_2/epoxy composites was investigated. The presence of silica reduced the μ and improved tribological properties composites [112].

The effects (SiC and TiC) and (TiO_2 and ZrO_2) at 0.5, 1, and 2 vol.% on the tribological properties of PF resin were studied by Atarian et al. [113] at various conditions. Results showed that carbide-based ceramic particles had better enhancing effect on tribological properties of PF resin as compared to the oxide particles.

The nanocomposites with 1 vol.% SiC nanoparticles showed the best tribological properties among the investigated samples [8].

A model based on movable cellular automata (MCA) is described and applied for simulating the sliding behavior of a nanocomposite consisting of an epoxy matrix and 6 vol.% of uniformly dispersed nano-SiO_2 particles. Sliding simulations revealed a significant impact of nanoparticles on the interface structure and smoothness of sliding mechanism. Furthermore, assuming both possibilities, bond breaking and rebinding of automata pairs can explain different friction levels of polymer materials [114].

Effects of SiO_2 nanoparticles on tribological properties of epoxy/SiO_2 nanocomposites were studied by Zhang et al. The w_s of epoxy matrix was improved when the of nano-SiO_2 content exceeded a critical value of 10 wt.%, while the static mechanical properties of the nanocomposites found to be proportionally increased with increasing the nano-SiO_2 content. Continuous transfer film on counterface played a key role in boosting wear performance of epoxy resin [115]. Nanocomposites, consisting of silane-functionalized nano-TiO_2 reinforced epoxy matrix, were elaborated using vacuum-assisted casting technique. Under sliding conditions, the w_s and μ were reduced by adding only 5 wt.% of nano-TiO_2. The sliding wear behavior showed that nano-TiO_2 nanoparticles impeded the propagation of cracks [116].

The tribological behaviors of polydicyclopentadiene/polystyrene@silica (poly (DCPD)/PS@SiO_2) nanocomposites with different PS@SiO_2 loading have been investigated details in the work of Peng et al. [117]. The results revealed that by adding PS@SiO_2 nanospheres to poly(DCPD), this was significantly lowered friction coefficient and wear rate of their resulted nanocomposites, especially as the loading of PS@SiO_2 nano-spheres attained 4 wt.%. The good interfacial compatibility between the PS shell layer of PS@SiO_2 nano-spheres and poly(DCPD) matrix should be responsible for improving the wear properties of poly(DCPD)/SiO_2 nanocomposites. Yan et al. [118] studied the effect of surface-modified nano-SiO_2 particles using a silane coupling agent on the friction and wear properties of the related BMI composites, and the results showed that the addition of nano-SiO_2 particles could decrease the μ and the w_r of the BMI composites.

5.7 COMPRESSION AND CREEP PROPERTIES

The creep and creep rupture behavior of a PF/Al_2O_3 composite was studied in an aqueous environment by Service et al. [119]. The composites exhibited power-law creep behavior in which the steady-state creep rate was a power function of the initial applied elastic stress. The creep exponent was found to be 5.3. The creep rupture behavior can be explained using a modified Monkman-Grant relationship, which provides a failure criterion that was independent of applied stress and stress state.

The result showed great enhancement of compressive modulus was achieved as filler size and filler content increased. For compressive strength, finer particles had shown better value than the courser filler at similar filler loading. The correlation between hardness and compressive strength of all samples was found to be lower than 3 [120]. Hassan et al. synthesized new bio-based nano-$CaCO_3$ were from eggshell and used to prepare UPE nanocomposites where addition of only 2 wt.% $CaCO_3$ to

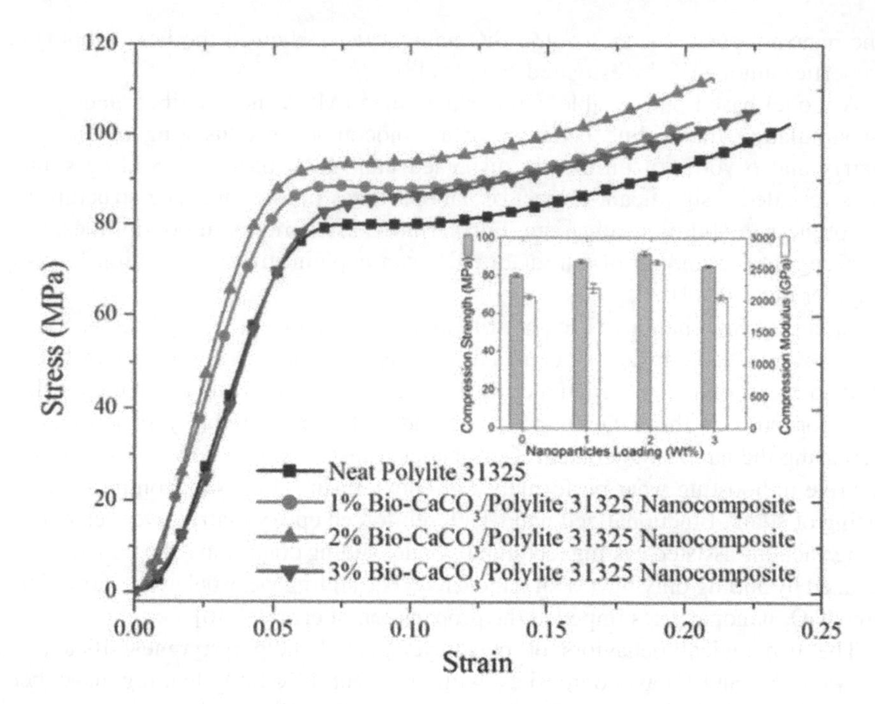

FIGURE 5.14 Quasi-static Compression Test Results for Neat Polylite 31325 and Bio-$CaCO_3$/polylite 31325 Nanocomposite. Inset Is a Bar Graph of Compression Results with Error Bars [121].

UPE thermoset resulted in an improvement of 14% and 27% in compressive strength and modulus, respectively, as shown in Figure 5.14 [121].

Rodrigues et al. treated B_4C with GPS silane to improve their adhesion with the epoxy resin. The mechanical strength was significantly improved mechanical strength for the treated composite. Initial observation on the surface chemistry

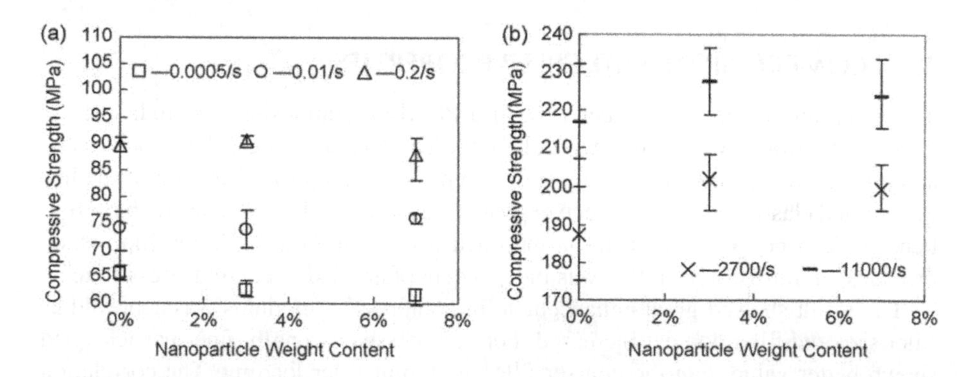

FIGURE 5.15 Measured Compressive Strength as a Function of Nanoparticle Weight Content of Epoxy/SiO_2 Composites. (a) Quasi-static Experiment Results and (b) Dynamic Experiment Results Correspond to Quasi-static and Dynamic Experiment Results, Respectively [124].

showed the presence of hydroxyl groups and B–O bonds, which justified the resulted improved adhesion through silane condensation and the formation of covalent bonds [122]. Rostamiyan et al. found that ternary nanocomposite had an improved ultimate compression and impact strength up to 45% and 414% compared to those of the pure epoxy matrix, respectively, but compression strain did not change considerably [123]. Nanoparticles at 0, 3, and 7 wt.% under uniaxial compression at different loading rates (10^{-4} to 10^{4} s^{-1}) are investigated by Guo and Li. The performance of the composite depended on loading rate and nanoparticle dispersion. The compressive strength of the composites is higher than that of pure epoxy at high strain-rates as shown in Figure 5.15 [124].

5.8 CONCLUSIONS

To sum up, the mechanical properties of the various polymers including thermoplastics, thermoplastic elastomers, and thermosets can be improved significantly by adding ceramic micro- and nanoparticles. Results also showed that the good selection of processing techniques and preparation conditions can promote these improvements dramatically. Another important issue is the using of surface-treatment and silane molecules, which can provide an extra improvement especially in the bending, tensile, and microhardness properties. In addition, the reinforcing of polymers with ceramic nanoparticles, even at lower nanofillers' loading provides significant enhancements in mechanical properties of polymeric matrices, however, these improvements were related to the dispersion states of nano-ceramics. With the rapid advancement in the processing technique and equipment, polymer/ceramic composites will gain promising industrial applications in the near future.

REFERENCES

1. Davidge, R. W. 1969. Mechanical properties of ceramic materials. *Journal Contemporary Physics* 10: 105–124.
2. Ma, X. Y., Yuan, L., Jia, Q. Y. et al. 2003. Composites of epoxy resin and BMI/BA reinforced with aluminum borate whiskers. *Journal of Material Sciences and Engineering* 21: 789–792.
3. Li, S. J., Tian, J. X., Gan, W. J. et al. 2005. Synthesis and characterization of bismaleimide-polyetherimide-silica hybrid by sol-gel process. *Polymers for Advanced Technologies* 16: 133–138.
4. Pinto D, Bernardo L, Amaro A, Lopes S. 2015. Mechanical properties of epoxy nanocomposites using titanium dioxide as reinforcement—A review. *Construct Build Mater* 95: 506–524.
5. McGrath, L. M., Parnas, R. S., King, S. H. et al. 2008. Investigation of the thermal, mechanical, and fracture properties of alumina-epoxy composites. *Polymer* 49: 999–1014.
6. Watanabe, R., Kunioka, M., Sato, H. et al. 2018. Management of both toughness and stiffness of polypropylene nanocomposites using poly(5-hexen-1-ol-co-propylene) and silica nanospheres. *Polymers for Advanced Technology* 29: 417–423.
7. Hasan, S. A. B., Dimitrijević, M. M., Kojović, A. et al. 2014. The effect of the size and shape of alumina nanofillers on the mechanical behavior of PMMA matrix composites. *Journal of the Serbian Chemical Society* 79: 1295–1307.

8. Wang, W., Wu, J. 2018. Interfacial influence on mechanical properties of polypropylene/polypropylene-grafted silica nanocomposites. *Journal of Applied Polymer Science* 135: 45887.
9. Johnsen, B. B., Stavnes, S. M. N., Olsen, T. et al. 2012. Preparation and characterisation of epoxy/alumina polymer nanocomposites. ECCM15—15TH European Conference on Composite Materials, Venice, Italy, 24–28.
10. Marković, G., Marinović-Cincović, M.S., Jovanović, V. et al. 2012. Gamma irradiation aging of NBR/CSM rubber nanocomposites. *Composites Part B: Engineering.* 43: 609–615.
11. Rong, M. Z., Zhang, M. Q., Zheng, Y. X. 2001. Structure–property relationships of irradiation grafted nano-inorganic particle filled polypropylene composites. *Polymer* 42: 167–183.
12. Chee, C. Y., Song, N. L., Abdullah, L. C. et al. 2012. Characterization of mechanical properties: low-density polyethylene nanocomposite using nanoalumina particle as filler. *Journal of Nanomaterial*: 1–6.
13. Rong N.Z., Zhang M.Q., Zheng Y.X. et al. 2001. Improvement of tensile properties of nano-SiO_2/PP composites in relation to percolation mechanism. *Polymer* 42: 3301–3304.
14. Lee, C. H., Hwang, S. Y., Sohn, J. Y. et al. 2006. Water-stable crosslinked sulfonated polyimide–silica nanocomposite containing interpenetrating polymer network. *Journal of Power Sources* 163: 339–348
15. Rajulu, A. V., Devi, L. G., Devi, R. R. 2008. Effect of Orientation of Fibers and Coupling Agent on Tensile Properties of Phenolic/Silica Composites. *Journal of Reinforced Plastics and Composites* 27: 599–603.
16. Yao, H. Y., Pan, H., Li, R. R. et al. 2015. Unsaturated polyester resin/epoxy functionalised nano-silica composites constructed by in situ polymerisation. *Micro & Nano Letters* 10: 427–431.
17. Moreira, D. C., Sphaier, L. A., Reis, J. M. L. et al. 2012. Determination of Young's modulus in polyester-Al_2O_3 and epoxy-Al_2O_3 nanocomposites using the Digital Image. *Composites Part A: Applied Science and Manufacturing* 43: 304–309.
18. Baskaran, R., Sarojadevi, M., Vijayakumar, C. T. 2011. Unsaturated polyester nanocomposites filled with nano alumina. *Journal of Materials Science* 46: 4864–4871.
19. Lim, S. H., Zeng, K. Y., He, C. B. Morphology, tensile and fracture characteristics of epoxy-alumina nanocomposites. *Materials Science and Engineering: A* 2010; 527: 5670–5676.
20. Kim, K. S., Jakubinek, M. B., Martinez-Rubi, Y. et al. 2015. Polymer nanocomposites from free-standing, macroscopic boron nitride nanotube assemblies. *RSC Advances* 5: 41186–41192.
21. Zhao, R. G., Luo, W. B. 2008. Fracture surface analysis on nano-SiO_2/epoxy composite. *Materials Science and Engineering: A* 483-484: 313–315.
22. Lin, J., Wu, X., Zheng, C. et al. 2014. Synthesis and properties of epoxy-polyurethane/silica nanocomposites by a novel sol method and in-situ solution polymerization route. *Applied Surface Science* 2014; 303: 67–75.
23. Hsieh, T. H., Kinloch, A. J., Masania, K. et al. 2010. The mechanisms and mechanics of the toughening of epoxy polymers modified with silica nanoparticles. *Polymer* 51: 6284–6294.
24. Peng, S., Chang, K., Ma, J., et al. 2018. Silica/polydicyclopentadiene nanocomposites: Preparation, characterization, and mechanical properties. *Ferroelectrics* 523: 32–40.
25. Ulus, H., Üstün, T., Eskizeybek, V. et al. 2014. Boron nitride-MWCNT/epoxy hybrid nanocomposites: Preparation and mechanical properties. *Applied Surface Science* 318: 37–42.
26. Trivedi, S., Sharma, S. C., Harsha, S. P. 2014. Evaluations of young's modulus of boron nitride nanotube reinforced nano-composites. *Procedia material science* 6: 1899–1905.

27. Al-Turaif, H. A. 2010. Effect of nano TiO_2 particle size on mechanical properties of cured epoxy resin. *Progress in Organic Coatings* 69: 241–246.
28. Fu, J.F., Chen, L.Y., Yang, H. et al. 2012. Mechanical properties, chemical and aging resistance of natural rubber filled with nano-Al_2O_3. *Polymer Composites* 33: 404–411.
29. Balachandran, M., Bhagawan, S. S. 2012. Mechanical, thermal, and transport properties of nitrile rubber–nanocalcium carbonate composites. *Journal of Applied Polymer Science* 126: 1983–1992.
30. Chen, D., Liu, Y., Hu, C. 2012. Synergistic effect between POSS and fumed silica on thermal stabilities and mechanical properties of room temperature vulcanized (RTV) silicone rubbers. *Polymer Degradation Stability* 97: 308–315.
31. Pelin, G., Pelin, C. E., Stefan, A. et al. 2016. Influence of nanometric silicon carbide on phenolic resin composites Properties. *Bulletin of Materials Science* 39: 769–775.
32. Hiremath V., Singh, M., Shukla, D. K. 2014. Effect of post curing temperature on viscoelastic and flexural properties of epoxy/alumina polymer nanocomposites. *Procedia Engineering* 97: 479–487.
33. Yan, H. X., Li, P. B., Zhang, J. P. et al. 2010. Mechanical properties of novel bismaleimide nanocomposites with Si_3N_4 nanoparticles. *Journal of Reinforced Plastics and Composites* 29: 1515–1522.
34. Wu, G. L., Kou, K. C., Chao, M. et al. 2013. Preparation and properties of nano-SiO_2/TDE-85/BMI/BADCy composites. *Journal of Wuhan University of Technology-Materials Science Edition* 28: 261–264.
35. Derradji, M., Ramdani, N., Zhang, T. et al. 2015. Mechanical and thermal properties of phthalonitrile resin reinforced with silicon carbide particles. *Materials & Design* 71: 48–55.
36. Li, X., Wang, Z., Wu, L. 2015. Preparation of a silica nanospheres/graphene oxide hybrid and its application in phenolic foams with improved mechanical strengths, friability and flame retardancy. *RSC Advances* 5: 99907–99913.
37. Yuan, J., Zhang, Y., Wang, Z. 2015. Phenolic foams toughened with crosslinked poly (n-butyl acrylate)/silica core-shell nanocomposite particles. *Journal of Applied Polymer Science* 2015; 132: 42590.
38. Zhou, Y., White, E., Hosur, M. et al. 2010. Effect of particle size and weight fraction on the flexural strength and failure mode of TiO_2 particles reinforced epoxy. *Materials Letters* 64: 806–809.
39. Petrović, Z. S., Javni, I., Waddon, A. et al. 2000. Structure and properties of polyurethane–silica nanocomposites. *Journal of Applied Polymer Science* 76: 133–151.
40. Li, L., Chen Y., Stachurski, Z. H. 2013. Boron nitride nanotube reinforced polyurethane composites. *Progress in Natural Science: Materials International*; 23(2): 170–173.
41. Mohandas, M., Ayyavu, C., Muthukaruppan, A. 2014. Nanoindentation studies of nano alumina-reinforced ether-linked bismaleimide toughened epoxy-based nanocomposites. *Polymer-Plastics Technology and Engineering* 53: 975–989.
42. Nhuapeng, W., Thamjaree, W., Kumfu, S. et al. 2008. Fabrication and mechanical properties of silicon carbide nanowires/epoxy resin composites. *Current Applied Physics* 8: 295–299.
43. Demian, C., Liao, H., Lachat, R. et al. 2013. Investigation of surface properties and mechanical and tribological behaviors of polyimide based composite coatings. *Surface and Coatings Technology* 235: 603–610.
44. Mohammed, R. D., Taylor, A. C., Sprenger, S. 2007. Toughening mechanisms of nanoparticle-modified epoxy polymers. *Polymer* 48: 530–541.
45. Rosso, P., Ye, L., Friedrich, K. et al. 2006. A Toughened Epoxy Resin by Silica Nanoparticle Reinforcement. *Journal of Applied Polymer Science* 100: 1849–1855.
46. Ragosta, G., Abbate, M., Musto, P. et al. 2005. Mascia: Epoxy-silica particulate nanocomposites: Chemical interactions, reinforcement and fracture toughness. *Polymer* 46: 10506–10516.

47. Xu, N., Zhou, W., Shi, W. 2004. Preparation and enhanced properties of poly(propylene)/silica-grafted-hyperbranched polyester nanocomposites. *Polymers for Advanced Technologies* 15: 654–661.
48. Xiong, L., Lian, Z. Y., Liang, H. B. et al. 2013. Influence of silica nanoparticles functionalized with poly(butyl acrylate-co-glycidyl methacrylate)-g-diaminodiphenyl sulfone on the mechanical and thermal properties of bismaleimide nanocomposites. *Polymer Composites* 34: 2154–2159.
49. Liang, G. Z., Hu, X. L. 2004. Aluminum-borate-whiskers-reinforced bismaleimide composites. 1: Preparation and properties. *Polymer International* 53: 670–674.
50. Suprapakorn, N., Dhamrongvaraporn, S., Ishida, H. 1998. Effect of $CaCO_3$, on the mechanical and rheological properties of a ring-opening phenolic resin: polybenzoxazine. *Polymer Composites* 19: 126–132.
51. Rusmirović, J. D., Trifković, K. T., Bugarski, B. et al. 2016. High performance unsaturated polyester based nanocomposites: Effect of vinyl modified nanosilica on mechanical properties. *Express Polymer Letters* 10: 139–159.
52. Chaeichian, S., Wood-Adams, P. M., Ho, S.V. 2015. Fracture of unsaturated polyester and the limitation of layered silicates. *Polymer Engineering & Science* 55: 1303–1309.
53. Kong, X. W., Wang, J. H., Li, S. X. et al. Et al. 2015. Interfacial structures and properties of granular SiO_2/unsaturated polyester resin composites. *Materials Research Innovations* 19: S1-226–S1-232.
54. Wang, Y., Guo, Y., Cui, R. et al. 2014. Preparation and mechanical properties of nano-silica/UPR polymer composite. *Science and Engineering of Composite Materials* 21: 471–477.
55. Kim, D. J., Kang, P. H., Nho, Y. C. 2004. Characterization of mechanical properties of γ-Al_2O_3 dispersed epoxy resin cured by γ-Ray radiation. *Journal of Applied Polymer Science* 91: 1898–1903.
56. Yao, X. F., Zhou, D., Yeh, H. Y. 2008. Macro/microscopic fracture characterizations of SiO_2/epoxy nanocomposites. *Science and Technology of Advanced Materials* 12: 223–230.
57. Patnaik, A., Bhatt, A. D. 2011. Mechanical and dry sliding wear characterization of epoxy–TiO_2 particulate filled functionally graded composites materials using Taguchi design of experiment. *Materials & Design* 32: 615–627.
58. Opelt, C. V., Beckera, D., Lepienski, C. M. et al. 2015. Reinforcement and toughening mechanisms in polymer nanocomposites Carbon nanotubes and aluminum oxide. *Composites Part B: Engineering* 75: 119–126.
59. Li, H., Zhang, Z., Ma, X. et al. 2007. Synthesis and characterization of epoxy resin modified with nano-SiO_2 and γ-glycidoxypropyltrimethoxy silane. *Surface and Coatings Technology* 201: 5269–5272.
60. Vijayan, P., Puglia, D., Dąbrowsk, A. et al. 2015. Mechanical and thermal properties of epoxy/silicon carbide nanofiber composites. *Polymers for Advanced Technologies* 2015; 26: 142–146.
61. Gao, J. N., Li, J. T., Zhao, S. et al. 2013. Effect of graft density and molecular weight on mechanical properties of rubbery block copolymer grafted SiO_2 nanoparticle toughened epoxy. *Polymer* 54: 3961–3973.
62. Bittmann, B., Haupert, F., Schlarb, A. K. 2011. Preparation of TiO_2/epoxy nanocomposites by ultrasonic dispersion and their structure property relationship. *Ultrasonics Sonochemistry* 8: 120–126.
63. Dittanet, P., Pearson, R. A. 2013. Effect of bimodal particle size distributions on the toughening mechanisms in silica nanoparticle filled epoxy resin. *Polymer* 54: 1832–1845.
64. Kothmann, M. H., Zeiler, R., De Anda, A. R. et al. 2015. Fatigue crack propagation behaviour of epoxy resins modified with silica-nanoparticles. *Polymer* 60: 157–163.

65. Rubab, Z., Afzal, A., Siddiqi, H. M. et al. 2014. Augmenting thermal and mechanical properties of epoxy thermosets: The role of thermally-treated versus surface-modified TiO_2 nanoparticles. *Materials Express* 4: 54–64.
66. Alhare, A.O., Akil, H. B. M., Ahmad, Z. A. B. 2015. Poly(methyl methacrylate) denture base composites enhancement by various combinations of nitrile butadiene rubber/treated ceramic fillers. *Journal of Thermoplastic Composite Materials* 30: 1–22.
67. Cai, P., Wang, Y., Wang, T. et al. 2015. Effect of resins on thermal, mechanical and tribological properties of friction materials. *Tribology International* 87: 1–10.
68. García, M., de Rooij, M., Winnubst, L. et al. 2004. Friction and wear studies on nylon-6/SiO_2 nanocomposites. *Journal of Applied Polymer Science* 92: 1855–1862.
69. Tanaka, K. Kawakami, S. 1982. Effect of various fillers on the friction and wear of polytetrafluoroethylene-based composites. *Wear* 79: 221–234.
70. Sawyer, W. G., Freudenberg, K. D., Bhimaraj, P. et al. A Study on the Friction and Wear Behavior of PTFE Filled with Alumina Nanoparticles. *Wear* 254: 573–580.
71. Burris, D. L., Sawyer, W. G. 2006. Improved Wear Resistance in Alumina-PTFE Nanocomposites with Irregular Shaped Nanoparticles. *Wear* 260: 915–918.
72. Wang, Q., Xue, H., Shen, W.C. et al. The effect of particle size of nanometer ZrO_2 on the tribological behaviour of PEEK. 1996. *Wear* 198: 216–219.
73. Schwartz, C.J., Bahadur, S. 2000. Studies on the tribological behavior and transfer film–counterface bond strength for polyphenylene sulfide filled with nanoscale alumina particles. *Wear* 237: 261–273.
74. Ng, C.B., Schadler, L.S., Siegel, R.W. 1999. Synthesis and mechanical properties of TiO_2-epoxy nanocomposites. *Nanostructure Materials* 12: 507–510.
75. Cai, H., Yan, F. Y., Xue, Q. J. et al. 2003. Investigation of tribological properties of Al_2O_3-polyimide nanocomposites. *Polymer Testing* 22: 875–882.
76. Lai, S.Q., Li, T.S., Wang, F.D. 2007. The effect of silica size on the friction and wear behaviors of polyimide/silica hybrids by sol–gel processing. *Wear* 262: 1048–1055.
77. Su, F.H., Zhang, Z.Z., Liu, W.M. 2007. Tribological and mechanical properties of Nomex fabric composites filled with polyfluo 150 wax and nano-SiO_2. *Composite Science and Technology* 67: 102–110.
78. Luo, Y., Yu, X. Y., Dong, X. M. et al. 2010. Effect of nano-Si_3N_4 surface treatment on the tribological performance of epoxy composite. eXPRESS Polymer Letters 4: 131–140.
79. Yan, H., Ning, R., Liang, G. et al. 2005. The performances of BMI nanocomposites filled with nanometer SiC. *Journal of Applied Polymer Science* 95: 1246–1250.
80. Yan, H., Ning, R., Liang, G., Huang, Y., Lu, T. 2007. The effect of silane coupling agent on the sliding wear behavior of nanometer ZrO_2/bismaleimide composites. *Journal of Material Science* 42: 958–965.
81. Krick, B. A., Ewin, J. J., McCumiskey, E. J. 2014. Tribofilm formation and run-in behavior in ultra-low-wearing polytetrafluoroethylene (PTFE) and alumina nanocomposites. *Tribology Transactions* 57: 1058–1065.
82. McElwain, S. E., Blanchet, T. A., Schadler, L. S. et al. 2008. Effect of particle size on the wear resistance of alumina-filled PTFE micro-and nanocomposites. *Tribology Transactions* 51: 247–253.
83. Burris, D. L., Sawyer, W. G. 2005. Tribological sensitivity of PTFE/alumina nanocomposites to a range of traditional surface finishes. *Tribology Transactions*, 48: 147–153.
84. Mandhakini, M., Chandramohan, T. L., Alagar, M. 2014. Effect of nanoalumina on the tribology performance of C4-ether-linked bismaleimide-toughened epoxy nanocomposites. *Tribology Letters* 54: 67–79.
85. Durand, J.M., Vardavoulias, M., Jeandin, M. 1995. Role of reinforcing ceramic particles in the wear behaviour of polymer-based model composites. *Wear* 181/183: 833–839.
86. Shi, G., Zhang M.Q., Rong, M.Z. et al. 2003. Friction and wear of low nanometer Si_3N_4 filled epoxy composites. *Wear* 254: 784–796.

87. Burris, D. L., Sawyer, W. G. 2006. Improved wear resistance in alumina-PTFE nanocomposites with irregular shaped nanoparticles. *Wear* 260: 915–918.
88. McCook, N.L., Hamilton, M.A., Burris, D.L., Sawyer, W.G. 2007. Tribological results of PEEK nanocomposites in dry sliding against 440C in various gas environments. *Wear* 262:1511–1515.
89. Krishna, K.G., Divakar C., Venkatesh, K. et al. 2009. Tribological studies of polymer based ceramic–metal composites processed at ambient temperature. *Wear* 266: 878–883.
90. Zhang, X. R., Pei, X. Q., Wang, Q. H. 2009. Friction and wear studies of polyimide composites filled with short carbon fibers and graphite and micro SiO_2. *Materials & Design* 30(10): 4414–4420.
91. Bahadur, S., Sunkara, C. 2005. Effect of transfer film structure, composition and bonding on the tribological behavior of polyphenylene sulfide filled with nano particles of TiO_2, ZnO, CuO and SiC. *Wear* 258: 1411–1421.
92. Kurahatti, R. V., Surendranathan, A. O., Srivastava, S. 2011. Role of zirconia filler on friction and dry sliding wear behaviour of bismaleimide nanocomposites. *Materials & Design* 32: 2644–2649.
93. Chang, L., Zhang, Z., Ye, L. et al. 2007. Tribological properties of high temperature resistant polymer composites with fine particles. *Tribology International* 40: 1170–1178.
94. Llorente, A., Serrano, B., Baselg, J. et al. 2016. Nanoindentation and wear behavior of thermally stable biocompatible polysulfone–alumina nanocomposites. *RSC Advances* 6: 100239–100247.
95. Yan, X. H., Ning, R. C., Liang, G. Z. et al. 2007. The effect of silane coupling agent on the sliding wear behavior of nanometer ZrO_2/bismaleimide composites. *Journal of Materials Science* 42: 958–965.
96. Yan, H. X., Jia, Y., Li, M. L. et al. 2013. The tribological properties of benzoxazine-bismaleimides composites with functionalized nano-SiO_2. *Journal of Applied Polymer Science* 129: 3150–3155.
97. Yan, H. G., Li, P. B., Ning, R. C. et al. 2008. Tribological properties of bismaleimide composites with surface-modified SiO_2 nanoparticles. *Journal of Applied Polymer Science* 110: 1375–1381.
98. Yan, H. X., Ning, R. C., Ma, X. Y. et al. 2004. Research on tribology of bismaleimide materials filled with graphite and namometer Si_3N_4. *Journal of Materials Engineering and Performance* 2: 29–31.
99. Yan, H. X., Ning, R. C., Liang, G. Z. et al. 2005. The performances of BMI nanocomposites filled with nanometer SiC. *Journal of Applied Polymer Science* 95: 1246–1250.
100. Yan, H. X., Ning, R. C., Song, C. W. et al. 2008. Properties of novel bismaleimide resin and its composites with nanometer Si_3N_4. *China Plastics* 22: 23–26.
101. Xiao, F., Wang, K., Zhan, M. 2010. Atomic oxygen erosion resistance of polyimide/ZrO_2 hybrid films. *Applied Surface Science* 256: 7384–7388.
102. Song, H. J., Zhang, Z. Z. 2008. Study on the tribological behaviors of the phenolic composite coating filled with modified nano-TiO_2. *Tribology International* 41: 396–403.
103. Taheri-Behrooz, F., Maher, B. M., Shokrieh, M. M. 2015. Mechanical properties modification of a thin film phenolic resin filled with nano silica particles. *Computational Materials Science* 96: 411–415.
104. Song, H. J., Zhang, Z. Z., Men, X. 2006. Effect of nano-Al_2O_3 surface treatment on the tribological performance of phenolic composite coating. *Surface and Coatings Technology* 201: 3767–3774.
105. Manocha, L. M., Prasad, G., Manocha, S. 2014. Carbon-ceramic composites for friction applications. *Mechanics of Advanced Materials and Structures* 21: 172–180.
106. Jin, H., Wu, Y., Hou, S. et al. 2013. The effect of spherical silica powder on the tribological behavior of phenolic resin-based friction materials. *Tribology Letters* 2013; 51: 65–72.

107. Wu, Y., Jin, H., Hou, S. et al. 2013. Effects of glass-to-rubber transition on the temperature, load and speed sensitivities of nano-ZrO_2 reinforced polybenzoxazine. *Wear* 297: 1025–1031.
108. Wu, Y., Zeng, M., Jin, H. et al. 2012. Effects of glass-to-rubber transition on the friction properties of ZrO_2 reinforced polybenzoxazine nanocomposites. *Tribology Letters* 2012; 47: 389–398.
109. Ji, Q. L., Zhang, M. Q., Rong, M. Z. et al. 2005. Friction and wear of epoxy composites containing surface modified SiC nanoparticles. *Tribology Letters* 20: 115–123.
110. Kang, Y., Chen, X., Song, S. et al. 2012. Friction and wear behavior of nanosilica-filled epoxy resin composite coatings. *Applied Surface Science* 258: 6384–6390.
111. Shi, G., Zhang, M. Q., Rong, M. Z. et al. 2004. Sliding wear behavior of epoxy containing nano-Al_2O_3 particles with different pre-treatments. *Wear* 256: 1072–1081.
112. Shen, J. T., Top, M., Pei, Y. T. et al. 2015. Wear and friction performance of PTFE filled epoxy composites with a high concentration of SiO_2 particles. *Wear* 322/323: 171–180.
113. Atarian, M., Salehi, H. R., Atarian, M. et al. 2012. Effect of oxide and carbide nanoparticles on tribological properties of phenolic-based nanocomposites. *Iranian Polymer Journal* 21: 297–305.
114. Dmitriev, A. I., Österle, W., Wetzel, B. et al. 2015. Mesoscale modeling of the mechanical and tribological behavior of a polymer matrix composite based on epoxy and 6 vol.% silica nanoparticles. *Computational Materials Science* 110: 204–214.
115. Zhang, J., Chang, L., Deng, S. et al. 2015. Some insights into effects of nanoparticles on sliding wear performance of epoxy nanocomposites. *Wear* 304: 138–143.
116. Srivastava, S., Tiwari, R. K. 2012. Synthesis of epoxy-TiO_2 nanocomposites: a study on sliding wear behavior, thermal and mechanical properties. *Progress in Organic Coatings* 61: 999–1010.
117. Peng, S., Chang, K., Ma, J. et al. 2018. In-situ polymerization of silica/polydicyclopentadiene nanocomposites: Improved tribological properties and evaluation of interfacial compatibility. *Ferroelectrics* 523: 22–31.
118. Yan, H., Li, P., Ning, R. et al. 2008. Tribological properties of bismaleimide composites with surface-modified SiO_2 nanoparticles. *Journal of Applied Polymer Science* 110: 1375–1381.
119. Service, T. H. 1993. Creep rupture of a phenolic-alumina particulate composite. *Journal of Materials Science* 28: 6087–6090.
120. Mahshuri, Y., Amalina, M. A. 2014. Hardness and compressive properties of calcium carbonate derived from clam shell filled unsaturated polyester composites. *Materials Research Innovations* 18: S6-291–S6-294.
121. Hassan, T. A., Rangari, V. K., Jeelani, S. 2013. Mechanical and thermal properties of bio-based $CaCO_3$/soybean-based hybrid unsaturated polyester nanocomposites. *Journal of Applied Polymer Science* 130: 1442–1452.
122. Rodrigues, D. D., Broughton, J. G. 2013. Silane surface modification of boron carbide in epoxy composites. *International Journal of Adhesion and Adhesives* 46: 62–73.
123. Rostamiyan, Y., Mashhadzadeh, A. H., SalmanKhani, A. 2014. Optimization of mechanical properties of epoxy-based hybrid nanocomposite: Effect of using nano silica and high-impact polystyrene by mixture design approach. *Materials & Design* 56: 1068–1077.
124. Guo, Y. Z., Li, L. Y. 2007. Quasi-static/dynamic response of SiO_2–epoxy nanocomposites. *Materials Science and Engineering: A* 458: 330–335.

6 Thermal Properties of Polymer/Ceramic Composites

6.1 INTRODUCTION

Determining thermal properties is necessary for any materials that change with temperature. They are investigated by several thermal analysis techniques, which include differential-scanning calorimetry (DSC), thermogravimetric analysis (TGA), differential thermal analysis (DTA), thermomechanical analysis (TMA), dynamic mechanical analysis (DMA), dielectric thermal analysis, etc. As is well-known, TGA/DTA and DSC are the two most extensively used techniques to evaluate the thermal properties of polymer/ceramic nanocomposites.

TGA can demonstrate the thermal stability, the onset of degradation, char yield, inflammability, and the percent ceramic incorporated in the polymer matrix. DSC can be effectively utilized to determine the thermal transition phenomena of polymer/ceramics composites. Furthermore, the CTE, which is the criterion for the dimensional stability of materials, can be tested with TMA. In addition, thermomechanical properties measured by DMA/DMTA are very important to understand the viscoelastic behavior of the composites. The storage modulus (G'), loss modulus (G''), and tan δ (G''/G') are three key parameters of dynamic mechanical properties that can be used to study the occurrence of molecular mobility transitions, such as the glass transition temperature (T_g). Dielectric thermal analysis is also useful to explain the viscoelastic behavior of such composites.

6.2 THERMAL STABILITY

Generally, the incorporation of ceramic particles into the various polymeric matrices increased the thermal stability of their resulted materials by acting as a superior thermal insulator and mass transport barrier to the volatile degradation products generated during the thermal decomposition [1]. The improvement in the thermal stability of polymer/ceramic micro- and nanocomposites was generally attributed to the formation of polymer/ceramic networks through various types of physical cross-linking of polymeric chains and ceramic particles, which can stabilize the integrated system by reducing the thermal motions of polymer chains. An example of such system was that reported in polyurethane/TiO_2 nanocomposites [2]. A decrease in the thermal stability, usually found in some other polymer/ceramic materials, was linked to the catalytic-effect metal-based ceramics oxidative degradation phenomena.

The incorporation of the ceramic fillers into flexible polymers having inferior thermal resistance can markedly enhance both their stiffness values and thermal stabilities [3]. The SiO_2 or Al_2O_3 particles without surface modification can be included into the polymer matrices such as PEEK. The involved mechanisms of thermal stability improvement for polymers/SiO_2 were related to their molecular dynamics [4], physical cross-linking, and specific interactions between the hybrids' constituents. The experiments on the polymaleimide/SiO_2 nanocomposites revealed that the ceramic networks were chemically joined to polymaleimides, which enhanced the thermal resistance of these nanocomposites. In several cases, the inclusion of ceramic nanofillers into polymer matrices causes physical or chemical cross-linking responsible for increasing the initial decomposition temperature of their resulted composites.

The extensive literature related to the thermal stability of polymer/ceramic composites provides inconsistent conclusions on the ceramic efficiency in enhancing thermal stability. For example, the thermal resistance of the boron-phenolic resin (B-PF)/KH-550-Si_3N_4 was improved by the well-dispersed and well-adhered KH-550-Si_3N_4 in these novel thermoset/ceramic composites. Improved grinding quality and binding media were showed by the new composites compared to pure B-PF [5].

The thermal decomposition temperatures (T_d) of polyimide (PI)/SiO_2 hybrids was determined by the TGA technique [6]. The pure PI exhibited a T_d of 561°C while for the 30 wt.% of SiO_2 loading composite reached 592°C. The T_d of a hybrid increased with increasing SiO_2 content. The thermal stabilities of the hybrids with coupling agent–modified nano-SiO_2 were slightly inferior to those of raw ones due to the low thermal stability of the alkyl chains of γ-glycidyloxypropyltrimethoxysilane, but these hybrids retained higher thermal stability compared to the pure PI.

The thermal stability of the PMMA/SiO_2 nanocomposites in certain cases did not improve by filling with silica ceramic particles [7]. For instance, TGA tests conducted on a series of PMMA/SiO_2 nanocomposites revealed such a trend, and even showed a slight decrease in the thermal stability of these hybrids at low degradation temperatures while slightly retarding the random thermal degradation along the PMMA backbone. Epoxy/SiO_2 and epoxy/TiO_2 hybrids were synthesized via a sol–gel process from a DGEBA prepolymer with a GPTMS silane-coupling agent to improve interfacial interaction between the epoxy resin and these ceramic nanoparticles. The addition of inorganic nanoparticles significantly enhanced the thermal stability of the epoxy matrix [8].

The DTG data displayed in Table 6.1 concludes that adding nano-SiO_2 particles into the PMMA polymer did not change the degradation mechanisms of the polymer, thus the thermal stability of PMMA was not improved by incorporating this ceramic nanofiller [9]. However, the char yield of the resulting PMMA nanocomposites monotonically increases with increasing the loading of nano-SiO_2.

Thermal stability can be defined as the protection of materials from the deterioration of properties by thermal stress [10]. Many published papers address the influence of the ceramics on polymer thermal stability. Some of the literature suggests that nanoparticles have no obvious effect on thermal stability; others report a small to marked improvement, while the last group found acceleration in the thermal decomposition of polymers. Low-temperature treatment of γ-Al_2O_3, produced

TABLE 6.1
Thermal Degradation Data of PMMA and Nanocomposite Materials in Nitrogen [9].

	Degradation Temperatures (°C)			Char Yield (wt.%)	Calculated Silica Content (wt.%)
Sample	Head-to-head Structure	Unsaturated End	Polymer Backbone		
PMMA	165,212	296	401	0	0
HM-40	177	296	388	39	43
HM-50	174	299	395	49	50
HM-60	177	297	388	60	63
HM-70	179	294	378	69	70
HM-80	180	299	378	75	76

by the electrical explosion of wires (EEW) method [11, 12], was utilized as fillers for polypropylene matrices. All additives showed good oxidation resistance under heating up to 400°C, where PP/Al_2O_3 hybrids released water under an endothermic decomposition, except γ-Al_2O_3. The results of these studies indicated that the oxidation rate reduced when polypropylene was reinforced with gibbsite and barite at concentration of 0.5 to 10 wt.%.

The possibility of using AlN nano-powders as a flame-retardant additive was investigated [13]. The AlN rations introduced into a polyethylene matrix was 0.1, 0.25, 0.75, 1.5, and 3 wt.%. The addition of 1.5 wt.% AlN in a polyethylene matrix caused a significant increase in the initial oxidation temperature by almost 33°C (to 183°C) compared to the unfilled polyethylene (150°C). On the other hand, the onset temperature of the intensive weight loss for these nanocomposites attained as high as 375°C compared to only 360°C for the pure polyethylene. This increase was related to the role of AlN nanofillers that acted as crystallization centers and contributed in the formation of fine-grained structure.

Chayan et al., in 2012, reported on the thermal decomposition of nitrile rubber (NR)/SiO_2 nanocomposites prepared using surface-treated nano-SiO_2. They observed that the first weight-loss temperature appeared at the range 350 to 490°C as the consequence of rubber component degradation. The second weight loss at the temperature range 520 to 640°C was attributed to the thermal decomposition of carbonaceous by-products. The temperature at maximum weight loss (T_{max}) was found to be higher for the prepared nanocomposites filled with surface-modified SiO_2 nanoparticles, indicating improvement in the thermal stability over that of composites containing raw SiO_2 nanofillers [14, 15]. In the presence of silane molecules, nano-SiO_2 can thus effectively enhance the thermal stability of NR matrix [14].

The effect of adding $CaCO_3$ nanoparticles on the thermal stability of polybutadiene rubber (PBR) was evaluated by Shimpi and Mishra [16]. They reported that the addition of nano $CaCO_3$ in PBR resulted in enhanced thermal stability. Since, at 12 wt.% of nano $CaCO_3$ (21, 15, and 9 nm)-reinforced PBR, decomposition temperatures of 491, 483, and 472°C, respectively, were recorded. This enhancement in thermal stability was traced to the uniform dispersion of nano-$CaCO_3$ throughout the matrix, which retained the rubber chains intact on cross-linking, so preventing out the diffusion of the volatile decomposition by-product [17]. It was also revealed inorganic nanofillers afforded improved thermal stability as compared with commercial microsized fillers. This trend was studied by Shimpi et al. [16], where they synthesized via a solution spray three different types of $CaCO_3$ nanoparticles at diameters 9, 15, and 21 nm to reinforce the polybutadiene rubber (PBR) matrix. They observed that the incorporation of nano-$CaCO_3$ in PBR with reduced size imparted better thermal stability as compared to commercial microsized-$CaCO_3$.

In order to achieve improved thermal stability and fire retardancy of PMMA and PS nanocomposites, ethylene glycol methacrylate phosphate (EGMP) monomers on grafted nano-Al_2O_3 were prepared. Thermal stability and decomposition routes of pure monomer, polymer reinforced with grafted Al_2O_3, and octylsilane-treated Al_2O_3 were evaluated by TGA technique. Due to strong interactions between EGMP and Al_2O_3 surface, the thermal stability of the monomer supported by the nanoparticles is higher than that of an unmodified monomer. As shown in Figure 6.1, the incorporation of 5 wt.% of surface-modified Al_2O_3 in both PMMA and PS led to an enhancement of thermal stability with respect to the unfilled matrices or their nanocomposites filled with pristine Al_2O_3. In addition, the surface modification of nano-Al_2O_3 significantly reduced the peak of HRR measured by cone calorimetry in the case of PMMA nanocomposites [18].

PF/SiO_2 core-shell composite hollow microspheres were produced by spray-coating process and exhibited low density and high-temperature resistance [19]. TiO_2 nanotube filled epoxy composite was prepared by ultrasonication followed by mechanical stirring as reported by El Saeed et al. [20]. The TiO_2 nanotubes significantly improved the thermal properties of epoxy resin compared to those resulted by using microsized TiO_2. The TiO_2 crystalline shape increased the surface area and led to reducing the shrinkage of the epoxy matrix at a higher temperature, and acted as laminar barrier, which can decelerate the evolution of volatiles during the degradation process. The thermal-degradation mechanism of polymer/ceramic nanocomposites is related to the kind of filler used and its ratio, the structure of the char formed during polymer degradation, the gas impermeability of ceramic nanoparticles, which inhibit the formation and escape of volatile by-products during degradation, and the interactions between polymer-reactive groups and inorganic nanoparticles [21].

ZrB_2/carbon/phenolic hybrid composites were prepared by immersion, and their ablative and thermal insulation property were evaluated by Chen et al. The linear ablation rate of carbon/phenolic composite decreased with the increasing of ZrB_2 content to attain its minimum value up to 9 wt.% filler loading, at which also an optimal insulation performance was obtained [22]. Micro-ZrO_2 was used to improve the thermal stability and the ablation properties of asbestos fiber/PF composites. The thermal stability of the asbestos/PF composites was enhanced by adding ZrO_2 through the formation of a thin melted layer of ZrO_2. The results showed that the linear and mass ablation rates, as well as the back surface temperature of the composites after adding 14 wt.% ZrO_2, decreased by 58%, 92%, and 49%, respectively [23].

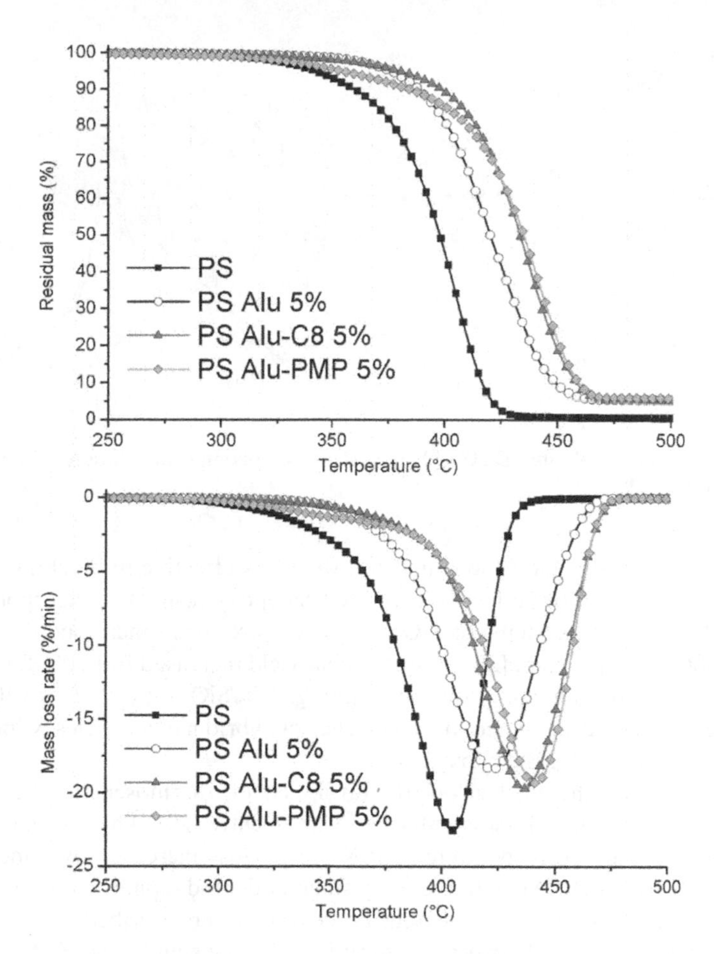

FIGURE 6.1 TGA and DTG Curves Under Air for PMMA Nanocomposites. Alu: Al_2O_3; C_8: Octasilane; PMP: Polymerization Step of EGMP [18].

B_4C/PF nanocomposites were first prepared and then used to fabricate CF-based BMC. The B_2O_3 produced by the conversion of B_4C generated a substantial improvement of the dimensional stability of the BMC, which also exhibited considerable residual structural integrity after burning [24]. An opposite effect for the thermal stability of irradiated PP was observed in the presence of TiO_2 nanoparticles, which promotes oxidation initiated by radiation processing [25]. The incorporation of other types of nanoparticles, such as TiO_2 and silsequioxanes, was found to influence the thermal degradation of the thermosetting materials by changing the degradation mechanism of the epoxy resin [26].

The thermal stability of epoxy resin/TiO_2 nanocomposites was found to depend on the nanoparticles' content, as well as on their dispersion state [27]. By adding a very low TiO_2 fraction, the nanoparticles were uniformly dispersed and formed a barrier to heat and oxygen, due to their ceramic nature. As the loading of the nanoceramics was increased, they tended to agglomerate into lumps, which were not well-dispersed

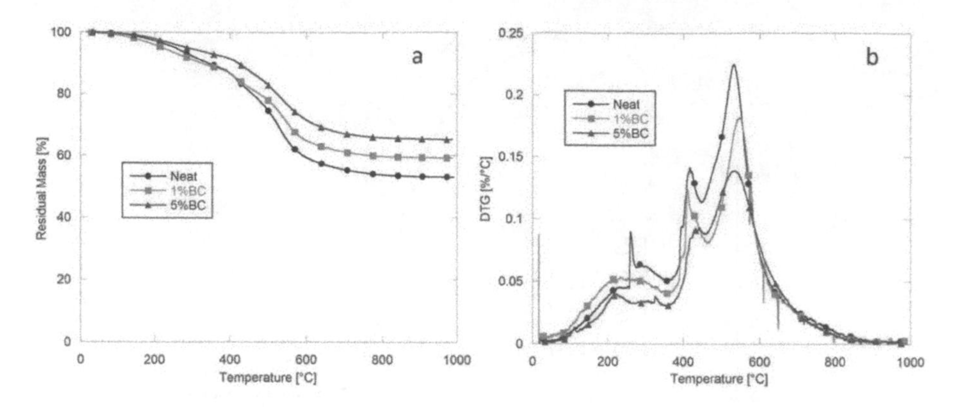

FIGURE 6.2 TGA (a) and DTG (b) Traces of Nanocomposites Studied in Nitrogen. Redrawn from [24].

into the matrix. In this case the nano-TiO_2 were less effective in blocking the heat and oxygen, but still more effective than the neat epoxy resin. The incorporation of the hybrid nanofillers' systems of TiO_2/SiO_2 into epoxy resin increased the thermal stability of the unfilled resin [28]. Also, the char yield increased from 0% for the neat resin to 25% for the nanocomposites containing TiO_2/SiO_2 ratio of 2.5 to 10 wt.% . These phenomena are attributed to the presence of hybrid nanoparticles, which acted as thermal stabilizers for the epoxy resin.

As the SiO_2 was included in poly(furfuryl alcohol) thermoset, the thermal stability of the furanic matrix was enhanced significantly [29]. The thermal decomposition temperature corresponding to 10% weight loss increased by almost 30°C in the presence of SiO_2. Also, in the early stage of degradation, the rate of thermal decomposition was markedly decreased. These results were ascribed to the interconnections, which occurred between the furanic end groups and nano-SiO_2, reducing the possibility of unzipping the furanic chain. This was justified by that the predomination of chain scission and the formation of high molecular-weight compounds exhibiting low volatility. Above 400°C, the rate of degradation was fixed for all the systems with or without SiO_2, suggesting that the SiO_2 network had no effect on the latter stages of poly(furfuryl alcohol) thermal degradation.

Schutz et al. studied the thermal properties of a phenolic nanocomposite containing silsesquioxanes [30]. The thermal oxidative stability of the nanocomposites was improved, as compared to that of the pure resin due to the formation of a protective layer of SiO_2 during silsesquioxanes' pyrolysis at the surface of nanocomposites. These ceramic layers decelerated the thermal oxidative degradation process. The thermal stability of nanocomposite films based on poly(triazole-imide) reinforced with raw and modified-SiC nanoparticles was investigated [31]. The nanocomposites' decomposition temperatures increased with increasing the SiC content, indicating an enhancement of the thermal stability. In addition, the char yield increased from 58% for the poly(triazole-imide) to 85% for the nanocomposites containing 10 wt.% SiC. The nanocomposites containing SiC nanoparticles functionalized with epoxide-end groups showed a lower thermal stability compared to those reinforced with neat SiC.

The significant improvements in thermal stability of polymer/ceramic micro- and nanocomposites could be related to the many factors of the ceramic fillers, such as the high surface volume of the filler, enhancing the thermal barrier properties in the presence ceramic which contributes to

- The formation of numerous tortuosity path,
- Reducing the organic matrix molecular mobility,
- Decreasing permeability and reduction in the rate of evolution of the resulted volatiles by-products,
- Constructing of high-performance carbonaceous char which insulates the underline composites layers and decelerates the escape of volatiles during the decomposition process,
- Adsorption of formed gasses into ceramic/polymer membrane.

By incorporating a low ratio of ceramic filler into the polymeric matrix, the ceramic particles are uniformly dispersed, which predominated the heat barrier effect, while with a further increase of filler loading, the promoter effect instantly increased and changed to an impressible, impermeable membrane; thus, the thermal stability of such composites was lowered. Such an accelerating effect depends on several factors including the presence of hydroxyl groups on the outer surface of ceramics that could catalyze the matrix thermal decomposition; the existence of active catalytic sites affects the matrix degradation as well. Several ceramic nanoparticles, such as TiO_2, MgO, $CaCO_3$, Al_2O_3, etc., were also reported to exhibit great stabilization effects on polymers' thermal degradation process [32].

Thermal degradation of HDPE occurred mainly in two mechanisms. The first one is related to small mass loss and is *n*th-reaction order, while the second is ascribed to the main mass loss and is autocatalysis *n*th-order. The incorporation of nano-SiO_2 particles increased the thermal stability of the HDPE matrix. However, both mechanisms exhibited increased activation energy values (201 and 266 kJ/mol for HDPE/SiO_2 5 wt.% compared to140 and 260 kJ/mol for the neat matrix), which demonstrated the thermal stabilization effect of nano-SiO_2.

The activation energy, E_a, of thermal decomposition can be computed by the isoconversional method of Ozawa, Flynn, and Wall (OFW), a "model free" method which considers that the conversion function $f(\alpha)$ does not change with the alteration of the heating rate for all values of α. It is based on the determination of the temperatures related to fixed values of heating rates β. Thus, we can plot $\ln(\beta)$ against 1/T in the form of:

$$\ln(\beta) = \ln\left[\frac{Af(\alpha)}{d\alpha / dT}\right] - \frac{E_a}{RT} \tag{6.1}$$

This plot should provide straight lines, and its slope is monotonically proportional to the activation energy ($-E_a$/R), where A is the pre-exponential factor, that is considered to be independent of temperature, E_a the activation energy, T the absolute temperature, and R is the gas constant. If the determined activation energy is the same for the various values of α, the existence of a single-step reaction can be concluded with certainty. On the contrary, a change of E_a with increasing degree of conversion indicates a complex reaction mechanism that invalidates the separation of variables involved in the OFW analysis [32].

To identify the mechanisms of thermal degradation, a comparison between the experimental and theoretical data was done, since initially it was found that the degradation of the specimens can be described only by one mechanism, without presuming the exact mechanism. If this mechanism failed to explain the relationship between the experimental and theoretical data, other mechanisms can be used. To determine the conversion function *f(α)*, a method known as the "model-fitting method" was used. This method was utilized simultaneously on the experimental data recorded at the heating rates of β = 5, 10, and 15°C/min, respectively. Results showed that the model-fitting method applied to multiple heating rate data afforded activation energies similar to those estimated by the isoconversional methods. In Figure 6.3 the results of this fitting for HDPE/SiO_2 5 wt.% is observed. The form of the conversion function, given by fitting, is the mechanism of autocatalysis n-order written as follow:

$$f(\alpha) = (1-\alpha)^n (1 + K_{cat}.X) \tag{6.2}$$

where, K_{cat} is a constant and X the reactants. The parameters of the mechanism were: the pre-exponential factor log $A(s^{-1}) = 15.2$, the activation energy $E_a = 249.9$ kJ/mol, the exponent value n is equal to 0.8, and the correlation coefficient was 0.9997. The value of the activation energy is between the limits of the calculated values from the OFW method.

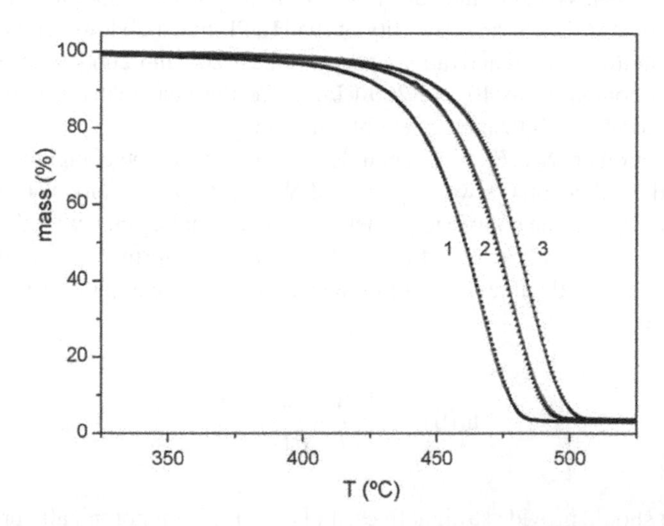

FIGURE 6.3 Fitting and Experimental Data TG Curves of HDPE/SiO_2 5 wt.% for One Reaction Mechanism and Different Heating Rates. 1: β = 5°C/min, 2: β = 10°C/min and 3: β = 15°C/min [33].

The influence of the presence of SiO_2 nanoparticles on the thermal stability of low-density polyethylene (LDPE) has been scrutinized [34]. Various concentration of nano-SiO_2 were included within a LDPE matrix by means of high-energy ball milling (HEBM) to afford uniform dispersion of nanofillers. Thermal characterization of the resulted nanocomposites showed insignificant changes at low filler content, reflecting that in the case of reinforced-LDPE, weak interactions occurred between the matrix and nano-SiO_2.

The impact of adding a nano-TiO_2 and SiO_2/TiO_2 hybrid filler system on the thermal decomposition of a styrene-cross-linked UPE resin was evaluated by Gorelov et al. The thermal resistivity of the UPE/TiO_2 nanocomposites increased at small nano-TiO_2 ratios of about 0.5 wt.%, while in the PUE/SiO_2-TiO_2 nanocomposites, a decrease in the parameters at loadings of up to 1.5 wt.% was replaced by some increases at higher loadings of up to 5 [35].

The thermal stability of the PBZ/SiO_2 composite was increased with decreasing the particle sizes of nano-SiO_2, suggesting better reinforcement of the smaller particles. The $T_{5\%}$ of PBZ/SiO_2 specimens filled with various particle sizes of SiO_2 nanoparticles shifted from the value of 325°C for the unfilled PBZ to about 340°C for nanocomposites reinforced with smallest nano-SiO_2 size, due to their efficient barrier effect resulting from larger surface area [36].

PU/SiO_2 nanocomposites were prepared with a good dispersion of nano-SiO_2 at a loading amount below 5 wt.%. Compared with the pure PU, the T_d of the PU/SiO_2 nanocomposites were respectively enhanced by 43.5°C at 3 wt.% nano-SiO_2 loading [37]. Jin et al. synthesized a new cyclotriphosphazene/h-BN hybrid and used it to reinforce bismaleimide resin. The results revealed that the filling with only a small amount of cyclotriphosphazene-BN substantially enhanced the thermal stability of the pure bismaleimide thermoset through forming protective char layers, which played the role of a barrier for heat transfer [38]. Lu et al. reported that the nano-SiO_2 networks chemically bound to polymaleimides can greatly improve the thermal stability of these nanocomposites [39]. Phosphorous materials were introduced into UPE resin to prepare flame-retardant UPE/SiO_2 hybrid materials by sol–gel method and curing process. The TG results indicate that the FR-UPR/SiO_2 hybrid materials possess higher thermal stability and increased Y_c than those of pure UPR [40].

The addition of n-SiC facilitated improved thermal stability and fire resistance of vinyl-ester matrix [41]. A series of the silane-functionalized nano-SiO_2/PBZ composites was produced, and their thermal stabilities were evaluated. The $T_{5\%}$ values were increased with the increasing of nano-SiO_2 loading and was shifted from 368°C or 405°C (of the neat PBZ) to 379°C and 426°C (of the nanocomposite containing 3 wt.% nano-SiO_2) under air or nitrogen, respectively, while the Y_c of the nanocomposite exceeded 50% [42].

Morgan and Putthanarat found that the addition of Al_2O_3 particles combined with carbon nanomaterials and/or nano-SiO_2 results in a substantial decrease in the thermal stability and heat release rate of polyimide matrix. For instance, the combination of 40 wt.% Al_2O_3 with 10 wt.% SiO_2 and 5 wt.% carbon nanofiber results in a 95% reduction in total heat generated for this composite compared to the base polyimide. Such composite materials will likely perform well in vigorous fire conditions or do not contribute to fire spread [43].

As reported by Ghezelbash et al., Al_2O_3 nanoparticles can be homogeneously dispersed in a polyimide matrix. The results of TG analysis showed that the addition of nano-Al_2O_3 improved thermal stability of the obtained hybrids [44].

According to Toiserkani [45], comparison between the TGA thermograms of pure PI and PI/TiO_2 nanocomposite revealed that there is a relatively large gap between the thermal stabilities of pure polymer and their nano-hybrids containing 5 wt.%, 10 wt.%, and 15 wt.% of nano-TiO_2. Furthermore, a linear relationship between the concentration of TiO_2 and thermal stability was also observed. The improved thermal stability of these nanocomposites is traced to the high heat resistance exerted by the thermal-stability of TiO_2 nanoparticles.

Using thermogravimetric analysis, the thermal stability of PI/SiO_2 nanocomposite films was evaluated by Akhter et al [46]. Results showed that the thermal decomposition temperature $T_{10\%}$ reached 558°C for pristine PI, while it was increased by 20°C and 34°C for the PI/SiO_2 nanocomposites filled with 30 wt.% nano-SiO_2. The surface-treatment of nano-SiO_2 provided much-improved values of $T_{10\%}$ compared to unmodified nano-SiO_2.

Periadurai and his co-workers described the preparation of SiO_2/PF nanocomposites through the in situ polymerization process with the presence of chemical linkage between the nano-SiO_2 and matrix. The addition of nano-SiO_2 enhanced the T_d and LOI value of the PF matrix by almost 70°C and 5, respectively, due to flame retardancy of SiO_2 nanofiller and also the nanocomposite structures [47].

Novel thermally-conductive bismaleimide-triazine resin (BT)/Si_3N_4 composites have been developed by Zeng et al. It was found that the new composites exhibited $T_{5\%}$ higher than 370°C as the Si_3N_4 content increased [48]. Novel PF/SiO_2 hybrid ceramers were synthesized via the sol–gel process, and their thermal degradation properties were studied by Chiang and Ma [49]. The char yields of the hybrids increased as the TEOS content increased. In addition, $T_{5\%}$ of the hybrid increased from 290°C for the specimen contained 20 wt.% TEOS to 312°C at 80 wt.% of TEOS content as illustrated in Figure 6.4. This result means that the TEOS can effectively improve the thermal stability and flame retardancy of these hybrids as proved by the LOI test [50].

Cenosphere (SiO_2/Al_2O_3) microparticle-filled PF-based composites were prepared, and their thermal, thermos-oxidation, and ablative behaviors were studied by Balaji et al [51]. The result showed that the thermal stability of the filled composites had improved, and reduction of mass loss was obtained with addition of cenosphere. Ablation results showed that the addition of cenosphere content exhibited the favorable ablation resistance. Chiang et al. [52] described the preparation of high flame-retardant PF/GPTS-SiO_2 nanocomposites by a sol–gel technique. The GPTS silane molecules reacted with the PF resin chains to form covalent bonds. The size of the particles of SiO_2 in the hybrids was below 100 nm. The temperature $T_{5\%}$ increased from 281°C to 350°C, while the LOI of the nanocomposite attained 37.

The addition of B_4C into a PF matrix efficiently increased the heat resistance of PF resin by seizing the active phenolic hydroxyl and methylol groups of the PF thermoset and holding the main components of some volatile, such as CO, H_2O, phenol, and its derivations in the matrix by chemical reactions. Similarly, the reaction product B_2O_3-oxide closed and mended the micro-cracks for its satisfactory wettability

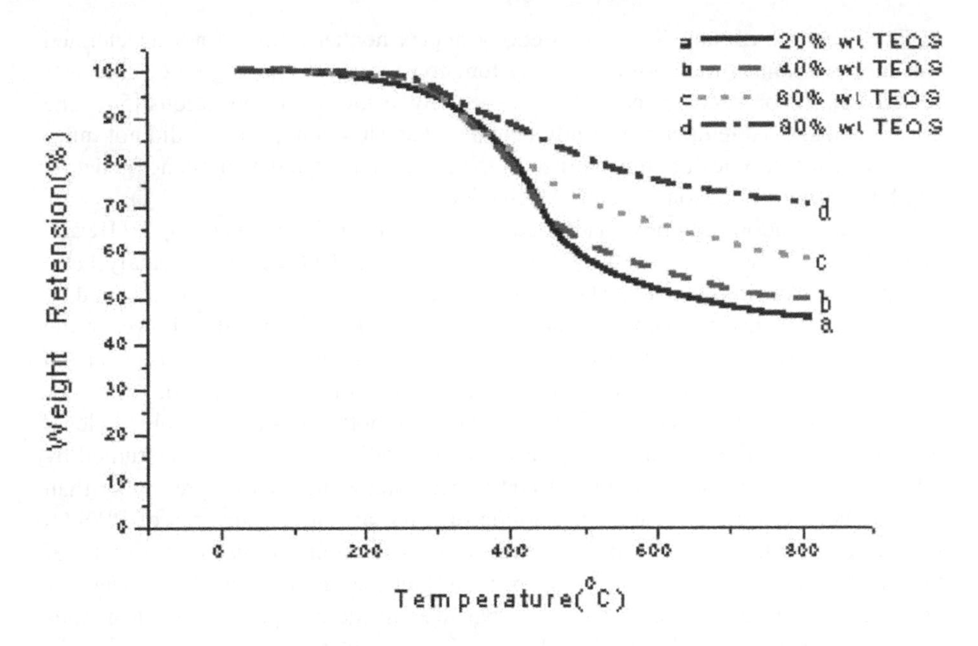

FIGURE 6.4 TGA Curves of Phenolic Resin/Silica Hybrid Ceramer. Reprinted with Permission from Elsevier [50].

and adhesive capacity. The bond strength and thermal stability of the PF matrix were also significantly elevated [53].

The thermal decomposition behavior of PF/SiO_2 composite was studied by a solar radiant-heating experiment coupled with one-sided heat flux. The composite exhibited an excellent thermal insulation during thermal exposure. The reduction in radius of the silica fillers with fewer pits were observed at the heated surface [54]. The higher the level of SiO_2, the better the thermal stability of the BMI/KH-921 silane-treated SiO_2 composite particles. The silane-modified SiO_2 particles significantly improved their dispersion capability within the continuous BMI oligomeric matrix [55].

6.3 KINETICS OF CURE

New polybenzoxazine (PBZ)/SiO_2 nanocomposites were developed by high shear mixing and compression molding as a function of filler content ranging from 0 to 30 wt.%. The curing process of the PBZ/nano-SiO_2 composites was found to be autocatalytic in nature with average E_a of 79-92 kJ.mol^{-1} [56]. The curing kinetics and the E_a of an epoxy resin reinforced with B_4C at various ratios (6 and 12 wt.%) and particle size (60 nm, and 7 and 23 μm) were evaluated by DSC. The E_a was calculated assuming Arrhenius' equation revealed a significant influence of the temperature on the curing-reaction mechanism. In addition, the curing kinetics were affected by the ratios and sizes of B_4C ceramic filler [57].

The presence of nano-TiO_2 in an epoxy matrix neither retarded nor accelerated the cure reaction of the epoxy/amine system considerably. The degree of cure was reduced in the presence of nano-TiO_2, especially at lower concentrations [58]. The conversion against temperature study revealed that TiO_2 nanoparticles did not influence the autocatalytic cure mechanism of the epoxy matrix and also did not generate any huge barrier effect on the curing behavior.

The cure reaction of epoxy-TiO_2 nanocomposites was investigated by FTIR and DSC. The FTIR showed that at the initial curing stage, TiO_2 acted as a catalyst and facilitated the curing reaction. The catalytic effect of TiO_2 was further confirmed by the decrease in maximum exothermal peak temperature (DSC results); however, the addition of TiO_2 reduced the overall degree of cure, as the heat of reaction for the cured nanocomposites was reduced compared to that of pure matrix [59].

The kinetics of non-isothermal crystallization of both high-density polyethylene/nano-silica (HDPE/SiO_2) and polypropylene/nano-silica (PP/SiO_2) were studied by DSC [60]. The activation energies of HDPE/SiO_2 nanocomposites were higher than that of the unfilled HDPE matrix. By contrast, the activation energies of PP/SiO_2 were inferior to that of the neat PP. However, the activation energies of both HDPE/SiO_2 and PP/SiO_2 increased with the increasing of loading of nano-SiO_2. The heterogeneous nucleation role of SiO_2 nanoparticles in the composite was clear, thus the crystallization of both HDPE and PP was not completed. The cure kinetics of a highly-reactive PU/SiO_2 nanocomposite were DSC as a function of preparation time and SiO_2 concentration. By adding nano-SiO_2 with a loading range of 0.5 to 2 wt.%, a reduction in reactivity and a delay of the maximum cure rate was detected. In addition, the kinetics of the reaction significantly depended on the preparation time [61].

6.4 GLASS TRANSITION TEMPERATURE (T_g)

Studies on the role of adding SiO_2 nanofiller on T_g of polymer showed a wide variety of trends. Also, they reported that T_g of those polymer nanocomposites changed for a variety of reasons, including filler size, filler content, and dispersion state [62]. In some cases, the polymer nanocomposites revealed an enhancement of the T_g [63]. In other cases, a decrease in the T_g values of polymer nanocomposites was observed [64]. An initial increase in T_g followed by a depression in T_g at a higher filler content was recorded due to the fact that nano-ceramics had insignificant influence on the T_g of the matrix, and in some cases a disappearance of T_g was also reported [65]. However, The T_g of the epoxy/SiO_2 composites with nanometer and micrometer-sized fillers showed different trends [66].

The micrometer-sized filler did not provoke any significant effect on T_g of the composites, whereas the nanofiller had a substantial impact. With increasing nano-SiO_2 loading, the resulting nanocomposites first exhibited a slight enhancement in the T_g values, and then T_g was decreased significantly at much higher nano-SiO_2 contents. Compared with the T_g of pure matrix, the 40 wt.% nano-SiO_2 specimen had a drop in T_g of almost 30°C. It was found that the T_g depression of the nanocomposite was closely attributed to the matrix/filler interfacial features. The improved matrix/filler interface generated extra free volume and thus controlled the large-scale segmental motion of the thermoset; this provoked a steady decrease in the T_g values when the ceramic filler increased. But, in the sub-T_g transition, local movement of the chain was involved needing a lower free volume. Thus, the improved interface did not impact the sub-T_g transition temperature.

The addition of nonpolar fumed SiO_2 to nonpolar Teflon produced no measurable change in T_g. The nanocomposite films filled with different ratios of SiO_2 (10, 25, and 40 wt.%) revealed no significant change in T_g values; those ranged between 242 and 246°C. This result was justified by the weak interaction between the Teflon chains and a polar fumed SiO_2, which resulted in low mobility [67]. The PMMA/SiO_2 nanocomposites had a T_g of 115°C, while all of its nanocomposites did not show any T_g from both DSC and DMA tests. The disappearance of T_g implied that the motion of the PMMA chains was significantly restricted by the SiO_2 nanoparticles. This phenomena could also result from the cross-linking bonding between PMMA chains and nano-SiO_2, since a reduction in the T_g was detected with the PMMA/SiO_2 hybrid nano-materials without interphase bonding [68, 69].

Highly-filled polybenzoxazine (PBZ)/SiO_2 nanocomposites were investigated for their thermal properties as a function of filler loading. A significant increment up to 16°C in the T_g of the PBZ/nano-SiO_2 nanocomposites was detected at a loading of 30 wt.% nano-SiO_2. The resulting PBZ/nano-SiO_2 nanomaterials can be used as coating material in micro-electronic packaging or other similar applications [70].

Studies on of the glass transition temperature T_g of polymer/ceramic showed that different effects can be recorded according to the type of ceramic used, the involved matrix, and the method of mixing the ceramics with polymer matrix, and the surface-modification of ceramic fillers. The change in T_g of such nanocomposites is controversial since its value is related to several parameters including the degree of polymerization, chemical structure of polymeric matrix, filler size, filler content, dispersion state, etc. In many cases, polymer/ceramic nanocomposites exhibited an improvement in the T_g, however, in other cases a decrease in T_g resulted. Typical thermograms of total specific heat flow versus temperature for polystyrene and five filled nanocomposites with silica particles are presented in Figure 6.5. All the values of T_g, are calculated using middle-point technique [71].

All DSC curves exhibit the similar curvature: a slow reduction of heat flow followed by a steeper decrease, generating a step corresponding to T_g. With increasing the nano-SiO_2 content, T_g irises from 94°C (for the unfilled matrix) to 99°C (for sample filled with 30 wt.% of SiO_2 nanoparticles). However, the magnitude and the width of the T_g remained fix as the nano-SiO_2 concentration increased. With the addition of silica, the determined positive T_g shift value by DSC can be correlated with strong interactions between polymer and nanoparticles. In order to determine a correlation between T_g data and the nano-SiO_2 content of the polystyrene, Eq. (6.3) is given:

$$T_g(x) = T_{g0} + \frac{x}{1+kx} \tag{6.3}$$

Where, T_g is the glass transition temperature of the composites, T_{g0} is the glass transition of the pure matrix and x is the SiO_2 fraction. The constant k depends on the type of nanofiller and it is calculated by the method of least squares ($k = 0.16$).

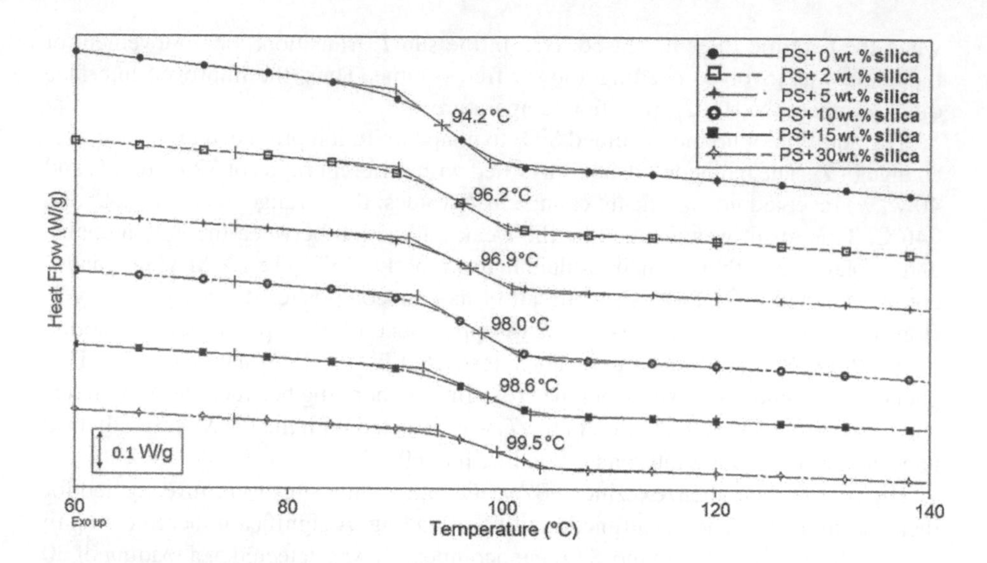

FIGURE 6.5 DSC Curves of Prepared Polystyrene Samples with Different Nano-silica Content [71].

The rate of change of polystyrene T_g is not strongly affected by large agglomerates, whose presence is characteristic for the samples reinforced with higher silica loading, which was a good agreement with the proposed fitted equation (Eq. 6.3) as Figure 6.6 reveals.

Polybenzoxazine/TiC nanocomposites containing 0 to 10 wt.% filler contents were produced by a solution-blending technique exhibiting improved T_g values compared to that of unfilled resin. The incorporation of nano-TiC was markedly increased the char yield which increase by 50% at 10 wt.% TiC content [72]. Effects of post-curing temperature on the T_g of epoxy/Al_2O_3 nanocomposite was investigated

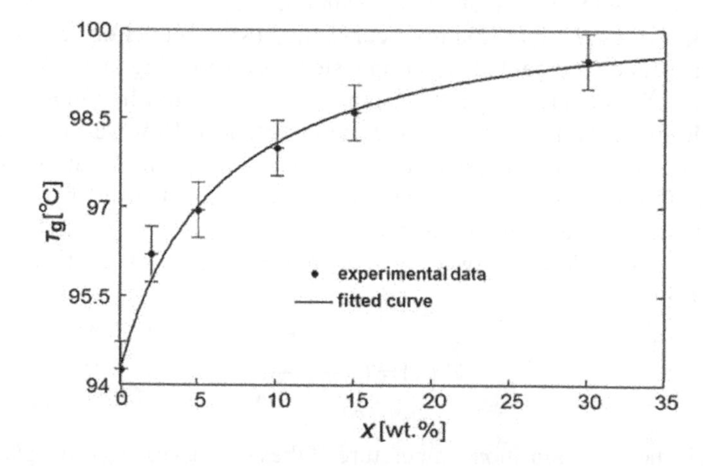

FIGURE 6.6 Dependence of T_g of Polystyrene on Silica Content, *x*. Vertical Errors Bars Are Standard Deviations from Measurements (±0.5 °C) [71].

by Hiremath et al. The post-curing at a temperature below the T_g of epoxy improved the viscoelastic and flexural properties of PNCs, while above the T_g had a detrimental effect on the properties of the pure epoxy and its nanocomposites. As the post-curing temperature rose, T_g of both of the neat epoxy and nanocomposite increased [73].

C_4-ether-linked BMI-toughened epoxy-reinforced with silane-functionalized Al_2O_3 nanocomposites were prepared by Mandhakini et al. Results revealed that incorporating at 5 wt.% nano-Al_2O_3, C_4-ether-linked-BMI-toughened epoxy nanocomposite improved both the viscoelastic and thermal properties because of the functionalization and uniform dispersion of Al_2O_3 nanofiller in the thermoset's matrix [74]. PAMAM: BF_3 grafted SiO_2 were used to cure epoxy resin to produce novel epoxy/silica nanocomposite. The T_g of nanocomposites in rubbery region of the new EP/SiO_2 nanocomposite were considerably higher than those using ethylene-diamine (EDA) as a curing agent with SiO_2. In addition, the T_g increased with increasing content of SiO_2-PAMAM: BF3 contents [75].

The ZrW_2O_8 nanoparticles with a smaller particle size and larger surface area led to a more significant reduction in gel-time and T_g of the epoxy nanocomposites, while a higher initial viscosity and significant shear-thinning behavior was observed in prepolymer suspensions containing ZrW_2O_8 with larger particle sizes and aspect ratios [76].

Yu et al. reported that the addition of hyperbranched aromatic polyamide-treated boron nitride (BN-HBP) to epoxy matrix resulted in a strong interface, and thus the composites exhibited higher T_g value, while the addition of BN nanoplatelets showed the least improvements of T_g [77].

Choi et al. proposed a sequential multiscale modeling approach to describe the thermoelastic behavior of nanocomposites, and they reported that the T_g values of epoxy/SiC nanocomposites having different particle sizes were changed dramatically in the temperature range of –23°C to 277°C [78]. The addition of GPTMS to epoxy/SiO_2 improved the interface interactions by forming chemical cross-links between the organic and ceramic constituents, enhanced bi-continuity of the phases, affected the size and distribution of silica, mitigated the phase separation, and increased the T_g even at higher concentrations of SiO_2 due to profound negative effects of free-volume increased, homopolymerization, and phase separation [79].

Milder processing procedures were applied to produce spherical SiO_2/epoxy nanocomposites. The results showed that even at higher nano-SiO_2 loadings, good filler dispersion was obtained. The T_g was slightly decreased with the increasing of nano-SiO_2 content. This T_g can be further minimized as nano-SiO_2 loading was increased [80]. Epoxy/SiO_2 nanocomposites exhibited improved T_g and higher flame retardancy than those of pure epoxy system as reported by Lin et al [81]. The incorporation of nano-SiO_2 into a DGEBA epoxy matrix was found to increase its T_g of cold-cured systems by more than 10°C. The study also offered new insights for the events taking place during gelation, vitrification, and curing to the equilibrium state [82].

6.5 COEFFICIENT OF THERMAL EXPANSION (CTE)

Under extreme service conditions, the toughness and the coefficient of thermal expansion (CTE) of polymers play a critical role in tailoring the durability of composite materials. The high-positive CTE in polymers is one of the major shortcomings in applying polymers in wide range of applications where the specimen dimensions must be considered. The use of ceramic fillers is a promising approach to produce isotropic-low CTE polymeric composite materials suitable for electronic circuit board applications.

Low CTE of ceramic particles were introduced into several polymeric matrices for this purpose. The CTE studies showed that the dispersion state of ceramic fillers into the matrix and the nature of interfacial bonding can affect the decrease in the CTE of the composites. TMA is usually a highly-sensitive technique used to determine thermal expansion and contraction of cross-linked or filled polymer micro- and nanocomposite materials [83]. Several compounds were used to prepare epoxy composite materials exhibiting lower CTE. The CTE of the developed epoxy systems reinforced with crystalline and amorphous fillers was evaluated and studied as a function of temperature [84].

The epoxy/SiO_2 nanocomposites suitable for cryogenic applications were developed by the sol–gel process. The average CTE value in the range of temperature from 298 K to 77 K was significantly reduced [85]. This is because SiO_2 nanofillers exhibit much lower CTE than epoxy resins do. The decrease in the CTE by the inclusion of nano-SiO_2 nanoparticles is recommended for cryogenic application. The relationship between the silica loading and the in-plane CTE of the PI matrix and PI/SiO_2 nano-hybrid films was established by Huang et al. [86]. As the nano-SiO_2 was incorporated, the CTE value was reduced, since it was decreased by almost 28.6% from 31.1 ppm for a pure PI to 22.2 ppm for a nanocomposite reinforced with 10 wt.% nano-SiO_2. The CTE was further reduced to 19.2, 16.2, and 14.9 ppm for the nanocomposites loaded with 20, 30, and 40 wt.% nano-SiO_2, respectively.

Badrinarayanan and Kessler [87] explored the potential for development of bulk polymer nanocomposites with appropriate thermal expansivity. The authors added zirconium tungstate (ZrW_2O_8) nanofiller, which has a negative CTE value, into a low-viscosity bisphenol E cyanate ester (BECy) matrix. They found that the dimensional stability of the composites can be increase by reducing the CTE discrepancies between polymer matrix and filler, which generally lead to the matrix micro-cracking. A best strategy to mitigate these residual stresses involves the elaboration of polymer nanocomposites containing well-dispersed nano-ZrW_2O_8 that decrease the extent of disparity in CTE. They confirmed that the addition of up to 10 vol.% ZrW_2O_8 whisker-like nanoparticles resulted in a 20% decrease in the CTE of BECy thermoset.

The influence of zirconium tungstate (ZrW_2O_8) surface-modification on the properties of polyimide nanocomposites reinforced with nano- ZrW_2O_8 was studied by Sharma et al. [88]. These nanocomposites showed a reduced CTE value while exhibiting improved mechanical properties. For example, the CTE of PI was reduced by almost 22% by adding 15 vol.% of nano-ZrW_2O_8. The surface functionalization of the nanofiller improved only the tensile properties. According to Sullivan and Lukehart, increasing the ceramic content generated a controlled decrease in CTE value of polymer/ceramic composite. They found that the CTE value of ZrW_2O_8/PI nanocomposites containing 0 to 50 wt.% ceramic loading reduced gradually with increasing ceramic content. For example, the incorporation of 22 vol.% ZrW_2O_8 loading into the PI matrix resulted in a CTE value as low as 10 ppm/K [89].

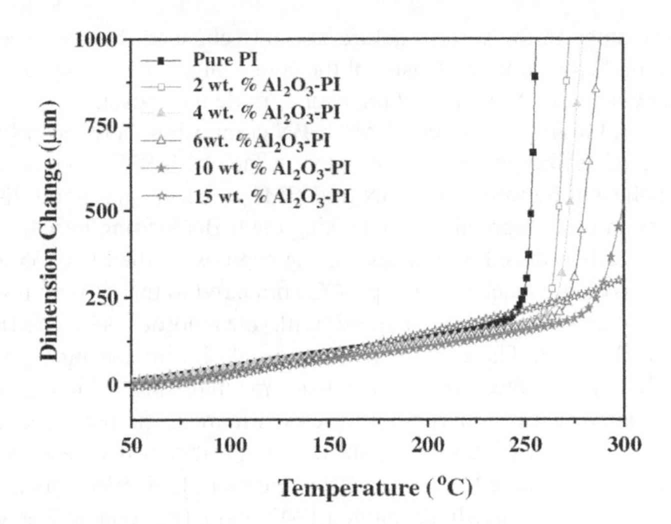

FIGURE 6.7 The Coefficient of Thermal Expansion (CTE) Curves of Pure Polyimide and the Hybrid Films with Various Al_2O_3 Content [90].

Tsai et al. [90] studied the influence of Al_2O_3 loading on the CTE of the PI matrix and the result was illustrated in Figure 6.7. According to their findings the CTE of the PI matrix decreased gradually with the loading Al_2O_3 filler. The 15 wt. % Al_2O_3/PI composite exhibited a CTE of 41 ppm/°C, in the temperature domain of 30 to 240°C, which was 24% smaller than that of an unfilled PI matrix. This substantial reduction in the CTE value was attributed to the low inherent CTE value of ceramic filler. In addition, the complex network structure of Al_2O_3 fillers led to a reduction in the segmental mobility of the PI molecular chains. The low CTE value of PI/Al_2O_3 hybrids can enhance the resulting thermal stress between Si_3N_4 barrier film and PI/Al_2O_3 film, which decreased the separation of Si/Al_2O_3 hybrid.

The effects of MWCNTs@MPSA on CTE of epoxy composite was evaluated by Chung et al. Due to the polar SiO_2 shell, the MWCNTs@MPSA exhibited uniform dispersion into epoxy matrix. By increasing the MWCNTs@MPSA loadings, the CTE value of the composites was gradually reduced due to the dipole–dipole interactions between the SiO_2 and epoxy polymer and to the confinement space of the mesoporous structure, which decreased the thermal mobility of the thermoset within the mesoporous space [91].The addition of 3 wt.% of SiO_2 microfiller along with 3 wt.% of CNF to an epoxy improved damping loss factors by up to 15.6% at room temperature while reducing CTE by almost 15%. It was noted that there is good agreement between CTE experimental data and the predicted ones [92].

A newly organized hexagonal boron nitride (Oh-BN) with significantly increased amounts of amine groups was prepared [93]. The proportion of amine groups for Oh-BN is about 5 times that for original hexagonal boron nitride (h-BN). In the Oh-BN/BD composite with 15 wt.% Oh-BN, its CTEs are about 0.92 times of the corresponding values of hBN/bismaleimide composite. The integrated performance of these composites closely depended on the surface feature of the ceramic fillers because the change in the surface nature not only changed the chemical structure, free volume, and cross-linking density of the composite, but also improved the interfacial nature between BN fillers and the bismalimide thermoset.

A new kind of high-performance BMI/h-BN composites were recently produced. The presence of amino groups on the outer surface of h-BN provides the positive interfacial adhesion between the h-BN and BMI matrices, and good dispersion of filler in the resin was observed. By increasing the h-BN loading into the matrix, the CTE value linearly reduced. In the case of the composite filled with 35 wt.% h-BN, its CTE in glassy state reached 0.63 ppm/°C compared to that of pure resin [94].

New composites of BMI resin filed with mesoporous silica (MPSA) were produced by Hu et al. These composites showed similar curing temperature as pure resin, but they showed different curing mechanisms, which gave different cross-linked networks. Compared with the neat matrix, the produced composites containing a suitable MPSA loading showed a significant decrease in their CTE values, whereas they gained several improvements in their mechanical properties [95]. The CTE values of newly developed PMMA/$PbTiO_3$ composites were determined. It was found that the CTE of the prepared composites can be controlled by tailoring the proportion of polymer to ceramic filler [96]. The SEM analysis revealed that the ceramic filler was homogeneously dispersed into the PMMA matrix if the composite is obtained by the polymerization of the suspension of $PbTiO_3$ in the monomer MMA.

The high CTE of the epoxy composite generally engendered the CTE-mismatch problem in semiconductor packaging. However, when the epoxy resins are used for composite, the CTE-mismatch problem is inevitable even at highly-filled conditions. In the Chun et al. study [97], a functionalized-epoxy was synthesized for the ultra-low CTE epoxy composite containing up to 85 wt.% of SiO_2 loading, which had ultra-low CTE values of 3.2 ppm/°C and 6.0 ppm/°C at the temperature ranges of $T < T_g$ and $T > T_g$, respectively. In comparison, a conventional epoxy/SiO_2 composite was prepared under the same conditions, and it exhibited higher CTE values of 8 ppm/°C and 40 ppm/°C at the temperature ranges of $T < T_g$ and $T > T_g$, respectively. The ZrW_2O_8/epoxy nanocomposites were produced, and their thermal expansion data were measured at the temperature range of 4 to 300 K. Relative to unfilled epoxy matrix, the prepared composites exhibited lower CTE values [98].

The effect of morphology and thermal expansivity of ZrW_2O_8 nanoparticles on the rheological, thermo-mechanical, dynamic-mechanical, and dielectric properties of ZrW_2O_8/epoxy nanocomposites was studied by Wu et al [99]. As shown in Figure 6.8a, the pure epoxy resin had a CTE of 87.8 ppm/°C. The incorporation of 10 vol.% ZrW_2O_8 decreased the CTE values of the nanocomposites by different percentages according to the types of the used nanoparticles. For instance, the Type-1, Type-2, and Type-3 nano-ZrW_2O_8 decreased the CTE of epoxy resin by almost 18%, 17%, and 11%, respectively. At 20 vol.% nanofillers loading the reduction in CTE was more relevant. Since, the Type-1 nano-ZrW_2O_8 showed a decrease in the CTE by 29%, whereas Type-2 and Type-3 nanoparticles decreased it by 27% and 23%, respectively. As illustrated in Figure 6.8b, in the rubbery domain, the CTE of the ZrW_2O_8/epoxy nanocomposites reduced with enhancing the nano-filler loadings for all three types of nanofillers. Nanocomposites filled with Type-1 ZrW_2O_8 had the

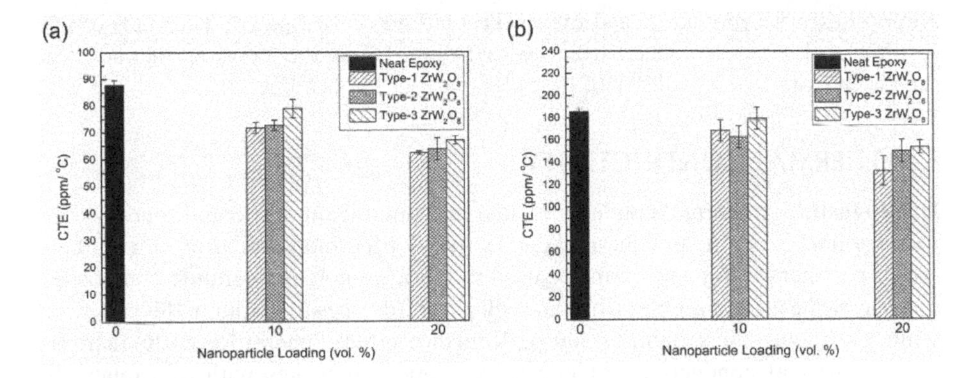

FIGURE 6.8 Comparison of the CTE Values of Neat Epoxy Resin and ZrW_2O_8/epoxy Nanocomposites in the (a) Glassy Region and (b) Rubbery Region. The Error Bars Represent Standard Deviations, Based on Five Samples. Type-1(Large, Bundle-like Rods); Type-2 (Large, Rectangular Rods); Type-3 (Small, Capsule-like, Short Rods) [99].

most pronounced CTE decrease with 29% at a 20 vol.% ZrW_2O_8 content, followed by those filled with Type-2 and Type-3 nanofillers which showed CTE reductions of 21% and 19%, respectively.

Komarov et al. [100] reported that the predicted CTE data using *molecular dynamics simulations* for PI/SiO_2 composite showed good agreement with the experimental data. The CTE of the composite model reduced with decreasing the SiO_2 ratio in agreement with many experimental studies. The results also showed that there is a threshold for the nano-SiO_2 beyond which the material model exhibited ultra-low CTE values. The addition of up to 10 vol.% whisker-like ZrW_2O_8 nanoparticles provoked a decrease of 20% in the CTE of the cyanate ester resin. However, at high-filler loadings, the nanoparticles exert a dramatic catalytic effect on the cure reaction of the matrix and subsequently decrease the onset temperature of the T_g for the cured polymer network [101].

The 32- to 40-nm barium strontium titanate (BST) ceramic fillers were formulated as $Ba_{0.77}Sr_{0.23}TiO_3$ using the sol–gel technique and used to reinforce PI matrix. It was found that the BST nanoparticles sized were dispersed homogeneously in the polyimide matrix without aggregation. The CTE of PI/BST containing 33 vol.% of BST was tested at the range of 36°C and 600°C. CTE values of the prepared nanocomposites showed a decrease of about 80% in the CTE relative to that of the unfilled PI matrix [102]. The change in the thermal expansion behavior of a cyanate ester/SiO_2 was studied by thermomechanical analysis. All nanocomposites exhibited reduced CTE values compared to that of pure matrix. However, at a fixed nano-SiO_2-loading, 12-nm nano-SiO_2-based nanocomposites had lower CTE values than those prepared from 40-nm nano-SiO_2 [103].

Zirconium tungstate (ZrW_2O_8) is a unique ceramic filler showing an isotropic-negative thermal expansion behavior over a wide range of temperature [104]. The addition of ZrW_2O_8 to polymeric matrices is expected to enhance their dimensional stabilities by decreasing the overall CTE values. Thermomechanical analysis revealed that the inclusion of ZrW_2O_8 engendered a reduction in the CTE value of cyanate ester at temperatures above and below its T_g. Phenolic resin/ZrW_2O_8

composites were produced, and their CTE data was investigated. The CTE of the resulting composites reduced from 46 to 14 ppm/°C as the ZrW_2O_8 volume ratio increased from 0 to 52 vol.% [105].

6.6 THERMAL CONDUCTIVITY

Due to their high thermal conductivity and insulation features, ceramics are widely used for improving the low thermal conductivity of thermoplastic, rubber, and thermoset polymers. They are normally used to fabricate polymer/ceramic composites to dissipate heat from high-performance electronic devices to maintain their lifetime while enhancing the reliability and performance of equipment. Recently, increasing the thermal conductivity of polymer/ceramic composites with at a relatively low fraction of fillers while flowing heat in a desired direction has similarly been explored [106]. In part, we summarized thermal features some ceramics that are normally used for high thermal conductivity, alignment methods of anisotropic ceramic fillers in the polymer matrix. As most research so far has been focused on enhancing heat spreading through a single in-plane direction, realization of vertical alignment formation for vertical heat conduction is very limited.

On the other hand, thermal conductivity of polymer composites in the vertical direction is still much lower than that recorded for the in-plane direction. Due to their higher thermal conductivities, many ceramics types, such as aluminum nitride (AlN), alumina (Al_2O_3), silicon carbide (SiC), silicon nitride (Si_3N_4), and boron nitride (BN), have been widely used as reinforcing agents for polymers to produce high-performance composites [107]. As displayed in Table 6.2, these inorganic fillers overall exhibit superior thermal conductivity that ranged between 30 and 390 W/m.K, low dielectric constant at 1MHz in the range of 4.5 to 9, and high electrical resistivity that exceeded 10^{13} W.cm. It was reported that the AlN ceramic filler is not chemically stable in the presence of moisture in the air since it can be easily decomposed into $Al(OH)_3$ and NH_3, which can harm the electrodes and devices. Thus, the industry is rarely uses composite materials of AlN nano-ceramics.

TABLE 6.2
Summary of Thermal Conductivity, Dielectric Constant, and Electrical Resistivity of Several Ceramics

Ceramics		Thermal Conductivity [W/m.K]	Dielectric Constant (at 1 MHz)	Electrical Resistivity [Ω.cm]
AIN	Spherical	200–320	8.5–8.9	$>10^{14}$
Al_2O_3	Spherical	30–42	6.0–9.0	$>10^{14}$
	Spherical	85–390	–	-
SiC	Nanowire	90	40	-
Si_3N_4	Spherical	86–155	8.3	$>10^{13}$
	Nanosheet	29–600	4.5	$>10^{13}$
BN	Nanotube	200–300	-	-

In contrast, Al_2O_3 exhibits good chemical stability and relatively low cost, but it still has lower thermal conductivity ranges between 30 and 42 W/m.K than other ceramics, like AlN or BN ceramics. SiC is a potential candidate for high-temperature applications due to its high-thermal conductivity and stability. However, it has limitations for some applications especially in highly integrated devices because it exhibits a high dielectric constant of nearly 40 at 1 MHz compared to other ceramics (4.5–9) at the same frequency.

As shown in Table 6.2, the Si_3N_4 ceramic filler has a thermal conductivity lower than those of AlN or BN. But, it has a higher chemical stability, improved resistance, and low cost compared to those of AlN and BN. On the other hand, BN nanoparticles show an anisotropic heat flow thanks to its honeycomb molecular structure. When the BN nanosheets are aligned in-parallel to the c-axis, the thermal conductivity is nearly 20 times (600 W/m.K) greater than that (30 W/m.K) when BN nanosheets are perpendicularly aligned to the c-axis. Therefore, special procedures are required for producing BN nanocomposites to attain high-thermal conductivity.

A highly thermally-conductive polymer nanocomposite filled with a nano-porous α-Al_2O_3 sheet was prepared. The thermal conductivity of these nanocomposites along the radial surface reached 12 W/m.K at 41 vol.% Al_2O_3, due to the formation of continuous interconnected thermally-conductive chains. The results confirmed the important role of structure of the filler for designing of thermally-conductive materials [108]. Comparatively, the through-plane thermal conductivity of nanocomposites prepared from a 1,200°C-anealed Al_2O_3 filler reached only 1.13 W/m.K at 39 vol.% Al_2O_3 loading. By adding annealed Al_2O_3 ceramic, low-refractive index mismatch is resulted between matrix and this ceramic filler by forming nano-optical fibers through the nanocomposite membrane. This study showed the potential of such nanocomposites in "heat dissipation" of lightening devices [109].

Highly thermally-conductive PI/BN-nanosheets (BNNSs)-based composites films with BNNSs filler were produced by solution blending process. The in-plane thermal conductivity of PI nanocomposites filled with 7 wt.% BNNSs attained up to 2.95 W/m.K, which increased by almost 1080% compared to that of the neat PI matrix. In contrast, the out-of-plane thermal conductivity of these composites reached only 0.44 W/m.K, which represents an enhancement of 76%. The alignment of the BNNSs can determine the increase in the thermal conductivity [110].

Ramdani et al. prepared new thermally-conductive polybenzoxazine/KH550-teated Si_3N_4 nanocomposite, since by adding 0 to 70 wt.% nano-Si_3N_4, thermal conductivities of the resulted nanocomposites increased from 0.18 to 5.78 W/m.K due to the formation of continuous conductive chains within the polymeric matrix [111] (Figure 6.9). The thermal conductivity of unsaturated polyester resin UPE/Al_2O_3 nanocomposites was evaluated by Moreira et al. An augmentation of more than 100% was recorded for all the loaded systems at maximum loading compared to that of the unfilled resin, while there was insignificant dependency between the thermal conductivity and temperature [112]. Improved thermal conductivities were recorded for UPE/Al_2O_3 composites [113].

The heat conduction of TiB_2-Al_2O_3/epoxy composites was discussed by Wu et al. results showed that TiB_2 had an optimistic potential in improving the thermal conductivity of epoxy matrix. By including TiB_2-Al_2O_3 binary filler system, the thermal

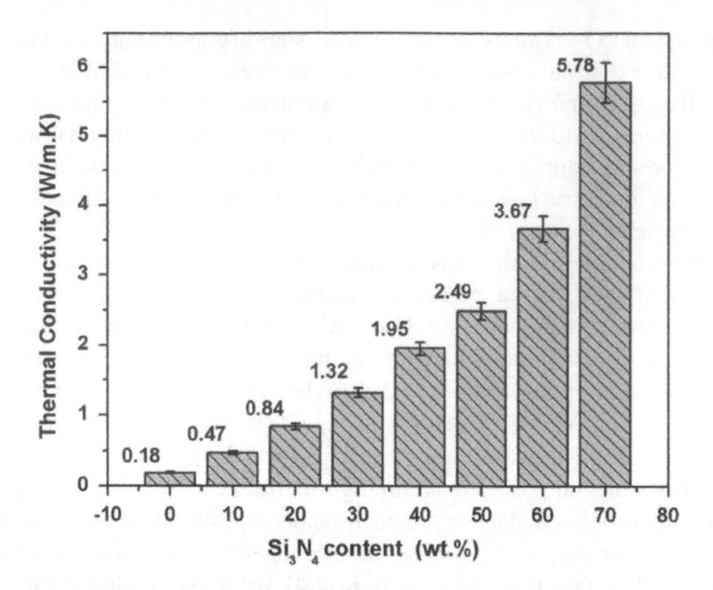

FIGURE 6.9 Thermal Conductivity of the Cured Nanocomposites at Various Contents of Nano-Si_3N_4. Reprinted from Ref. [111]. Copyright (2015), with permission from Elsevier.

conductivity of the composites was markedly enhanced at higher fillers ratios. The surface-treatment of ceramic fillers by silane molecules was further increased the heat diffusion in these composites [114].

An investigation conducted by Dai et al. showed that as-prepared polyimide composites with SiC nanowires grown on graphene hybrid fillers (PI/SiC_{NWs}-GSs) films exhibited synchronously relatively good thermal conductivity. Moreover, the heat-conducting properties of PI/SiC_{NWs}-GSs films perform a good reproducible in the temperature range from 25 to 175°C. The maximum value of thermal conductivity of PI composite reached 0.577 W/m. K at 7 wt.% fillers loading, presenting an increase of 138% in comparison to that of pure PI matrix [115].

Heo et al. found that the thermal conductivities of 80 wt.%-Al_2O_3/epoxy composites containing 5 wt.% GO and $Al(OH)_3$-coated GO attained 3.5 and 3.1 W/m.K, respectively. On the other hand, the $Al(OH)_3$-coated GO/Al_2O_3/epoxy composites exhibited the more retained electrical resistivity compared with GO/Al_2O_3/epoxy composite. Thus, the $Al(OH)_3$-coated GO composites showed simultaneously improvements in the thermal conductivity and retention of electrical resistivity [116].

Gao et al. studied the effects of adding several ceramic fillers on the thermal properties of epoxy resin, including Al_2O_3 micro-particles, which maximally enhanced the thermal conductivity and thermal stability of epoxy matrix at 45 wt.% micro-Al_2O_3 particles content. In these composites, the thermal conductivity was controlled by the inherent properties of Al_2O_3 fillers and the formation of conductive-paths within the matrix [117].

Effect of backbone moiety of liquid crystalline epoxy (LCE) resin on thermal conductivity of epoxy/Al_2O_3 composite was investigated by Giang et al. It was found that the thermal conductivity of LCE/Al_2O_3 composite strongly depended on the backbone structure of LCE, where, higher thermal conductivity was obtained as the ordering in LCE backbone increased [118].

The non-equilibrium molecular dynamics simulation was performed to obtain the effective thermal conductivity on composites of epoxy/α-Al_2O_3 fillers, which generally applied as a heat dissipation material. Effects of surface-coupling agent were also scrutinized by adding model molecules to the polymer subsystem [119]. ZrB_2/Al_2O_3 hybrid fillers were functionalized with epoxide modified with GPTS functionalized via the covalent bonding method to improve the interfacial compatibility of hybrids in epoxy matrix. The thermal conductivities of functionalized composites containing 3 vol.% and 5 vol.% ZrB_2/Al_2O_3 loadings were enhanced by 8.3% and 12.5% compared to those of unmodified composites, respectively [120].

The soft SiO_2 intermediate nanolayer on Ag nanowires (Ag_{NWs}) alleviated the mismatch between Ag_{NWs} and epoxy, increased their interfacial interaction, and enhanced the thermal conductivity of an epoxy/Ag_{NWs}@SiO_2 composites. In addition, the insulating SiO_2 nanolayer inhibited the formation of electrically-conductive Ag_{NWs} networks within the matrix, leading to better electrical insulation and the dielectric properties of the composite [121]. Dang et al. used functionalized β-SiC whiskers (β-SiCw) to elaborate β-SiCw/BMI composites by powder blending-casting process. The thermal conductivity of the β-SiCw/BMI composites reached nearly 1 W/m.K at 40 wt.% of treated-β-SiCw, which was five times higher than that of pure matrix [122].

Hong et al. used mixtures of thermally-conductive AlN and BN ceramic fillers to scrutinize the effects of the particle size and the relative composition on the thermal conductivity of epoxy matrix. Optimal thermal conductive chains were significantly affected by the packing efficiency and interfacial resistance of these particles, and the highest thermal conductivity was attained up to 8 W/m.K in the 1:1 volume ratio of AlN:BN fillers. The optimization of contact resistance and contact areas of these fillers was done to maximize the thermal conductivity [123]. Agrawal et al. evaluated the thermal properties of the epoxy filled with AlN ceramic filler, suitable for electronic packaging and printed circuit boards. By increasing the AlN loading, thermal conductivity and T_g of epoxy were gradually increased [124].

Agrawal et al. stated that increasing AlN concentrations in the epoxy matrix increased the thermal conductivity of their composite, while the rate of enhancement was intensified at higher volume fraction up to 35 vol.% AlN. Such epoxy/AlN composites are more suitable for some applications, including electronic packages, communication device, thermal grease, thermal interface material and electric cable insulation [125]. Choi et al. used both AlN at different sizes alone and in combination with Al_2O_3 to elaborate new thermally-conductive epoxy composites. Using these hybrid fillers was reported to be more effective for enhancing the thermal conductivity of the matrix due to the increased connectivity formed by the binary filler [126].

The thermal conductivity of an AlN/epoxy resin was evaluated as a function of the fillers content. AlN filled-epoxy composite exhibited thermal conductivity as high as 7.15 W/m.K at 68.5 vol.%. The improvements were controlled by the AlN particle size and the number of resin layers on the surface of an AlN particle. The presence of Al_2O_3 as impurity in AlN filler decreased the thermal conductivity of composites [127].

In another study by Ma et al., MWCNTs/AlN hybrid fillers functionalize by KH-560 silane was used to reinforce epoxy resin by casting process. The thermal conductivity of epoxy matrix was markedly increased by filling with MWCNTs/AlN hybrid filler to reach 1.04 W/m.K at 29 wt.% MWCNTs/AlN hybrid filler (4 wt.% MWCNTs +25 wt.% AlN). The thermal stability of MWCNTs/AlN/epoxy composite enhanced with increasing the hybrid filler contents, and surface-treatment by KH-560 had a positive effect on the thermal conductivity of epoxy composite [128].

Three different interface structures of AlN, γ-aminopropyl-triethoxysilane treated-AlN (γ-APS), and AlN- hyperbranched aromatic polyamide (HBP) were used as reinforcing agents for epoxy resin. AlN-HBP nanoparticles showed a strong interface and thus the incorporation of the AlN-HBP nanoparticles enhanced both the dispersion of the AlN nanoparticles in the matrix and increased the thermal conductivity [129]. Agrawal et al. developed epoxy/AlN micro-composites for microelectronics applications, at the filler concentration range of 0 to 25 vol.%. The incorporation of AlN in epoxy thermoset increased the thermal conductivity whereas it favorably decreased the CTE of the composite [130].

KH560 treated-AlN micro-particles were used to reinforce EP matrix. At lower AlN loadings, the thermal conductivity of the composites increased with increasing AlN loadings. The λ of AlN/EP composites improved with the addition of AlN to reach 0.98 W/m.K at 70 wt.% treated AlN. For a fixed AlN fraction, the surface-treatment of AlN had an extra improvement of thermal conductivity of these composites [131]. Thermally-conductive PF/SiC composites were produced in the presence of graphene nanoplatelets (G_{np}) to further improve their thermal-conduction properties. In the composite containing 50 vol.% filler, its thermal conductivity increased to 5.5 W/m.K due to a synergetic effect between the SiC filler and the binder induced by the graphene nanoplatelets [132].

Compact α-Si_3N_4/PF composites exhibiting high thermal conductivity were produced. The enhancements in thermal conductivity were dependent on the good compatibility and strong interfacial adhesion between compact α-Si_3N_4 nano-spheres and PF thermoset, as large amount of α-Si_3N_4 nano-spheres were uniformly disperse into the thermosets forming thermally-conductive networks, which accelerate heat diffusion process [133].

A solvent-free method for the fabrication of thermally-conductive EP/BN nanocomposite was developed by Wang et al. The maximum λ reached 5.24 W/m.K, which represented 1,600% improvement in comparison to that of pristine EP matrix [134].

A relevant study on several factors influencing the thermal conductivity of EP/BN composites were investigated by Donnay et al. The thermal conductivity of the pure resin significantly increased by only adding 20 w.% of BN-concentrated GPTMS had a significant positive effect on the thermal conductivity of the composites while ball milling was an efficient method to break the large aggregates of BN powder [135].

Raza et al. developed a thermally-conducting but electrically-insulating epoxy composite reinforced by hybrids of BN and vapor-grown carbon nanofiber (VGCNF) using by 3-roll milling. The thermal conductivity of the prepared epoxy hybrid composites was enhanced with increasing the VGCNF and BN contents. SEM revealed that BN inhibited the VGCNF contacts, which further increased the electrically-insulating properties of these composites. The thermal contact resistance of 6 wt.% BN/8wt.%VGCNF/rubbery hybrid reached 3.36 10^5 m^2.K/W at a bond-line thickness of 18 mm [136].

Control of the BN anisotropy was dispersed in the epoxy without damaging the composite films or requiring any surface modification of the BN. The degree of BN orientation perpendicular to the nanocomposite film plane, which was parallel to the electric flux, could be controlled by applying the nanosecond pulse for different lengths of time before cross-linking. The resulting composite films with oriented BN nanosheets manifested improved thermal diffusivity compared to a composite prepared without orientation [137].

The h-BN nano-sheet is a novel high thermally-conductive and high aspect-ratio filler that has the potential to substantially enhance the thermal conductivity of polymer composites. The addition of h-BN nano-sheet to epoxy resin demonstrated an enhanced thermal conductivity of their resulted composite. Lin et al. found that a significant increase of thermal conductivity of epoxy resin, when filled with a 5 wt.% of exfoliated h-BN, represents an increase of almost 113%. This effect was insignificant at high h-BN filler contents due to the large thermal boundary resistance. The prepared h-BN nano-sheet/epoxy nanocomposite also reduced the thermal resistance between Cu–Si interfaces [138].

BN was treated with different surface-modifiers to elaborate thermally-conductive epoxy composites to allow high BN filler loadings as reported by Kim et al. The functionalized BN filler tailored the thermal properties of the polymeric matrix by reducing its free energy. Surface-curing agents interrupted the interaction between the filler and matrix, and did not always enhanced thermal conductivity, while the maximum thermal conductivity was recorded at 30 wt.% fraction [139].

Kim et al. investigated the thermal conductivities of epoxy-terminated dimethylsiloxane (ETDS) filled with two 3-glycidoxypropyltrimethoxysilane (KBM-403)-treated-BN and 3-chloropropyltrimethoxysilane (KBM-703)-treated-BN fillers by a solvent blending method. The thermal conductivities of the composites containing 70 wt.% BN fraction functionalized with KBM-403 and KBM-703 reached 4.11 and 3.88 W/m.K, respectively, compared to 2.92 W/m.K for the composite reinforced with raw BN due to the formation of conductive pathways was maximized while minimizing the thermal interface resistance along the heat-flow path [140].

Teng et al. studied the synergistic effect on thermally conductivity of an EP matrix reinforced with combined system filler of MWCNTs and BN flakes, which were functionalized to form covalent bonds. The hybrid filler imparted a marked increase of thermal conductivity, since by adding 30 vol.% of modified-BN and 1 vol.% functionalized-MWCNTs resulted in an enhancement of 743% in thermal conductivity compared to that of neat epoxy [141]. Wattanakul et al. reported that the adsorption of the surfactants changed the property of the BN surface from partial ionic to a hydrophobic surface with better wetting by the epoxy resin. This change has led to an increase in the thermal conductivity and mechanical properties of the surfactant-treated BN-epoxy composites [142].

Cho et al. demonstrated that the orientation of untreated BN nanosheets can be controlled in epoxy matrix either perpendicular or parallel to the nanocomposite film surface with high anisotropy triggered by a rotating superconducting magnetic field.

The resulting polymer nanocomposite had outstanding thermal anisotropy, and the increased thermal diffusivity was proportional to the anisotropic orientation of the BN nanosheets [143].

Fu et al. measured the thermal conductivity of different kinds of thermal adhesives filling the epoxy resin, including BN, which improved the thermal conductivity of the epoxy matrix. The layer-shape filler was more favorable than the ball-shape filler and the sharp-corner-shape filler in enhancing the thermal conductivity of thermoset [144]. The thermal conductivity of the liquid-crystalline epoxy composites was significantly enhanced even at lower volume fraction of BN due to the formation of thermally-conductive paths by the BN filler through exclusion of the BN filler from the epoxy matrix during the curing process [145].

The increase in thermal conductivity of BN-filled epoxy composites was related to the decrease in the filler/matrix thermal contact resistance by improving the matrix/particles interface after surface modifications by acetone, acids (nitric and sulfuric), and silane, which had the most relevant impact especially for a larger BN volume fraction. For example, at 57 vol.% BN, the thermal conductivity reached 10.3 W/m.K. Silane-treatments also was effectively change the thermal conductivity AlN/EP which reached 11 W/m.K at 60 vol.% AlN [146].

Zeng et al. developed new thermal-conductive epoxy composites reinforced by Si_3N_4 and SiO_2 multi-fillers. Si_3N_4 particles were most effective in improving the thermal conductivity and reduced cost compared to SiO_2, since a thermal conductivity of 2.51 W/m.K was obtained for 80 vol.% of Si_3N_4/EP. At a fixed Si_3N_4 content, the thermal conductivity increased for a hybrid Si_3N_4/SiO_2/EP with increasing nano-SiO_2 contents [147].

Shi et al. produced epoxy/Si_3N_4 composites suitable for electronic packaging application, where the Si_3N_4 filler has a great influence on thermal conductivity, which markedly increased for Si_3N_4 > vol.40%, whereas the $Al(OH)_3$ filler greatly improved the flame resistance of the composites. The simultaneous presence of both Si_3N_4 and $Al(OH)_3$ fillers in the composites positively affected both properties. At 60 vol.% and a volume ratio of Si_3N_4:$Al(OH)_3$ of 3 exhibited a thermal conductivity of 2.15 W/m.K [148].

To improve the thermal conductivity of epoxy resins without losing their inherent insulation features, epoxy/β-Si_3N_4 nanowire composites were produced by Kusunose et al. The epoxy composite filled with 60 vol.% β-Si_3N_4 nanowires had a thermal conductivity as high as 9.2 W/m.K along the preferred orientation of the nanowires [149]. Zhu et al. observed and modeled the pre- and post-percolation transition in EP/β-Si_3N_4 heterogeneous composites. The surface modification was found capable of increasing the filler dispersion. The competition between the bridging effect and the interface thermal resistance was deduced as the main cause of the turning point in the thermal conductivity [150].

SiC/epoxy composites were prepared by a wetting method. The SiC particles were treated to increase wettability of the epoxy resin and improved the dispersion of fillers via a secondary matrix/ceramic interaction. The maximum thermal transport properties were recorded at 70 wt.% of SiC [151]. The AlN- and BN-hybrid ceramic-filler reinforced-composite systems were designed to identify the effect of different size and relative composition on thermal-conducting paths. The highest thermal conductivity was obtained at a composition ratio of AlN to BN of 1:1 with similar particle sizes. Results also showed that any change in the relative composition of AlN to BN can affect the interfacial thermal resistance and inhibit the formation of conductive networks. Moreover, the influence of the relative filler dimension on thermally-conductive paths can be defined using particle-size ratio (R_D) of the AlN and BN ceramics, and can also be schematically expressed by the bimodal distribution curves. As the bimodal distribution become a continuous valley ($R_D \approx 1$), this leads to the formation of increased conducting networks and the contact area is optimized, engendering high thermal conductivity of these composites [152].

h-BN@GO hybrids were prepared by electrostatic self-assembly of h-BN and graphene oxide (GO), and their epoxy composites were fabricated by a solution-free curing process. The thermal conductivity of the h-BN@GO composites reaches 2.23 W/m.K, at 40 wt.% hybrid filler content, which is approximately two times larger than that of the composites filled with only h-BN at the same loading. This enhancement was attributed to the existence of GO, which improves the compatibility of h-BN with epoxy resin [153].

Novel poly(cyclotriphosphazene-co-bisphenol A)-coated boron nitride (PCB-BN) was designed and synthesized by in situ condensation polymerization on the surfaces of boron nitride (BN) particles based on the reaction of hexachlorocyclotriphosphazene (HCCP) with bisphenol A (BPA). Incorporating PCB-BN particles in the epoxy matrix markedly increased thermal conductivity of the epoxy matrix. By adding 20 wt.% PCB-BN, the thermal conductivity of the EP/PCB-BN composites was 0.708 W/m.K, which was 3.7 times higher than that of the unfilled epoxy [154].

The γ-glycidoxy propyl trimethoxy silane/polyhedral oligomeric silsesquioxane (KH-560/POSS) grafted nano-sized boron nitride (POSS-g-BN) were used to fill BMI matrix. The BMI nanocomposite reinforced with 15.4 vol.% POSS-g-nBN showed improved thermal conductivity of 0.607 W/m.K, which represents an increase of 266% relative to that of neat BMI matrix [155]. The modified h-BN by 3-glycidyloxypropyltrimethoxy silane (γ-MPS-BN) micro-particles were used to reinforce PI matrix where an enhanced thermal conductivity was resulted. The composite filled with 40 wt.% γ-MPS-BN ratio exhibited a thermal conductivity of 0.748 W/m.K, which is 4.6 times higher than that of neat matrix [156]. 3-aminopropyl triethoxy silane (APTES) modified h-BN were used to develop thermally-conductive epoxy/APTES-BN composites. At 30 wt.% APTES-BN loading, the thermal conductivity of the composites reached 1.178 W/m.K, which is 6.14 times higher than that of the neat resin [157].

Fang et al. [158] designed a nano–micro structure composed from hyperbranched aromatic polyamide (HBP) treated both 2-D h-BN micro-particles and 0-D α-Al_2O_3 nanoparticles hybrid systems for preparing epoxy composites exhibiting high thermal conductivity and breakdown strength. Importantly, the nano–micro structure provided noticeable synergistic effects on both thermal conductivity and breakdown strength. The composite specimen containing 26.5 vol.% of fillers showed a high thermal conductivity of 0.808 W/m.K (4.3 times that of epoxy). The thermal conductivity of epoxy was enhanced by up to 69% at only 5 wt.% boron nitride nanotube (BNNTs) [159].

The thermal conductivity of the composites containing 60 wt.% of KH550-BN BN attained 1.052 W/m.K, which is five times higher than that of virgin matrix

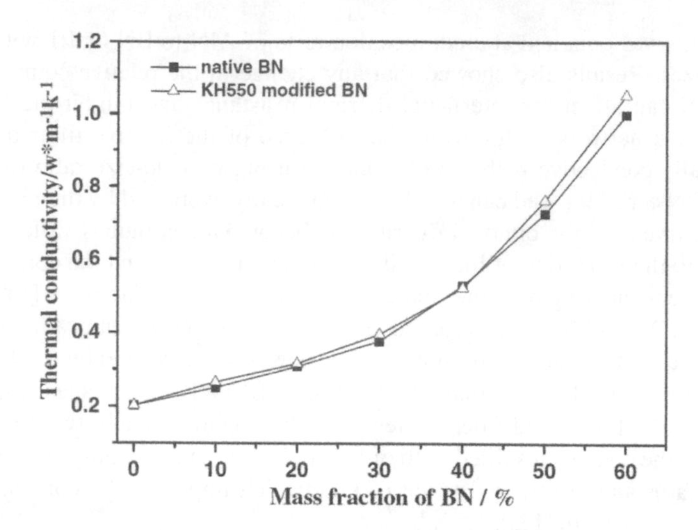

FIGURE 6.10 Effects of BN Content on the Thermal Conductivity of the BN/EP Composites [160].

(0.202 W/m.K). For a fixed-BN loading, the surface-treatment of BN by KH550 exhibited a positive effect on the thermal conductivity of such composites as shown in Figure 6.10 [160]. Compared to the surface-treatment of BN particles, the addition of maleic anhydride grafted high-density polyethylene (PE-g-MAH) was more effective to enhance the thermal conductivity of the polyethylene/BN composites [161]. PU/BN nanocomposites were elaborated at different weight ratios of the alkoxysilane-coated BN using shear mixing and ultrasonication. Thermal conductivity of nanocomposite at 5 wt.% indicated that raw BN nanoparticles increased the thermal conductivity by almost 50%, while the silane-treated BN particles enhanced it by only 20% [162].

6.7 THERMOMECHANICAL PROPERTIES

Dynamic mechanical analysis (DMA) over a wide range of temperatures and frequencies allows the determination of the viscoelastic behavior of polymeric materials. The stiffness and damping characteristics of polymer/ceramic composites can be analyzed using DMA which could be influenced by their morphologies. The dynamic mechanical properties analysis is a suitable method for determining the polymeric composites transition temperatures that can affect the fatigue and impact properties of the composite. The use of various polymer/ceramic combination to tailor and enhance the dynamic mechanical performance of these composites has been widely studied in recent years [163].

The thermomechanical properties of polymer/ceramic composites were significantly influenced by their polymer/filler wetting phenomena. Bansal et al. found that low molecular weight (MW) PS melts with lengths <880 wet the graphed nano-SiO_2 particles, which improved the viscoelastic properties of these nanocomposite increases [164]. Yuan et al. found that the storage modulus of epoxy/SiC nanocomposites enhanced with the addition of treated-SiC whiskers to the matrix. For example, at the glassy state, generally taken at 50°C, the unfiled epoxy and the filled nanocomposites 10 wt.% and 20 wt.% SiC whisker-filled epoxy exhibited storage moduli of 1825, 2356, and 2852 MPa, respectively [165].

The PEEK/stearic acid treated-SiO_2 nanocomposites were produced by the compression-molding technique. The TMA tests on PEEK/stearic acid treated-SiO_2 hybrids on showed that then dynamic modulus of the PEEK nanocomposites increased by almost 40% at elevated the temperature range of 100-250°C [166]. In PI/SiO_2, nanocomposites revealed that interphase chemical links between the SiO_2 network and the polyimide chain avoid the agglomeration of nanoparticles, reducing the particle size to nano-level thus resulting in improved distribution and better homogenous in the matrix. The improved interfacial interaction between the phases improved the thermomechanical particularly at high temperatures [167].

The fumed micro- and nano-SiO_2 originated form the silicon and ferrosilicon industry were used to fill epoxy thermoset at different filler content. Up to 30 wt.%, the composite behavior was explained as a predominantly agglomeration effect. For 30 wt.% and higher filler loadings, single particles seem to play a more important role [168].

The effect of the particle size, loading, and size distribution of AlN on the thermomechanical properties of the composites were studied. There was a total change trend for T_g of epoxy/AlN nanocomposites was lower than that of that of micro-composites, especially at high nano-AlN ratios. The cross-link density of the epoxy matrix reduced for epoxy/AlN nanocomposites. The DMA analysis revealed that E' and E'' increased with enhancing the AlN content or rising nano-AlN fraction at a fixed AlN fraction due to the increase in the interfacial areas between AlN and the epoxy matrix (Figure 6.11) [169].

A new series of polylactic acid/SiO_2 nanocomposites have been prepared by a melt-mixing procedure. The storage modulus is always higher for PLA/nanocomposites compared to that of pure polylactic acid, while the reinforcing effect is higher for PLA/SiO_2 nanocomposites probably due to the finer dispersion of nanoparticles. The evolution of storage modulus with aging time is monotonically decreasing because different type of applied deformation, showing to be different than that of the Young's modulus [170].

Hybrid silica/poly(butyl acrylate-methyl methacrylate-acrylic acid) (SiO_2/P(BA-MMA-AA)) were synthesized via in situ semi-batch emulsion polymerization. The results showed that this process was produced with high monomer conversion and low formation of agglomerates. The nanocomposite films display significantly improved thermomechanical properties over its pure polymer film, and also presents almost the same high transparency [171]. Vassileva et al. observed that the addition of Al_2O_3 to the epoxy thermoset enhanced the moduli of nanocomposites to higher values in the glassy state than the rubbery region, and broadened the damping peaks, which were slightly shifted to higher temperatures [172]. The epoxy–TiO_2-nanocomposite was manufactured by an ultrasonic cavitation's process and studied in detail by Chatterjee and Islam. The addition of nano-TiO_2 increased the storage

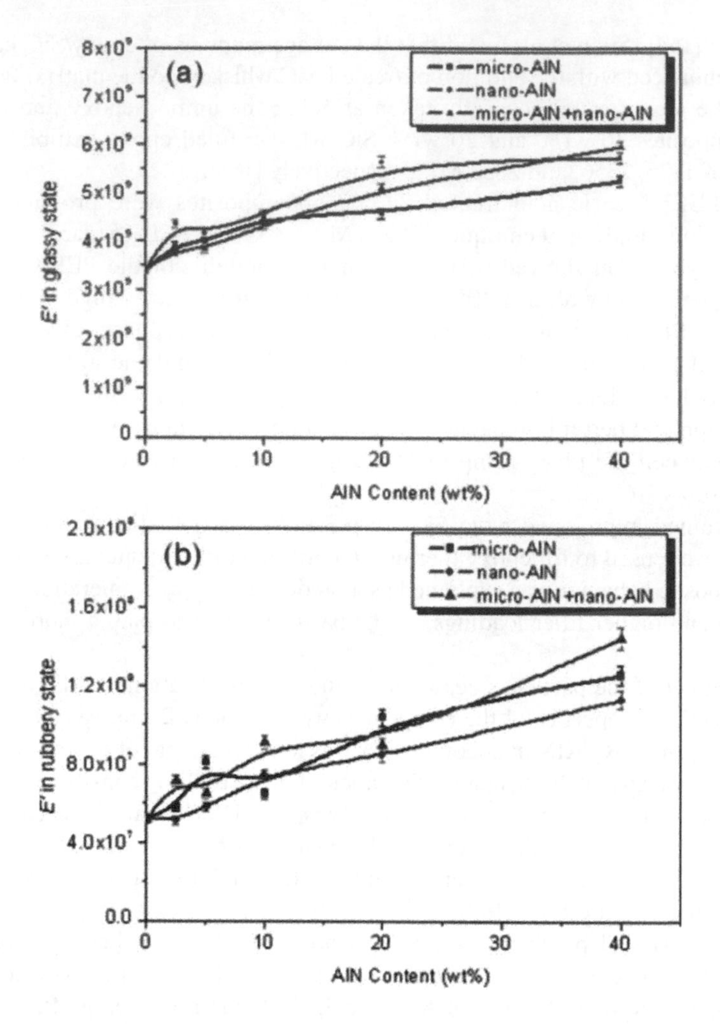

FIGURE 6.11 *E'* in the (a) Glassy and (b) Rubbery Regions of the Composites as a Function of AlN Content [169].

modulus and the T_g the pure epoxy matrix, while the extent of improvement was related to the dispersion state of nanofillers (Figure 6.12) [173].

Duraibabu et al. observed significant improvements in the thermomechanical properties of functionalized nano-Al_2O_3 reinforced 1,4′-bis (4-amine-phenoxy) sulphone benzene epoxy resin (TGBAPSB) nanocomposites, which appear to be ideal material for advanced high-performance applications when compared to those of neat epoxy matrices [174]. Morselli et al. observed that the addition of TiO_2 in epoxy matrices significantly enhanced both T_g and modulus (in the rubbery region) due to both the presence of inorganic nanofillers and, most importantly, a higher cross-linking density of the composite material with respect to the pristine epoxy matrix [175]. The dynamics in DGEBA epoxy/Al_2O_3 nanocomposites was studied using various techniques as reported by Kyritsis et al. and established a

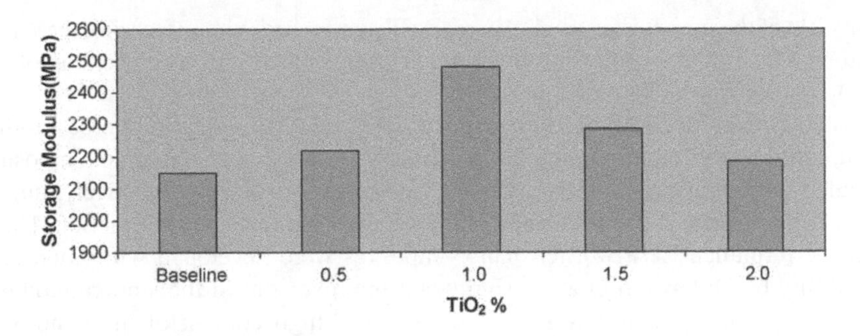

FIGURE 6.12 Variation of Storage Modulus with % of Loading of the TiO_2 [173].

correlation between viscosity (of the nano-dispersions), storage modulus, and T_g of the nanocomposites [176].

The effect of adding nano-Al_2O_3 particles having fiber-like and spherical shapes on the thermomechanical properties of an epoxy resin was explored by Johnsen et al. DMA tests revealed that significant improvements in the stiffness of the epoxy matrix was obtained at only small nano-Al_2O_3 ratios, even when agglomerates presented [177]. EP/SiO_2 composite films based on DGEBA/DDM type epoxy were produced by sol–gel processes exhibiting superior dynamic storage modulus and higher T_gs. The effects were more intense in the presence of APTES as it had the role of binding agent, which enhanced interphase compatibility and allowed better dispersion of SiO_2 in the matrix [178]. Highly-filled micro-composites based on PBZ/Al_2O_3 were prepared with excellent processability. The storage modulus was significantly increased from 5.93 GPa of the neat PBZ to 45.27 GPa for composite containing 83 wt.% micro-Al_2O_3 as illustrated in Figure 6.13. The T_g, degradation temperature and

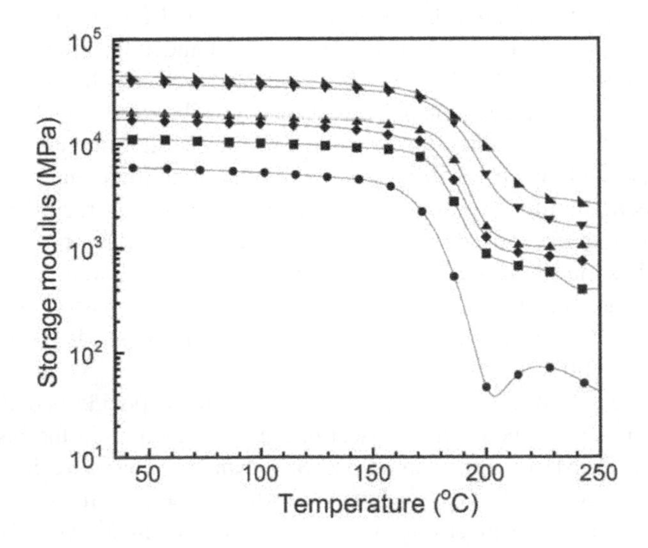

FIGURE 6.13 Storage Modulus of PBA-a/Al_2O_3 Composites at Various Alumina Contents: (●) wt.%, (■) 50 wt.%, (◊) 60 wt.%, (▲) 70 wt.%, (▼) 80 wt.%, and (▶) 83 wt.% [179].

char yield of the composites were also markedly enhanced with increasing the Al_2O_3 contents. SEM of the composite fracture surface revealed a good dispersion of the Al_2O_3 particles in the PBZ matrix [179].

Polybenzoxazine/Si_3N_4 nanocomposites were prepared with good processability via a compression molding technique. The DMA analysis of the nanocomposites revealed an increase of 2 GPa and 47°C in stiffness and T_g, respectively, due to uniform nanoparticle dispersion and to the strong matrix-particle interactions [180]. High-performance EP/PBZ/SiO_2 nanocomposites were developed with good processability by Okhawilai et al. A significant improvement in the storage modulus of the prepared nanocomposites was observed at high content of the nano-SiO_2 due to the strong bonding between EP/PBZ matrix and the nano-SiO_2 through ether-linkage as bonds [181].

Storage modulus (E') of polybenzoxazine/Al_2O_3 at room temperature had an increase of 663% compared to that of the neat matrix at maximum Al_2O_3 content of 83 wt.%. The T_g of the composites significantly increased by adding Al_2O_3 filler implying the presence of a substantial interfacial interaction between the Al_2O_3 particle and the P(BA-a) matrix [182].

The effect of Al_2O_3 nanoparticles on the thermo-physical properties of EP and UPE matrices were evaluated. However, the obtained T_g showed no significant change with nanoparticle concentration. The cross-link density studied by DMA for each nanocomposite showed that Al_2O_3 fillers are responsible for the decrease in cross-link density [183]. Using casting technique UPE/$CaCO_3$ nanocomposites were produced by Baskaran et al. with good dispersibility of nanofiller in the matrix. The DMA of the nanocomposites clearly indicated a progressive increase in T_g with the gradual addition of nano-$CaCO_3$ up to a loading of 5 wt.% of nanofiller [184].

Phthalonitrile/KH550-Si_3N_4 was prepared by a hot compression-molding technique. The nanocomposites exhibited improved modulus and T_g values due to the good dispersion and adhesion between the particles and the resin in the presence of KH550 silane [185]. Phthalonitrile/KH550-TiO_2 nanocomposites were prepared by Derradji et al. At 6 wt.% KH550-TiO_2 content, the T_g and the storage modulus were considerably enhanced reaching 364°C and 3.2 GPa, respectively (Figure 6.14) [186]. Geng et al. studied the effect of nano-SiO_2 on the thermomechanical properties of bismaleimide resin by DMA and reported that as the content of nano-SiO_2 increased, more significant improvement in the T_g value was recorded due to the steric hindrance effect caused by nano-SiO_2 particles [187]. The introduction of the TiO_2 nanofillers into the 4,4′-bismaleimidodiphenyl methane modified with novolac resin did not improve the T_g of the nanocomposites, but it increased the modulus of the composite at lower temperatures (< 200°C), while decreased it at higher temperatures (>250°C) [188].

SiO_2 at different sizes and surface modifications (of expoxide and BMI groups) showed to some extent the catalytic effect of the cure reaction on the bismaleimide/diamine (BMI/DDM) matrix. Unmodified SiO_2 nanoparticles were interacted with the BMI/DDM matrix through hydrogen bonding, whereas the surface-modified SiO_2 nanoparticles showed strong adherence to the resin by covalent bonding. Nanocomposites reinforced with surface-modified SiO_2 nanoparticles revealed significant improvements over raw SiO_2, and improvements were also noted in

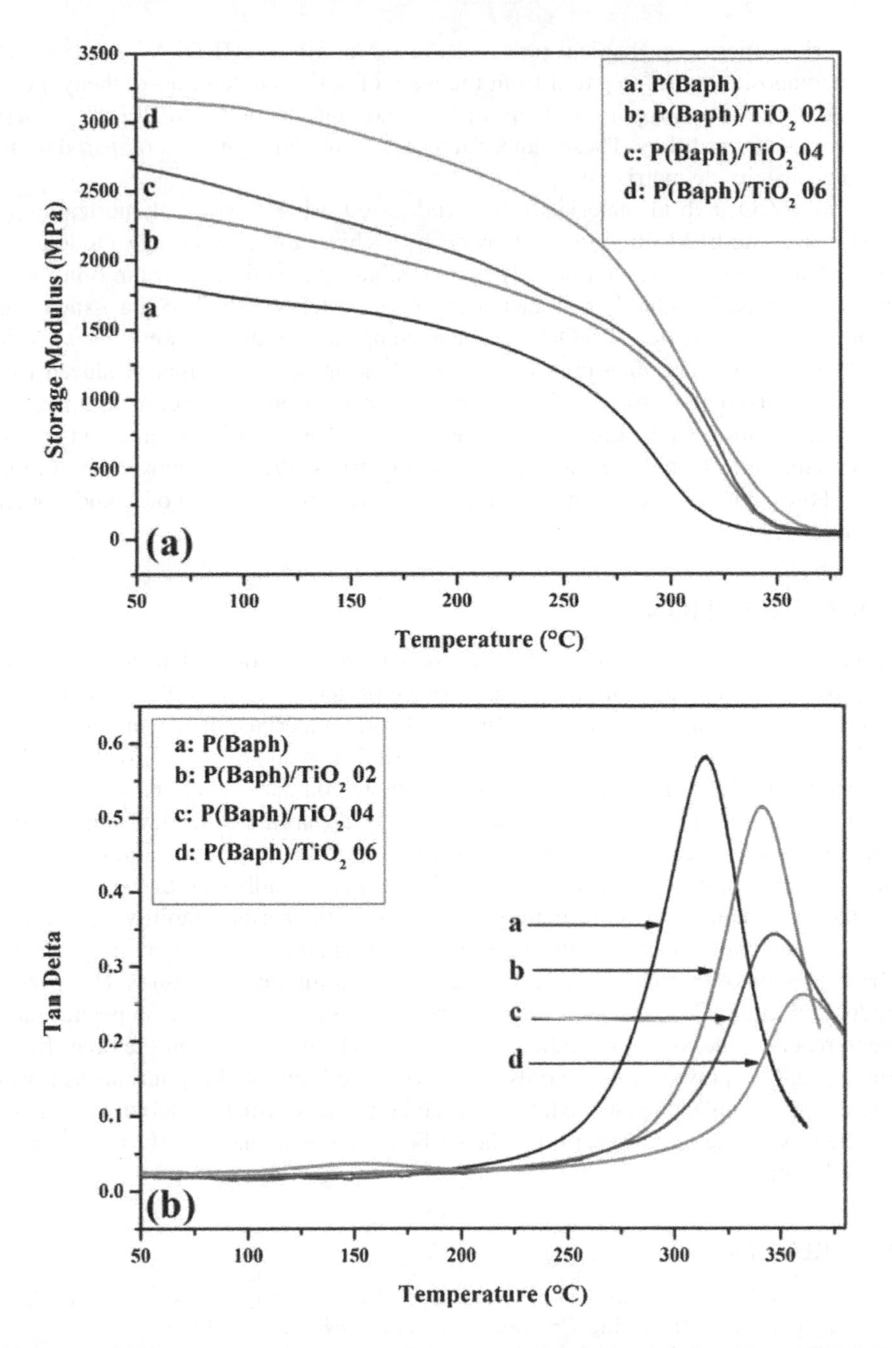

FIGURE 6.14 Evolution of Storage Modulus (a) and Tan Delta (b) of the Cured P(Baph)/ TiO_2 Nanocomposites [186].

the various thermomechanical properties of the unfilled BMI/DDM matrix [189]. Nanocomposites were prepared from the resin of 4,4′-bismaleimido-diphenylmethane (BDM) with dipropargyl others of hexafluorobisphenol A (DPBPF) as matrix and nano-SiO_2 as fillers. These nanocomposites exhibited higher T_g compared to the pure bismaleimide matrix [190].

Epoxy/ZrO_2 hybrid materials were synthesized using in situ polymerization of acetic acid-modified zirconium alkoxide by Ochi et al. The storage modulus in the rubbery region was significantly increased and the peak area of tan δ in the T_g region decreased gradually with increasing ZrO_2 contents [191]. The quasi-static and dynamic characteristics of SiO_2/epoxy nanocomposites with SiO_2 were investigated by Hölck et al. since they proposed a cross-linking algorithm that produced true 3D-cross linked atomistic packing models within a reasonable time. After an extensive equilibration procedure, the validated bulk models were investigated on epoxy/SiO_2 composites with respect to their elastic properties [192]. The epoxy matrix reinforced the cellular structure of SiO_2 imparted a great enhancement of T_g and storage modulus [193].

6.8 CONCLUSIONS

This chapter has proved the efficiency of ceramic micro- and nanoparticles in improving the flame retardancy, glass transition temperature, CTE, and thermal conductivity of polymeric matrices. In these hybrid materials, the ceramic fillers act as barriers against the polymer degradation gases' flow and oxygen diffusion within a condensed phase. Thus, generating strong, more compact, crack-free coating surface layers during combustion is the most critical parameter in developing more effective polymer/ceramic composite showing improved thermal stability. Several studies also confirmed that the ceramic filler dispersion adhesion to the matrix are both determinant factors contributing to the synergistic thermal stability effects. The addition of ceramic particles also improved the thermal conductivity of polymers by the formation of thermally-conductive pathways within their structures. However, a reduction in the CTE values of polymer/ceramic micro- and nanocomposites have been resulted due to the low inherent CTE values of the fillers. On the other hand, the trend T_g of polymers in the presence of ceramic fillers still mysterious as it was increased for some cases while it was reduced for others, which was found to depend on the type of the selected ceramic, the surface-treatment, and also the type of reinforced matrix.

REFERENCES

1. Ray, S. S., Okamoto, M. 2003. Polymer/layered silicate nanocomposites: a review from preparation to processing. *Progress in Polymer Science* 28: 1539-1641.
2. Chen, J., Zhou, Y., Nan, Q. et al. 2007. Preparation and properties of optically active polyurethane/TiO_2 nanocomposites derived from optically pure 1,1'-binaphthyl. *European Polymer Journal* 43: 4151–4159.
3. Goyal, P. K., Negi, Y. S., Tiwari, A. N. 2005. Preparation of high performance composites based on aluminum nitride/poly(ether-ether-ketone) and their properties. *European Polymer Journal* 41: 2034–2044.

4. Cassagnau, P. 2003. Payne effect and shear elasticity of silica-filled polymers in solutions and in molten state. *Polymer* 44: 2455–2462.
5. Lin, C. T., Lee, H. T., Chen, J. K. 2015. Preparation and properties of bisphenol-F based boron-phenolic resin/modified silicon nitride composites and their usage as binders for grinding wheels. *Applied Surface Science* 330: 1–9.
6. Shang, X. Y., Zhu, Z. K., Yin, J. et al. 2002. Compatibility of soluble polyimide/silica hybrids induced by a coupling agent. *Chemistry Materials* 14: 71–77.
7. Kashiwagi, T., Morgan, A. B., Antonucci, J. M. et al. 2003. Thermal and flammability properties of a silica–poly(methylmethacrylate) nanocomposite. *Journal of Applied Polymer Science* 89: 2072–2078.
8. Wu, C. C., Hsu, S. L. C. 2010. Preparation of epoxy/silica and epoxy/titania hybrid resists via a sol–gel process for nanoimprint lithography. *The Journal of Physical Chemistry C* 114: 2179–2183.
9. Liu, Y. L., Hsu, C. Y., Hsu, K. Y. 2005. Poly (methyl methacrylate)-silica nanocomposites films from surface-functionalized silica nanoparticles. *Polymer* 46: 1851–1856.
10. Pandey, J. K., Reddy, K. R., Kumar, A. P. et al. 2005. An overview on the degradability of polymer nanocomposites. *Polymer Degradation and Stability* 88: 234–250.
11. Kumara, A. P., Depana, D., Tomerb, N. S. et al. 2009. Nanoscale particles for polymer degradation and stabilization—Trends and future perspectives. *Progress in Polymer Science* 34: 479–515.
12. Kwon, Y. S., Kim, J. C., Ilyin, A. P. et al. 2012. Electroexplosive technology of nano powders production: current status and future prospect. *Journal of Korean Powder Metallurgy Institute* 19: 40–48.
13. Nazarenko, O. B., Amelkovich, Y. A., Ilyin, A. P. et al. 2014. Prospects of using nanopowders as flame retardant additives. *Advanced Materials Research* 872: 123–127.
14. Chayan, D., Bharat, K. 2012. Preparation and studies of nitrile rubber nanocomposites with silane modified silica nanoparticle. *Research Journal of Recent Sciences* 1: 357–360.
15. Chaichua, B., Prasassaraki, P., Poompradub, S. 2009. In situ silica reinforcement of natural rubber by sol-gel process via rubber solution. *Journal of Sol–Gel Science and Technology* 52: 219–227.
16. Shimpi, N. G., Mishra, S. 2010. Synthesis of nanoparticles and its effect on properties of elastomeric nanocomposites. *Journal of Nanoparticle Research* 12: 2093–2099.
17. Chen, G., Liu, S., Chen, S. et al. 2001. FTIR spectra, thermal properties, and dispersibility of a polystyrene/montmorillonite nanocomposites. *Macromolecular Chemistry and Physics* 202, 1189 1193.
18. Cinausero, N., Azema, N., Cuesta, J. M. L. et al. 2011. Impact of modified alumina oxides on the fire properties of PMMA and PS nanocomposites. *Polymers for Advanced Technologies* 22: 1931–1939.
19. Fan, J., Yuan, F., Zhao, P. et al. 2015. Study on preparation and property of phenolic-SiO_2 core-shell composite hollow microspheres. *Procedia Engineering* 102: 435–442.
20. El Saeed, A. M., El-Fattah, M. A., Dardir, M. M. 2015. Synthesis and characterization of titanium oxide nanotubes and its performance in epoxy nanocomposite coating. *Progress in Organic Coatings* 78: 83–89.
21. Chrissafis, D. B. 2011. Can nanoparticles really enhance thermal stability of polymers? Part I: An overview on thermal decomposition of addition polymers. *Thermochimica Acta* 523: 1–24.
22. Chen, Y. X., Chen, P., Yu, J. L. et al. 2013. Ablation performance and insulation property of carbon/phenolic composites with ZrB_2 particles. *Rare Metal Materials and Engineering* 42: 445–448.
23. Mirzapour, M. A., Haghighat, H. R., Eslami, Z. 2013. Effect of zirconia on ablation mechanism of asbestos fiber/phenolic composites in oxyacetylene torch environment. *Ceramics International* 39: 9263–9272.

24. Rallini, M., Torre, L., Kenny, J. M. et al. 2017. Effect of boron carbide nanoparticles on the thermal stability of carbon/phenolic composites. *Polymer Composites* 8:1819–1827.
25. Zaharescu, T., Jipa, S., Adrian, M., et al. 2008. Nanostructured isotactic polypropylene—TiO_2 systems. *Journal of Optoelectronic. Advanced Materials* 10: 2205–2209.
26. Zabihi, O., Ghasemlou, S. 2012. Nano-CuO/epoxy composites: Thermal characterization and thermo-oxidative degradation. *International Journal of Polymer Analysis and Characterization* 17: 108–121.
27. Chatterjee, A., Islam, M.S. 2008. Fabrication and characterization of TiO_2–epoxy nanocomposite. *Materials Science and Engineering: A* 487: 574–585.
28. Omrani, A., Afsar, S., Safarpour, M.A. 2010. Thermoset nanocomposites using hybrid nano TiO_2–SiO_2. *Materials Chemistry and Physics* 122: 343–349.
29. Guigo, N., Mija, A., Zavaglia, et al. 2009. New insights on the thermal degradation pathways of neat poly(furfuryl alcohol) and poly(furfuryl alcohol)/SiO_2 hybrid materials. *Polymer Degradation and Stability* 94: 908–913.
30. Schutz, M. R., Sattler, K., Deeken, S., et al. 2010. Improvement of thermal and mechanical properties of a phenolic resin nanocomposite by in situ formation of silsesquioxanes from a molecular precursor. *Journal of Applied Polymer Science* 117: 2272–2277.
31. Bazzar, M., Ghaemy, M. 2013. 1,2,4-Triazole and quinoxaline based polyimide reinforced with neat and epoxide-end capped modified SiC nanoparticles: Study thermal, mechanical and photophysical properties. *Composites Science and Technology* 86: 101–108.
32. Chrissafis, K., Bikiaris, D. 2011. Can nanoparticles really enhance thermal stability of polymers? Part I: an overview on thermal decomposition of addition polymers. *Thermochimica Acta* 523: 1–24.
33. Chrissafis, K., Paraskevopoulos, K.M., Pavlidou, E. et al. 2009. Thermal degradation mechanism of HDPE nanocomposites containing fumed silica nanoparticles. *Thermochimica Acta* 485: 65–71.
34. Olmos, D., Gonzalez-Gaitano, G., Gonzalez-Benito, J. 2015. Effect of a silica nanofiller on the structure, dynamics and thermostability of LDPE in LDPE/silica nanocomposites. *RSC Advances* 5: 34979–34984.
35. Gorelov, B., Gorb, A., Korotchenkov, O. et al. 2015. Impact of titanium and silica/titanium fumed oxide nanofillers on the elastic properties and thermal decomposition of a polyester resin. *Journal of Applied Polymer Science* 132: 42010.
36. Liawthanyarat, N., Rimdusit, S. 2015. Effects of particles size of nanosilica on properties of polybenzoxazine nanocomposites. *Key Engineering Materials* 659: 394–398.
37. Wu, G., He, X., Xu, L. et al. 2015. Synthesis and characterization of Biobased polyurethane/SiO_2 nanocomposites from natural *Sapium sebiferum oil. RSC Advances* 5: 27097–27106.
38. Jin, W. Q., Yuan, L., Liang, G. Z. et al. 2014. Multifunctional cyclotriphosphazene/hexagonal boron nitride hybrids and their flame retarding bismaleimide resins with high thermal conductivity and thermal stability. *ACS Applied Materials & Interfaces* 6: 14931–14944.
39. Lu, G. T., Huang, Y. 2002. Synthesis of polymaleimide/silica nanocomposites. *Journal of Materials Science* 37: 2305–2309.
40. Bai, Z. M., Jiang, S. D., Tang, G. et al. 2014. Enhanced thermal properties and flame retardancy of unsaturated polyester-based hybrid materials containing phosphorus and silicon. *Polymers for Advanced Technologies* 25: 223–232.
41. Alhuthali, A., Low, I. M. 2013. Multi-scale hybrid eco-nanocomposites: synthesis and characterization of nano-SiC-reinforced vinyl-ester eco-composites. *Journal of Materials Science* 48: 3097–3106.

42. Yan, C., Fan, X., Li, J. et al. 2011. Study of surface-functionalized nano-SiO_2/polybenzoxazine composites. *Journal of Applied Polymer Science* 120: 1525–1532.
43. Morgan, A. B., Putthanarat, S. 2011. Use of inorganic materials to enhance thermal stability and flammability behavior of a polyimide. *Polymer Degradation and Stability* 96: 23–32.
44. Ghezelbash Z., Ashouri, D., Mousavian, S. et al. 2012. Surface modified Al_2O_3 in fluorinated polyimide/Al_2O_3 nanocomposites: Synthesis and characterization. *Bulletin of Materials Science* 35: 925–931.
45. Toiserkani, H. 2015. Polyimide/nano-TiO_2 hybrid films having benzoxazole pendent groups: In situ sol–gel preparation and evaluation of properties. *Progress in Organic Coatings* 88: 17–22.
46. Akhter, T., Park, O. O., Siddiqi, H. M. et al. 2014. An investigation of physicochemical properties of a new polyimide–silica composites. *RSC* Advances 4: 46587–46594.
47. Periadurai, T., Vijayakumar, C. T., Balasubramanian, M. 2010. Thermal decomposition and flame retardant behaviour of SiO_2-phenolic nanocomposite. *Journal of Analytical and Applied Pyrolysis* 89: 244–249.
48. Zeng, X. L., Yu, S. H., Sun, R. et al. 2011. High thermal conductive BT resin/silicon nitride composites. *IEEE International Conference on Electronic Packaging Technology & High Density Packaging* 346–348.
49. Chiang, C. L., Ma, C. C. M. 2006. *Phenolic resin/SiO_2 organic-inorganic hybrid nanocomposites*. Ed: Mai, Hall, Abington, Cambridge CB1 6AH, England, p. 485–507.
50. Chiang, C. L., Ma, C. C. M. 2004. Synthesis, characterization, thermal properties and flame retardance of novel phenolic resin/silica nanocomposites. *Polymer Degradation and Stability* 83: 207–214.
51. Balaji, R., Sasikumar, M., Elayaperumal, A. 2015. Thermal, thermos-oxidative and ablative behavior of cenosphere filled ceramic/phenolic composites. *Polymer Degradation and Stability* 114: 125–132.
52. Chiang, C. L., Ma, C. C. M., Wu, D. L. et al. 2003. Preparation, characterization, and properties of novolac-type phenolic/SiO_2 hybrid organic–inorganic nanocomposite materials by sol–gel method. *Journal of Polymer Science Part A: Polymer Chemistry* 41: 905–913.
53. Jiang, H., Wang, J., Wu, S. et al. 2014. Study on the property of boron carbide-modified phenol-formaldehyde resin for silicon carbide bonding. *Russian Journal of Applied Chemistry* 87: 904–908.
54. Shi, S., Liang, J., He, R. 2015. Thermal decomposition behavior of silica-phenolic composite exposed to one-sided radiant heating. *Polymer Composites* 36:1557–1564.
55. Su, H. L., Hsu, J. M., Pan, J. P. et al. 2007. Silica nanoparticles modified with vinyltriethoxysilane and their copolymerization with N,N'-bismaleimide-4,4'-diphenylmethane. *Journal of Applied Polymer Science* 103: 3600–3608.
56. Dueramae, I., Jubsilp, C., Takeichi, T. et al. 2014. High thermal and mechanical properties enhancement obtained in highly filled polybenzoxazine nanocomposites with fumed silica. *Composites Part B: Engineering* 2014; 56: 197–206.
57. Abenojar, J., Encinas, N., Real, J. C. D. et al. 2014. Polymerization kinetics of boron carbide/epoxy composites. *Thermochimica Acta* 575: 144–150.
58. Rajabi, L., Kaihani, S., Kurdian, A. R., et al. 2010. Dynamic Cure Kinetics of Epoxy/TiO_2 Nanocomposites. *Science of Advanced Materials* 2: 200–209.
59. Parameswaranpillai, J., George, A., Pionteck, J., et al. 2013. Investigation of Cure Reaction, Rheology, Volume Shrinkage and Thermomechanical Properties of Nano-TiO_2 Filled Epoxy/DDS Composites. *Journal of Polymers* 2013:1–17.

60. Qian, J. S., Yang, H.Y., He, P. S. 2004. Effects of nano-SiO_2 on the kinetics of non-isothermal crystallization of HDPE and PP. *Polymeric Materials Science and Engineering* 20:161–164.
61. Chiacchiarelli, L. M., Puri, I., Puglia, D. et al. 2012. Cure kinetics of a highly reactive silica-polyurethane nanocomposite. Thermochimica Acta 549: 172–178.
62. Sun, Y. Y., Zhang, Z. Q., Moon, K. S. et al. 2004. Glass transition and relaxation behavior of epoxy nanocomposites. *Journal of Polymer Science Part B: Polymer Physics* 42: 3849–3858.
63. Charpentier, P. A., Xu, W. Z., Li, X. S. 2007. A novel approach to the synthesis of SiO_2-PVAc nanocomposites using a one-pot synthesis in supercritical CO_2. *Chemistry* 9: 768–776.
64. Jain, S., Goossens, H., van Duin, M. et al. 2005. Effect of in situ prepared silica nano-particles on non-isothermal crystallization of polypropylene. *Polymer* 46: 8805–8818.
65. Liu, Y. L., Hsu, C. Y., Hsu, K. Y. 2005. Poly(methyl methacrylate)-silica nanocomposites films from surface-functionalized silica nanoparticles. *Polymer* 46: 1851–1856.
66. Sun, Y., Zhang, Z., Moon, K. S. et al. 2004. Glass transition and relaxation behavior of epoxy nanocomposites. *Journal of Polymer Science Part B: Polymer Physics* 42: 3849–3858.
67. MerkeL, T. C., He, Z., Pinnau,I. 2003. Sorption and transport in poly(2,2-bis(trifluoromethyl)-4,5-difluoro-1,3-dioxole-co-tetrafluoroethylene) containing nanoscale fumed silica. *Macromolecules* 36: 8406–8414.
68. Liu, Y. L., Hsu, C. Y., Hsu, K. Y. 2005. Poly (methyl methacrylate)-silica nanocomposites films from surface-functionalized silica nanoparticles. *Polymer* 46: 1851–1856.
69. Yu, Y. Y., Chen, C. Y., Chen, W. C. 2003. Transparent organic–inorganic hybrid thin films prepared from acrylic polymer and aqueous monodispersed colloidal silica. *Polymer* 44: 593–601.
70. Dueramae, I., Jubsilp, C., Takeichi, T. et al. 2014. High thermal and mechanical properties enhancement obtained in highly filled polybenzoxazine nanocomposites with fumed silica. *Composites Part B: Engineering* 56: 197–206.
71. Bera, O., Pilić, B., Pavličević, J. 2011. Preparation and thermal properties of polystyrene/silica nanocomposites. *Thermochimica Acta* 515: 1–5.
72. Ramdani, N., Wang, J., Liu, W. B. 2014. Preparation and thermal properties of polybenzoxazine/TiC hybrids. *Advanced Materials Research* 887-888: 49–52.
73. Hiremath, V., Singh, M., Shukl, D. K. 2014. Effect of post curing temperature on viscoelastic and flexural properties of epoxy/alumina polymer nanocomposites. *Procedia Engineering* 97: 479–487.
74. Mandhakini, M., Lakshmikandhan, T., Chandramohan, A. et al. 2014. Effect of nanoalumina on the tribology performance of C_4-ether-linked bismaleimide-toughened epoxy nanocomposites. *Tribology Letters* 54: 67–79.
75. Ukaji, M., Takamura, M., Shirai, K. et al. 2008. Curing of epoxy resin by hyperbranched poly(amidoamine)-grafted silica nanoparticles and their properties. *Polymer Journal* 40: 607–613.
76. Wu, H., Rogalski, M., Kessler, M. R. 2013. Zirconium tungstate/epoxy nanocomposites: Effect of nanoparticle morphology and negative thermal expansivity. *ACS Applied Materials & Interfaces* 5: 9478–9487.
77. Yu, J. H., Huang, X. Y., Wu, C. et al. 2012. Interfacial modification of boron nitride nanoplatelets for epoxy composites with improved thermal properties. *Polymer* 53: 471–480.
78. Choi, J., Yang, S., Yu, S. et al. 2012. Method of scale bridging for thermoelasticity of cross-linked epoxy/SiC nanocomposites at a wide range of temperatures. *Polymer* 53: 5178–5189.

79. Afzal, A., Siddiqi, H. M. 2011. A comprehensive study of the bicontinuous epoxy-silica hybrid polymers: I. Synthesis, characterization and glass transition. *Polymer* 52: 1345–1355.
80. Chen, C., Morgan, A. B. 2009. Mild processing and characterization of silica epoxy hybrid nanocomposite. *Polymer* 50: 6265–6273.
81. Lin, C. H., Feng, C. C., Hwang, T. Y. 2007. Preparation, thermal properties, morphology, and microstructure of phosphorus-containing epoxy/SiO_2 and polyimide/SiO_2 nanocomposites. *European Polymer Journal* 43:725–742.
82. Lionetto, F., Mascia, L., Frigione, M. 2013. Evolution of transient states and properties of an epoxy–silica hybrid cured at ambient temperature. *European Polymer Journal* 49: 1298–1313.
83. Sonje, P. U., Subramanian, K. N., Lee, A. 2004. Characterization of Polymeric Composites with Low CTE Ceramic Particulate Fillers. *Chemical Engineering and Materials Science* 36: 22–29.
84. Glavchev, I., Petrova, K., Ivanova, M. 2002. Determination of the coefficient of thermal expansion of epoxy composites. *Polymer Testing* 21: 177–179.
85. Huang, C. J., Fu, S. Y., Zhang, Y. H. et al. 2005. Cryogenic properties of SiO_2/epoxy nanocomposites. *Cryogenics* 45: 450–454.
86. Huang, J. C., Zhu, Z. K., Yin, J. et al. 2001. Preparation and properties of rigid-rod polyimide/silica hybrid materials by sol–gel process. *Journal of Applied Polymer Science* 79: 794–800.
87. Badrinarayanan, P., Kessler, M.R. 2011. Zirconium tungstate/cyanate ester nanocomposites with tailored thermal expansivity. *Composites Science and Technology* 71: 1385–1391.
88. Sharma, G. R., Lind, C., Coleman, M. R. 2012. Preparation and properties of polyimide nanocomposites with negative thermal expansion nanoparticle filler. *Materials Chemistry and Physics* 137: 448–457.
89. Sullivan, L. M., Lukehart CM. 2005. Zirconium Tungstate (ZrW_2O_8)/Polyimide Nanocomposites Exhibiting Reduced Coefficient of Thermal Expansion. *Chemistry of Materials* 17: 2139–2141.
90. Tsai, M. H., Wang, H. Y., Lu, H. T. et al. 2011. Properties of polyimide/Al_2O_3 and Si_3N_4 deposited thin films. *Thin Solid Films* 519: 4969–4973.
91. Chung, M. H., Chen, L. M., Wang, W. H. et al. 2014. Effects of mesoporous silica coated multi-wall carbon nanotubes on the mechanical and thermal properties of epoxy nanocomposites. *Journal of the Taiwan Institute of Chemical Engineers* 45: 2813–2819.
92. Jang, J. S., Varischetti, J., Lee, G. W. et al. 2011. Experimental and analytical investigation of mechanical damping and CTE of both SiO_2 particle and carbon nanofiber reinforced hybrid epoxy composites. *Composites Part A: Applied Science and Manufacturing* 42: 98–103.
93. Jin, W. Q., Zhang, W., Gao, Y. et al. 2013. Surface functionalization of hexagonal boron nitride and its effect on the structure and performance of composites. *Applied Surface Science* 270: 561–571.
94. Gao, Y. W., Gu, A. J., Jiao, Y.C. et al. 2012. High-performance hexagonal boron nitride/bismaleimide composites with high thermal conductivity, low coefficient of thermal expansion, and low dielectric loss. *Polymers for Advanced Technologies* 23: 919–928.
95. Hu, J., Gu, A., Liang, G. et al. 2011. Preparation and properties of mesoporous silica/bismaleimide/diallylbisphenol composites with improved thermal stability, mechanical and dielectric properties. *Express Polymer Letters* 5: 555–568.
96. Chandra, A., Meyer, W. H., Best, A. et al. 2007. Modifying Thermal Expansion of Polymer Composites by Blending with a Negative Thermal Expansion Material. *Macromolecular Materials and Engineering* 292: 295–301.

97. Chun, H., Kim, Y.J., Tak, S. Y. 2018. Preparation of ultra-low CTE epoxy composite using the new alkoxysilyl-functionalized bisphenol A epoxy resin. *Polymer* 135: 241–250.
98. Shan, X., Huang, C., Yang, H. 2015. The Thermal Expansion and Tensile Properties of Nanofiber-ZrW_2O_8 Reinforced Epoxy Resin Nanocomposites. *Physics Procedia* 67: 1056–1061.
99. Wu, H., Rogalski, M., Kessler, M. R. 2013. Zirconium tungstate/epoxy nanocomposites: effect of nanoparticle morphology and negative thermal expansivity. *ACS Applied Materials and Interfaces*. 2013 5: 9478–9487.
100. Komarov, P. V., Chiu, Y. T., Chen, S. M., et al. 2010. Investigation of Thermal Expansion of Polyimide/SiO_2 Nanocomposites by Molecular Dynamics Simulations. *Molecular Theory and Simulation* 19: 64–73.
101. Badrinarayanan, P., Kessler, M.R. 2011. Zirconium tungstate/cyanate ester nanocomposites with tailored thermal expansivity. *Composites Science and Technology* 71: 1385–1391.
102. Enhessari, M., Ozaee, K., Karamali, E., et al. 2012. Fabrication and Thermal Properties of Polyimide/Ba0.77Sr0.23TiO_3 Nanocomposites. *International Journal of Polymeric Materials and Polymeric Biomaterials* 61: 89–98.
103. Goertzen, W. K., Kessler, M. R. 2008. Thermal expansion of fumed silica/cyanate ester nanocomposites. *Journal of Applied Polymer Science* 109: 647–653.
104. Badrinarayanan, P., Murray, B.M., Kessler, M. R. 2009. Zirconium tungstate reinforced cyanate ester composites with enhanced dimensional stability. *Journal of Materials Research* 24: 2235–2242.
105. Tani, J., Kimura, H., Hirota, K., et al. 2007. Thermal expansion and mechanical properties of phenolic resin/ZrW_2O_8 composites. *Journal of Applied Polymer Science* 106: 3343–3347.
106. Wong, C. P., Bollampally, R. S. 1999. Thermal conductivity, elastic modulus, and coefficient of thermal expansion of polymer composites filled with ceramic particles for electronic packaging. *Journal of Applied Polymer Science* 74: 3396–3403.
107. Hong He, H., Fu, R., Han,Y. et al. 2007. Thermal conductivity of ceramic particle filled polymer composites and theoretical predictions. *Journal of Materials Science* 42: 6749–6754.
108. Shimazaki, Y., Hojo, F., Takezawa, Y. 2009. Highly thermoconductive polymer nanocomposite with a nanoporous α-alumina sheet. *ACS Applied Materials & Interfaces* 1(2): 225–227.
109. Poostforush M.D., Azizi, H. 2014. Superior thermal conductivity of transparent polymer nanocomposites with a crystallized alumina membrane. *EXPRESS Polymer Letters* 8: 293–299.
110. Wang, T., Wang, M., Fu, L. et al. 2018. Enhanced thermal conductivity of polyimide composites with boron nitride nanosheets. *Scientific Reports*, 2018, 8:1557–1565.
111. Ramdani, N., Derradji, M., Feng, T. T. et al. 2015. Preparation and characterization of thermally-conductive silane-treated silicon nitride filled polybenzoxazine nanocomposites. *Materials Letters* 155: 34–37.
112. Moreira, D. C., Sphaier, L. A., Reis, J. M. L. et al. 2012. Experimental analysis of heat conduction in UPR-alumina nano-composites. *High Temperatures-High Pressures* 41: 185–195.
113. Moreira, D. C., Sphaier, L. A., Reis, J. M. L. et al. 2011. Experimental investigation of heat conduction in polyester-Al_2O_3 and polyester-CuO nanocomposites. *Experimental Thermal and Fluid Science* 35: 1458–1462.
114. Wu, Y., Yu, Z., He, Y. et al. 2015. Heat conduction models of interfacial effects for TiB_2-Al_2O_3/epoxy composites. *Materials Chemistry and Physics* 162:182–187.

115. Dai, W., Yu, J., Liu, Z. et al. 2015. Enhanced thermal conductivity and retained electrical insulation for polyimide composites with SiC nanowires grown on graphene hybrid fillers. *Composites Part A: Applied Science and Manufacturing* 76:73–81.
116. Heo, Y. S., Im, H. G., Kim, J. W. et al. 2012. The influence of $Al(OH)_3$-coated graphene oxide on improved thermal conductivity and maintained electrical resistivity of Al_2O_3/epoxy composites. *Journal of Nanoparticle Research* 14:1196–1206.
117. Gao, Z. F., Zhao, L. 2015. Effect of nanofillers on the thermal conductivity of epoxy composites with micro-Al_2O_3 particles. *Materials & Design* 66: 176–182.
118. Giang, T., Kim, J. 2015. Effect of backbone moiety in diglycidylether-terminated liquid crystalline epoxy on thermal conductivity of epoxy/alumina composite. *Journal of Industrial and Engineering Chemistry* 30: 77–84.
119. Tanaka, K., Ogata, S., Kobayashi, R. et al. 2015. A molecular dynamics study on thermal conductivity of thin epoxy polymer sandwiched between alumina fillers in heat-dissipation composite material. *International Journal of Heat and Mass Transfer* 89: 714–723.
120. Yu, Z. Q., Wu, Y. C., Wei, B. et al. 2015. Boride ceramics covalent functionalization and its effect on the thermal conductivity of epoxy composites. *Materials Chemistry and Physics* 164: 214–222.
121. Chen, C., Tang, Y., Ye, Y. S. et al. 2014. High-performance epoxy/silica coated silver nanowire composites as underfill material for electronic packaging. *Composites Science and Technology* 105: 80–85.
122. Dang, J., Wang, R. M., Yang, L. et al. 2014. Preparation of β-SiCw/BDM/DBA composites with excellent comprehensive properties. *Polymer Composites* 35: 1875–1878.
123. Hong, J. P., Yoon, S. W., Hwang, T. et al. 2012. High thermal conductivity epoxy composites with bimodal distribution of aluminum nitride and boron nitride fillers. *Thermochimica Acta* 537: 70–75.
124. Agrawal, A., Satapathy, A. Thermal and dielectric behavior of epoxy composites filled with ceramic micro particulates. *Journal of Composite Materials* 2014; 48: 3755–3769.
125. Agrawal, A., Satapathy, A. 2013. Development of a heat conduction model and investigation on thermal conductivity enhancement of AlN/epoxy composites. *Procedia Engineering* 51: 573–578.
126. Choi, S., Kim, J. 2013. Thermal conductivity of epoxy composites with a binary-particle system of aluminum oxide and aluminum nitride fillers. *Composites Part B: Engineering* 51: 140–147.
127. Nagai, Y., Lai, G. C. 1997. Thermal conductivity of epoxy resin filled with particulate aluminum nitride powder. *Journal of the Ceramic Society of Japan* 105: 197–200.
128. Ma, A. J., Chen, W., Hou, Y. 2012. Enhanced thermal conductivity of epoxy composites with MWCNTs/AlN hybrid filler. *Polymer-Plastics Technology and Engineering* 2012; 51: 1578–1582.
129. Qian, R., Yu, J., Xie, L. et al. 2013. Efficient thermal properties enhancement to hyperbranched aromatic polyamide grafted aluminum nitride in epoxy composites. *Polymers for Advanced Technologies* 24: 348–356.
130. Agrawal, A., Satapathy, A. 2015. Epoxy composites filled with microsized AlN particles for microelectronic applications. *Particulate Science and Technology* 33: 2–7.
131. Zhang, J., Qi, S. 2015. Mechanical, thermal and dielectric properties of aluminum nitride/epoxy resin composites. *Journal of Elastomers & Plastics* 47: 431–438.
132. Kim, T., Bae, J. C., Cho, K.Y. et al. 2013. Thermal conductivity behaviour of silicon carbide fiber/phenolic resin composites by the introduction of graphene nanoplatelets. *Asian Journal of Chemistry* 25: 5625–5630.
133. Hou, G., Cheng, B., Ding, F. et al. 2015. Synthesis of uniform α-Si_3N_4 nanospheres by RF induction thermal plasma and their application in high thermal conductive nanocomposites. *ACS Applied Materials & Interfaces* 7: 2873–2881.

134. Wang, Z. F., Fu, Y. Q., Meng, W. J. et al. 2014. Solvent-free fabrication of thermally conductive insulating epoxy composites with boron nitride nanoplatelets as fillers. *Nanoscale Research Letters* 9: 643–650.
135. Donnay, M., Tzavalas, S., Logakis, E. 2015. Boron nitride filled epoxy with improved thermal conductivity and dielectric breakdown strength. *Composites Science and Technology* 110: 152–158.
136. Raza, M. A., Westwood, A. V. K., Stirling, C. et al. 2015. Effect of boron nitride addition on properties of vapour grown carbon nanofiber/rubbery epoxy composites for thermal interface applications. *Composites Science and Technology* 120: 9–16.
137. Cho, H. B., Tu, N. C., Fujihara, T. et al. 2011. Epoxy resin-based nanocomposite films with highly oriented BN nanosheets prepared using a nanosecond-pulse electric field. *Materials Letters* 65: 2426–2428.
138. Lin, Z. Y., Mcnamara, A., Liu, Y. et al. 2014. Exfoliated hexagonal boron nitride-based polymer nanocomposite with enhanced thermal conductivity for electronic encapsulation. *Composites Science and Technology* 90: 123–128.
139. Kim, K., Kim, J. 2014. Fabrication of thermally conductive composite with surface modified boron nitride by epoxy wetting method. *Ceramics International* 40: 5181–5189.
140. Kim, K., Kim, M., Hwang, Y. et al. 2014. Chemically modified boron nitride-epoxy terminated dimethylsiloxane composite for improving the thermal conductivity. *Ceramics International* 40: 2047–2056.
141. Teng, C. C., Ma, C. C. M., Chiou, K. C. et al. 2011. Synergetic effect of hybrid boron nitride and multi-walled carbon nanotubes on the thermal conductivity of epoxy composites. *Materials Chemistry and Physics* 126: 722–728.
142. Wattanakul, K., Manuspiya, H., Yanumet, N. 2010. The adsorption of cationic surfactants on BN surface: Its effects on the thermal conductivity and mechanical properties of BN-epoxy composite. *Colloids and Surfaces A: Physicochemical and Engineering Aspects* 369: 203–210.
143. Cho, H. B., Mitsuhashi, M., Nakayama, T. et al. 2013. Thermal anisotropy of epoxy resin-based nano-hybrid films containing BN nanosheets under a rotating superconducting magnetic field. *Materials Chemistry and Physics* 139: 355–359.
144. Fu, Y. X., He, Z. X., Mo, D. C. 2014. Thermal conductivity enhancement with different fillers for epoxy resin adhesives. *Applied Thermal Engineering* 66: 493–498.
145. Harada, M., Hamaura, N., Ochi, M. et al. 2013. Thermal conductivity of liquid crystalline epoxy/BN filler composites having ordered network structure. *Composites Part B: Engineering* 55: 306–313.
146. Xu, Y. S., Chung, D. D. L. 2000. Increasing the thermal conductivity of boron nitride and aluminum nitride particle epoxy-matrix composites by particle surface treatments. *Composite Interfaces* 2000; 7: 243–256.
147. Zeng, J., Fu, R. L., Shen, Y. et al. 2009. High thermal conductive epoxy molding compound with thermal conductive pathway. *Journal of Applied Polymer Science* 113: 2117–2125.
148. Shi, Z., Fu, R., Agathopoulos, S. et al. 2012. Thermal conductivity and fire resistance of epoxy molding compounds filled with Si_3N_4 and $Al(OH)_3$. *Materials & Design* 34: 820–824.
149. Kusunose, T., Yagi, T., Firoz, S. H. et al. 2013. Fabrication of epoxy/silicon nitride nanowire composites and evaluation of their thermal conductivity. *Journal of Materials Chemistry A* 1: 3440–3445.
150. Zhu, Y., Chen, K. X., Kang, F. Y. 2013. Percolation transition in thermal conductivity of β-Si_3N_4 filled epoxy. *Solid State Communications* 158: 46–50.
151. Hwang, Y., Kim, M., Kim, J. 2014. Fabrication of surface-treated SiC/epoxy composites through a wetting method for enhanced thermal and mechanical properties. *Chemical Engineering Journal* 246: 229–237.

152. Hong, J.P., Yoon, S.W., Hwang, T., et al. 2012. High thermal conductivity epoxy composites with bimodal distribution of aluminum nitride and boron nitride fillers. *Thermochimica Acta* 537:70–75.
153. Huang T, Zeng X, Yao Y, et al. 2016. Boron nitride@graphene oxide hybrids for epoxy composites with enhanced thermal conductivity. *RSC Advances* 42: 35144–36217.
154. Qu, T., Yang, N., Hou, J. et al. 2017. Flame retarding epoxy composites with poly(phosphazene-co-bisphenol A)-coated boron nitride to improve thermal conductivity and thermal stability. *RSC Advances* 7: 6140–6151.
155. Gu, J., Liang, C., Dang, J. et al. 2016. Ideal dielectric thermally conductive bismaleimide nanocomposites filled with polyhedral oligomeric silsesquioxane functionalized nanosized boron nitride. *RSC Advances* 6: 35809–35814.
156. Yang, N., Xu, C., Hou, J. et al. 2016. Preparation and properties of thermally conductive polyimide/boron nitride composites. *RSC Advances* 6: 18279–18287.
157. Hou, J., Li, G., Yang, N. et al. 2014. Preparation and characterization of surface modified boron nitride epoxy composites with enhanced thermal conductivity. *RSC Advances* 4: 44282–44290.
158. Fang, L., Wu, C., Qian, R. et al. 2014. Nano–micro structure of functionalized boron nitride and aluminum oxide for epoxy composites with enhanced thermal conductivity and breakdown strength. *RSC Advances* 4: 21010–21017.
159. Zhi, C.Y., Bando, Y., Terao, T. et al. 2010. Dielectric and thermal properties of epoxy/boron nitride nanotube composites. *Pure Applied Chemistry* 82(11): 2175–2183.
160. Gu, J., Zhang, Q., Dang, J. et al. 2012. Thermal conductivity epoxy resin composites filled with boron nitride. *Polymer Advanced Technologies* 23: 1025–1028.
161. Zhang, X., Wu, H., Guo, S. 2015. Effect of Interfacial Interaction on Morphology and Properties of Polyethylene/Boron Nitride Thermally Conductive Composites. *Polymer-Plastics Technology and Engineering* 54: 1097–1105.
162. Costa, J. V., Ramotowski, T., Warner, S. et al. 2010. High Thermal Conductivity Polyurethane-Boron Nitride Nanocomposite Encapsulants. Proceedings of the SEM Annual Conference June 7-10 Indianapolis, Indiana, 337-342.
163. Menard, K. P. 2008. *Dynamic Mechanical Analysis: A Practical Introduction*. Boca Raton: CRC Press.
164. Bansal, A., Yang, H., Li, C. et al. Controlling the thermomechanical properties of polymer nanocomposites by tailoring the polymer–particle interface. *Journal of Polymer Science Part B: Polymer Physics* 44: 2944–2950.
165. Yuan, Z., Yu, J., Rao, B. et al. 2014. Enhanced thermal properties of epoxy composites by using hyper branched aromatic polyamide grafted silicon carbide whiskers. *Macromolecular Research* 22: 405–411.
166. Lai, Y. H., Kuo, M. C., Huang, J. C. et al. 2007. Thermomechanical Properties of Nanosilica Reinforced PEEK Composites. *Key Engineering Materials* 351: 15–20.
167. Al Arbash, A., Ahmad, Z., Al-Sagheer, F. et al. 2006. Microstructure and Thermomechanical Properties of Polyimide-Silica Nanocomposites. *Journal of Nanomaterials*: 2006: 1–9.
168. Tarrío-Saavedra, J., López-Beceiro, J., Naya, S. et al. 2010. Controversial effects of fumed silica on the curing and thermomechanical properties of epoxy composites. *eXPRESS Polymer Letters* 4(6): 382–395.
169. Yung, K. C., Zhu, B. L., Yue, T. M. et al. 2010. Effect of the filler size and content on the thermomechanical properties of particulate aluminum nitride filled epoxy composites. *Journal of Applied Polymer Science* 116: 225–236.
170. Georgiopoulos, P., Kontou, E. Meristoudi, A. et al. 2014. The effect of silica nanoparticles on the thermomechanical properties and degradation behavior of polylactic acid. *Journal of Biomaterials Applications* 29: 662–674.

171. Fuentes-Miranda, A., Campillo-Illanes, B., Fernández-Garcia, M. et al. 2017. Thermomechanical properties of silica-polyacrylic nanocomposites. *MRS Advances* 2017: 2745–2750.
172. Vassileva, E., Friedrich, K. 2003. Epoxy/alumina nanoparticle composites. I. dynamic mechanical behavior. *Journal of Applied Polymer Science* 89: 3774–3785.
173. Chatterjee, A., Islam, M. S. 2008. Fabrication and characterization of TiO_2–epoxy nanocomposite. *Materials Science and Engineering: A* 487: 574–585.
174. Duraibabu, D., Alagar, M., Kumar, S. A. 2014. Studies on mechanical, thermal and dynamic mechanical properties of functionalized nanoalumina reinforced sulphone ether linked tetraglycidyl epoxy nanocomposites. *RSC Advances* 4: 40132–40140.
175. Morselli, D., Bondioli, F., Sangermano, M. et al. 2012. Photo-cured epoxy networks reinforced with TiO_2 in-situ generated by means of non-hydrolytic sole gel process. *Polymer* 53: 283–290.
176. Kyritsis, A., Vikelis, G., Maroulas, P. et al. 2011. Polymer dynamics in epoxy/alumina nanocomposites studied by various techniques. *Journal of Applied Polymer Science* 121: 3613–3627.
177. Johnsen, B. B., Frømyr, T. R., Thorvaldsen, T. et al. 2013. Preparation and characterisation of epoxy/alumina polymer nanocomposites. *Composite Interfaces* 20: 721–740.
178. Nazir, T., Afzal, A., Siddiqi, H. M. et al. 2010. Thermally and mechanically superior hybrid epoxy–silica polymer films via sol–gel method. *Progress in Organic Coatings* 2010; 69:100–106.
179. Kajohnchaiyagual, J., Jubsilp, C., Dueramae, I. et al. 2014. Thermal and mechanical properties enhancement obtained in highly filled alumina-polybenzoxazine composites. *Polymer Composites* 35: 2269–2279.
180. Ramdani, N., Wang, J., Wang, H. et al. 2014. Mechanical and thermal properties of silicon nitride reinforced polybenzoxazine nanocomposites. *Composites Science and Technology* 105: 73–79.
181. Okhawilai, M., Dueramae, I., Jubsilp, C. et al. 2017. Effects of high nano-SiO_2 contents on properties of epoxy-modified polybenzoxazine. *Polymer Composites* 38:2261–2271.
182. Kajornchaiyakul, J., Jubsilp, C., Rimdusit, S. 2013. Thermal and mechanical properties of highly-filled polybenzoxazine/alumina composites. *Key Engineering Materials* 545: 211–215.
183. De Souza, J. P. B., Dos Reis, J. M. L. 2015. Influence of Al_2O_3 and CuO nanoparticles on the thermal properties of polyester- and epoxy-based nanocomposites. *Journal of Thermal Analysis and Calorimetry* 119:1739–1746.
184. Baskaran, R., Sarojadevi, M., Vijayakumar, C. T. 2011. Mechanical and thermal properties of unsaturated polyester/calcium carbonate nanocomposites. *Journal of Reinforced Plastics and Composites* 30: 1549–1556.
185. Derradji, M., Ramdani, N., Zhang, T. et al. 2015. High thermal and thermomechanical properties obtained by reinforcing a bisphenol-A based phthalonitrile resin with silicon nitride nanoparticles. *Materials Letters* 149: 81–84.
186. Derradji, M., Ramdani, N., Zhang, T. et al. 2016. Effect of silane surface modified titania nanoparticles on the thermal, mechanical, and corrosion protective properties of a bisphenol-A based phthalonitrile resin. *Progress in Organic Coatings* 90: 34–43.
187. Geng, D. B., Zeng, L. M., Hu, B. et al. Dynamic mechanical analysis of nano-SiO_2/bismaleimide composite. Proceedings of the 3rd IEEE Int. Conf. on Nano/Micro Engineered and Molecular Systems January 6-9, 2008, Sanya, China 626-629.
188. Lu, G. T., Huang, Y., Yan, Y. H. et al. 2006. 4,4'-Bismaleimidodiphenyl methane modified novolac resin/titania nanocomposites: preparation and properties. *Journal of Applied Polymer Science* 102: 52–57.

189. Sipaut, C. S., Mansa, R. F., Padavettan, V. et al. 2015. The effect of surface modification of silica nanoparticles on the morphological and mechanical properties of bismaleimide/diamine matrices. *Advances in Polymer Technology* 34: 21492–21502
190. Huang, F. W., Huang, F. R., Zhou, Y. et al. 2011. Nanocomposites of a bismaleimide resin with octaphenylsilsesquioxane or nano-SiO_2. *Polymer Composites* 32: 125–130.
191. Ochi, M., Nii, D., Harada, M. 2010. Effect of acetic acid content on in situ preparation of epoxy/zirconia hybrid materials. *Journal of Materials Science* 45: 6159–6165.
192. Hölck, O., Dermitzaki, E., Wunderle, B. et al. 2011. Basic thermo-mechanical property estimation of a 3D-crosslinked epoxy/SiO_2 interface using molecular modelling. *Microelectronics Reliability* 51: 1027–1034.
193. Daud, N., Shanks, R. 2014. Epoxy–silica composites replicating wood cell structure. *Composites Part A: Applied Science and Manufacturing* 62: 11–15.

7 Piezoelectric and Ferroelectric Polymer/ Ceramic Composites

7.1 INTRODUCTION

The piezoelectric effect was discovered by electrification of certain crystals, such as tourmaline, quartz, topaz, cane sugar, and Rochelle salt under mechanical pressure. When these materials are subjected to mechanical stress, this produces an electric charge proportional to the applied stress, which is known as the "direct piezoelectric effect." However, the inverse or converse piezoelectric effect results when piezoelectric materials generate a mechanical strain under an electric field. The phenomenon of piezoelectricity is a reversible process in which a linear electromechanical-coupling trend between the mechanical strain and electrical field exists. The piezoelectric phenomenon is detected in non-centro symmetric crystalline materials [1–3]. Piezoelectric materials are utilized to fabricate many electrical devices suitable for a variety of applications, such as in communication systems, defense, industrial automation, medical diagnostics, and energy-harvesting applications.

Ceramics are the most extensively investigated material because of their tremendous physical properties in terms of chemical-resistance, high pressure and temperature resistance, high hardness, and good mechanical properties and wear [4–8]. Since their discoveries, these inorganic fillers find numerous applications in many industries including aerospace, nuclear power plants, mining equipment, and cutting tools [9, 10]. The most useful ceramic particles are categorized into oxides (silica, alumina, and zirconia), non-oxides (carbides, borides, nitrides, and silicides), and combinations of oxides and non-oxides. These fillers are usually treated by introducing primary and secondary phase (fibers and whiskers, surfactants), and multiple phases (nanoparticles) to increase their durability and flaw resistance [11].

Ferroelectric ceramics and piezoelectric polymer/ceramic composites are widely used in many applications, such as sensors, transducers, and actuators used in MEMS devices. Intermetallic ceramic fillers, such as lead zirconate titanate (PZT), are one of the most interesting; they have been used in the development of many piezoelectric devices [12]. Due to their high capacity to transform the mechanical deformations into electrical energy, such energy harvesters can be very useful in several potential applications, like medical diagnostic devices, security systems, space vehicles, and other nondestructive testing [13, 14]. The generated energy aspect shows tremendous possibilities as it can be fitted with almost all dynamic movements in nature, including human motion.

The present chapter deals with the recent development in polymer/ceramic composites showing piezoelectric and ferroelectric responses. The mechanisms of harvesting energy and various key parameters affecting the piezoelectric and ferroelectric performance of such composite materials, including the filler ratio, particle size, and the type of polymer structure are discussed.

7.2 PIEZOELECTRIC PROPERTIES OF POLYMER/CERAMIC COMPOSITES

Numerous types of ceramic fillers are investigated for their piezoelectric properties, which means that they are able to convert mechanical energy to electrical energy and vice versa. These materials also exhibit pyroelectric effects (converting thermal energy into electrical energy); it is well-known that all pyroelectric materials are piezoelectric. This characteristic can also be related to the ferroelectricity, by which electric dipoles can be specifically oriented under an electric field [15].

Examples of ferroelectric ceramics are lead titanate (PT), lead zirconate titanate (PZT), calcium-doped lead titanate (PTCa), and lead manganese niobate–lead titanate/barium titanate (PMN-PT/$BaTiO_3$). The piezoelectric properties of ceramic particles are applied in the design of several electronic devices such as actuators, transducers, high-frequency loudspeakers, etc. The most widely utilized ceramic particles for these purposes are $BaTiO_3$ and PZT ceramics as shown in Table 7.1 [16].

Aiming to surpass the inferior compliance of ceramics and the reduced piezoelectricity of polymers, researchers produced piezoelectric polymer composites which combined passive polymer matrixes with piezoelectric ceramic fillers, using the inherent outstanding characteristics of each component [17–21]. Researchers identified a distinction in performance between ceramic particle connectivity patterns in these composites and established nomenclature to differentiate these composites. These composites are annotated in the format of *x-y*, where *x* is the number of dimensions in which the particles are continuously connected to each other, and *y* is the number of dimension in which the matrix is connected to itself. For instance, a randomly oriented composite in which the ceramic particles are not disconnected in a defined pattern was denoted 0 to 3, whereas a pillar type composite was denoted 1 to 3. Polymer/ceramic composites having a connectivity in the range of 0 to 1 have

TABLE 7.1
Piezoelectric and Dielectric Constants of Common Piezoelectric Material [16]

Material	Relative Dielectric Constant, ε_r	Piezoelectric Charge Constant, d_{33} (pC/N)	Piezoelectric Voltage Constant, g_{33} (10^{-3} Vm/N)
$BaTiO_3$	1,700	191	12.6
Quartz	4.5	2.3 (d 11)	50.0
PVDF	13	–33	–339.0
PZT-4	1,300	289	25.1

TABLE 7.2
Significant 0 to 1 Polymer/Ceramic Composites

Ceramic	% Volume	Host	d_{33}	Source
PZT	67	PVDF	48	[22]
PTCa	50	Epoxy	22	[23]
PTCa	65	P(VDF-TrFE)	28	[24]
PTCa	20–50	Rubber	17–44	[25]
Pt	70	Acrylic	32	[26]
(Pb, Bi)TiO_3	35	Epoxy	41	[27]

been widely explored in the past few decades to enhance flexibility, improve dielectric properties, and increase piezoelectricity. Table 7.2 summarizes the various 0 to 1 polymer/ceramic composites recently investigated.

Numerous groups have investigated the inclusion of many piezoelectric ceramics into both passive and active polymeric matrices. Table 7.2 summarizes the various 0 to 1 polymer/ceramic composites recently investigated. Nonpiezoelectric polymeric matrices such as rubbers, thermosetting epoxies, and acrylics have been successfully combined with piezoelectric ceramics such as lead titanate (PT), calcium modified lead titanate (PTCa), lead titanium oxide ($PbTiO_3$) and lead zirconate titanate (PZT). These composites have produced piezoelectric *d* values roughly equal to that of PVDF matrix. However, the best results have been reached when a piezoelectric ceramic and piezoelectric polymer are used, since a 50% enhancement in the d_{33} was attained by incorporating PZT ceramic into PVDF [22].

Dias and Das Gupta produced new composites by varying piezoelectric ceramics and polymers (both passive and active) [24]. Composites prepared from PTCa and poly(vinylidene difluoride–tri-fluoroethylene) P(VDF-TrFE) were investigated and results showed that a d_{33} value of 55 pC/N could occur. The piezoelectricity of these composites was controlled by the electrical breakdown of the raw material, which was near the coercive field of the ceramic particles. They deduced that although the d_{33} enhanced remarkably, the g_{33} was naturally decreased. However, these composites afforded many advantages, such as being polarized easily, and they exhibit a good ability to be self-supporting without requiring any substrate.

Another composite system was explored using various polymers with PZT particles. In that work, the 0 to 3 composites have the average properties of the polymer and ceramic, where Y is the Young's modulus measured in MPa, and ε is the elastic constant [28, 29]. As this model was applied to reproduce the *d* and permittivity values for a piezoelectric composite by other researchers, it was detected that both values were generally higher than those expected. Aiming to account for the deviations (n) this modeling of 0 to 3 connectivity, other researchers established a new model that consider that the composite has a certain amount of mixed connectivity, 0 to 3, 1 to 3, 2 to 3 [30]. Zewdie and Brouers developed a similar calculation technique based on local field coefficients, and both models have been validated as reasonably accurate [31].

7.3 CERAMIC-FILLED SEMI-CRYSTALLINE AND CRYSTALLINE POLYMER COMPOSITES

A review conducted by Ramadan et al. deals with the significant parameters affecting the piezoelectric response of many types of polymers [32]. Other than the nature of the polymer matrices and ceramic fillers used to elaborate their composites, the selected fabrication technique, alignment of polymer crystallites, poling conditions, etc., can also influence the piezoelectric performance of a typical composite specimen. In addition, several polymer/ceramic nanocomposites exhibited a piezo- and thermoelectric properties [33]. Many polymers revealed this energy-generating characteristic, and here, a classification of polymer nanocomposites based on the polymer crystallinity is given for increase comprehension of these phenomena. Table 7.3 contains various polymer/ceramic composites showing piezoelectric property and their production techniques.

Sundaram et al. reviewed the different polymer/ceramic composites for both their dielectric and piezoelectric properties in terms of their piezoelectric strain coefficient (d_{33}) and voltage coefficient (g_{33}). They found that g_{33} is directly proportional to d_{33}, while it was inversely proportional to the permittivity. They also discovered that the d_{33} values depend on the volume fraction of the ceramic particles and the direction of polarization of the polymeric matrix. By mixing the ferroelectric ceramic, PZT, with PVDF matrix in 0.5 volume fraction using hot pressing technique, the resulted composite demonstrated a reasonable d_{33} value, in addition to the good stability and flexibility [34].

TABLE 7.3
Piezoelectric Properties of a Few Ceramic Particles and Its Polymer Composites

Ceramic Particle	Polymer	Fabrication Method	Piezoelectric Charge Constant, d_{33} (pC/N)
PZT	Thermoplastic	Cold isostatic pressing	44
PZT	Epoxy	1–3 composite commercially available	593
PMNT	Epoxy	Dice-and-fill method	1,20
PMNT	Epoxy	Modified dice-and-fill method	1,256
PLZT	PVDF	Hot press	102
PLZT	PVDF	Solution mixing followed by hot press	18
PZT	Polyester resin	Spinning PZT in the resin in a centrifuge	29
PZT	PVDF	Brabender mixing	33
PZT	PVC	Calendering and tape casting	29
PZT	PU	Blending	28
PZT	P(VDF-TrFE)	Solution casting followed by hot pressing	410
PZT	PVC	Solution mixing	31
PZT and carbon black	PVC	Hot pressing	20
PZT	Poly (3-hydroxybutyrate)	Powder mixing and hot pressing	6

Various dimension of $BaTiO_3$ particles, 500 nm (tetragonal), 100 nm (cubic), and 10 nm (cubic), were incorporated into polyvinylidene fluoride-trifluoroethylene (PVDF-TrFE), and its piezoelectric performance was compared with that of pure PVDF matrix. All the prepared specimens were formed as electrospun fibers. Results showed that the $BaTiO_3$ particles of lower size (<500 nm) were trapped inside the polymer fibers whereas those having a diameter of 500 nm were randomly dispersed in and outside the fibers [35]. The electrically-poled fibers were tested for the produced voltage, and the effect of frequency, time, and filler dimension on their performance is illustrated in Figure 7.1.

Shin et al. [36] studied the piezoelectric behavior of $BaTiO_3$-filled polymer composite and its possible use in manufacturing nanogenerators. The PVDF copolymer poly(vinylidene fluoride-co-hexafluoropropylene) or P(VDF-HFP) was adopted as the matrix, and a solvent-assisted film-formation technique was selected for the composite elaboration. An excellent output performance was obtained in the form of both voltage and current, respectively, 110 V and 221 A with corresponding yield power density reaching 0.48 Wcm^{-3}. An optimal ratio of the $BaTiO_3$ ceramic filler was also estimated by evaluating the effect of solvent ratios on ceramic clustering.

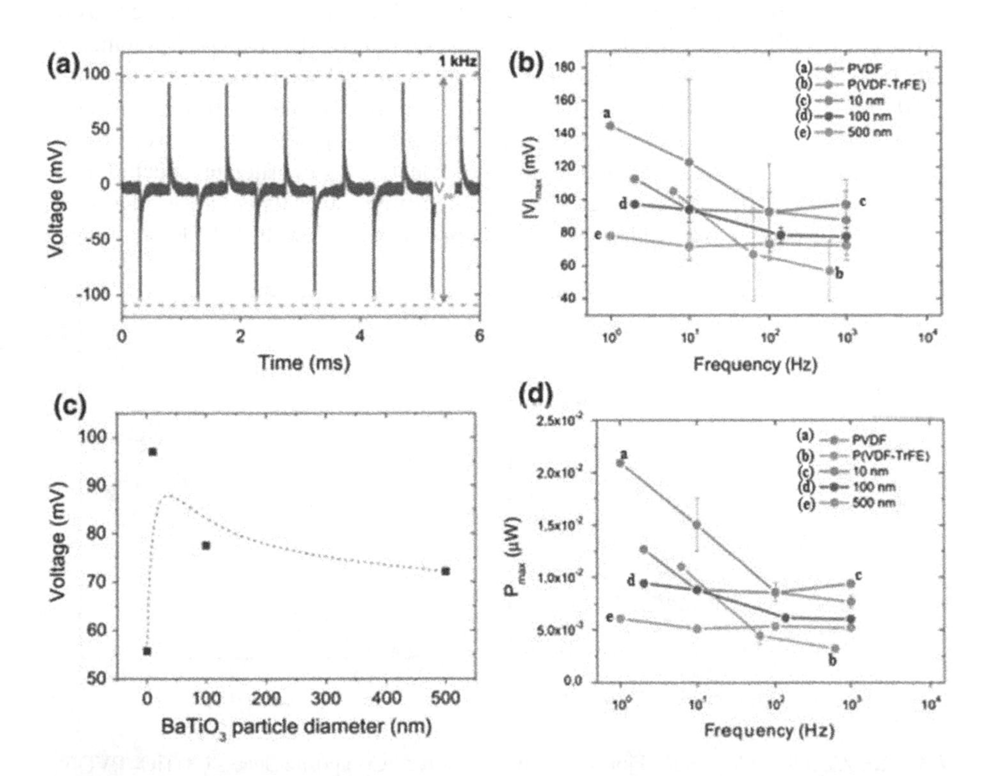

FIGURE 7.1 (a) Voltage Generated During 6 ms at a Frequency of 1 kHz, of Electrospun $BaTiO_3$/P(VDF-TrFE) at 20% filler; (b) Maximum Voltage Generated at Different Frequencies for All Samples; (c) Piezopotential Obtained at 1 kHz for P(VDF-TrFE) and $BaTiO_3$/P(VDF-TrFE) at 20% Filler; and (d) Maximum Power Generated by the Electrospun Membranes [35].

Many types of PVDF composites filled with 50 to 90 vol.% of PLZT were produced by hot pressing [37]. With increasing the PLZT volume fraction and its particle size, both d_{33} and relative permittivity values of the composites is significantly enhanced. For example, the composite specimen filled with 85% PLZT of 150-μm diameter, showed the best performance with a d_{33} of 101 pC/N and relative permittivity of 299. The energy-harvesting efficiency of fibrous $BaTiO_3$/PVDF composites based on longitudinal stretch movement was studied by Kakimoto et al. [38]. Because the fibrous $BaTiO_3$ fillers of 800-nm diameter were well-dispersed in PVDF by means of the extruder, the piezoelectric system can behave like a parallel-plate capacitor. The energy output resulted under an electric field of 2 MX is given in Figure 7.2.

A hybrid combination of PZT and polyaniline (PANI) for improving the properties of PVDF was reported by Sakamoto et al. [39]. A percolation at 20 to 30 vol.% was detected for the hybrid filler (PZT was coated on PANI) combination. At 30 vol.%, the prepared composite exhibited the best value for d_{33} constant, which was attributed to the conducting nature of the composite and the poling process that was employed (5 MV/m electric field for 15 min). This functional composite material is a candidate for applications in structural health-monitoring systems.

Polypropylene (PP) is also a polymer showing piezoelectric response, specifically in its fiber form. The cellular PP is generally produced using a modified extrusion process, in which voids of μm dimension are generated in the melted polymer by gas blowing before foam blowing. The cooling in the subsequent step maintain the voids within the PP polymer. Corona charging is similarly used to fill PP, other than the electrode charging and electron-beam charging [39]. Zhang et al. scrutinized the effect of a fabrication method on the piezoelectric d_{33} coefficient of cellular PP, determining that high-pressure exposure to long-time intervals enhanced the thickness and the d_{33} coefficient. However, static pressures bellow 10 kPa do not change

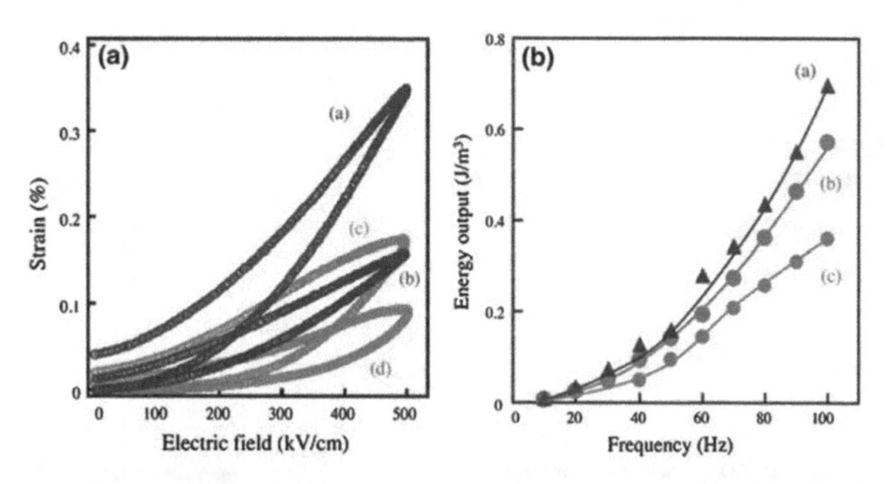

FIGURE 7.2 S–E Curves for Fibrous $BaTiO_3$-reinforced Composite Sheets ($BaTiO_3$/PVDF = 30/70 v/v%) Having Different $BaTiO_3$ Orientation Ratios of (a) 83% and (b) 38%, and Two Reference Samples of (c) Spherical $BaTiO_3$-reinforced Composite Sheets ($BaTiO_3$/PVDF = 30/70 v/v%) and (d) Pure PVDF; (b) Generated Power as a Function of Frequency Applied to Piezoelectric Energy Harvesting Based on Stretch Movement [38].

the value, whereas higher pressures provoke a reversible reduction. The d_{33} constant is also correlated with the reduction in Young's modulus and the enhancement in chargeability of the material due to expansion [40].

Epoxy thermosets are also found to have piezoresponses when ceramic particles are dispersed in them. In a study, lead magnesium niobate titanate (PMNT)-reinforced epoxy composites exhibiting high 1,3-piezoelectric coefficient was reported when compared to the PZT/epoxy composite studied by Li et al. [41]. Wang et al. [42] also explored the same system of PIMNT/epoxy 1–3 composite filled with 60 vol.% filler loading and produced by a modified dice-and-fill technique, where they suggested the applicability in ultrasonic transducers. Another PZT/epoxy composite showing good piezoresponse was explored using a freeze-casting process as reported by Xu and Wang [43]. The main recorded data from their studies are showed in Figure 7.3.

Xu and Wang studied both longitudinal and transverse piezoelectric strain coefficients (d_{33} and d_{31}) for the PZT/epoxy 3-1 composites [43]. These piezo-composites were produced by tert-butyl alcohol (TBA)–based freeze casting of PZT followed by the infiltration of epoxy. The volume fraction of PZT was changed from 0.36 to 0.69 by varying the initial solid loading in freeze-casting slurry. Figure 7.3 reveals the change of d_{33} with respect to the poling voltage and volume fraction of PZT. Generally, the saturation polarization phenomenon is exhibited by the PZT ceramics, when the poling voltage exceeded 2 kV/mm. However, for the studied composite, saturation was achieved at only 0.9 kV/mm itself (Figure 7.3a), which is related to the consistency of polarization direction and one-dimensional pore channel direction as well as the resultant easy deflection of electric domains along the polarization direction.

Figure 7.3b gives the variation in d_{33} and d_{31} constants with the volume fraction of PZT at a poling voltage of 0.9 kV/mm. Both coefficients reveal an enhancement as the PZT ratios increased till 0.57 vol.%. This is due to the weak structural constitution of these composites above this particular PZT volume fraction, confirming that the d_{33} and d_{31} coefficients depend on the piezoelectric phase of the composite

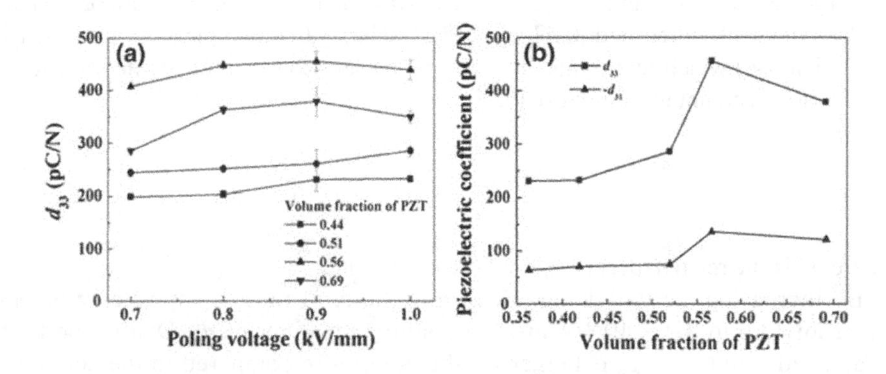

FIGURE 7.3 (a) Effect of Poling Voltage on Longitudinal Piezoelectric Strain Coefficient, d_{33} of 3-1–type PZT/epoxy Composites; (b) d_{33} and Transverse Piezoelectric Strain Coefficient (d_{31}) of 3-1–type PZT/epoxy Composites as a Function of PZT Volume Fraction [43].

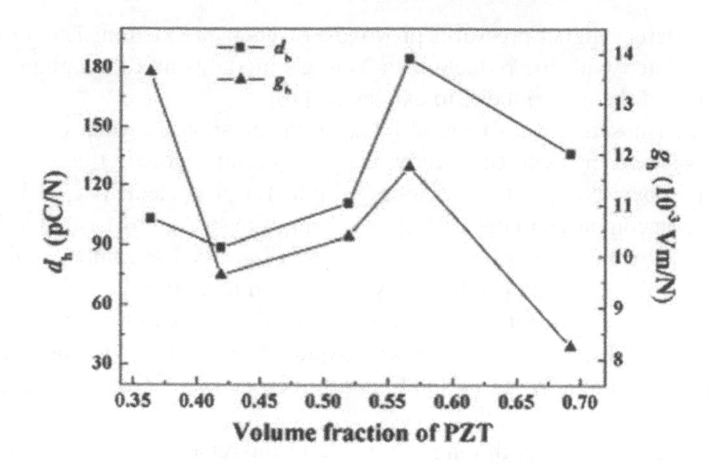

FIGURE 7.4 Variation in Hydrostatic Strain Coefficient (d_h) and Hydrostatic Piezoelectric Voltage Coefficient (g_h) with Volume Fraction of PZT [43].

and interfacial connectivity. As the volume fraction attains 0.69, the unidirectional porous structure ceases its directionality provoking more disorder, thereby lowering the piezoelectricity.

Figure 7.4 summarizes the influence of PZT volume fraction on the hydrostatic strain coefficient (d_h) and hydrostatic piezoelectric voltage coefficient (g_h) of epoxy/PZT composites. The d_h is calculated from the d_{33} and d_{31} constants as presented in Eq. (7.1)

$$d_h = d_{33} + 2d_{31} \tag{7.1}$$

As the PZT concentration was increased from 0.36 to 0.69, the d_h value enhanced from 103 to 136 pC/N. At 0.57 vol.%, the d_h reaches 184 pC/N, which is nearly three times higher than that of the neat PZT (56 pC/N). This behavior is likely ascribed to the large decrease detected in d_{31} respected to d_{33} and the porous structure of the particular composite filled with 0.57 vol.% PZT. The hydrostatic piezoelectric voltage coefficient (g_h) which represents the hydrophone sensitivity in ultrasonic transducer application is computed using Eq. (7.2)

$$g_h = \frac{d_h}{\varepsilon_0 \varepsilon_r} \tag{7.2}$$

where, ε_r is the relative permittivity of the composite.

By increasing the PZT volume fraction, the g_h of these composites was varied from 13.6 to 8.2 × 10^{-3} V m/N in a similar manner of d_h. Due to the high d_h and reduced ε_r, the g_h is bigger for the composite compared to the neat PZT (1.5 × 10^{-3} V m/N). Jain et al. exhaustively studied the relationship between the dielectric properties and piezoelectricity in their recent study of PVDF/PZT composites [44]. They well-established a good theoretical survey based on mathematical models and discussed the two models—Yamada model and Furukawa model—to reproduce the piezoelectric constants. According to the Yamada model, the g_{33} can be given by the Eq. (7.3) as follows:

$$g_{33} = \frac{V_f n a d_{33_2} / \varepsilon_0}{\varepsilon 33_2 + (n-1)\varepsilon_{33_1} \left\{ 1 + \frac{n V_f \left(\varepsilon_{33_2} - \varepsilon_{33_1}\right)}{n\varepsilon_{33_1} + \left(\varepsilon_{33_2} - \varepsilon_{33_1}\right)\left(1 - V_f\right)} \right\}} \tag{7.3}$$

Since a is the poling ratio of ceramic inclusions, ε_{33i} is the dielectric constant of the inclusion at the ith direction of poling, V_f is the filler volume fraction, and n is a dimensionless parameter depending on the shape and orientation of ceramic inclusions.

The Furukawa model can be written by Eqs. (7.4) and (7.5).

$$d = V_f L_T L_E d_2 \tag{7.4}$$

$$g = V_f L_T L_D g_2 \tag{7.5}$$

where d and g denote the piezoelectric coefficients of the composites, and d_2 and g_2 represent the coefficients of the ceramic particles. L_T, L_D, and L_E are the local field coefficients with respect to the stress, electrical displacement, and electric field.

Over the past few decades, the elaboration of flexible PVDF-based percolative nanocomposites with high piezoelectric performance is of great interest from both academia and industry. In their work, Fu et al. developed a sensitive flexible piezoelectric energy harvester by filling a PVDF polymer matrix with oriented $BaTi_2O_5$ nanorods. Out of all the compositions, the 5 vol.% composite sample exhibited the optimal energy-harvesting performance with a high-power density of 27.4 μW/cm^3 together with improved mechanical properties under a large acceleration of 10 g [45].

Piezoelectric nanogenerators (PENGs) having a good flexibility and improved outputs have several promising applications for harvesting mechanical energy and powering electronics. Dutta et al. reported on the design of a NiO@SiO_2/PVDF nanocomposite, a facile piezoelectric material possessing superior flexibility that is light in weight and has low cost, which is a better choice for the next-generation mechanical energy harvester and tactile e-skin sensors. The fabricated piezoelectric nanogenerator comprising nanocomposites exhibits a very promising output under application of the biomechanical force on it [46].

Other PENGs were developed using electrospun nanocomposite fiber mats comprised of barium titanate (BT) nanoparticles, graphene nanosheets, and PVDF. When the nanocomposite fiber mats are constituted of 0.15 wt% graphene nanosheets and 15 wt% BT nanoparticles, the open-circuit voltage and electric power of the PENG can attain as high as 11 V and 4.1 μW under a loading frequency of 2 Hz and a strain of 4 mm, and no apparent decrease of the open-circuit voltage was detected after 1,800 cycles in the durability test [47].

Flexible nanocomposite films of PVDF matrix and barium hexaferrite (BHF) nanoparticles with high dielectric constant were fabricated by the solution cast technique. The dielectric constant was significantly enhanced due to the improved electrostatics and interfacial interaction between the local electric field of the BHF nanoparticle and CH_2/CF_2 dipole of PVDF chain [48].

Kar et al. [49] reported on the prototype fabrication of a flexible, facile multiwalled carbon nanotube@SiO_2-incorporated PVDF nanocomposite-based piezoelectric energy harvester as a cheaper and cleaner source of alternative energy. The prepared flexible nanogenerator demonstrated high performance with a maximum recordable output voltage of 45 V, current density of 1.2 $\mu A/cm^2$, and power density of 5,400 W/m^3 under periodically vertical compression and release operations via biomechanical force.

Highly flexible biocompatible nanocomposites comprising of polyvinylpyrrolidone (PVP)-modified barium calcium zirconate Titanate (BCT-BZT)/PVDF were produced. The optimal device with 60 wt.% PVP-modified BCT-BZT powders showed maximum peak to peak voltage of 23 V when tested for harnessing waste biomechanical energy (human hand palm force) [50]. A tri-phasic filler system comprised of one-dimensional TiO_2 nanotubes, two-dimensional reduced graphene oxide, and three-dimensional $SrTiO_3$, included into a PVDF semi-crystalline polymer was developed. The results demonstrated that the piezoelectric constant has risen due to the change in the filler loading and attained a value of 7.52 pC/N at 1:2 filler combination. The output voltage for a fixed-filler composition was nearly 10.5 times that of the voltage generated by the unfilled polymer [51].

Sun et al. [52] designed and fabricated a wireless piezoelectric device consisting of a piezoelectric pressure sensor based on electrospun PVDF/$BaTiO_3$ nanowire (NW) nanocomposite fibers, and a wireless circuit system integrated by a data-conversion control module, a signal acquisition and amplification module, and a Bluetooth module. Polyamide11 (PA11)/$BaTiO_3$/carbon nanotube (CNT) ternary nanocomposites (PBCNCs-3D) with 3D segregated percolation routes were produced by selective laser sintering (SLS) of CNT-coated PA11/$BaTiO_3$ nanocomposite powder. The ternary composites exhibited an enhanced piezoelectric performance by constructing 3D segregated percolation routes [53].

7.4 PIEZOELECTRIC AMORPHOUS POLYMER/CERAMIC COMPOSITES

The search for flexible piezoelectric materials based on polymer composites has been started from the reports of Kitayama and Sugawara in 1972 [54]. Piezoelectric flexible composites can be particularly used in hydrophones were developed by Banno and Saito in 1983 [55]. These composites were produced form a synthetic rubbers as matrix and ceramic particles such as lead titanate and PZT. The rubber/ceramic composites produced by hot rolling were poled in silicone oil by applying 100 kV/cm at 60°C for 1 hour. The composite sample produced from pure lead titanate exhibited the highest piezo-properties and was able to be used in producing hydrophones. The experimental data were in good agreement with the theoretical "modified cubes model" as well. Mamada et al. studied a set of pseudo-1-3 piezoelectric ceramic/polymer composites consisting of linearly organized piezoelectric ceramic particles reinforced silicone gel, poly-methyl-methacrylate, urethane rubber, and silicone rubber matrices for their piezoelectric behavior [56].

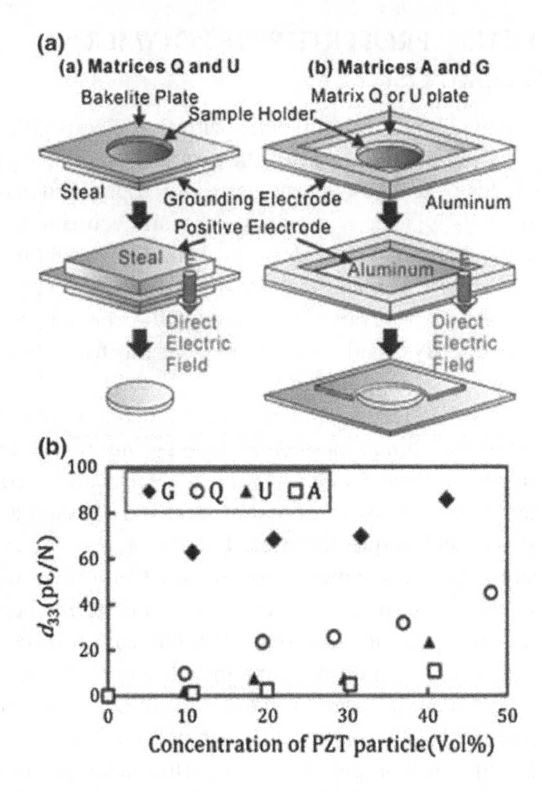

FIGURE 7.5 (a) Fabrication Process of Aligned-type Piezoelectric Composite: (a) Matrices Q and U, (b) Matrices A and G; (b) Relationship Between PZT Particle Concentration and d_{33} for Aligned-type Piezoelectric Composite [56].

The polymer matrices were categorized into two sets—matrices Q and U not adhered to the electrodes (Q is the silicone rubber and U is the urethane rubber) while matrices A and G adhered to the electrodes (A represents PMMA and G represents silicone gel). Figure 7.5a shows the elaboration steps followed for the two sets of ceramic/polymer composites and Figure 7.5b their d_{33} constant with compared to the filler volume ratios. At a fixed PZT ratio, the d_{33} values exhibit a variation in the order of $A < U < Q < G$. As the PZT concentration reaches 10 vol.%, the d_{33} for silicone gel composite was exceeded 60 pC/N and at 40 vol.%, the value attained 80 pC/N. However, for the PMMA/PZT composite, the d_{33} reached only 10 pC/N even at 40 vol.% PZT. This change in piezoelectric coefficient is attributed to the discrepancy in the relative permittivity and Young's modulus of the composites. The difference in dielectric constant is related to the different area of exposure of each polymeric host where PZT particle alignment is exposed out of the composite surface. In addition, the d_{33} values increased with decreasing the Young's modulus of the matrix because of the enhancement in force applied on the aligned PZT particles in these composites having inferior Young's modulus and the increased force with the increased area of exposure.

7.5 FERROELECTRIC PROPERTIES OF POLYMER/CERAMIC COMPOSITES

The elaboration of new polymeric composites with ferroelectric fillers and the evaluation of their properties are urgent tasks of the growing practical application of these materials in many fields of technology, such as radio-, opto- and acousto-electronics, nonlinear optics, etc. [57–59]. The selection of the manufacturing method of the composite (dielectric matrix/nanodispersed ferroelectric filler) would markedly eliminate uncertainty in the microstructure and the size of the ferroelectric grains. By tailoring the properties of the matrix, it is possible to change the characteristics and functionality of composites in a fairly broad range. Today, various ferroelectric/polymer composite coatings are known as shown in Table 7.4.

Recently, ferroelectric materials have attracted much attention due to their large industrial applications, including nonvolatile memory and transducers in sensors and actuators, electrostriction, electric energy storage, electrocaloric cooling, fuel injectors for high efficiency-low emission diesel engines, and ultrasonic rotary inchworm motors with high power and torque densities. These materials are essentially divided into two groups: ceramics and polymer ferroelectrics. Ceramic ferroelectrics demonstrate a highly desirable properties; however, they are brittle and heavy materials. By contrast, the ferroelectric polymers are light, flexible, easy to process and low cost; but, their polar properties are an order of magnitude lower. Among these polymers, PVDF and its composites are a widely utilized class of polymer ferroelectrics showing marked ferroelectric properties due to their b polymorph characteristic [70].

The elaboration and development of a high-performance piezoelectric and ferroelectric material based on poly(vinylidene fluoride)-reduced graphene oxide-barium titanate (PVDF-RGO-BTO($BaTiO_3$)) nanocomposites have been recently studied [71]. The nanocomposites were produced by the polymerization of the graphene oxide (GO) into the PVDF matrix to provide an insulator-conductor-insulator sandwich structure, and BTO was then carefully added to the composite. The as-prepared nanocomposite exhibited a dielectric constant of 98 and relatively low dielectric loss of 0.081 at a fixed frequency of 1 MHz. A very low leakage current density (1.29×10^{-7} A/cm^2) was measured at 80 MV/m, revealing the high breakdown strength of this nanocomposite. Moreover, it showed a high energy density of 4.5 J/cm^3 at an electric field of 50 MV/m.

TABLE 7.4
Various Ferroelectric/Polymer Composite Coatings

Matrix	Ceramic Filler	References
PVDF	$BaTiO_3$, $Ba(Fe_{0.5}Nb_{0.5})O_3$	[60, 61]
PVDF copolymer with hexafluoropropylene	$BaTiO_3$	[62]
Polyimide	SiO_2@GO	[63]
Polyurethane	$BaTiO_3$	[64]
Polytetrafluoroethylene	hBN	[65]
polyvinyl chloride	$CaCu_3Ti_4O_{12}$, $BaTiO_3$	[66]
Epoxy	$Pb(Zr,Ti)O_3$ (PZT), $BaTiO_3$	[67–69]

Ishaq et al. [72] designed a new hybrid composite based on polyvinylidene fluoride as a dielectric polymer matrix reinforced with graphene platelets as a conductive and barium titanite as ceramic ferroelectric fillers. The results revealed that a combination of graphene and ferroelectric ceramic are an excellent approach to significantly improve the performance of dielectric and ferroelectric properties of piezoelectric polymers for broad applications including energy storage. The introduction of high-ε nanomaterials into ferroelectric polymers has proven to be a promising strategy for the fabrication of high-ε nanocomposites.

Zhang et al. introduce nanosized $BaSrTiO_3$ with systematically varied morphologies into a ferroelectric polymer matrix to produce ferroelectric polymer nanocomposites. The solution-processed polymer nanocomposites show an extraordinary room-temperature electrocaloric effect via the synergistic combination of the high breakdown strength of a ferroelectric polymer matrix and the large change of polarization with temperature of ceramic nanofillers [73]. Wang et al. [74] elaborated on four kinds of nanowires ($Na_2Ti_3O_7$, TiO_2, $BaTiO_3$, and $SrTiO_3$) with different inherent characteristics that are used to produce high-ε ferroelectric polymer nanocomposites. Dopamine functionalization eases the homogenous dispersion of these nanowires in the ferroelectric polymer matrix due to the strong polymer/nanowire interfacial adhesion. They deduced that among the four kinds of nanowires, $BaTiO_3$ NWs exhibit the best potential in enhancing the energy-storage capability of the proposed nanocomposites, generating from the most significant enhancement of ε while keeping the rather low dielectric loss and leakage current.

Concurrent improvements in dielectric constant and breakdown strength are reached in a solution-processed ternary ferroelectric polymer nanocomposite incorporated with two-dimensional BN nanosheets and zero-dimensional $BaTiO_3$ nanoparticles that synergistically interact to provide a remarkable energy-storage capability, such as large discharged energy density, high charge–discharge efficiency, and great power density [75]. Solution-processable ferroelectric polymer nanocomposites are developed by Zhang et al. as a new form of electrocaloric materials that can be effectively worked under both modest and high electric fields at ambient temperature [76].

Prateek et al. [77] give a comprehensive review of the use of ferroelectric polymers, especially PVDF and PVDF-based copolymers/blends as potential components in dielectric nanocomposite materials for high-energy density capacitor applications. Different parameters like dielectric constant, dielectric loss, breakdown strength, energy density, and flexibility of the polymer nanocomposites have been thoroughly studied. Li et al. [78] prepared uniform $BaTiO_3$ nanofibers (BT-nfs) having a large aspect ratio via the electrospinning technique, surface modified by poly(vinyl pyrrolidone) (PVP) and used them as fillers for elaborating the PVDF nanocomposites. They reported that the nanocomposite reinforced with 3 vol.% BT-nfs has a marked enhancement in the discharged energy density of 8.55 J/cm^3 at an applied electric field of 300 MV/m, which represents 43% higher than the value recorded for the pure polymer matrix and more than four times that that of the commercial biaxial-oriented polypropylene dielectric (2 J/cm^3 at over 600 MV/m).

Zhu et al. [79] used three types of core–shell structured polymer@$BaTiO_3$ nanoparticles with polymer shells having different electrical properties as reinforcing agent to produce ferroelectric polymer nanocomposites via a surface-initiated

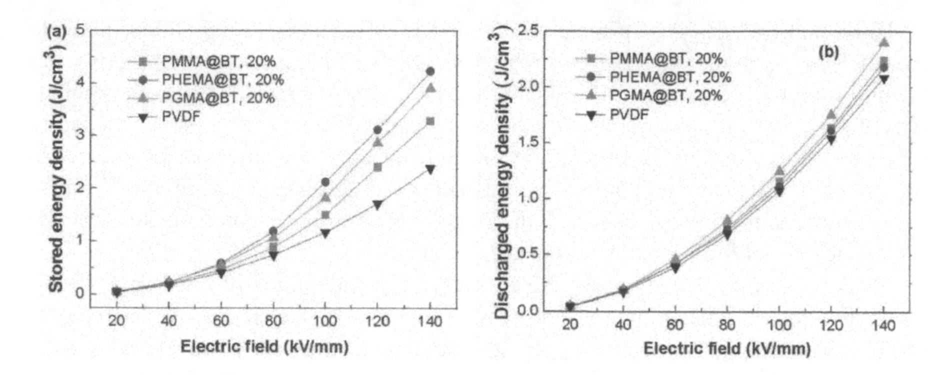

FIGURE 7.6 Stored (a) and Discharged (b) Energy Densities of PVDF and Polymer@BT-based PVDF Nanocomposites [79].

reversible addition–fragmentation chain transfer (RAFT) polymerization. They found that to develop nanocomposites with high discharged energy density, the core-shell nanoparticle filler should simultaneously exhibit high dielectric constant and small electrical conductivity as shown in Figure 7.6.

7.6 CONCLUSIONS

The accelerated developments in personal electronics and sensor networks have increased urgent and challenging requirements for portable and sustainable power sources. Flexible piezoelectric materials have attracted much attention due to their efficient route to scavenge energies from the living environment to power personal electronics and nanosystems. To deal with the limitations of piezoelectric polymers, which have a low dielectric constant, and to improve their dielectric and ferroelectric efficiency for energy storage applications, the addition of ceramic filler is one promising solution. This chapter reviewed the various piezoelectric characteristics of polymer/ceramic composites. The effect of adding ceramic to both semi-crystalline and amorphous polymeric matrices on the piezoelectric performance of composites were studied for their piezoelectric properties. The synthesis routes of ceramic particles and polymer/ceramic composites was first introduced. By analyzing the data published in literature, it appeared that the piezoelectricity can be influenced by several parameter such as the dielectric properties, Young's modulus of the matrix, poling conditions, filler orientation, etc. The high-performance piezoelectric-ferroelectric polymer/ceramic material can be a promising building block for its application in energy harvesting and high-frequency capacitors.

REFERENCES

1. Jing, Q., Kar-Narayan, S. 2018. Nanostructured polymer-based piezoelectric and triboelectric materials and devices for energy harvesting applications. *Journal of Physics D: Applied Physics* 51: 303001.
2. Jaffe, B., Cook, W., Jaffe, H. 1971. *Piezoelectric Ceramics*. Academic Press: New York, p. 115.

3. Moulson, A. J., Herbert, J. M. 2003. *Piezoelectric Ceramics*. In *Electroceramics*, John Wiley & Sons, Ltd, pp. 339–410.
4. Carroll, L., Sternitzke, M., Derby, B. 1996. Silicon carbide particle size effects in alumina-based nanocomposites. *Acta Materialia* 44: 4543–4552.
5. Peigney, A., Flahaut, E., Laurent, C. et al. 2002. Aligned carbon nanotubes in ceramic-matrix nanocomposites prepared by high-temperature extrusion. *Chemical Physics Letters* 352: 20–25.
6. Yuan, J., Galetz, M., Luan. X. et al. 2016. High-temperature oxidation behavior of polymer-derived SiHfBCN ceramic nanocomposites. *Journal of the European Ceramic Society* 36: 3021–3028.
7. Peigney, A., Laurent, C., Flahaut, E. et al. 2000. Carbon nanotubes in novel ceramic matrix nanocomposites. *Ceramics International* 26: 677–683.
8. Yoon, S., Byun, H., Yun, Y. 2015. Characterization and photocatalytic properties of ceramics TiO_2 nanocomposites. *Ceramics International* 41: 8241–8246.
9. Saheb, N., Mohammad, K. 2016. Microstructure and mechanical properties of spark plasma sintered Al_2O_3-SiC-CNTs hybrid nanocomposites. *Ceramics International* 42: 12330–12340.
10. Hirvonen, A., Nowak, R., Yamamoto, Y. et al. 2006. Fabrication, structure, mechanical and thermal properties of zirconia-based ceramic nanocomposites. *Journal of the European Ceramic Society* 26: 1497–1505.
11. Ionescu, E., Linck, C., Fasel, C. et al. 2010. Polymer-derived SiOC/ZrO_2 ceramic nanocomposites with excellent high-temperature stability. *Journal of the American Ceramic Society* 93: 241–250.
12. Sundaram, S., Sampathkumar, P., Gowdhaman, P. et al. 2014. Dielectric and piezoelectric properties of various ferroelectric ceramic-polymer composites. *Journal of Environmental Nanotechnology* 3: 27–31.
13. Patel, I. 2011. *Ceramic based intelligent piezoelectric energy harvesting device*. INTECH Open Access Publisher, Croatia.
14. Patel, I., Siores, E., Shah, T. 2010. Utilisation of smart polymers and ceramic based piezoelectric materials for scavenging wasted energy. *Sensors and Actuators A* 159: 213–218.
15. Pullar, R. C. 2012. Hexagonal ferrites: a review of the synthesis, properties and applications of hexaferrite ceramics. *Progress in Materials Science* 57: 1191–1334.
16. Priya, S., Inman, D. J. 2008. *Energy Harvesting Technologies*. Springer: New York.
17. Senthilkumar, R., Sridevi, K., Venkatesan, J. et al. 2005. Investigations on ferroelectric PZT-PVDF composites of 0–3 connectivity. *Ferroelectrics* 325(1): 121–130.
18. Nhuapeng, W., Tunkasiri, T. 2002. Properties of 0–3 lead zirconate titanate-polymer composites prepared in a centrifuge. *Journal of the American Ceramic Society* 85(3): 700–702.
19. Cai, X., Zhong, C., Zhang, S. 1997. A surface treating method for ceramic particles to improve the compatibility with PVDF polymer in 0-3 piezo-electric composites. *Journal of Materials Science Letters* 16: 253–254.
20. Cui, C., Baughman, R. H., Iqbal, Z., Kazmar, T. R., Dahlstrom, D. K. 1997. Improved piezoelectric ceramic/polymer composites for hydrophone applications. *Synthetic metals* 85: 1391.
21. Venkatragavaraj, E., Satish, B., Vinod, P. R., Vijaya, M. S. J. 2001. Piezoelectric properties of ferroelectric PZT-polymer composites. *Journal of Physics D: Applied Physics* 34: 487.
22. Yamada, T., Ueda, T., Kitayama, T. 1982. Piezoelectricity of a high-content lead zirconate titanate/polymer composite. *Journal of Applied Physics* 53: 4328.
23. Garner, G., Shorrocks, N., Whatmore, R., Goosey, M., Seth, P., Ainger, F. 1986. 0–3 piezoelectric composites for large area hydrophones. *Ferroelectrics* 93: 169–176.

24. Dias, C. J., Das-Gupta, D. K. 1996. Inorganic ceramic/polymer ferroelectric composite electrets. *IEEE Transactions on Dielectrics and Electrical Insulation* 3: 706–734.
25. Banno, H., Ogura, K., Sobue, H., Ohya, K. 1987. Piezoelectric and acoustic properties of piezoelectric flexible composites. *Japanese Journal of Applied Physics* 26: 153.
26. Hanner, K., Safari, A., Newnham, R., Runt, J. 1989. Thin film 0–3 polymer/piezoelectric ceramic composites: Piezoelectric paints. *Ferroelectrics* 100: 255–260.
27. Han, K., Safari, A., Riman, R. E. 1991. Colloidal Processing for Improved Piezoelectric Properties of Flexible 0–3 Ceramic–Polymer Composites. *Journal of the American Ceramic Society* 74: 1699–1702.
28. Furukawa, T., Fujino, K., Fukada, E. 1976. Electromechanical properties in the composites of epoxy resin and PZT ceramics. *Japanese Journal of Applied Physics* 15: 2119.
29. Furukawa, T., Ishida, K., Fukada, E. 1979. Piezoelectric properties in the composite systems of polymers and PZT ceramics. *Journal of Applied Physics* 50: 4904.
30. Pardo, L., Mendiola, J., Alemany, C. 1988. Theoretical treatment of ferroelectric composites using Monte Carlo calculations. *Journal of Applied Physics* 64: 5092.
31. Zewdie, H., Brouers, F. 1990. Theory of ferroelectric polymer-ceramic composites. *Journal of Applied Physics* 68: 713.
32. Ramadan, K.S., Sameoto, D., Evoy, S. 2014. A review of piezoelectric polymers as functional materials for electromechanical transducers. *Smart Materials and Structures* 23: 033001.
33. Ponnamma, D., Ogunleye, G. J., Sharma, P. et al. 2016. Piezo- and thermoelectric materials from biopolymer composites. In Sadasivuni, K. K. et al. (ed.) *Biopolymer Composites in Electronics*, Amsterdam: Elsevier, p. 333–352.
34. Sundaram, S., Sampathkumar, P., Gowdhaman, P. et al. 2014. Dielectric and piezoelectric properties of various ferroelectric ceramic-polymer composites. *Journal of Environmental Nanotechnology* 3: 27–31.
35. Nunes-Pereira, J., Sencadas, V., Correia, V. et al. 2013. Energy harvesting performance of piezoelectric electrospun polymer fibers and polymer/ceramic composites. *Sensors and Actuators A* 196: 55–62.
36. Shin, S. H., Kim, Y. H., Jung, J. Y. 2014. Solvent-assisted optimal $BaTiO_3$ nanoparticles-polymer composite cluster formation for high performance piezoelectric nanogenerators. *Nanotechnology* 25(48): 485401.
37. Han, P., Pang, S., Fan, J. 2013. Highly enhanced piezoelectric properties of PLZT/PVDF composite by tailoring the ceramic Curie temperature, particle size and volume fraction. *Sensors and Actuators A* 204: 74–78.
38. Kakimoto, K., Fukata, K., Ogawa, H. 2013. Fabrication of fibrous $BaTiO_3$-reinforced PVDF composite sheet for transducer application. *Sensors and Actuators A* 200: 21–25.
39. Lindner, M., Bauer-Gogonea, S., Bauer, S. et al. 2002. Dielectric barrier microdischarges: Mechanism for the charging of cellular piezoelectric polymers. *Journal of Applied Physics* 91: 5283–5528.
40. Zhang, X., Hillenbrand, J., Sessler, G. M. 2004. Piezoelectric d33 coefficient of cellular polypropylene subjected to expansion by pressure treatment. *Applied Physics Letters* 85: 1226–1228.
41. Li, G., Luan, G. D., Qu, H. 2014. Study on novel relaxor ferroelectric single crystal PMNT/epoxy composite. *Applied Mechanics and Materials* 475–476: 1257–1261.
42. Wang, W., Or, S.W., Yue, Q. 2013. Cylindrically shaped ultrasonic linear array fabricated using PIMNT/epoxy 1-3 piezoelectric composite. *Sensors and Actuators A* 192: 69–75.
43. Ponnamma, D., Chamakh, M. M., Deshmukh, K. et al. 2017. Ceramic-based polymer nanocomposites as piezoelectric materials. Ponnamma, D. et al. (Eds.), *Smart Polymer Nanocomposites,* Springer Series on Polymer and Composite Materials. New York: p. 397.

44. Jain, A., Prashanth, K. J., Sharma, A. K. et al. 2015. Dielectric and piezoelectric properties of PVDF/PZT composites: A review. *Polymer Engineering & Science* 55(7): 1589–1616.
45. Fu, J., Hou, Y., Gao, X., Zheng, M., Zhu, M. Highly durable piezoelectric energy harvester based on a PVDF flexible nanocomposite filled with oriented $BaTi_2O_5$ nanorods with high power density. *Nano Energy* 52: 391–401.
46. Dutta, B., Kar, E., Bose, N., and Mukherjee, S. 2018. NiO@SiO_2/PVDF: A flexible polymer nanocomposite for a high performance human body motion-based energy harvester and tactile e-skin mechanosensor. *ACS Sustainable Chemical Engineering* 6(8): 10505–10516.
47. Shi, K., Sun, B., Huang, X., Jiang, P. 2018. Synergistic effect of graphene nanosheet and BaTiO3 nanoparticles on performance enhancement of electrospun PVDF nanofiber mat for flexible piezoelectric nanogenerators. *Nano Energy* 52: 153–162.
48. Kumar, S., Supriya, S., Kar, M. 2018. Enhancement of dielectric constant in polymer-ceramic nanocomposite for flexible electronics and energy storage applications. *Composites Science and Technology* 157: 48–56.
49. Kar, E., Bose, N., Dutta, B., Mukherjee, N., and Mukherjee, S. 2018. MWCNT@SiO_2 heterogeneous nanofiller-based polymer composites: a single key to the high-performance piezoelectric nanogenerator and x-band microwave shield. *ACS Applied Nano Materials* 1(8): 4005–4018.
50. Patra, A., Pal, A., Sen, S. Polyvinylpyrrolidone modified barium zirconate titanate/polyvinylidene fluoride nanocomposites as self-powered sensor. *Ceramics International* 44(10): 11196–11203.
51. Ponnamma, D., Erturk, A., Parangusan, H. et al. 2018. Stretchable quaternary phasic PVDF-HFP nanocomposite films containing graphene-titania-$SrTiO_3$ for mechanical energy harvesting. *Emergent Materials* 1–11. https://doi.org/10.1007/s42247-018-0007-z.
52. Sun, B., Guo, W. Z., Tan, C. et al. 2018. Wireless piezoelectric device based on electrospun PVDF/BaTiO3 NW nanocomposite fibers for human motion monitoring. *Nanoscale*. DIO: 10.1039/C8NR05292A
53. Qi, F., Chen, N., Wang, Q. 2018. Dielectric and piezoelectric properties in selective laser sintered polyamide11/$BaTiO_3$/CNT ternary nanocomposites. *Materials & Design* 143: 72–80.
54. Kitayama, T., Sugawara, S. 1972. Flexible Piezoelectric Materials. Report in Proceeding of Institute of Electrical Communication Engineers of Japan. *CPM* 27, 17.
55. Banno, H., Saito, S. 1983. Piezoelectric and dielectric properties of composites of synthetic rubber and PbTiO3 or PZT. *Japanese Journal of Applied Physics* 22: 67–69.
56. Mamada, S., Yaguchi, N., Hansaka, M. et al. 2015. Matrix influence on the piezoelectric properties of piezoelectric ceramic/polymer composite exhibiting particle alignment. *Journal of Applied Polymer Science* 132(15): 41817.
57. Singh, P., Borkar, H., Singh, B.P., Singh, V.N., Kumar, A. 2014. Ferroelectric polymer-ceramic composite thick films for energy storage applications. *AIP Advances* 8: 087117.
58. Barber, P., Balasubramanian, S., Anguchamy, Y. et al. 2009. polymer composite and nanocomposite dielectric materials for pulse power energy storage. *Materials* 2: 1697–1733.
59. Tanaka, T. 2005. Dielectric nanocomposites with insulating properties. *IEEE Transactions on Dielectrics and Electrical Insulation* 12: 914–928.
60. Zhang, X., Zhao, S., Fang, W. et al. 2017. Improving dielectric properties of $BaTiO_3$/poly (vinylidene fluoride) composites by employing core-shell structured $BaTiO_3$@Poly(methylmethacrylate) and $BaTiO_3$@Poly(trifluoroethyl methacrylate) nanoparticles. *Applied Surface Science* 403: 71–79.
61. Joseph, N., Singh, S. K., Sirugudu, R. K., Murthy, V. R. K., Ananthakumar, S., Sebastian, M. T. 2013. Effect of silver incorporation into PVDF-barium titanate composites for EMI shielding applications. *Materials Research Bulletin* 48: 1681–1687.

62. Shin, S.-H., Kim, Y.-H., Jung, J.-Y., Hyung Lee, M., Nah, J. 2014. Solvent-assisted optimal $BaTiO_3$ nanoparticles-polymer composite cluster formation for high performance piezoelectric nanogenerators. *Nanotechnology* 25: 485401.
63. Liu, L., Lv, F., Zhang, Y. et al., 2017. Enhanced dielectric performance of polyimide composites with modified sandwich-like SiO_2@GO hybrids. *Composites Part A: Applied Science and Manufacturing* 99: 41–47.
64. Khan, M.N., Jelania, N., Li, C., Khaliq, J. 2017. Flexible and low cost lead free composites with high dielectric constant. *Ceramics International* 43: 3923–3926.
65. Pan, C., Kou, K., Jia, Q., Zhang, Y., Wu, G., Ji, T. 2017. Improved thermal conductivity and dielectric properties of hBN/PTFE composites via surface treatment by silane coupling agent. *Composites Part B: Engineering* 111: 83–89.
66. Singh, A. P., Singh, Y .P. 2017. Dielectric behavior of $CaCu_3Ti_4O1_2$: Poly vinyl chloride ceramic polymer composites at different temperature and frequencies. *Modern Electronic Materials* 2: 121–126.
67. Khaliq, J., Deutz, D. B., Frescas, J. F. A. S. et al., 2017. Effect of the piezoelectric ceramic filler dielectric constant on the piezoelectric properties of PZT-epoxy composites. *Ceramics International* 43: 2774–2779.
68. Wu, M., Yuan, X., Luo, H. et al. 2017. Enhanced actuation performance of piezoelectric fiber composites induced by incorporated $BaTiO_3$ nanoparticles in epoxy resin. *Physics Letters A* 381: 1641–1647.
69. Kim, D. S., Baek, C., Ma, H. J., Kim, D. K. 2016. Enhanced dielectric permittivity of $BaTiO_3$/epoxy resin composites by particle alignment. *Ceramics International* 42: 7141–7147.
70. Yang, L., Li, X., Allahyarov, E. et al. 2013. Novel polymer ferroelectric behavior via crystal isomorphism and the nanoconfinement effect. *Polymer*, 54: 1709–1728.
71. Yaqoob, U., Uddin, A. S. M. I., Chung, G.S. 2016. The effect of reduced graphene oxide on the dielectric and ferroelectric properties of PVDF-$BaTiO_3$ nanocomposites. *RSC Advances* 36: 29863–30768.
72. Ishaq, S., Kanwal, F., Atiq, S. et al. 2018. Advancing dielectric and ferroelectric properties of piezoelectric polymers by combining graphene and ferroelectric ceramic additives for energy storage applications. *Materials* 11(9): 1553.
73. Zhang, G., Zhang, X., Yang, T. et al. 2015. Colossal room-temperature electrocaloric effect in ferroelectric polymer nanocomposites using nanostructured barium strontium titanates. *ACS Nano* 9(7): 7164–7174.
74. Wang, G., Huang, X., Jiang, P. et al. 2015. Tailoring dielectric properties and energy density of ferroelectric polymer nanocomposites by high-k nanowires. *ACS Applied Materials & Interfaces* 7(32): 18017–18027.
75. Li, Q., Han, K., Gadinski, M. R., Zhang, G., Wang, Q. 2014. High energy and power density capacitors from solution-processed ternary ferroelectric polymer nanocomposites. *Advanced Materials* 26(36): 6244–6249.
76. Zhang, G., Li, Q., Gu, H. et al. 2015. Ferroelectric polymer nanocomposites for room-temperature electrocaloric refrigeration. *Advanced Materials* 27(8): 1450–1454.
77. Prateek, V. K. T., Gupta, R. K. 2016. Recent progress on ferroelectric polymer-based nanocomposites for high energy density capacitors: Synthesis, dielectric properties, and future aspects. *Chemical Reviews* 116(7): 4260–4317.
78. Li, Z., Liu, F., Yang, G. et al. 2018. Enhanced energy storage performance of ferroelectric polymer nanocomposites at relatively low electric fields induced by surface modified $BaTiO_3$ nanofibers. *Composites Science and Technology* 164: 214–221.
79. Zhu, M., Huang, X., Yang, K. et al. 2014. Energy storage in ferroelectric polymer nanocomposites filled with core–shell structured polymer@$BaTiO_3$ nanoparticles: Understanding the role of polymer shells in the interfacial regions. *ACS Applied Materials & Interfaces* 6(22): 19644–19654.

8 Electrical Properties of Polymer/Ceramic Composites

8.1 INTRODUCTION

Ceramic micro- and nanofillers like $BaTiO_3$, $SrTiO_3$, alumina (Al_2O_3), silica (SiO_2) and titania (TiO_2) ameliorate the dielectric properties of polymeric matrices remarkably, and the treatment of their outer surface can provide an extra-improvement on these properties. Ceramics like SiO_2 or TiO_2 nanoparticles generally exist in from of powder or colloidal and are produced by sol–gel or the micro-emulsion technique. The presence of hydroxyl groups on the surfaces of these fillers ensure the good grafting of these particles on the various polymers. This kind of treatment enhances significantly their dispersion and decreases the formation of agglomerates.

Ceramics modification includes treating them with different surfactants or chemical grafting of organic molecules on their outer surfaces. Surfactants generate wrapping phenomena on the ceric particles and thus reduce their tendency to agglomerate. They can equally reduce the surface tension of the ceramic suspension, which improves the filler segregation and good dispersion. Chemical treatments aim to add functional groups on ceramic surfaces and contribute to making covalent bonds between filler and organic matrices. Although these techniques afford improved bonding, they can result in some filler defects, which in turn can alter the final electrical properties of their polymer composites.

Polymer/ceramic composites are potential candidate dielectrics for several other electronic systems, such as embedded capacitors. Because of the poor adhesion and poor thermal stress reliability at high-filler loadings, commercially available polymer/ceramic composites can only attain a maximum dielectric constant (ε) of 30. But a higher ε of 50 to 200 is recommended to elaborate the layout area dense enough for embedding electronics applications.

Electrical properties of any material include several electrical characteristics that are generally associated with dielectric properties and conductivity characteristics. Electrical properties of reinforced polymers' nanocomposites are intended to be different when the fillers reach the nanoscale for many reasons. First, quantum effects start to be more significant as the electrical properties of nanoparticles can be modified compared with the microparticles. Second, as the particle size reduces, the interparticles' spacing decreases at a fixed-volume fraction. Thus, percolation can take place at lower volume ratios. Moreover, the decrease rate of resistivity is inferior to that detected in micrometer-scale fillers because of the large interfacial area and high interfacial resistance.

The dielectric spectroscopy revealed a reduction in dielectric permittivity for the nanocomposite compared to the unfilled polymer. The most remarkable difference between micrometer scale and nanoscale fillers was the significant increase in the interfacial area in nanocomposites. The increased interfacial zone, in addition to particle/polymer bonding, played an important role in evaluating the dielectric behavior of nanocomposites.

Xu et al. [1] systematically studied the new formulations of polymer/ceramic composites in order to obtain a high dielectric constant ($\varepsilon \approx 50$) at the lowest ceramic filler loading. They reported that composite design and the involved processing technique were critical parameters.

By modifying the epoxy matrix with a chelating agent and using bimodal fillers and a proper amount of dispersing agent, a dielectric constant of 50 was attained at moderate filler loadings. Al-Bayer reported that the applied frequency and filler fractions can influence the AC electrical conductivity and dielectric behavior of the epoxy/Al_2O_3 composites. They also found that both the optical energy gap (E_{opt}) and tail widths (D_E) vary with the Al_2O_3 content dispersed in the epoxy matrix [2]. Al-Shabander et al. [3] investigated the effect of ZrO_2 on dielectric properties of epoxy resin. They found that by adding ZrO_2 to epoxy, the electrical performance of their resulted composites improved.

Mohanty et al. [4] reported a remarkable improvement in the breakdown voltage and breakdown time of epoxy/Al_2O_3 nanocomposites as compared to neat epoxy due to uniform dispersion of nano-Al_2O_3 particles. At 2 wt.% of Al_2O_3 nanoparticles content, the breakdown voltage and breakdown time were increased by 91% and 155%, respectively, compared to unfilled thermoset. Das et al. [5] developed various types of $BaTiO_3$/epoxy nanocomposites based on thin-film capacitors. They highlighted high capacitance, large area, thin film passives, their integration in printed wire board (PWB) substrates, and the reliability of the embedded capacitors.

New insulation materials for heavy electric equipment were prepared by adding spherical Al_2O_3 to epoxy resin at different fractions and sizes as reported by Park [6]. The insulation breakdown strength for a composite containing 50 wt.% Al_2O_3 reached 44.0 kV/1 mm, and this value was gradually reduced with increasing filler loading. The good dispersion of larger particle sizes of Al_2O_3 had effectively interrupted the propagation of the electrical trace. Al_2O_3 nanoparticles having three different interface structures were selected to reinforce epoxy resin as reported by Yu et al. [7]. The study found that the inclusion of the Al_2O_3-APS and Al_2O_3-HBP nanoparticles improved the dispersion of the nano-Al_2O_3 in the epoxy matrix and significantly affected the dielectric properties of the epoxy nanocomposites as compared to those filled with native nano-Al_2O_3 ones.

Maity et al. [8] stated that epoxy nanocomposite filled with 40 nm nano-Al_2O_3 contained an interphase region nearly 200 nm thick, and they exhibited a permittivity lower than that of the matrix. In addition, heating the nano-Al_2O_3 prior to reinforcing the epoxy matrix thickened the interphase region and slightly lowered its permittivity.

Singha et al. prepared new epoxy/Al_2O_3 nanocomposites, which exhibited lower values of permittivity/tan delta over a wide range of frequency compared to that of pure resin. This was attributed to the interaction between the epoxy chains and the

Al_2O_3 nanofillers. The DC resistivity and AC dielectric strength of the nanocomposites were also lowered compared to that of the neat epoxy, whereas the electrical discharge resistant was significantly increased [9].

The effects of the AlN contents on the physical and dielectric properties of epoxy/AlN composites are studied by Wu et al. They reported that increasing AlN loading in the epoxy increased the crystal intensity as well as enhanced the dielectric constant, while slightly reducing the loss tangent of epoxy/AlN composites as the measured frequency increased. The results also indicated that the elaboration method had a strong effect on reducing porosities, which in turns lead to a low-loss tangent [10]. New mesoporous silica containing amine groups (MPSA) were synthesized and employed to reinforce bismaleimide-diallyl bisphenol (BD)/cyanate ester (CE) resin. The MPSA/BD/CE hybrids had a very low dielectric constant and loss as well, compared to that of BD/CE pure resin blend due to the different structure between MPSA/BD/CE hybrids and BD/CE resin [11].

In a study by Li et al. [12], small-scale ZrO_2 nanocrystals were incorporated into polyimide (PI) thin film by grafting oleic acid-capped ZrO_2 nanocrystals on the main chain of polyamic acid via ring-opening reaction of epoxides, thus ZrO_2 nanocrystals were homogeneously embedded in the PI films. The optimal weight ratio of PAA and ZrO_2 nanocrystals and the sintering temperature are 2:3 and 320°C, respectively. They found that the dielectric constant (k) of PI/ZrO_2-nanocrystals hybrid thin films is much higher than that of pristine PI film, and a maximum value of k reached 6.1.

Multi-doped $BaTiO_3$/epoxy composites were developed by Kuo et al. [13], using various ceramic fillers treated at different temperatures. The ceramic/epoxy composite filled with the 900°C treated ceramic filler exhibited the highest k value. The increase of dielectric loss at frequencies exceeding 1 MHz was attributed to the mechanism of domain-wall motion.

An enhancement in $BaTiO_3$ particle size and volume fraction in the PMMA film tended to rise the dielectric constant while maintaining the dissipation factor around 5%. The dielectric constant of the film produced from a particle size of 24 nm at 39 vol.% reached a value of 19.8 that was nearly four times higher than that of the unfilled PMMA film [14].

Effects of the filler ratio and testing temperature as well as frequency on the dielectric properties of self-synthesized $BaTiO_3$/epoxy composites were discussed. At 40 wt.% $BaTiO_3$-loaded composite, k reached 44. The large $BaTiO_3$ aggregates formed within the epoxy matrix were responsible for this improvement. The k of composites was proportional to the volume ratio of $BaTiO_3$ filler while remaining fixed at different test temperatures and frequencies. In addition, the processing method of slurry-casting or screen-printing affected the porosity as low porosity lead to a low dielectric loss tangent [15].

Dielectric properties and relaxation phenomena of epoxy/$BaTiO_3$ composites, prepared by direct mixing or in tetrahydrofuran (THF), were investigated as a function of filler content as published by Ramajo et al. Both the k and dielectric losses of composites depended on the testing temperature and the frequency, while k only was influenced by filler content [16]. The effect of silane-coupling agents on the microstructure and dielectric behavior of epoxy/$BaTiO_3$ composites was explored by Ramajo et al. [17]. Methoxy silane was applied on the outer surface of ceramic particles at different silane ratios of 0.25, 0.35, and 0.50 wt.%. Composites presented good dielectric properties and a strong dependence with the silane concentration. The effect of combined temperature and voltage aging on the conduction mechanism of epoxy-$BaTiO_3$ composite was investigated by Alam et al. [18]. It was observed that the value of leakage current for all the tested composites did not increase during temperature and voltage aging unlike pure $BaTiO_3$ dielectrics.

High-dielectric permittivity of EP/$BaTiO_3$ composites with different $BaTiO_3$ particle sizes were produced, and their dielectric properties were studied. By using KH-550 silane coupling agent, the interaction between $BaTiO_3$ and EP improved and induced a high-dielectric permittivity and a low-loss tangent. The dielectric properties of the composites with different size of $BaTiO_3$ particles gave different temperature dependency due to the existence of phase transition of large-size $BaTiO_3$ particles at its Curie temperature [19]. Preetha et al. observed that the AC breakdown strength epoxy/Al_2O_3 nanocomposites containing 1 wt.% Al_2O_3 was marginally decreased up to 1 wt.% loading, while increased at 5 wt.% loading as compared to the neat epoxy [20]. Yuen et al. used TiO_2-coated-(aminopropyl)triethoxysilane (ATS)-MWCNT particles to fill the epoxy resin where the electrical resistivity of the TiO_2-ATS-MWCNT/epoxy composite exceeded that of the unmodified MWCNT/epoxy composites at the same filler loading [21].

Spherical $BaTiO_3$ nanoparticles having three different diameters were synthesized by the alkoxid route, and they were mixed with thermoplastic polyvinylidene fluoride (PVDF) or polyimide (PI) thermoset to produce nanocomposite films. Effects of nanoparticle size and polymer matrix on the dielectric properties were studied at frequencies ranging from 10^2 Hz to 10^6 Hz. The roles of nanoparticles (size and crystal phase) and properties of polymers (chemistry and chain structure) are used to explain the difference in dielectric behaviors [22]. It can be seen that the permittivity of two kinds of BT/polymer nanocomposite films increases with temperature as shown in Figure 8.1.

The effects of PATP-treatment and filler fraction of $BaTiO_3$ on dielectric properties of epoxy thermoset were investigated. Even if the degree of dispersion of ceramic nanofillers was low, the reduced dielectric loss of their resulted composites was recorded. The composites had high k and very low dielectric loss, which can be controlled through surface-modification of nano-$BaTiO_3$ and varying the filler concentration [23].

The effects of filler size on the structure and filler/matrix interface on the properties of epoxy/$BaTiO_3$ were discussed. It was found that the tetragonal phase formation in the $BaTiO_3$ can increase the permittivity of these composites. The conduction mechanism is identified as barrier hopping and the tunneling effect of electrons. Based on the electrical properties of $BaTiO_3$–epoxy composite with 100 nm $BaTiO_3$ at GHz, a compact band-pass filter with significant area reduction was produced. The measured results show that the $BaTiO_3$/epoxy composite is promising for advanced wireless device applications [24].

The evolution of ε for epoxy/$BaTiO_3$ composites reinforced by different $BaTiO_3$ particle sizes and fractions were studied. The best $\varepsilon \approx 90$ was recorded using two different-sized $BaTiO_3$-powders mixture. Typically, capacitors of 12-nm-thick film with 8 nF/cm^2 with less than ±5% capacitance tolerance and low-leakage current

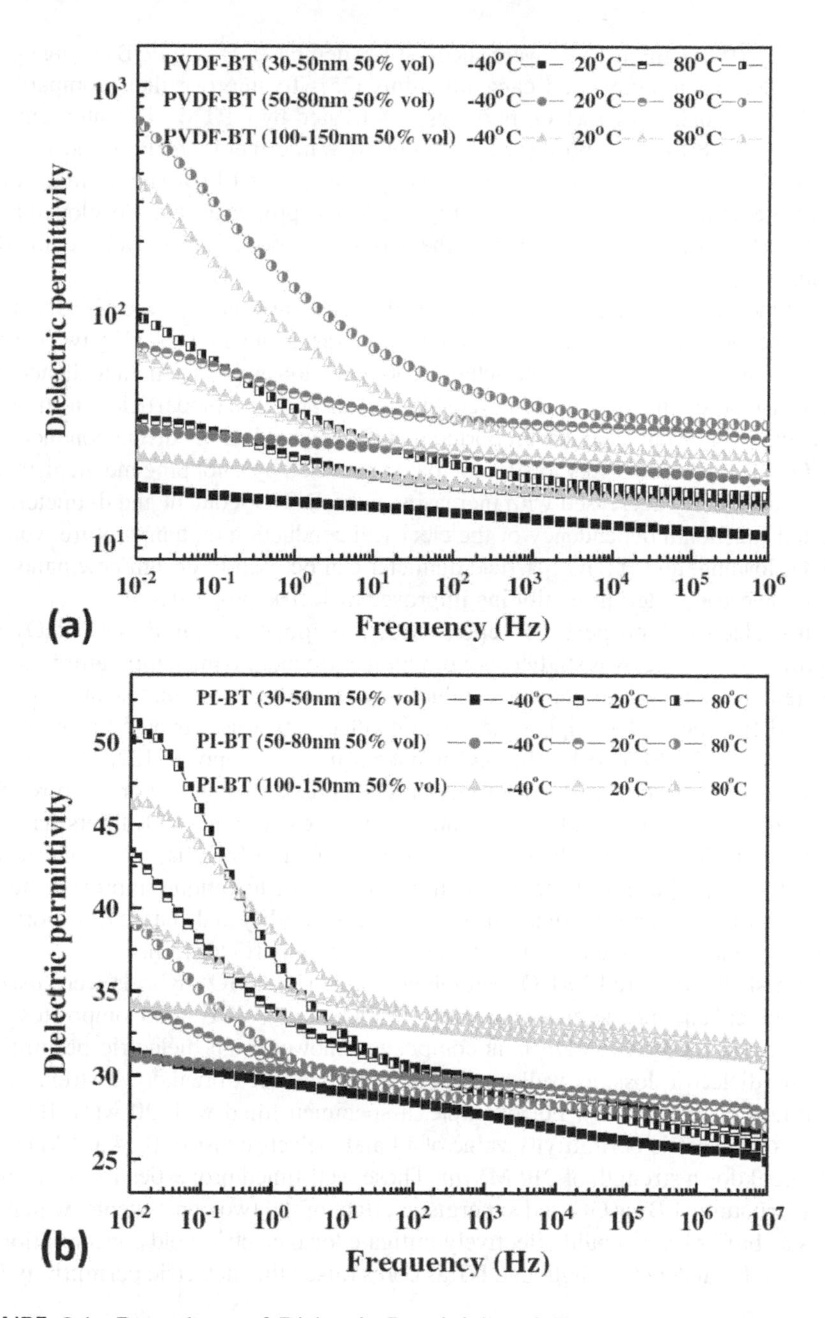

FIGURE 8.1 Dependence of Dielectric Permittivity of Polymer Nanocomposite Films Loaded with Three Kinds of BT Nanoparticles at 50 vol.% in the Frequency Range from 10^2 Hz to 10^6 Hz, Measured at −40°C, 20°C, and 80°C, respectively, (a) for BT/PVDF; (b) for BT/PI [22].

(less than 10^{-7} A/cm^2 at 10 V) were successfully demonstrated on PCBs using epoxy/$BaTiO_3$-composite embedded capacitor films [25]. To improve their compatibility with EP, the surface of $BaTiO_3$ particles was treated by GPTMS in water, ethanol, and xylene at 80°C as reported by Iijima et al. While there was no trend between the suspension viscosity and the dielectric properties of EP/$BaTiO_3$ composites, a composite with a smooth surface and high dielectric properties was developed using the $BaTiO_3$ treated with xylene where the epoxy groups were effectively retained on the surface [26].

Tsekmes et al. [27] proposed that the dielectric response of the EP/BN nanocomposites produced by three different techniques was mainly influenced by two possible mechanisms, which were polymer-chain reorganization and water-uptake, hence high levels of reproducibility was achieved for well-dispersed nanoparticles in the polymeric matrices. Alam et al. [28] reported that the mechanism of current conduction in $BaTiO_3$/EP nanocomposites was found to be controlled by a hopping mechanism, and its E_a was linearly decreased with increasing nano-$BaTiO_3$ content and diameter. The resulted functional dependency of the electrical conduction on temperature, voltage, $BaTiO_3$ loading, and $BaTiO_3$ particle diameter can be used to design new nanocomposite dielectric materials exhibiting improved dielectric properties.

This electrical property of epoxy/SiC_w composites coated with TiO_2 and poly(divinylbenzene) was studied as a function of frequency and temperature. Results revealed that alternating-current conductivity increased with increasing temperature and frequency. The higher electrical conductivity was attributed to numerous conduction paths formed by the SiC whiskers in the composite [29]. A comparative study on the dielectric permittivity and AC breakdown behavior of epoxy/SiO_2 nanocomposites loaded with both treated and untreated nano-SiO_2 fillers has been conducted by Wang et al. The surface-treatment of nano-SiO_2 has a strong effect on the dielectric properties of the nanocomposites as this technique improved the dispersions, which generally hinder the epoxy-chain mobility and charge transportation through the interface region between epoxy and nanoparticles [30].

Core shell-structured $BaTiO_3$@graphene oxide (BT@GO) hybrids were used as fillers for enhancing the energy density of dielectric PVDF nanocomposites. The as-fabricated BT@GO/PVDF nanocomposites showed high dielectric permittivity and low dielectric loss, as well as significantly increased breakdown strength and maximum energy density. For example, a specimen filled with 20 wt.% BT@GO displays a dielectric permittivity value of 14 and dielectric loss of 0.04 at 1 kHz with high breakdown strength of 210 MV/m. These well-tuned properties result from the novel structure of BT@GO and synergistic effect of the two constituents, which GO shells as buffer layers could effectively mitigate local electric-field concentration for increased breakdown strength and BT as cores raised the dielectric permittivity [31].

8.2 ELECTRICAL CONDUCTIVITY OF POLYMER/CERAMIC COMPOSITES

Electrical conductivity in a polymer composite is often formed as a single conductive filament. However, each percolated linkage works as a series of resistors where each particle and each contact point participates to the total resistance in that filament.

Contact resistance can be explained by constriction resistance and tunneling resistance parameters. In order that any polymer/ceramic composites become electrically conductive, the concentration of the conducting ceramic agents has to exceed the percolation threshold as given by the percolation theory. The correlation that links the electrical conductivity of ceramic particles in different types polymeric matrices and filler ratios generally is related to the volume fraction of the constituents and the filler morphology.

It is known that as filler loadings increased in the matrix, more continuous conductive chains are formed until a certain critical volume fraction called the percolation threshold. Beyond the percolation threshold, the conductive network increased as the filler concentrations increased until the electrical conductivity is fixed. Recent studies [32] reported that several factors can influence the electrical conductivity of the polymer/ceramic composite, including filler dispersion, shape and size, aspect ratio, inherent electrical conductivity of filler and matrix, the type of polymer matrix, wettability, shape, orientation, surface energy, and processing technique.

8.2.1 AC Electrical Conductivity by Dielectric Spectroscopy

Dielectric properties in polymers are due to electronic, ionic, molecular, and interfacial polarization. These properties are associated with the physical and chemical structure of the polymers. Dielectric spectroscopy is a non-invasive, very sensitive technique to investigate complex systems and is frequently used to study relaxation processes and conductivity. In dielectric experiments, when an AC voltage is applied to a polymer, it produces an alternating electric polarization.

The complex field and displacement are given by Eqs. (8.1) and (8.2). Complex permittivity, $\varepsilon^*(\omega)$ can be expressed as a complex number as shown in Eq. (8.3), where ε' and ε'' are the real and imaginary parts of the complex dielectric constant (complex permittivity). The relation between the complex field and displacement is shown in Eq. (8.4).

$$E^* = E_o\, exp^{(i\omega t)} \tag{8.1}$$

$$D^* = D_o\, exp^{(i\omega t-\delta)} \tag{8.2}$$

$$\varepsilon^*(\omega) = \varepsilon'(\omega) - i\varepsilon''(\omega) \tag{8.3}$$

$$D^* = \varepsilon^* E^* = (\varepsilon' + i\varepsilon'')E^* \tag{8.4}$$

The electrical properties of UV-curable *co*-polyacrylate/SiO_2 nanocomposite resins produced via the in situ sol–gel technique were studied [33]. Experimental data showed that adding nano-SiO_2 particles in polymeric matrix effectively enhanced the electrical properties of thermosetting specimens. The intercalation of ceramic particles stopped the migration of charge carriers, thus decreasing the leakage current density of thermosetting nanocomposite and enhancing their electrical properties. In addition, friction was created between the matrix/silica functional groups such as $-O-CH_2CH_3$, $-OH$, and $-Si-O-Si-$ groups capping SiO_2 nanofillers and the *co*-polyacrylate chains. This hindered the molecular-chain mobility and increased the dielectric properties of the resulted nanocomposites. It was reported that the curing process and the formation of SiO_2 nanoparticles finely dispersed into matrix, the leakage current density of the nanocomposite films were reduced from 235 to 1.3 $nA{\cdot}cm^{-2}$ at an electrical field of 10 $kV{\cdot}cm^{-1}$. The prepared hybrid films exhibited improved dielectric properties (dielectric constant ε of 3.93 and tangent loss tan δ of 0.0472).

Wong et al. [34] reported that at low frequency, epoxy/SiO_2 nanocomposite showed a higher dielectric loss because of the increased ionic conductivity generated by the by-products from the sol–gel synthesized nano-SiO_2. Due to the extra free volume at the nano-SiO_2/epoxy interface, the relaxation temperature of these nanocomposite was inferior to those of the microcomposite and that of the pure matrix. In a study published by Petrović et al., [35] dielectric tests were conducted on polyurethane/SiO_2 nanocomposites. Experimental results revealed that although the nano-SiO_2 showed an improved interaction with the matrix, there were no significant changes in the dielectric behavior between the two series of composites of PU/SiO_2 nanocomposites and PU/SiO_2 microcomposites.

The effect of structural and electrical parameters of the lead zirconate titanate (PZT) on the pyroelectric properties of composites based on poly(vinylidene fluoride), vinylidene fluoride–tetrafluoroethylene copolymer (F-42) was investigated. It was found that the pyroelectric coefficient of the composite can be determined by the reorientation polarization and the mobility of domains in the PZT pyroelectric ceramic agent, which was related to the homogeneous parameter of the spontaneous strain of the perovskite cell [36].

Han and Armes [37] found that the electrical conductivity of a conducting poly(3,4-ethylenedioxythiophene)/SiO_2 nanocomposite decreased with the increasing of SiO_2 loading. This was attributed to the inhibition in the transport of carriers between different polymer chains in the presence of nano-SiO_2 and also to strong interaction at the interface of matrix/filler that could resulted in a reduction of the conjugation length of conductive polymer. To enhance the electrical conductivity of such composites, the nano-SiO_2 should be directly coated on polymer chains. The sunflower-like SiO_2/PPy nanocomposites achieved high electrical conductivity of nearly 8 $S{\cdot}cm^{-1}$ at room temperature owing to their special morphologies [38].

The electrical conductivity of the pure poly(2-chloroaniline) and its SiO_2 nanocomposites were reached the values of 4.6×10^{-7} and 1.3×10^{-5} S cm^{-1}, respectively. The increase in electrical conductivity of the prepared nanocomposites can be justified by the high efficiency of charge transfer between SiO_2 and the organic matrix chains, and the nano-SiO_2 could also enhance their protonation [39]. PU/TiO_2 nanocomposites showed an excellent piezoelectric behavior, hence their dielectric constant and electrical conductivity were modified significantly with the changing in the applied stress [40].

8.3 DIELECTRIC PROPERTIES

A dielectric material is defined as an insulator material that under the effect of an external electric field can be polarized by separating their charges within the material. As these charges are bound within its structure, it can only pass limited movements, called as charge redistribution under applied electric field. Charge redistribution takes place either by positive and negative charges moving toward opposite electrodes to construct induced dipoles or rotation and orientation of permanent dipoles in the direction of electric force. This phenomenon is known as polarization of the dielectric material. In a dielectric material containing n molecules per cubic meter, the polarization P is provided by Eq. (8.5) [41].

$$P = nP_{av} = n\alpha E' \tag{8.5}$$

where E' is the local electric field in the vicinity of the dipole, and α is defined as the polarizability constant.

Dielectric materials are generally subdivided according to their polar and apolar groups. Polar dielectrics contain molecules with permanent dipole moments in the absence of any electric field. Apolar dielectrics do not have permanent dipoles and only have induced dipoles generated by one or more of polarization mechanisms that can appear when an electric field is applied to the dielectric. Polarization mechanisms include electronic, atomic, orientational (dipolar), and interfacial polarizations [41].

The overall dielectric permittivity of a material as a result of adding all these relaxation mechanisms is generally presented as the dielectric permittivity at room temperature and a specific frequency, for example 1 kHz, while under static electric field, the dielectric permittivity of the material will be called dielectric constant. Dielectric permittivity values for some ceramics and polymers dielectrics are listed in Table 8.1.

The dielectric constant of a material reflects its ability to store energy (e.g., charges) when it is subjected to an electric field, while the dielectric loss tangent represents the ratio of dissipated energy to the stored energy by a dielectric in a cyclic field. Figure 8.2 reveals the frequency dependency of dielectric constants and dielectric loss tangents of poly (lactic acid)(PLA)/$BaTiO_3$–based nanocomposites at various filler concentrations [43]. All the prepared nanocomposites showed a slight frequency dependence of dielectric constant trend. Respect to the unfilled PLA, their nanocomposites exhibited improved dielectric constant, which were gradually increased with the increasing of $BaTiO_3$ nanofillers' content. When the nan-$BaTiO_3$ loading attained 20 vol.%, the dielectric constants of the reinforced nanocomposites with $BaTiO_3$, $BaTiO_3$@ polydopamine (PDA), and $BaTiO_3$@PDA @PLA at 1 kHz reached 7.52, 8.10, and 8.74, respectively, which are all higher than two times the inherent dielectric constant of neat PLA matrix (3.04 at 1 kHz). This increase in dielectric constant value was mainly traced to the electric field increase in the PLA polymer, which was influenced by the large dielectric constant mismatch between $BaTiO_3$ nanoparticles and PLA matrix.

Ciuprina et al. [44] reported that by applying a dose of 50 kGy on LDPE-g-AM filled with 5 wt.% nano-Al_2O_3 led to a relative permittivity lower than that

TABLE 8.1
Dielectric Permittivity of Several Dielectric Materials, Ceramic, and Polymers [42]

Composition	Dielectric Permittivity	Polymer	Dielectric Permittivity
$BaTiO_3$	1,700	Nonfluorinated aromatic polyimides	32–3.6
PMN-PT (65/35)	3,640	Fluorinated polyimide	2.6–2.8
$PbNb_2O_6$	225	Poly(phenyl quinoxaline)	3.0
PLZT (7/60/40)	2,590	Poly(arylene ether oxazole)	2.6–2.8
SiO_2	39	Poly(arylene ether)	3.0
Al_2O_3	9	Polyquinoline	3.0
Ta_2O_5	22	Silsesquioxane	2.8–3.0
TiO_2	80	Poly(norborene)	2.4
$SrTiO_3$	2,000	Perfluorocyclobutane polyether	2.4
ZrO_2	25	Fluorinated poly(arylene ether)	3.7
HfO_2	25	Pol3maphthalene	2.2
$HfSiO_4$	11	Poly(tetrafluoroethylene)	2.9
La_2O_3	30	Polystyrene	3.6
Y_2O_3	15	Poly(vinylidene fluoride-co-hexafluoropropylene)	~12
α-$LaAlO_3$	30	Poly(ether ketone ketone)	~3.5

of the pure LDPE matrix. The γ-Radiation can provoke a reduction in the dielectric losses of LDPE/Al_2O_3 nanocomposites when suitable of dose/filler content was used. Equation 8.6 shows that the addition of high-ε fillers should directly lead to an enhanced ε value of their reinforced polymer composites. In addition, the coupling effect occurring at the filler/matrix interface areas in such nanocomposites can generate higher levels of interfacial polarization within the filler and matrix, which are responsible for advancing the energy density of these materials [45, 46].

$$\varepsilon = \varepsilon_2 \frac{2\varphi_1(\varepsilon_1 - \varepsilon_2) + \varepsilon_1 + 2\varepsilon_2}{2\varepsilon_2 + \varepsilon_1 + \varphi_1(\varepsilon_2 - \varepsilon_1)} \tag{8.6}$$

Indeed, experimental investigations confirmed that much improved ε values of polymer nanocomposites result from the inclusion of high-ε nanofillers like $BaTiO_3$, $Ba_xSr_{1-x}TiO_3$, $CaCu_3Ti_4O_{12}$, and Pb(Zr, Ti)O_3, providing enhanced energy densities [47, 48, 49–51]. Perry et al. reported phosphonic acid-modified $BaTiO_3$/polymer nanocomposites and found that the use of such modified-ceramic particles ligands improved the $BaTiO_3$ nanofillers' dispersion and exhibiting high ε [52]. The authors investigated a series of ligands as the modifier, each bearing an aliphatic octyl chain with a various terminal binding group. Results showed that the ligand bearing the phosphonic acid functional group could be tightly bound on $BaTiO_3$ nanoparticles in a tridentate form.

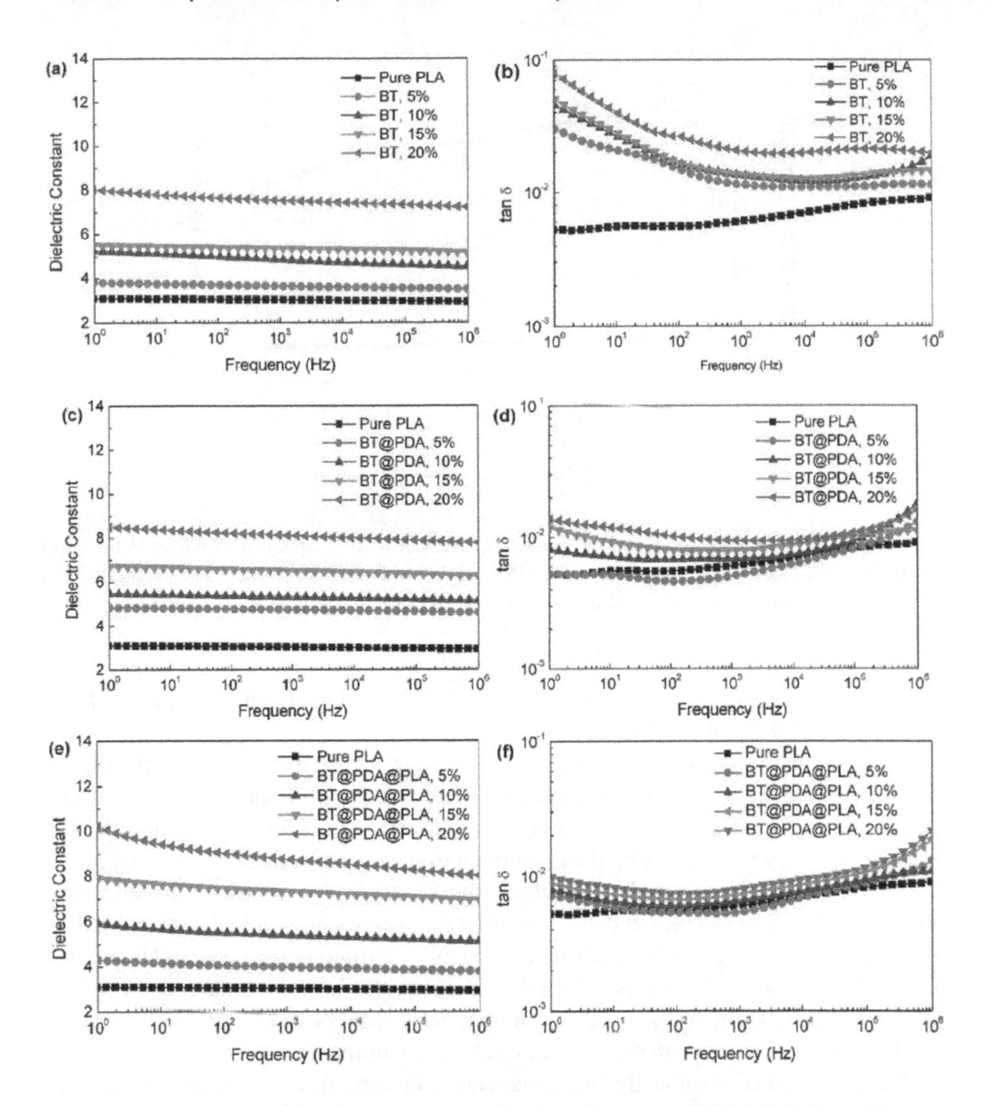

FIGURE 8.2 Frequency Dependence of Dielectric Constant and Dielectric Loss Tangent of (a, b) $BaTiO_3$/PLA, (c, d) $BaTiO_3$@PDA/PLA, and (e, f) $BaTiO_3$@PDA@ PLA/PLA Nanocomposites at Room Temperature [43].

To prove the effectiveness of the proposed technique, the authors gave two examples using these ligands to treat $BaTiO_3$ nanoparticles, and two polymer dielectrics were selected as matrices, i.e., the bisphenol-A-type polycarbonate (PC) and the poly(vinylidenefluoride-*co*-hexafluoropropylene) (P(VDF-HFP)). The prepared nanocomposites exhibited improved ε values in respect to the unfilled polymers with fairly low dielectric loss (Figure 8.3), even though the breakdown strength was also relatively decreased (~210 MV/m) compared with those of the polymer matrices (e.g., E_b of the P(VDF-HFP) is around 500 MV/m). By consequence, the stored energy

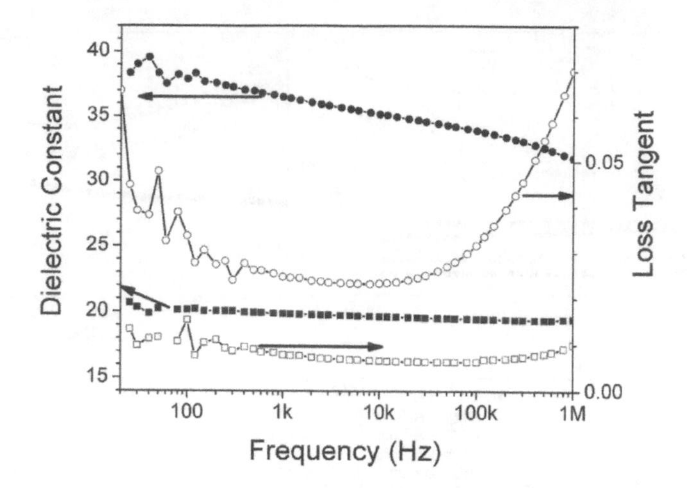

FIGURE 8.3 Frequency-dependent Dielectric Response of Capacitor Devices Fabricated from PEGPA-BT/PC (*squares*) and PFBPA-BT/P(VDF-HFP) (*circles*) (Reprinted from Ref. [52] with permission from John Wiley and Sons).

densities were enhanced, i.e., 3.9 J/cm^3 for the $BaTiO_3$/PC nanomaterials and 6.1 J/cm^3 for the $BaTiO_3$/P(VDF-HFP) nano-hybrids.

Using the same technique, several other ligands were applied to change the high-ε ceramics for dielectric polymer nanocomposites, such as ethylene diamine, dopamine [53], organic titanite–coupling agent [54], and silane-coupling agent [55, 56]. Another strategy for increasing the dispersion of ceramic filler particles is to directly join the particles to the molecular chain of the polymeric matrices via chemical bonding. This can be realized by two different approaches which are coined as "grafting from" [57, 58] and "grafting to" [59, 60] methods, respectively. The known as "grafting from" method is referred to present the initialization the growth of polymer chains from the surface of nanoparticles. Marks et al. applied the in-situ synthesis of high-energy-density metal oxide nanocomposites [61].

The approach of chemically functionalization of ceramics was extensively used to reinforce the various kinds of polymer matrices via different polymerizing techniques. For example, Jiang et al., [62] developed a $BaTiO_3$/poly(methyl methacrylate) (PMMA) nanocomposite that was prepared by in situ atom transfer radical polymerization (ATRP) of methyl methacrylate (MMA) from the surface of $BaTiO_3$ nanoparticles. By doing so, the ε value of the nanocomposite was improved to ~14.6 from 3.5 of the unfilled PMMA, and simultaneously, the dielectric loss was almost independent of the filler loadings.

The "grafting to" method is based on the linking of ceramic filler covalently with the polymeric matrix through functional groups bearded on the polymer chains and the outer surface of filler. As an example, Wang and coworkers [63] reinforce the ferroelectric polymers with phosphonic acid end groups and subsequently used the reactive terminal groups of the polymer for direct coupling with ceramic oxide particles. The formation of covalent coupling between the polymer matrix and filler particles make the resulted nanocomposites exhibit better stability and improve filler dispersion. Due to

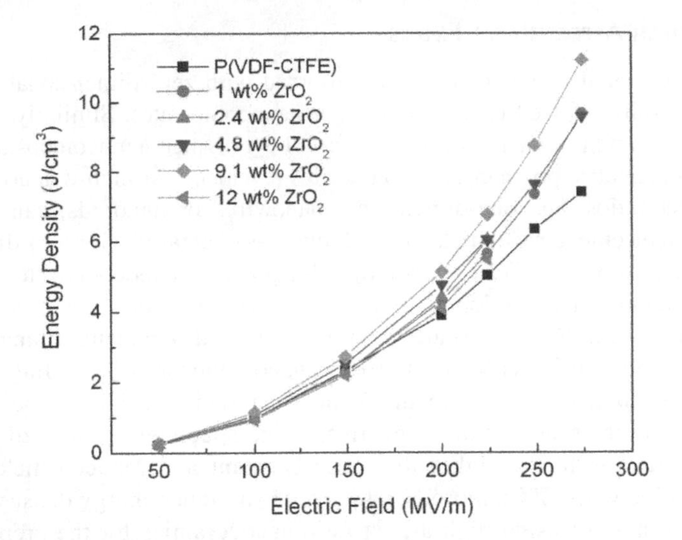

FIGURE 8.4 Stored Energy Density of the Polymer/ZrO_2 Nanocomposites as a Function of the Applied Field (Reprinted with Permission from Ref. [63], Copyright 2010 American Chemical Society).

the enhancement of the electric displacement resulting from the additional nanofillers, the energy storage capability of the nanocomposites was enhanced (Figure 8.4).

Another study prepared glycidyl methacrylate (GMA)–functionalized P(VDF-HFP) as the matrix to host $BaTiO_3$ nanoparticles, and the covalent linking of the matrix and filler reinforcing agent was attained via ring-opening reaction between the epoxy groups of GMA grafted onto P(VDF-HFP) and the amino groups on the surface of $BaTiO_3$ nanoparticles [64]. In addition, PVDF/$BaTiO_3$ composites were found to exhibit higher dielectric values as illustrated in Figure 8.5.

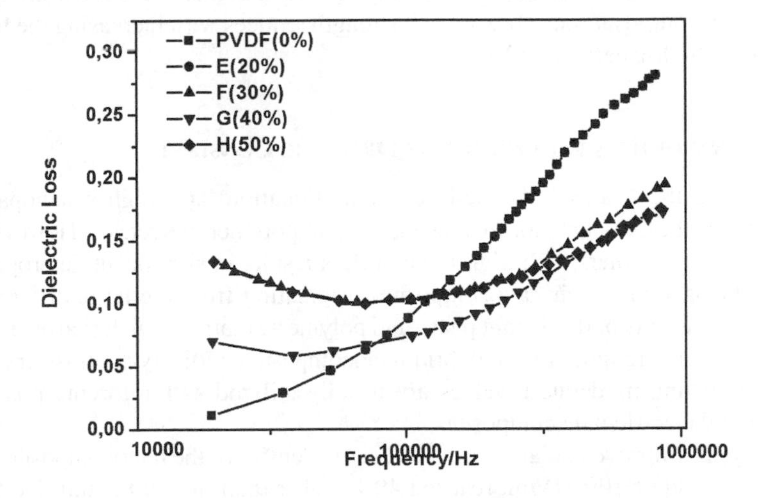

FIGURE 8.5 Frequency Dependence of Dielectric Loss for the Composite Films [41].

8.3.1 High Aspect Ratio Fillers

Three-dimensional polymer matrices reinforced with zero-dimensional nanoparticles is generally referred to as 0-3 polymer nanocomposites. Similarly, those containing one-dimensional incorporations are 1-3 polymer nanocomposites. Recent studies on dielectric polymer nanocomposites revealed that high-ε nanofillers with high aspect ratios, i.e., one-dimensional nanowires or nanorods, can provoke a marked enhancement in ε of dielectric polymers, as compared with zero-dimensional nanoparticles [65, 66]. Thus, elaborating 1-3 polymer nanocomposites became a promising approach to develop high-energy-density capacitors.

Sodano et al., [67] elaborated a series of lead zirconate titanate (PZT)/polyvinylidene fluoride (PVDF) dielectric nanocomposites by adding nanosized filler at various aspect ratios in the form of nanowires (higher aspect ratio) or nanorods (lower aspect ratio), respectively. The specimen reinforced with PZT nanowires demonstrated higher dielectric constant and reduced dielectric loss than that filled with PZT nanorods, which led to a higher energy density, showing the effectiveness of using high aspect ratio nanoceramics for the preparation of polymer nanocomposites for capacitive energy storage applications. By selecting $BaTiO_3$ nanowires as reinforcing agent, the energy density of a ferroelectric terpolymer poly(vinylidene fluoride-trifluoroethylene-chlorofluoroethylene) (P(VDFTrFE-CFE)) was enhanced by almost 45%, offering 10.48 J/cm^3 at the electric field of 300 MV/m [68].

Recently, $Ba_{0.2}Sr_{0.8}TiO_3$ nanowires were also utilized to fabricate a composite dielectric employing PVDF as the matrix (Figure 8.6) [69]. The discharged energy density of the prepared nanocomposite reached 14.86 J/cm^3, at the electric field of 450 MV/m, which was 43% larger than that of the unfilled polymer (Figure 8.7). As polymer dielectrics loaded with high-ε nanowires generally exhibit higher dielectric constant values than those reinforced with high-ε nanoparticles, lower filler ratio is needed for 1-3 systems to generate an equal electric displacement as compared with the 0-3 system. This could be automatically provoke a higher breakdown strength for the 1-3 nanocomposite since breakdown strength reduces with increasing the feeding ratio of the high-ε ceramic fillers.

8.3.2 Nanofillers with Moderate Dielectric Constant

By applying the Maxwell Garnett equation (Equation 8.6), high-ε nanoparticles would be the best dopants for increasing the ε of polymer dielectrics. However, filling dielectric polymers with high-ε nanofillers results in significant heterogeneous electric fields across their nanocomposites originating from the large difference in dielectric constants of the dopant phase and polymeric matrix, which sharply reduced the breakdown strength of such hybrid nanocomposites [70]. By contrast, the nanofillers exhibiting moderate ε values are usually utilized as reinforcing phases for producing dielectric nanocomposites.

Due to the improved polarization, the energy density of the nanocomposite, under an electric field of 200 MV/m, reached 40% higher than that of the unfilled terpolymer matrix. By using nanoparticles showing a moderate ε, the polarization could

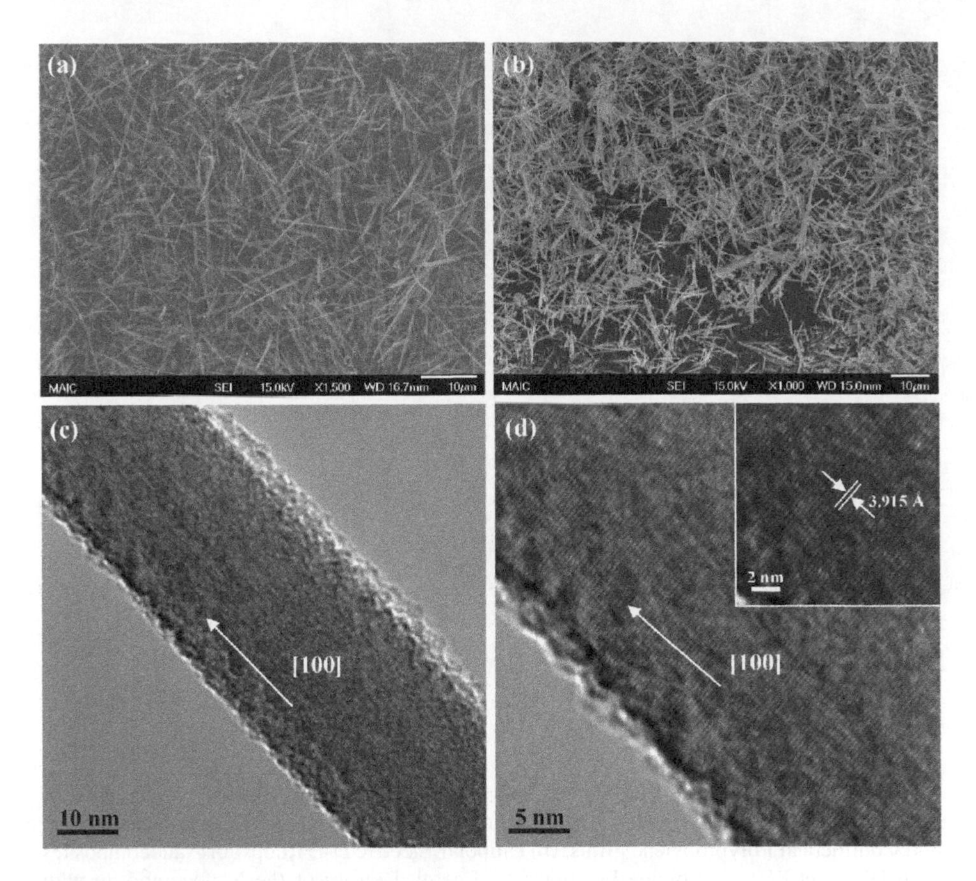

FIGURE 8.6 SEM Images of (a) Sodium Titanite NWs and (b) $Ba_{0.2}Sr_{0.8}TiO_3$ NWs. (c) TEM Image of a Single $Ba_{0.2}Sr_{0.8}TiO_3$ Nanowire. (d) Representative HRTEM Image Showing Clear Crystal Lattice Fringes of the $Ba_{0.2}Sr_{0.8}TiO_3$ Nanowire (Reprinted with Permission from Ref. [69], Copyright 2013 American Chemical Society).

be enhanced without sacrificing the breakdown strengths of host matrices and thus providing much increased capacitive energy storage capabilities. Other studies on polymer nanocomposites reinforced with several other moderate dielectric constant nanofillers were also evaluated [71–79]. This promising technique confirms that the using of high-ε nanofillers are not efficient for preparing high-energy-density dielectric polymer nanocomposites.

The percolation theory states that in any conductive inorganic/polymer hybrid system, the ε is enhanced sharply with increasing the filler loading till approaching the vicinity of the percolation threshold, which can be written by Eq. (8.7):

$$\frac{\varepsilon}{\varepsilon_m} = \left| \frac{f_c - f_f}{f_c} \right|^{-q} \tag{8.7}$$

in which ε is the dielectric constant of the composite, ε_m is the dielectric constant of the polymer matrix, fc is the percolation threshold, and f_f is the filler content [80].

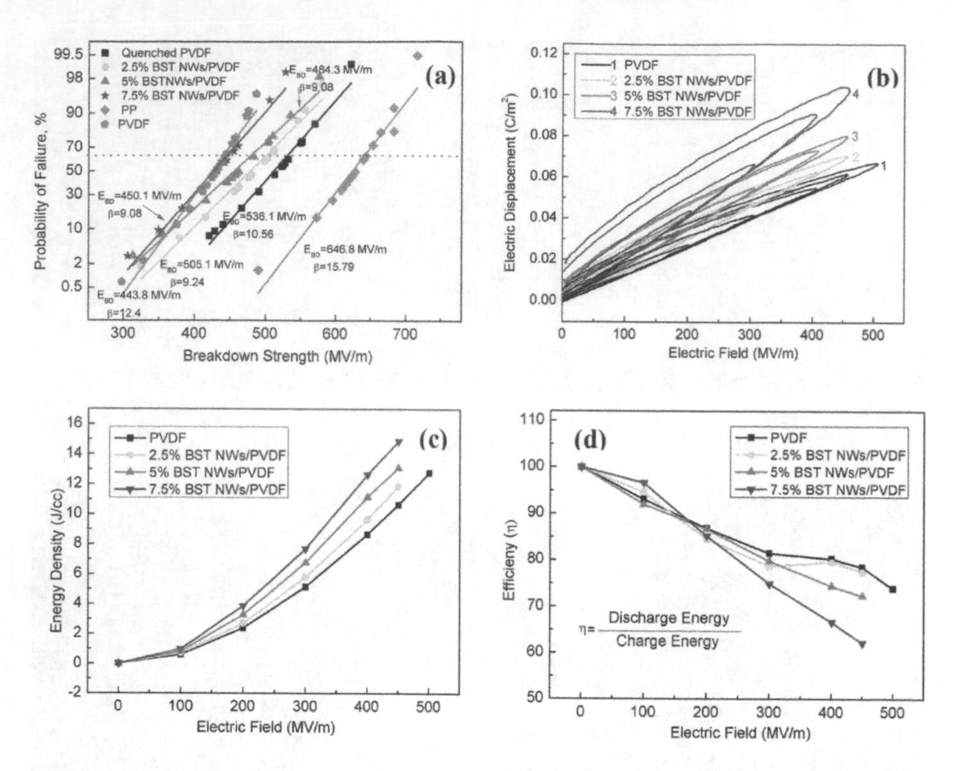

FIGURE 8.7 (a) Weibull Distribution and Observed Dielectric Breakdown Strength of Nanocomposites with Different Volume Fraction of BST, Quenched PVDF, Untreated PVDF, and Commercial Polypropylene Films; (b) Unipolar Electric *D-E* Loops for Nanocomposites with Different BST_{NWs} Volume Fractions; (c) Energy Density of the Nanocomposite with Different Volume Fraction as a Function of the Electric Field Calculated from *D-E* Loops; (d) Efficiencies of the Nanocomposites with Different Volume Fraction as a Function of the Electric Field (Reprinted with Permission from Ref. [69], Copyright 2013 American Chemical Society).

This significant rise in the ε value is caused by the construction of the so-called microcapacitor networks formed from numerous large localized capacitors [81]. Each of these microcapacitors is constructed by the neighboring conductive fillers as the two electrodes and a thin layer of dielectric polymer in their interface. These microcapacitors contribute to the intensified local electric fields and by consequence in the accumulation of charge carriers at the filler/matrix interface which is responsible for improving the interfacial polarization. One of the electrical properties that have been widely investigated is the resistance to partial discharges (PD).

Several research results showed that nanocomposites reinforced with small amount of ceramic like Al_2O_3, SiO_2, TiO_2, and SiC have a significant resistance to PD. The erosion depth decreased with respect to the unfilled specimens [82–84]. The damage generated by a PD on filled epoxy nanocomposites can be delayed if the filler particles played the role as thermal shields or heat sinks [85]. Superior resistance can be also related to the strong bonding between the epoxy matrix and the embedded ceramic filler [86, 87]. Polymer/ceramic nanocomposites were even more resistant to PD in the case where the used nanoparticle outer surfaces were treated [88]. The dielectric breakdown (BD) strength usually changed for better or worse as the nanocomposite subjected to an AC voltage [89]. The AC BD strength can also be improved by adding nanoparticles to a microcomposite [90].

Recent studies reported that an improvement in the direct current banding strength was resulted for various polymer composites reinforced with nanofillers such as AlN, BN, Al_2O_3 and SiO_2 [91–93]. The addition of a very small amount of Al_2O_3 nanoparticles in the epoxy can provide a significant enhancement of the long term BD degradation process by rising the treeing time to breakdown [94–96]. The addition of nanoparticles can also decrease the space charge accumulation in some nanocomposite systems. The amount of space charge accumulated by the epoxy-based composites filled with TiO_2, Al_2O_3 was considerably lower than that of the unfilled [97]. A study of the complex permittivity as a function of temperature or frequency is also recommended for characterizations of dielectrics. The inclusion of a small amount of nanofillers can influence the real and imaginary parts of the complex permittivity. A few percent of nanoparticles can modify the relative permittivity of a composite in such a way that it might be higher than any of its constituents [98, 99] or even lower [100–102].

Besides the amelioration of dielectric properties, the addition of ceramic nanoparticles leads to the improvement of thermal and mechanical properties, or to a change of the physical properties of a polymeric matrix, such as the glass transition temperature (T_g). The thermal conductivity of filled epoxy also can be increased using thermally-conductive but electrically insulating ceramic fillers such as Al_2O_3, BN, SiC, Si_3N_4 [103, 104]. The ε and dielectric loss of SCE-Al_2O_3/PES-MBAE composite reached 3.63 (at 100 Hz), and 1.52×10^{-3} (100 Hz) when the content of SCE-Al_2O_3 was 3 wt.% and PES was 5 wt.% [105].

The effects of adding h-BN on the dielectric properties bismaleimide-triazine (BT) resin was studied. The data showed that inclusion of h-BN in the BT resin both dielectric constant and dielectric loss did not exceed the values of 4.5 and 0.015, respectively [106].

Xia et al. [107] developed BT/Al_2O_3 supported on polyimide fiber core–shell structure (200 nm- Al_2O_3@PI) with. The resulting composites had a low dielectric constant (3.38 to 3.50, at 100 kHz) and dielectric loss (0.0102 to 0.0107, at 100 kHz). Yao et al. [108] prepared novel insulating composites by adding new hollow silica tubes (mHST) to a bismaleimide/diallylbisphenol A (BDM/DBA) matrix. In addition, the new composites exhibited a lower and stable ε, better frequency stability of dielectric loss, and significantly improved toughness and flame retardancy. Wu et al. [109] studied the effects of different curing conditions on the dielectric properties of nano-TiO_2/BMI/CE composite systems. Dielectric properties of microwave-curing nano-TiO_2/BMI/CE composites was significantly increased via controlling post-process temperatures and post-process time.

The silane treatment of $BaTiO_3$ nanoparticles resulted in excellent dispersion and enhanced the BT/$BaTiO_3$ interactions. The derived nanocomposite films exhibit improved ε and low dielectric loss. For the nanocomposite containing 70 wt.% of $BaTiO_3$, the effective dielectric constant at room temperature reached 23.63, which is about 7 times larger than that of neat BT matrix, and the dielectric loss was only 0.0212 at 100 Hz [110]. The interaction between $BaTiO_3$ and BT matrix changed both the phase transition of $BaTiO_3$ as well as the thermal behavior of BT resin.

The dielectric constant and dielectric loss tangent of microwave-cured BT resins composites reinforced with micro-and SiO_2 nanoparticles were lower than those of the thermally-cured ones. With rising micro-SiO_2 loadings, the ε of BT composites increased, whereas the dielectric loss tangent reduced. In the case of using nano-SiO_2 particles as reinforcing agents, the dielectric properties of their resulted nanocomposites were reduced due to the decrease of the matrix-free volume [111]. High ε of surface-modified $BaTiO_3$ beads was used to improve the affinity between $BaTiO_3$ particles and blend of PA/BMI which was selected as the matrix from a viewpoint of both the processability and the thermal stability. These technologies were combined and optimized for embedded capacitor materials [112].

A novel method to prepare high-performance copper-clad laminates (CCLs) was developed by using BMI/$Al_{18}B_4O_{33}$ whisker hybrid as the matrix. Results showed that the incorporation of surface-treated $Al_{18}B_4O_{33}$ whiskers at an optimal concentration can noticeably increase the mechanical and dielectric properties of the CCLs [113]. Experimental data of the ε of BMI/$BaTiO_3$ were fitted to several theoretical equations. The ε data of all the prepared composites were enhanced with increasing of $BaTiO_3$ contents. Whether BMI matrix or their prepared composites, their dielectric losses were improved by increasing the $BaTiO_3$ loading.

The predictions of the effective ε by the Lichterecker mixing rule well-reproduced the experiment data [114]. As reported by Wu et al., the PI hybrid films containing 10 wt.% of Al_2O_3 exhibited obviously enhanced electrical aging performance with the time to failure 3.4 times longer than that of pure PI film, which was traced to the nano-sized Al_2O_3 particles, which were highly dispersed in the hybrid composites [115]. The study of Kim et al., confirmed that dielectric properties of the Al_2O_3/polyimide composite thick films could be easily controlled by the mixing ratio of starting constituents [116].

The dielectric constant of pure polyimide changed from 2.40 to 2.60 and exhibited low-frequency dependence, whereas Li et al., [117] found that the ε of the nano-Al_2O_3/polyimide hybrids synthesized by in situ polymerization increased with the increasing Al_2O_3 contents and exhibited a gradual dependency with frequency. This was attributed to the fact that Al_2O_3 had a higher dielectric constant of 8 to 10 compared to that of pure matrix (2.5).

According to a study of Alias et al. [118], the polyimide/Al_2O_3 composite showed higher ε than polyimide nanocomposites. This was related to the replacement of significant polyimide parts which possess a ε of 3.5 with the Al_2O_3 particles, which have a much higher ε ranging from 8 to 10. The ε value increased with rising Al_2O_3 concentrations, and more parts in the matrix were replaced with Al_2O_3 particles. The interfacial polarization inside these composites in applied electric fields influenced the dielectric loss. As the frequency increased, the interfacial polarization gradually became incapable of following the oscillation of the applied electric field, so more energy is used, which is shown as the dielectric loss tangent. Wang et al. [119] found

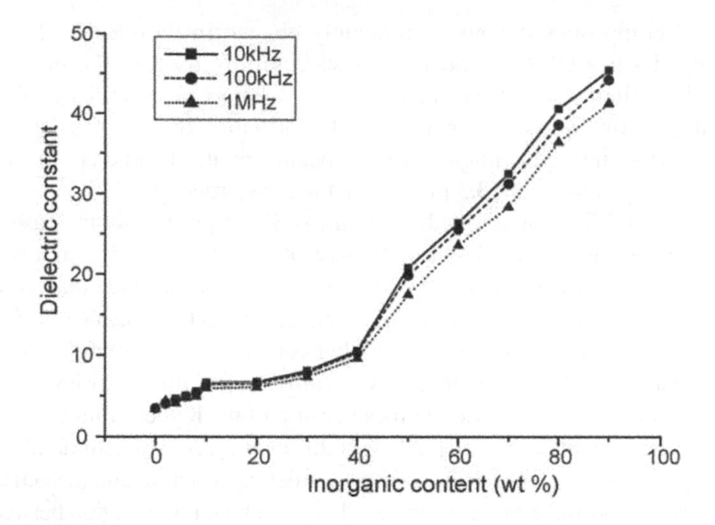

FIGURE 8.8 The Effects of $BaTiO_3$ Content on the Dielectric Constants and Dielectric Losses of Polyimide/$BaTiO_3$ Composites, at the Sweep Frequencies of 10 kHz, 100 kHz, and 1 MHz [119].

that the ε varies with the sweeping frequency and $BaTiO_3$ loading. Figure 8.8 reveals the effects of $BaTiO_3$ content on the ε and dielectric losses of polyimide/$BaTiO_3$ composite films at the sweep frequencies of 10 kHz, 100 kHz, and 1 MHz.

Süleyman et al., [120] synthesized a new class of polyimide/TiO_2 hybrids by the sol–gel technique. The ε value was significantly decreased but was found to be enhanced with further increasing TEOT loadings. However, the value of PI containing 30 wt.% of TiO_2 is still lower than that of the pure matrix due to the expanding the free volume and domains sizes of TEOT. It may be assumed that the Ti–O–Ti and the Ti–O bond in the cured network structures can provoke an increased ε with increasing the TEOT contents.

Panomsuwan et al. [121], fabricated new barium strontium titanate (BST)/PBZ composites with BST content ranging from 30 to 80 wt.%. It was found that composite ε increased with increasing ceramic contents. By adding 80 wt.% of BST fillers, the ε reached approximately 28, and they were still stable in the frequency range of 1 kHz to 10 MHz, with a temperature range of 20 to 130°C.

Dielectric properties of PBZ/$BaTiO_3$ composites with 0 to 3 connectivity were fabricated, where the ε of composites at frequency range of 1 kHz to10 MHz increased with increasing the fraction of $BaTiO_3$, and showed low-frequency dependence. At 70 wt.% of $BaTiO_3$, the ε at 10 MHz significantly increased from 3.56 of PBZ to 13.2 at room temperature, while dielectric losses of were below 0.016 [122]. Selvaraj et al. [123] synthesized PBZ/SiO_2 composites with an increased free volume. The thermal properties depended on the fraction of the SiO_2 filler, while dielectric tests proved that PBZ/SiO_2 hybrids had lower ε and dielectric loss values compared to those exhibited by the pure PBZ matrix.

The effect of absorbed moisture on electrical properties of SiO_2/UPE composites has been investigated by Sharma et al. [124]. The absorption of moisture in the prepared composites did not significantly change in the electrical properties. The developed silica-UPR composite was successfully used in the preparation of medium-voltage inductive transformers. Using a composite of UPE reinforced at a nano-$BaTiO_3$ loading of 74 wt.% permitted the formation of dielectric layers using a modified screen-printing technique. After capacitor mounting and composite curing an initial capacity density of 13.3 pF/mm(A^2) was recorded [125].

The filling of UPE thermoset with nano/micro SiO_2 significantly increased the surface, volume resistivity, dielectric strength, dissipation factor, and dry arc resistivity of their resulted composites as reported by Sharma et al. It was detected that the inclusion of nano-SiO_2 better enhanced the electrical properties respected to using micro-SiO_2 due to their better dispersion in the matrix [126]. For composites containing 55 wt.% $SrTiO_3$ previously treated at 1,000°C enhanced the permittivity by almost 25% while decreasing the loss-factor value by almost 10 times because of the higher-grain growth and the formation of larger crystallite sizes [127]. Gorelov et al., [128] studied the filler-loading effects of high-reactive-surface nano-sized SiO_2 filler on structure and dielectric properties of a styrene-cross-linked UPE resin have been evaluated. Interaction between active surface centers of SiO_2-nanoparticles and atoms of UPE chains and styrene molecules provides the non-monotonous influence on the structure of polyester chains, macromolecule polarity, dielectric parameters, and the number of positronium nanotraps.

8.3.3 Combined Improvements of Dielectric Constant and Breakdown Strength

They used new fillers, such as $BaTiO_3$–TiO_2 nanofibers, where $BaTiO_3$ nanoparticles are included into TiO_2 nanofibers to disperse into PVDF polymer. The discharged energy density of such nanocomposite was found to exceed the value 20 J/cm^3, which provoked an increase of 72% in the electric displacement and enhancement in breakdown strength by almost 8%. The improvement in the breakdown strength was attributed to the large aspect ratio and partial orientation of hybrid nanofibers, while that recorded for the electric displacement was related to the interfacial effect inside the integrated $BaTiO_3$–TiO_2 nanofillers. In addition to the interfacial polarization between the ceramic nanofillers and the PVDF matrix, the interfacial effect inside hierarchical multiphase ceramics inclusions could equally influence the dielectric properties of such polymer nanocomposites and thus could be used to attain high-energy density in dielectric capacitors [129].

8.3.4 Effect of Particle Size

Polymer/ceramic-based dielectric micro- and nanocomposites have the advantage of combining the high-dielectric permittivity of ceramic particles with lightweight, lower cost, easy processing and high dielectric BD of the polymeric matrices. This would impart significant improvements in stored energy density of the dielectric. But, when adding ceramics to conventional polymer composites, this would generally improve the dielectric permittivity of the polymer matrix, while the dielectric breakdown of such systems usually decreased with loading of ceramic fillers. Therefore, the overall stored energy density of such composite systems cannot be markedly enhanced. For examples, in a report published by Ma et al., the addition of 5 wt.% micro-TiO_2 particles to LDPE matrix led to an enhancement in the dielectric permittivity with an increment of 10%, whereas this was accompanied with a decrease in dielectric breakdown of 28% [130].

In Singha's work, the dielectric breakdown of an epoxy host was reduced by 26% and 38% after including TiO_2 and Al_2O_3 microparticles, respectively [131]. This trend of improving dielectric permittivity while lowering dielectric strength as ceramic microparticles were added to polymeric hosts was also reported by several other researchers. Some examples of such trends will be described and discussed in the next sections. To solve this contradictory trend many studies were conducted in the late 1980s [132, 133] and the first cited one was published by Fukushima and Inagaki [134]. But the most prominent study was that by Lewis [135], who conducted excellent research in this field, which increased interest in dielectrics polymer nanocomposites. Lewis found a promising technique for dramatic enhancement in dielectric nanocomposites by controlling the interface and its importance behavior in the heterogeneous systems containing nanofillers [136].

As Lewis stated, by reducing the size of fillers to reach nanoscale, the surface area of interaction within nanoparticles/matrices will be exponentially enhanced as shown in Figure 8.9. Hence, approaching the nanoscale, the proportion of the material at the interaction area will intensify in such a manner that, in an extremely small nanoscale regime, most of the material consists of such an interaction zone. Thus, the feature of the polymer host will change from polymer bulk properties and be characterized by properties of the interaction.

Lewis then postulated that it is theoretically conceivable to reach a synergistic impact of the nanocomposite effective properties through a suitable joining of particle and matrices based on studying the physics and chemistry of each compound. Lewis's proposed idea gives a real opportunity to develop specific nano-dielectric composites having effective properties surpassing those of each phase, which can

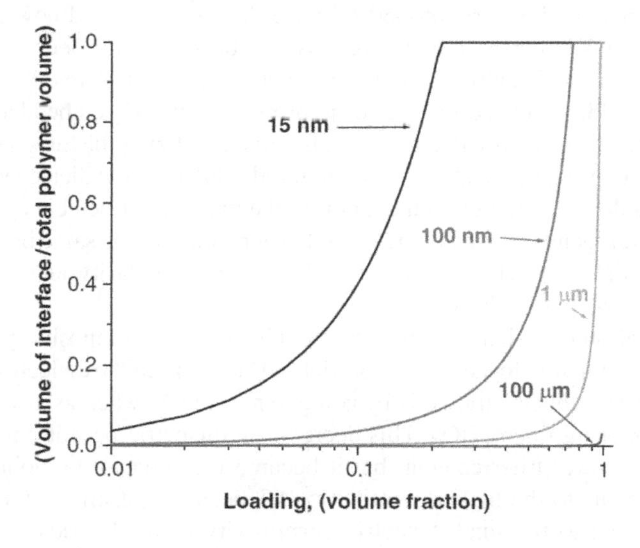

FIGURE 8.9 Surface-to-volume Ratios of Nanocomposites as a Function of Nanoparticle Loading [136].

be considered as new-generation materials for many applications, such as electrical insulation, capacitors, and smart materials [136].

A polymer nanocomposite consists of a multiphase system hosted in a polymer matrix, which represents the continuous host reinforced with nanoparticles of at least below 100 nm in one of the dimensions. This was in contrast to conventional microcomposites since higher contents of microparticles were required to give a significant improvements in dielectric properties. However, Lewis stated that by decreasing the particle size to nanoscale, the same improvement in dielectric properties can be achieved, if not much higher, due to marked enhancement in the interfacial surface area between ceramic nanoparticles and the polymeric matrix. Lewis's results extensively boosted the research on polymer/ceramic nanocomposites in the recent few years.

Several types of ceramic-oxide nanoparticles have been utilized for this purpose including TiO_2 [137–139], Al_2O_3 [140], SiO_2 [141], ZrO_2 [142], $BaTiO_3$ [143], as well as different types of polymers, such as polyolefin, phenolic, cyanate ester, polybenzoxazine, bismalimide, polyimides, fluoropolymers, and epoxies [144, 145]. In the following sections, many of these studies will be presented. Nelson et al. [146] scrutinized the dielectric properties of epoxy resin upon filling with TiO_2 ceramic filler. Expectedly, dielectric permittivity of the matrix increased by loading of TiO_2 microparticles, while it reduced when TiO_2 nanoparticles were used.

Nelson et al. [147] justified this decrease by the poor interfacial polarization and the reduced dipole mobility at interface nano-TiO_2/epoxy composites. They determined the relative value of free volume in neat epoxy systems and their micro- and nanocomposites, where a reduction in free volume for microcomposite was detected. However, there was an enhancement in the free volume for the nanocomposites. They ascribed the expansion of free volume in the case of nanocomposites to the increase in the mobile interlayer that somehow improved the interfacial polarization. Nelson's work showed the critical role of the interface in polymeric nanocomposite and its influence on dielectric properties since the work revealed how surface features that were insignificant in microcomposite hybrids played a determinant role in controlling the final dielectric properties of nanocomposite systems.

Roy et al. [148] evaluated the dielectric properties of SiO_2-embedded polyethylene composites. Results revealed that even by functionalizing the SiO_2 nanoparticles using silane-coupling agents, this did not markedly increase the dielectric permittivity. These results were related to the fact that the particle surface curvature and the quantity of hydrogen bonding among silanol groups was not the same between nano- and microparticles, which in turn changed the surface polarity and consequently affected dielectric permittivity.

Singha et al. studied dielectric properties of TiO_2/epoxy composite systems when micron-sized compared to nano-sized particles. Data showed the dielectric permittivity of composites reduced gradually by going to nano-TiO_2, whereas it was enhanced in the case of using micro-TiO_2. This decrease in dielectric permittivity was more pronounced at lower filler content, but it became insignificant by going to higher filler contents due to the higher contribution of dielectric permittivity of the TiO_2-reinforcing agent to the final dielectric permittivity of the hybrids. Nevertheless, adding micro-TiO_2 to an epoxy matrix resulted in the best enhancement in the dielectric permittivity. Singha believed that an innermost layer of polymer at interphase as a bound layer can be produced around each TiO_2 particle. Due to the strong surface interactions with the ceramic particle surface, polymer chains are infused and bound to each other; showing dipolar polarization of polymer chains at this layer could be restricted and thus a decrease in dielectric permittivity will result [131].

Dang et al. investigated the change in dielectric permittivity of 30- to 60-nm $BaTiO_3$/PVDF nanocomposites. The SEM photos of the resulting nano-hybrids filled by 20 vol.% (a) and 40 vol.% (b) are displayed in Figure 8.10. Micro-clusters of $BaTiO_3$ clearly appear on the SEM photos of the tested composite sample filled with 20 vol.%. In the case of high-volume fraction of 40 vol.%, the $BaTiO_3$ nanoparticles are interconnected to each other [149].

Figure 8.11 reveals the variation in the dielectric permittivity of $BaTiO_3$/PVDF as a function of filler ratios. This figure reflected that the experimental data mostly pursues

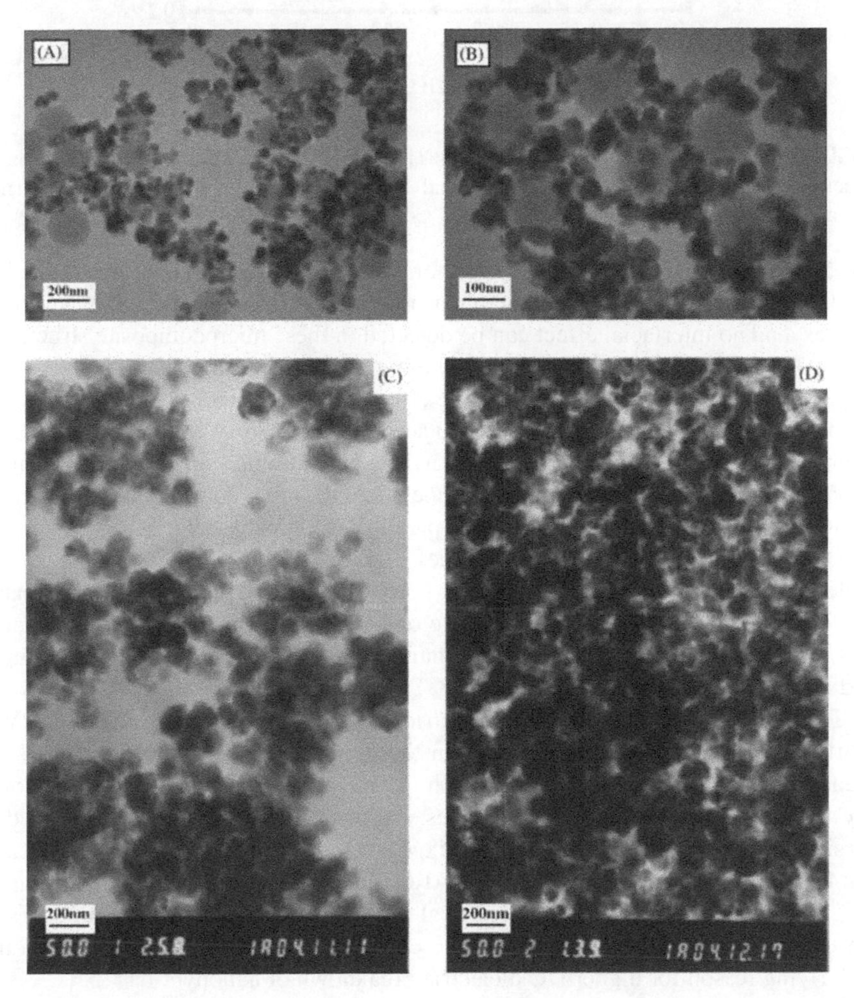

FIGURE 8.10 TEM Micrographs of the Dry BT/PVDF Mixtures with a (A) 0.20 and (B) 0.40 Volume Fraction of Nano-sized BT Particles; TEM Micrographs of the BT/PVDF Nanocomposites with a (C) 0.20 and (D) 0.40 Volume Fraction of Nanosized BT Particles [149].

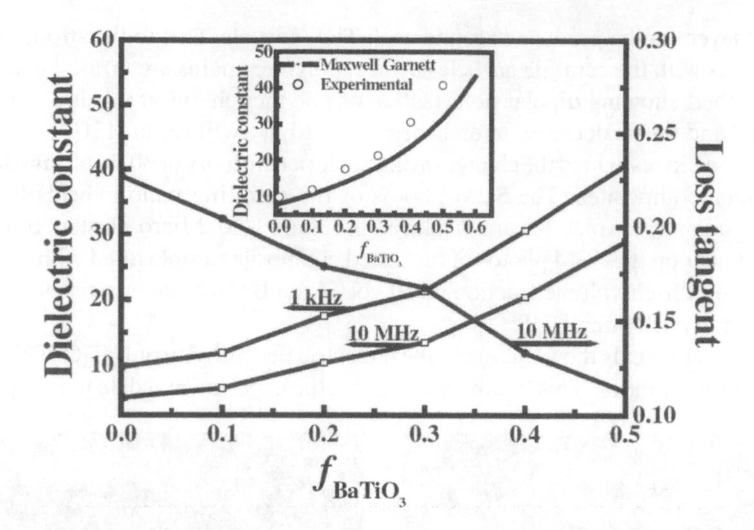

FIGURE 8.11 Dielectric Permittivity of $BaTiO_3$/PVDF as a Function of $BaTiO_3$ Volume Fraction. The Inset Shows the Experimental Data Mostly Follows the Maxwell-Garnett Behavior [149].

the Maxwell-Garnett trend. The reason for this good agreement could be related to the fact that these composites are acting as microcomposites in contrast to nanocomposites, and no interfacial effect can be detected in these microcomposite structures. The results of Dang et al. study showed the important role of nanoparticle dispersion in polymer nanocomposites. To study the effect of adding nanoceramic on dielectric properties of polymer nanocomposites and accomplish the so-called "nano-advantage" in both the nanoscale geometry of these particles and ultrahigh surface area, we might make sure that the prepared composite does not fall into a conventional microcomposite category and instead reaches each particle size in nanoscale.

Dou et al. studied dielectric properties of $BaTiO_3$/PVDF nanocomposites. The dielectric breakdown was reduced for the composite filled with 30 wt.% of untreated $BaTiO_3$ while increases up to 120% were obtained when functionalized nanoparticles were used. Figure 8.12 shows the change in the dielectric breakdown strength and dielectric constant as a function of $BaTiO_3$ contents [150].

Tuncer et al. [151] explored the dielectric breakdown of polyvinyl alcohol (PVA) reinforced with $BaTiO_3$ nanoparticles. In these composites, an increase of 25% in breakdown was recorded after filling with $BaTiO_3$. This enhancement in breakdown strengths of the prepared nanocomposites was mainly assigned to the uniform dispersion of nanoceramics within the matrix. Cao et al. also reported improved dielectric breakdown values after filing with Al_2O_3 and SiO_2 nano-powders to polyimide (PI). They attributed the inferior electrical conductivity of these nanocomposites to the different trapping and confined ionic conduction behavior at interphase as the underlying reason for higher DC dielectric breakdown of nanohybrids [152].

However, contradictory results have been found elsewhere where a decrease in breakdown was detected when micro and nanoparticles were included into polymer matrices. For instance, Tuncer et al. studied the dielectric breakdown of epoxy/

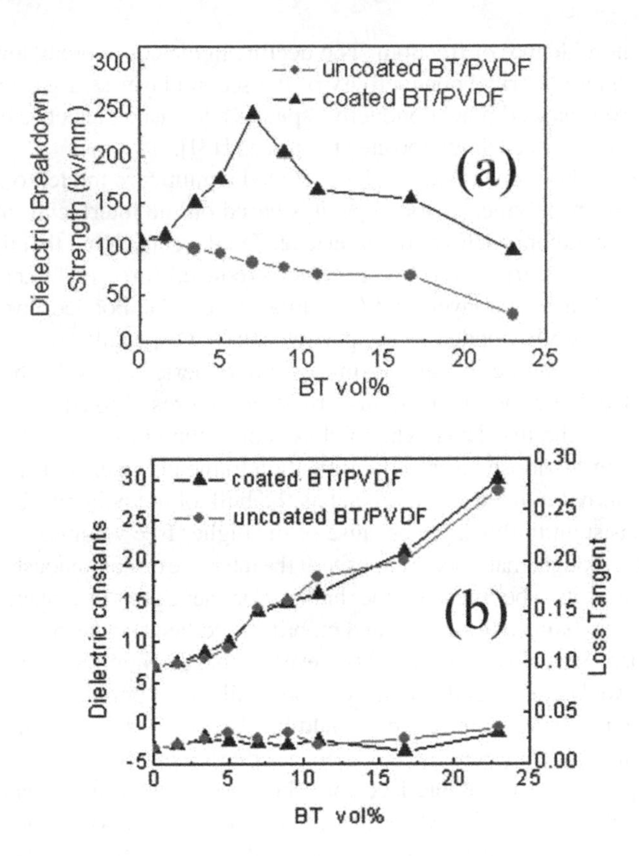

FIGURE 8.12 (a) Dielectric Strength of Nanocomposites as a Function of BT Volume Fraction; (b) Dielectric Constants and Loss Tangent of the Nanocomposites Measured at 1 kHz and 25°C [150].

$BaTiO_3$ micro- and nanocomposites. The prepared nanocomposite exhibited similar behavior as the unfilled resin while the microcomposite had a decreased dielectric breakdown but sharper slope, which revealed higher shape factor and more reliability for industrial applications [153].

Khalil found that DC dielectric breakdown of LDPE polymer decreased by almost 16% after reinforcing LDPE matrix with only 1 wt.% $BaTiO_3$. This trend was attributed to the enhancement in the carrier mobility after filling the matrix with $BaTiO_3$. In addition, the clusters of $BaTiO_3$ particles behave like field-increasing defect centers, resulting in a decrease in DC dielectric breakdown [154]. Singha reported that the dielectric breakdown of epoxy composites was reduced by adding both micro- and nano-sized TiO_2 and Al_2O_3 particles.

Based on multicore model approach, Singha et al., explained the decrease in dielectric behavior. In their methods, they considered that each particle in the polymer matrix is circled with various layers, namely an innermost layer, where polymer chains are joined to the surface of the particle and are completely stopped; and a second loosely bound layer formed free volume, constituted of loose chain ends and chains with more degree of freedom. The decline in dielectric breakdown was justified by the higher electrical conductivity of the second layer as it gave a conduction path for electrons as well other conductive species and eased dielectric breakdown in lower electric fields in such composite specimens [131].

As mentioned before, Tanaka et al. developed a multicore model to describe dielectric behavior of polymer nanocomposites based on the interfacial interactions of polymer and the nanoparticles at the interface. Tanaka considered that the interfacial region between the particle and the polymer produced three different layers: (1) a bonded layer, (2) a bound layer, and (3) a loose layer. The bonded layer consists of polymer chains tightly bonded to the particle surface especially when particles are chemically treated to interact and/or link with polymeric chains. In the second one, as the bounded layer, polymer chains are strongly cross-linked and entangled to the bonded layer, the first layer. The thickness of second layer can attain 2 to 9 nm. The third one in the loose layer is the transition to the polymer bulk where polymer chains are weakly bounded and more act as the bulk of the polymer. Chain mobility is usually consistent in this layer because of the higher free volume.

Tanaka also thought that since composite at the interface discontinuously transformed, and there is a transition between polymer bulk to polymer chains in contact with the filler surface, polymer chain conformation and mobility modified and deviated from the bulk. This phenomena was also investigated by several other researchers in particulate-filled composites as well as laminated polymer composite films and polymer-free surfaces [156] thus, the effective dielectric properties could be also affected, based on the change in polymer chain mobility where the polymer experienced heterogeneity. As a result, the alteration in polymer layers at interface could be markedly modified in nanocomposites filled with small nanoparticles where the interface becomes a dominant portion of the total volume and by consequence could influence the final dielectric properties.

The major role of the interface in polymer nanocomposites not only can increase from the physical and chemical structure of each constituent phase and from their interactions with each other, but also from unique electrical properties of interphase as well. Lewis reported a similar concept to that of electronic behavior of electrolytic colloidal systems and found that at the interface of particles/matrix there could be a multilayered charged media based on "reorganization" of a variety of various charged species upon application of an electric field [157].

The dielectric characteristics of the interface thus will be controlled by this set of charged layers. The charged species could either arise from any mobile ions existing in the matrix, such as ionic impurities, or they can be generated from polarization of heterogeneous dielectric media due to the contrast in dielectric permittivity or electrical conductivity values of different phases in the composite. It is further argued that the feature of the multilayer-charged media at the interface could be similar to the diffuse double layer theory in the electrolyte composites.

8.3.5 Effect of Dispersion State

Scrutinizing the effect of a dominating interface is not achieved without any difficulties; ceramics generally have high surface-tension values. On the other hand polymers and their organic solvents hold onto relatively low surface-tension values. Due to increased interfacial tension between the nanoparticles and the polymer matrix, nanoparticles tend to agglomerate in polymeric matrices and make it too difficult to reach a "nano" dispersion [158].

As a result of aggregation and heterogeneous dispersion, the resulted nanocomposites not only exhibited inferior electrical performance but also acted like microcomposites with micron-sized aggregates. Therefore, the full potential of the nanoscale approach has been limited to few papers where novel physical properties are explored due to the dominance of an interfacial phase in the boundaries of matrix/particle interface [159, 160], the so-called "interphase," with unique characteristics compared to both particles and matrix. The higher the difference between surface-tension values of the materials to be mixed, the lower the compatibility and miscibility between the two. This becomes particularly a critical problem in dispersing high surface-area nanoparticles into polymeric matrices, where nanoparticles tend to aggregate and coagulate with each other to reduce high-interfacial tension with the polymer matrix, hence making it difficult to achieve a nanodispersed system. Therefore, addressing the challenge of dispersion comes as the first essential step in order to achieve a nanodispersed polymer-composite systems.

There have been several studies on the polymer dielectric community to deal with the dispersion challenge. In addition to high-shear mechanical-mixing [161] chemical treatment and functionalization of particles [162], other methods such as in situ polymerization where particles are dispersed into the monomer before undergo polymerization [163–165], grafting nanoparticles to the polymer backbone using organometallic oligomers [166] or polymerization of monomer on particle surface [167, 168] and finally in situ synthesis of nanoparticles in polymer solution [169, 170]. Li et al. [171] treated the outer surface of rod-shaped TiO_2 nanoparticles with an aspect ratio of 3.5 to achieve uniform dispersion in PVDF terpolymer. Figure 8.13 shows the cross-sectional FE-SEM image of the TiO_2/PVDF copolymer nanocomposite [171].

FIGURE 8.13 Cross-sectional FE-SEM Image of the TiO_2/PVDF Copolymer Nanocomposite Prepared by Li et al. [171].

The evolution of dielectric permittivity of TiO_2/PVDF as a function of temperature at 1 kHz result is given in Figure 8.14. At 10 vol.% TiO_2, the dielectric permittivity was increased, in a study reported by Wang et al., in most of the temperature span under measurements. For example, dielectric permittivity grew by 11% at room temperature. Li et al. [171] also reported a 45% improvement on stored electrical energy density of 10 vol.% TiO_2-reinforced PVDF terpolymer nanocomposites at 200 MV/m. The stored energy density of 10 vol.% was the highest among all span-tested ratios.

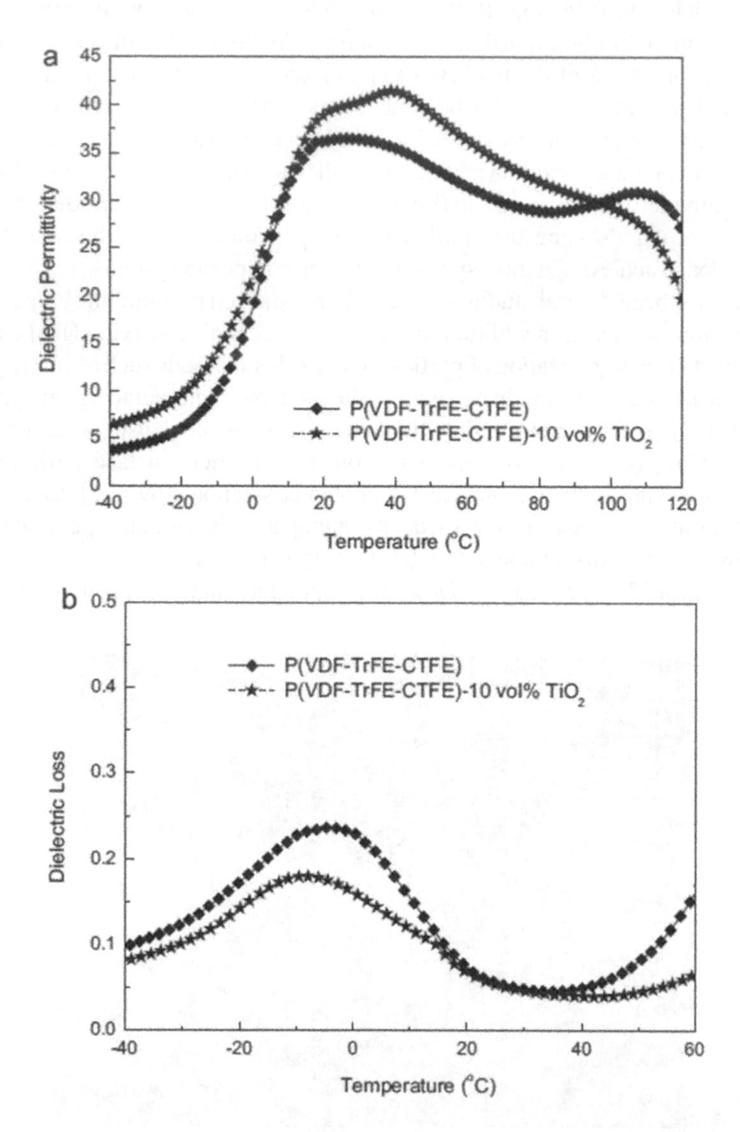

FIGURE 8.14 Temperature-dependence of the (a) Dielectric Constant and (b) Dielectric Loss of the Polymer and Nanocomposite Measured at 1 kHz. by Li et al [171].

Barber et al. [172] detected that the existence of such a sweet spot in energy density in Wang et al.'s work, as well as other similar works by other researchers, is the maximum interfacial area that can be attained for each particles size. This maximum interfacial surface area for TiO_2 nanoparticles in this study was achieved at the ratio of 10 vol.%. At this fraction, the ratio of interfacial area to the total of interfacial bulk phases reaches a maximum. Hence, if dielectric response in terms of dielectric permittivity and breakdown is affected by interfacial area, it can show itself as an alteration in dielectric behavior as a function of vol.%. $BaTiO_3$/polyimide nanocomposites were produced by Xie et al. by in situ polymerization of poly(amic acid) in the presence of the ceramic particles. The SEM images of their nanocomposites are illustrated in Figure 8.15 [173]. Even though the SEM images reveal a uniform dispersion of $BaTiO_3$ nanoparticles, it is difficult to certainly comment on each particle size in the nanocomposites.

Both the dielectric permittivity and loss of nanocomposites were enhanced by increasing particle loadings. Xie et al reported on nine times enhancement in dielectric permittivity and three times increase in dielectric loss for 50 vol.% $BaTiO_3$-filled nanocomposite as compared to those of neat polyimide. These significant improvements in dielectric permittivity depends on mixing laws predicted to the interfacial polarization and nano-enhancement effect.

Aiming to attain an improved dispersion state of nanoparticles into polymer matrices, the surface of the particles can be chemically treated via a variety of techniques

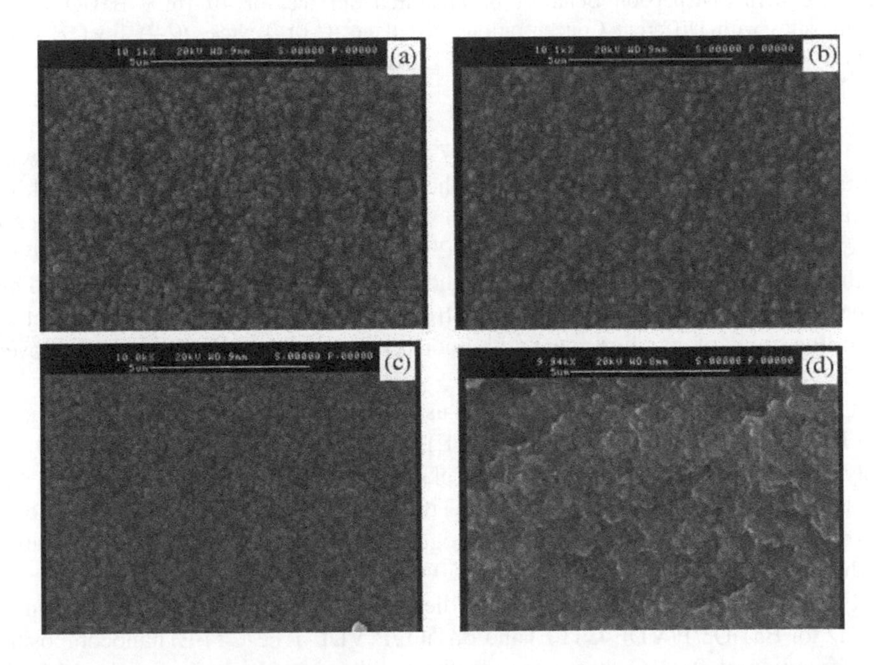

FIGURE 8.15 Surface SEM images of $BaTiO_3$/Polyimide Composites by Xie et al Using an In Situ Polymerization of Poly(amic acid) in the Presence of $BaTiO_3$ Particles. (a) 20 Vol.% (b) 33 Vol.% (c) 50 Vol.% and (d) the Fractured Section SEM of 20 vol.% $BaTiO_3$ [173].

FIGURE 8.16 Dispersion Behavior of Fractured Surfaces of 40 vol.% $BaTiO_3$/PVDF Composites with Different Concentrations of Cilane (C-0) 0 wt.%, (C-2) 1 wt.%, (C-4) 2 wt.%, and (C-6) 5 wt.%, Respectively [177].

[174–176]. For example, Dang et al., [177] treated the $BaTiO_3$ nanoparticles with aminopropyltriethoxy silane (KH550). The fractured cross section of their resulting films, without and after functionalization with various concentration of KH550, are given in Figure 8.16. They reported a 20% enhancement in dielectric permittivity without any change in dielectric loss at the range of 1 to 100 kHz frequencies. This is because in nanocomposites filled with treated nanoparticle, the dipolar interaction of amine group of coupling agent with fluorine atoms of the PVDF matrix improved the interface.

Li et al. successfully treated $BaTiO_3$ using ethylene diamine and then included the $BaTiO_3$ nanoparticle in P(VDF-CTFE) copolymer and P(VDFTrFe-CTFE) terpolymer. Figure 8.17 illustrates the TEM photo of 5 vol.% $BaTiO_3$/P(VDFCTFE) and cross-sectional FE-SEM of 23 vol.% $BaTiO_3$/P(VDF-TrFe-CTFE) [178]. As shown in this figure, the nanoparticles are homogenously dispersed in the polymer matrix with the average particle size of around 60 nm [178].

Figure 8.18 shows the evolution of dielectric permittivity and loss tangent at 1 kHz for BaTiO3/P(VDF-CTFE) and $BaTiO_3$/P(VDF-TrFe-CTFE) nanocomposites as a function of $BatTiO_3$ contents with the prediction of Lichtenecker logarithmic mixture law. As depicted in the Figure 8.18, the enhancement in dielectric permittivity is reproduced with the utilized model. On the other hand, Li et al. can keep

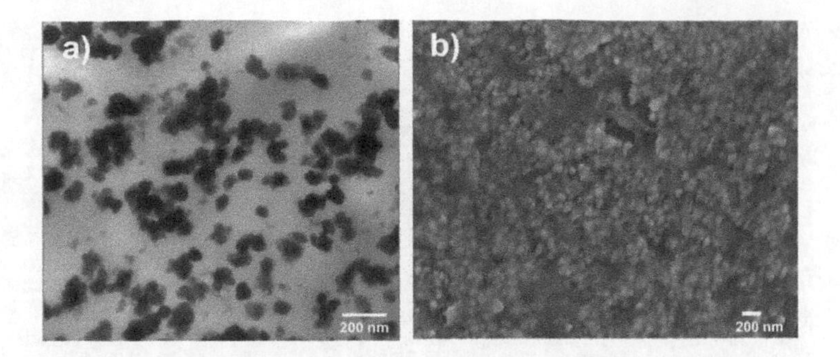

FIGURE 8.17 (a) TEM Image of 5 vol.% $BaTiO_3$/P(VDF-CTFE) and (b) Cross-sectional FE-SEM of 23 vol.% $BaTiO_3$/P(VDF-TrFe-CTFE) [178].

the loss tangent of their nanocomposites in an optimal low content. They concluded that the homogenous dispersion of $BaTiO_3$ nanoparticles caused the low-loss behavior by arguing that homogenous dispersion of nanoparticles kept the nanoparticles separated from each other and thus increased charge trapping, as opposed to an aggregated system in which a conduction path was established between nanoceramics with higher dielectric loss [178].

Choi et al. investigated the effect of dispersing $BaTiO_3$ into PI on the dielectric properties of their composites by treatment with two different coupling agents. INAAT[isopropyl tris(N-amino-ethyl aminoethyl)]titanite, also called also KR44

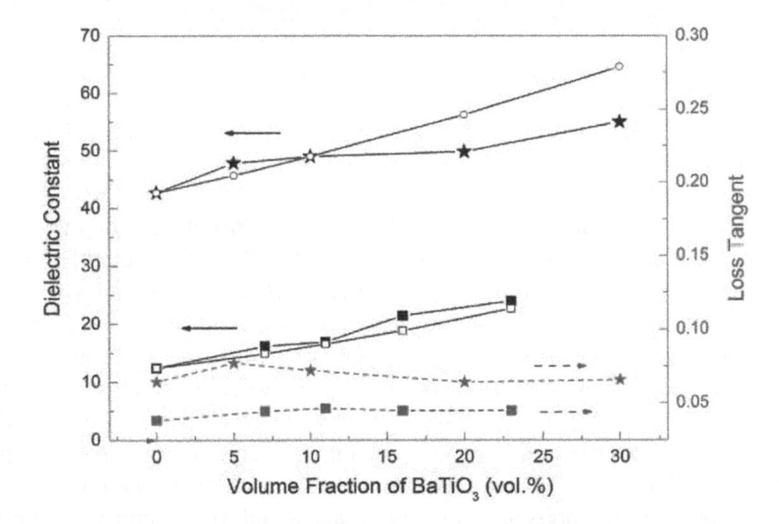

FIGURE 8.18 Dielectric Permittivity and Loss Tangent at 1 kHz for $BaTiO_3$/P(VDF-CTFE) (Square) and $BaTiO_3$/P(VDF-TrFe-CTFE) (Star) Nanocomposites as a Function of Vol.% of $BatTiO_3$ with the Prediction of Lichtenecker Logarithmic Mixture Law by Li et al. Open Squares and Circles are the Calculated Effective Permittivity from the Lichtenecker Law for $BaTiO_3$/P(VDF-CTFE) and $BaTiO_3$/P(VDF-TrFe-CTFE), Respectively [178].

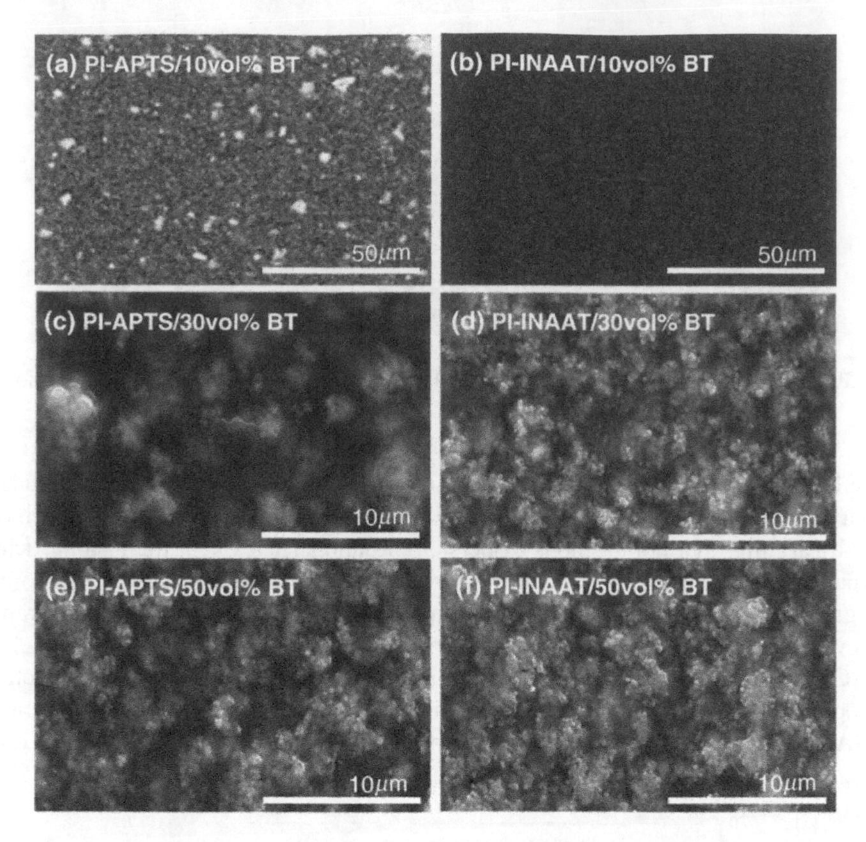

FIGURE 8.19 SEM Images of Effect of $BaTiO_3$/Polyimide Composites: (a) 10% $BaTiO_3$(APTS-modified), (b) 10% $BaTiO_3$(INAAT-modified), (c) 30% $BaTiO_3$(APTS-modified), (d) 30% $BaTiO_3$(INAAT-modified), (e) 50% $BaTiO_3$(APTS-modified), (f) 50% $BaTiO_3$(INAAT-modified) [179].

and APTS (3-amino-propyl-triethoxysilane) were chosen as coupling agents. Figure 8.19 shows the dispersion behavior of $BaTiO_3$/PI by Choi et al [179].

At 10 vol.%, the dispersion of INAAT-treated $BaTiO_3$ nanoparticles was clearly more homogenous with lower average aggregate size. However, by approaching 30 vol.% and 50 vol.%, the difference in dispersion between the two coupling agents disappeared. Figure 8.20 shows the comparison between the dielectric permittivity of INAAT versus APTS treated-$BaTiO_3$ in PI matrix. INAAT-treated $BaTiO_3$/Polyimide exhibited a higher dielectric constant in all nanofillers' range as reported by Choi et al. This improvement was ascribed to the better dispersion state of the reinforcing phase in the presence of coupling agent and to the improved interfacial polarization in such a case [179].

8.3.6 Sol–gel Processing of Nanocomposites

Controlled sol–gel synthesis of ceramic nanoparticles has been reported in the literature as an efficient technique to avoid aggregation and reach a uniform dispersion state

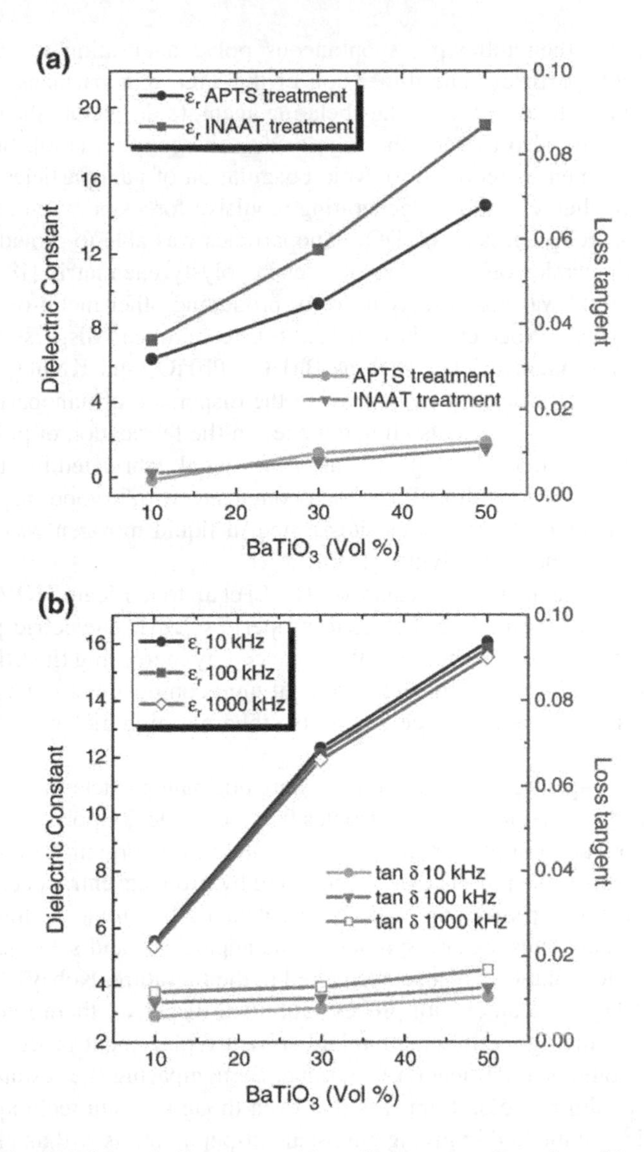

FIGURE 8.20 (a) Dielectric Constant and Loss Tangent as a Function of Loading Volume of $BaTiO_3$ Particles Treated by INAAT and APTS Coupling Agents and (b) Frequency Dependence for Dielectric Constant and Dielectric Loss of the Polyimide/INAAT-treated $BaTiO_3$ Composite [179].

of nanoparticles in the various polymeric matrices. The outstanding of the sol–gel process is the possibility of adding the polymeric phase at the time of sol–gel reaction; the presence of polymer chains in the sol–gel system gives steric hindrance to protect nanoparticles' coagulation after their formation and keeps them apart from each other.

Wang et al. synthesized TiO_2 nanoparticles in polystyrene solution via a sol–gel process [180]. Titanium butoxide, $TiBuO_4$ as TiO_2 precursor, was hydrolyzed using deionized water, then following spontaneous polycondensation reaction, it provides nano-TiO_2 particles. The dimension of the TiO_2 nanoparticles was tailored by using acetyl acetone, acacH as the chelating agent, to decelerate the polycondensation reaction and also to keep the high-acidity the solution [180]. Strong acidic or basic environment is required to avoid coagulation of nanoparticles by forming similar surface charges and thus generating repulsive forces between each particle. The in-situ sol–gel processing of TiO_2 nanoparticles was able to provide nanoscale homogenous dispersion of TiO_2 nanoparticles in polystyrene matrix [180].

Sol–gel method was also widely used for producing other metal-oxide particles reinforced polymers. Yogo et al. has utilized this technique to disperse the different kinds of ceramic nanoparticles, such as $BaTiO_3$, $PbTiO_3$ and $KNbO_3$, in polymer matrices. Due to its efficiency in improving the dispersion of nanoparticles within polymers, this technique was also implemented in the fabrication of polymer-based nanocomposites for dielectric applications. Tuncer et al. fabricated in situ TiO_2 nanoparticles in polyvinyl alcohol (PVA) nanocomposite with a good dispersion. The dielectric breakdown of nanocomposites tested in liquid nitrogen was remarkably higher than that of the neat polymer [151].

The in situ sol–gel method was also used by Li et al. to fabricate TiO_2/polyarylene ether nitriles and investigated dielectric properties [181]. Dielectric permittivity and loss factor were both monotonically enhanced by increasing the filler contents. They justified the improved dielectric loss of nanocomposites by the presence of impurities left from sol–gel process and also inherent high dielectric loss of TiO_2 particles.

The use of sol–gel technique to produce inorganic nanoparticles has not only been limited to in situ formation of nanoparticles in the presence of polymer solution, but also to the production of these nanofillers separately using organic or inorganic precursors. In contrast, the presence of residues and by-products entrapped in polymers in the case of in situ processing could be problematic by providing high-dielectric loss to the system. Thus, ex situ synthesis of nanoparticles and subsequently introducing a polymer solution has also been cited in the literature. Kobayashi et al. synthesized $BaTiO_3$ using an ex situ process and then dispersed them into a polymer solvent, by a centrifuge sedimentation and solvent replacement process, for a subsequent nanocomposite fabrication technique. By comparing the results for ex situ process and in situ process, it appears that even though ex situ technique is not as effective as in situ method in giving nanoscale dispersion, it is still an effective way to create a uniform dispersion of nanoparticles into polymer matrices.

In a new work [182], titanium carbide@boehmite (TiC@AlOOH) nanoparticle core–shell structure was elaborated by coating the AlOOH layer of nearly 10 nm on the surface of the titanium carbide (TiC) nanoparticle. These hybrid nanoparticles were used to fill a PVDF matrix using mechanical blending technique. The dielectric tests revealed that the ε of the nanocomposites significantly raised with an increase in filler loadings, since it reached 1.8×10^7 at 100 Hz for the specimen filled with 41 wt.% TiC@AlOOH, which represented 1.8 million times higher than that of pure PVDF. This result was explained according to the percolation theory, where the doped TiC@AlOOH nanoparticles could markedly induce the space-charge polarization in PVDF films by increasing charged carrier generation.

8.4 CONCLUSIONS

The current high-energy-density polymer/ceramic nanocomposites are specifically produced for ambient condition applications because of the limitation in thermal stabilities of some high-K polymers, which cannot be suitable for severe-condition operations usually recommended by the emerging applications likes electric-drive vehicles and deep oil and gas explorations. Such problems provoke new challenges in developing the new-generation organic-capacitor dielectrics. In addition, the matrix/filler interface and the ceramic-filler size from micro to nano can affect the dielectric properties of polymer/ceramic composites, thus, the good selection of ceramic type and size, as well as the coupling agent, is required for improving the dielectric polarization that represents the energy density. Moreover, the electrical conduction mechanisms in dielectric polymer/ceramic composites can help with the optimization of material design, preparation, and processing to ensure that the efficiency of charge–discharge could be reached. As presented in this chapter, dielectric polymer/ceramic micro- and nanocomposites emerged as promising candidates for enhancing the storage energy density through a synergistic coupling of high-dielectric permittivity issued from ceramic oxide nanoparticles and high dielectric breakdown from polymer matrices. More improvements in stored energy density was achieved by enhancing the dielectric permittivity and dielectric breakdown and controlling low-dielectric loss, which led to novel generation of polymer nano-dielectrics exhibiting superior properties and multifunctional ties suitable for many applications including energy storage in mobile electronics, hybrid cars, medical implants, aerospace, defense, and sporting goods.

REFERENCES

1. Xu, J. W., Bhattacharya, S., Pramanik, P. et al. 2006. High dielectric constant polymer-ceramic (epoxy varnish-barium titanate) nanocomposites at moderate filler loadings for embedded capacitors. *Journal of Electronic Materials* 35: 2009–2015.
2. Al-Bayer, R., Zihlif, A., Lahlouh, B. et al. 2013. AC electrical and optical characterization of epoxy–Al_2O_3 composites. *Journal of Materials Science: Materials in Electronics* 24: 2866–2872.
3. Al-Shabander, B. M., Kareem, A. A. 2015. The effect of zirconium dioxide on dielectric properties and thermal conductivity of polymers. *International Journal of Research in Mechanical and Materials Engineering* 1: 15–18.
4. Mohanty, A., Srivastava, V. K. 2013. Dielectric breakdown performance of alumina/epoxy resin nanocomposites under high voltage application. *Materials & Design* 47: 711–716.
5. Das, R. N., Lauffer, J. M., Markovich, V. R. 2008. Fabrication, integration and reliability of nanocomposite based embedded capacitors in microelectronics packaging. *Journal of Materials Chemistry* 18: 537–544.
6. Park, J. J. 2013. Electrical insulation breakdown strength in epoxy/spherical alumina composites for HV insulation. *Transactions on Electrical and Electronic Materials* 14: 105–109.
7. Yu, J. H., Huo, R. M., Wu, C. et al. 2012. Influence of interface structure on dielectric properties of epoxy/alumina nanocomposites. *Macromolecular Research* 20: 816–826.
8. Maity, P., Gupta, N., Parameswaran, V. et al. 2010. On the size and dielectric properties of the interphase in epoxy-alumina nanocomposite. *IEEE Transactions on Dielectrics and Electrical Insulation* 17: 1665–1675.

9. Singha, S., Thomas, M. J. 2010. Dielectric properties of epoxy–Al_2O_3 nanocomposite system for packaging applications. *IEEE Transactions on Components and Packaging Technologies* 33: 373–385.
10. Wu, C. C., Chen, Y. C., Yang, C. F. et al. 2007. The dielectric properties of epoxy/AlN composites. *Journal of the European Ceramic Society* 27: 3839–3842.
11. Hu, J. T., Gu, A. J., Liang, G. Z. et al. 2012. Synthesis of mesoporous silica and its modification of bismaleimide/cyanate ester resin with improved thermal and dielectric properties. *Polymers for Advanced Technologies* 23: 454–462.
12. Li, X., Wang, G., Huang, L. et al. 2015. Significant enhancement in dielectric constant of polyimide thin films by doping zirconia nanocrystals. *Materials Letters* 148: 22–25.
13. Kuo, D. H., Chang, C. C., Su, T. Y. et al. 2001. Dielectric behaviours of multi-doped $BaTiO_3$/epoxy composites. *Journal of the European Ceramic Society* 21: 1171–1177.
14. Kobayashi, Y., Kurosawa, A., Nagao, D. et al. 2009. Fabrication of barium titanate nanoparticles-polymethylmethacrylate composite films and their dielectric properties. *Polymer Engineering & Science* 49: 1069–1075.
15. Kuo, D. H., Chang, C. C., Su, T. Y. et al. 2004. Dielectric properties of three ceramic/epoxy composites. *Materials Chemistry and Physics* 85: 201–206.
16. Ramajo, L., Reboredo, M., Castro, M. 2005. Dielectric response and relaxation phenomena in composites of epoxy resin with $BaTiO_3$ particles. *Composites Part A: Applied Science and Manufacturing* 36: 1267–1274.
17. Ramajo, L., Castro, M. S., Reboredo, M. M. 2007. Effect of silane as coupling agent on the dielectric properties of $BaTiO_3$-epoxy composites. *Composites Part A: Applied Science and Manufacturing* 38: 1852–1859.
18. Alam, M. A., Azarian, M. H., Osterman, M. et al. 2011. Temperature and voltage aging effects on electrical conduction mechanism in epoxy-$BaTiO_3$ composite dielectric used in embedded capacitors. *Microelectronics Reliability* 51: 946–952.
19. Dang, Z. M., Yu, Y. F., Xu, H. P. et al. 2008. Study on microstructure and dielectric property of the $BaTiO_3$/epoxy resin composites. *Composites Science and Technology* 68: 171–177.
20. Preetha, P., Thomas, M. J. 2010. AC breakdown characteristics of epoxy alumina nanocomposites. 2010 *Annual Report Conference on Electrical Insulation and Dielectric Phenomena*, West Lafayette, IN, 17–20.
21. Yuen, S. M., Ma, C. C. M., Chuang, C. Y. et al. 2008. Preparation, morphology, mechanical and electrical properties of TiO2 coated multiwalled carbon nanotube/epoxy composites. *Composites Part A: Applied Science and Manufacturing* 39: 119–125.
22. Fan, B. H., Zha, J. W., Wang, D. R. et al. 2013. Preparation and dielectric behaviors of thermoplastic and thermosetting polymer nanocomposite films containing $BaTiO_3$ nanoparticles with different diameters. *Composites Science and Technology* 80: 66–72.
23. Wei, X. D., Xing, R. G., Zhang, B. W. et al. 2015. Preparation and dielectric properties of PATP-coated nano-$BaTiO_3$/epoxy resin composites. *Ceramics International* 41: S492–S4.
24. Yang, W. H., Yu, S. H., Luo, S. B. et al. 2015. A systematic study on electrical properties of the $BaTiO_3$–epoxy composite with different sized $BaTiO_3$ as fillers. *Journal of Alloys and Compounds* 620: 315–323.
25. Cho, S. D., Lee, J. Y., Hyun, J. G. et al. 2004. Study on epoxy/$BaTiO_3$ composite embedded capacitor films (ECFs) for organic substrate applications. *Materials Science and Engineering: B* 110: 233–239.
26. Iijima, M., Sato, N., Lenggoro, I. W. et al. 2009. Surface modification of $BaTiO_3$ particles by silane coupling agents in different solvents and their effect on dielectric properties of $BaTiO_3$/epoxy composites. *Colloids and Surfaces A: Physicochemical and Engineering Aspects* 352: 88–93.

27. Tsekmes, I. A., Kochetov, R., Morshuis, P. H. F. et al. 2015. The role of particle distribution in the dielectric response of epoxy–boron nitride nanocomposites. *Journal of Materials Science* 50: 1175–1186.
28. Alam, A. M., Azarian, M. H., Pecht, M. G. 2013. Modeling the electrical conduction in epoxy–$BaTiO_3$ nanocomposites. *Journal of Electronic Materials* 42: 1101–1107.
29. Elimat, M. Z. 2010. AC electrical properties of epoxy/silicon carbide whiskers composites coated with TiO_2 and poly(divinylbenzene). *Journal of Reinforced Plastics and Composites* 29: 331–342.
30. Wang, Q., Chen, G. 2014. Effect of pretreatment of nanofillers on the dielectric properties of epoxy nanocomposites. *IEEE Transactions on Dielectrics and Electrical Insulation* 21: 1809–1816.
31. Li, Y., Yang, W., Ding, S. et al. 2018. Tuning dielectric properties and energy density of poly(vinylidene fluoride) nanocomposites by quasi core–shell structured $BaTiO_3$@graphene oxide hybrids. *Journal of Materials Science: Materials in Electronics* 29: 1082–1092.
32. Thongruang, W., Spontak, R. J., Balik, C. M. 2002. Bridged double percolation in conductive polymer composites: an electrical conductivity, morphology and mechanical property study. *Polymer* 43: 3717–3725.
33. Wang, Y. Y., Hsieh, T. E., Chen, I. C. et al. 2007. Direct encapsulation of organic light-emitting devices (OLEDs) using photo-curable co-polyacrylate/silica nanocomposite resin. *IEEE Transactions on Advanced Packaging* 30: 421–427.
34. Sun, Y. Y., Zhang, Z. Q., Wong, C. P. 2005. Influence of interphase and moisture on the dielectric spectroscopy of epoxy/silica composites. *Polymer* 46: 2297–2305.
35. Zhou, S. X., Wu, L. M., Sun, J. et al. 2002. The change of the properties of acrylic-based polyurethane via addition of nano-silica. *Progress in Organic Coatings* 45: 33.
36. Kerimov, M. K., Kerimov, É. A., Musaeva, S. N. et al. 2007. The influence of structural and electrical parameters of the pyroelectric phase on the pyroelectric properties of a polymer–pyroelectric ceramic composite. *Physics of the Solid State* 49(5): 877–880.
37. Han, M. G., Armes, S. P. 2003. Synthesis of poly(3,4-ethylenedioxythiophene)/silica colloidal nanocomposites. *Langmuir* 19: 4523–4526.
38. Yang, X. M., Dai, T. Y., Lu, Y. 2006. Synthesis of novel sunflower-like silica/polypyrrole nanocomposites via self-assembly polymerization. *Polymer* 47: 441–447.
39. Gök, A., Şen, S. 2006. Preparation and characterization of poly(2-chloroaniline)/SiO_2 nanocomposite via oxidative polymerization: Comparative UV–vis studies into different solvents of poly(2-chloroaniline) and poly(2-chloroaniline)/SiO_2. *Journal of Applied Polymer Science* 102: 935–943.
40. Nayak, S., Sahoo, B., Chaki, T. K. et al. 2013. Development of polyurethane–titania nanocomposites as dielectric and piezoelectric material. *RSC Advances* 3: 2620–2631.
41. Li-jie, D., Chuan-xi, X., Juan, C. et al. 2004. Dielectric behavior of $BaTiO_3$/PVDF nanocompositesIn-situ synthesized by the sol-gel method. *Journal of Wuhan University of Technology* 19: 9–11.
42. Yuxing, R., and Davud, C. L. 2008. Properties and Microstructures of Low Temperature Processable Ultralow-Dielectric Porous Polyimide Films. *Journal of Electronic Materials* 37: 955.
43. Fan, Y., Huang, X., Wang, G. et al. 2015. Core–shell structured biopolymer@$BaTiO_3$ nanoparticles for biopolymer nanocomposites with significantly enhanced dielectric properties and energy storage capability. *The Journal of Physical Chemistry C* 119: 27330–27339.
44. Ciuprina, F., Zaharescu, T., Pleşa, I. 2013. Effect of γ-radiation on dielectric properties of LDPE Al_2O_3 nanocomposites. *Radiation Physics and Chemistry* 84: 145–150.
45. Wang, Q., Zhu, L. 2011. Polymer nanocomposites for electrical energy storage. *Journal of Polymer Science Part B: Polymer Physics* 49: 1421–1429.

46. Dang, Z. M., Yuan, J. K., Yao, S. H. et al. 2013. Flexible nanodielectric materials with high permittivity for power energy storage. *Advanced Materials* 25: 6334–6365.
47. Kim, P., Doss, N. M., Tillotson, J. P. et al. 2009. High energy density nanocomposites based on surface-modified $BaTiO_3$ and a ferroelectric polymer. *ACS Nano* 3: 2581–2592.
48. Dang, Z. M., Zhou, T., Yao, S. H. et al. 2009. Advanced calcium copper titanate/polyimide functional hybrid films with high dielectric permittivity. *Advanced Materials* 21: 2077.
49. Yao, J., Xiong, C., Dong, L. et al. 2009. Enhancement of dielectric constant and piezoelectric coefficient of ceramic–polymer composites by interface chelation. *Journal of Materials Chemistry* 19: 2817–2821.
50. Banerjee, S., Cook-Chennault, K. A. 2011. Influence of Al particle size and lead zirconate titanate (PZT) volume fraction on the dielectric properties of PZT-epoxy-aluminum composites. *Journal of Engineering Materials and Technology* 133: 041016.
51. Tang, H., Lin, Y., Sodano, H. A. 2012. Enhanced energy storage in nanocomposite capacitors through aligned PZT nanowires by uniaxial strain assembly. *Advanced Energy Materials* 2: 469–476.
52. Kim, P., Jones, S. C., Hotchkiss, P. J. et al. 2007. Phosphonic acid-modified barium titanate polymer nanocomposites with high permittivity and dielectric strength. *Advanced Materials* 19: 1001–1005.
53. Song, Y., Shen, Y., Liu, H. et al. 2012. Improving the dielectric constants and breakdown strength of polymer composites: effects of the shape of the $BaTiO_3$ nanoinclusions, surface modification and polymer matrix. *Journal of Materials Chemistry* 22: 16491–16498.
54. Yu, K., Wang, H., Zhou, Y. et al. 2013. Enhanced dielectric properties of $BaTiO_3$/poly(vinylidene fluoride) nanocomposites for energy storage applications. *Journal of Applied Physics* 113: 034105.
55. Tomer, V., Polizos, G., Manias, E. et al. 2010. Epoxy-based nanocomposites for electrical energy storage. I: effects of montmorillonite and barium titanate nanofillers. *Journal of Applied Physics* 108: 074116.
56. Xia, W. M., Xu, Z., Wen, F. 2012. Electrical energy density and dielectric properties of poly (vinylidene fluoride-chlorotrifluoroethylene)/$BaSrTiO_3$ nanocomposites. *Ceramics International* 38: 1071–1075.
57. Paniagua, S. A., Kim, Y. S., Henry, K. et al. 2014. Surface-initiated polymerization from barium titanate nanoparticles for hybrid dielectric capacitors. *ACS Applied Materials & Interfaces* 6: 3477–3482.
58. Yang, K., Huang, X. Y., Xie, L. Y. et al. 2012. Core-shell structured polystyrene/$BaTiO_3$ hybrid nanodielectrics prepared by in situ RAFT polymerization: a route to high dielectric constant and low loss materials with weak frequency dependence. *Macromolecular Rapid Communications* 33: 1921–1926.
59. Li, Z., Fredin, L. A., Tewari, P. et al. 2010. In situ catalytic encapsulation of core-shell nanoparticles having variable shell thickness: dielectric and energy storage properties of high-permittivity metal oxide nanocomposites. *Chemistry of Materials* 22: 5154–5164.
60. Jung, H. M., Kang, J. H., Yang, S. Y. 2010. Barium titanate nanoparticles with diblock copolymer shielding layers for high-energy density nanocomposites. *Chemistry of Materials* 22: 450–456.
61. Guo, N., DiBenedetto, S. A., Kwon, D. K. et al. 2007. Supported metallocene catalysis for in situ synthesis of high energy density metal oxide nanocomposites. *Journal of the American Chemical Society* 129: 766–769.
62. Xie, L., Huang, X., Wu, C. et al. 2011. Core-shell structured poly(methyl methacrylate)/$BaTiO_3$ nanocomposites prepared by in situ atom transfer radical polymerization: a

route to high dielectric constant materials with the inherent low loss of the base polymer. *Journal of Materials Chemistry* 21: 5897–5906.
63. Li, J., Khanchaitit, P., Han, K. et al. 2010. New route toward high-energy-density nanocomposites based on chain-end functionalized ferroelectric polymers. *Chemistry of Materials* 22: 5350–5357.
64. Xie, L. Y., Huang, X. Y., Yang, K. et al. 2014. "Grafting to" route to PVDF-HFPGMA/$BaTiO_3$ nanocomposites with high dielectric constant and high thermal conductivity for energy storage and thermal management applications. *Journal of Materials Chemistry A* 2: 5244–5251.
65. Tang, H. X., Sodano, H. A. 2013. High energy density nanocomposite capacitors using non-ferroelectric nanowires. *Applied Physics Letters* 102: 063901.
66. Zhou, Z., Tang, H. X., Lin, Y. R. et al. 2013. Hydrothermal growth of textured BaxSr1-xTiO_3 films composed of nanowires. *Nanoscale* 5: 10901–10907.
67. Tang, H. X., Lin, Y. R., Andrews, C. et al. 2011. Nanocomposites with increased energy density through high aspect ratio PZT nanowires. *Nanotechnology* 22: 015702.
68. Tang, H. X., Lin, Y. R., Sodano, H. A. 2013. Synthesis of high aspect ratio $BaTiO_3$ nanowires for high energy density nanocomposite capacitors. *Advanced Energy Materials* 3: 451–456.
69. Tang, H. X., Sodano, H. A. 2013. Ultra high energy density nanocomposite capacitors with fast discharge using $Ba_{0.2}Sr_{0.8}TiO_3$ nanowires. *Nano Letters* 13: 1373–1379.
70. Ducharme, S. 2009. An inside-out approach to storing electrostatic energy. *ACS Nano* 3: 2447–2450.
71. Zou, C., Kushner, D., Zhang, S. 2011. Wide temperature polyimide/ZrO_2 polyimide/ZrO_2 nanodielectric capacitor film with excellent electrical performance. *Applied Physics Letters* 98: 082905.
72. Balasubramanian, B., Kraemer, K. L., Reding, N. A. et al. 2010. Synthesis of monodisperse TiO_2-paraffin core-shell nanoparticles for improved dielectric properties. *ACS Nano* 4: 1893–1900.
73. Lin, S., Kuang, X., Wang, F. et al. 2012. Effect of TiO_2 crystalline composition on the dielectric properties of TiO_2/P(VDF-TrFE) composites. *Physica Status Solidi (RRL)* 6: 352–354.
74. Ouyang, G., Wang, K., Chen, X. Y. 2012. TiO_2 nanoparticles modified polydimethylsiloxane with fast response time and increased dielectric constant. *Journal of Micromechanics and Microengineering* 22: 074002.
75. Dang, Z. M., Xia, Y. J., Zha, J. W. et al. 2011. Preparation and dielectric properties of surface modified TiO_2/silicone rubber nanocomposites. *Materials Letters* 65: 3430–3432.
76. Zha, J. W., Dang, Z. M., Zhou, T. et al. 2010. Electrical properties of TiO_2-filled polyimide nanocomposite films prepared via an in situ polymerization process. *Synthetic Metals* 160: 2670–2674.
77. Zha, J. W., Fan, B. H., Dang, Z. M. et al. 2010. Microstructure and electrical properties in three-component (Al_2O_3–TiO_2)/polyimide nanocomposite films. *Journal of Materials Research* 25: 2384–2391.
78. McCarthy, D. N., Stoyanov, H., Rychkov, D. et al. 2012. Increased permittivity nanocomposite dielectrics by controlled interfacial interactions. *Composites Science and Technology* 72: 731–736.
79. Nan, C. W. 1993. Physics of inhomogeneous inorganic materials. *Progress in Materials Science* 37: 1–116.
80. Tanaka, T., Yazagawa, T., Ohki, Y. et al. 2007. Frequency accelerated partial discharge resistance of epoxy/clay nanocomposite prepared by newly developed organic modification and solubilization methods. *IEEE International Conference on Solid Dielectrics*, Winchester, UK, 337–340.

81. Nan, C. W., Shen, Y., Ma, J. 2010. Physical properties of composites near percolation. *Annual Review of Materials Research* 40: 131–151.
82. Kozako, M., Yamano, S., Kido, R. Y. et al. 2005. Preparation and preliminary characteristic evaluation of epoxy/alumina nanocomposites. *International Symposium on Electrical Insulating Materials*, 231–234.
83. Imai, T., Sawa, F., Ozaki, T. et al. 2006. Influence of temperature on mechanical and insulation properties of epoxy-layered silicate nanocomposite. *IEEE Transactions on Dielectrics and Electrical Insulation* 13: 445–452.
84. Chen, Y., Imai, T., Ohki, Y. et al. 2010. Tree initiation phenomena in nanostructured epoxy composites. *IEEE Transactions on Dielectrics and Electrical Insulation* 17: 1509–1515.
85. Henk, P. O., Korsten, T. W., Kvarts, T. 1999. Increasing the electrical discharge endurance of acid anhydride cured DGEBA epoxy resin by dispersion of nanoparticle silica. *High Performance Polymers* 11: 281–296, 1999.
86. Kozako, M., Kuge, S., Imai, T. et al. 2005. Surface erosion due to partial discharges on several kinds of epoxy nanocomposites. *IEEE Conference on Electrical Insulation and Dielectric Phenomena*, Nashville, TN, 162–165.
87. Tanaka, T., Matsuo, Y., Uchida, K. 2008. Partial discharge endurance of epoxy/SiC nanocomposite. *IEEE Conference on Electrical Insulation and Dielectric Phenomena*, Québec City, QC, Canada, 13–16.
88. Imai, T., Sawa, F., Ozaki, T. et al. 2005. Evaluation of insulation properties of epoxy resin with nano-scale silica particles. *International Symposium on Electrical Insulating Materials, Kitakyushu, Japan*, 239–242.
89. Li, S., Yin, G., Chen, G. et al. 2010. Short-term breakdown and long-term failure in nanodielectrics: a review. *IEEE Transactions on Dielectrics and Electrical Insulation* 17: 1523–1535.
90. Imai, T., Sawa, F., Nakano, T. et al. 2006. Effects of nano- and micro-filler mixture on electrical insulation properties of epoxy based composites. *IEEE Transactions on Dielectrics and Electrical Insulation* 13: 319–326.
91. Andritsch, T., Kochetov, R., Gebrekiros, Y. T. et al. 2009. Synthesis and dielectric properties of epoxy based nanocomposites. *IEEE Conference on Electrical Insulation and Dielectric Phenomena*, Virginia Beach, VA, 523–526.
92. Andritsch, T., Kochetov, R., Gebrekiros, Y. T. et al. 2010. Short term DC breakdown strength in epoxy based BN nano- and microcomposites. *IEEE International Conference on Solid Dielectrics*, Potsdam, Germany, 179–182.
93. Andritsch, T., Kochetov, R., Morshuis, P. H. F. et al. 2010. Short term DC breakdown and complex permittivity of Al_2O_3- and MgO-epoxy nanocomposites. *IEEE Conference on Electrical Insulation and Dielectric Phenomena*, West Lafayette, IN, 530–533.
94. Ding, H. Z., Varlow, B. R. 2004. Effect of nano-fillers on electrical treeing in epoxy resin subjected to AC voltage. *IEEE Conference on Electrical Insulation and Dielectric Phenomena*, Boulder, CO, 332–335.
95. Alapati, S., Thomas, M. J. 2009. Electrical tree growth in high voltage insulation containing inorganic nanofillers. *16th International Symposium on High Voltage Engineering*, Cape Town, South Africa, 145–150.
96. Danikas, M. G., Tanaka, T. 2009. Nanocomposites – a review of electrical treeing and breakdown. *IEEE Electrical Insulation Magazine* 25: 19–25.
97. Hajiyiannis, A., Chen, G., Zhang, C. et al. 2008. Space charge formation in epoxy resin including various nanofillers. *IEEE Conference on Electrical Insulation and Dielectric Phenomena*, Québec City, QC, Canada, 714–717.
98. Sarathi, R., Sahu, R. K., Rajeshkumar, P. 2007. Understanding the thermal, mechanical and electrical properties of epoxy nanocomposites. *Materials Science and Engineering A* 445–446: 567–578.

99. Zhang, C., Mason, R., Stevens, G. C. 2005. Dielectric properties of epoxy and polyethylene nanocomposites. *International Symposium on Electrical Insulating Materials*, Kitakyushu, Japan, 393–396.
100. Singha S., Thomas, M. J. 2008. Reduction of permittivity in epoxy nanocomposites at low nanofillers loadings. *IEEE Conference on Electrical Insulation and Dielectric Phenomena*, Quėbec City, QC, 726–729.
101. Singha, S., Thomas, M. J. 2008. Permittivity and tan delta characteristics of epoxy nanocomposites in the frequency range of 1 MHz-1GHz. *IEEE Transactions on Dielectrics and Electrical Insulation* 15: 2–11.
102. Kochetov, R., Andritsch, T., Morshuis, P. H. F. et al. 2012. Anomalous behaviour of the dielectric spectroscopy response of nanocomposites. *IEEE Transactions on Dielectrics and Electrical Insulation, IEEE Transactions on Dielectrics and Electrical Insulation* 19: 107–117.
103. Stevens, G. C., Herman, H., Han, J. et al. 2009. The role of nano and micro fillers in high thermal conductivity electrical insulation systems. *11th Insucon Conference*, Birmingham, UK, 286–291.
104. Kochetov, R., Korobko, A. V., Andritsch, T. 2011. Modelling of the thermal conductivity in polymer nanocomposites and the impact of the interface between filler and matrix. *Journal of Physics D: Applied Physics* 44: 395401.
105. Chen, Y. F., Dai, Q. W., Zhang, X. W. et al. 2014. Microstructure and properties of SCE-Al_2O_3/PES-MBAE composite. *Journal of Nanomaterials* 2014; 2014–2022.
106. Zeng, X. L., Yu, S. H., Sun, R. 2013. Thermal behavior and dielectric property analysis of boron nitride-filled bismaleimide-triazine resin composites. *Journal of Applied Polymer Science* 128: 1353–1359.
107. Xia, J. W., Li, J. H., Zhang, G. P. et al. 2016. Highly mechanical strength and thermally conductive bismaleimide–triazine composites reinforced by Al_2O_3@polyimide hybrid fiber. *Composites Part A: Applied Science and Manufacturing* 80: 21–27.
108. Yao, W., Gu, A. J., Liang, G. Z. et al. 2012. Preparation and properties of hollow silica tubes/bismaleimide/diallylbisphenol A composites with improved toughness, dielectric properties, and flame retardancy. *Polymers for Advanced Technologies* 23: 326–335.
109. Wu, G. L., Kou, K. C., Chao, M. et al. 2011. Investigation on curing kinetics and dielectric properties of nano-TiO_2/BMI/CE composites. *Journal of Aeronautical Materials* 31: 79–84.
110. Zeng, X. L., Yu, S. H., Sun, R. et al. 2011. Microstructure, thermal and dielectric properties of homogeneous bismaleimide-triazine/barium titanate nanocomposite films. *Materials Chemistry and Physics* 131: 387–392.
111. Yuan, L. X., Yan, H. X., Jia, Y. et al. 2014. Microwave curing bismaleimide-triazine composites filled with micro-silica and nano-silica. *Journal of Reinforced Plastics and Composites* 33: 1743–1750.
112. Takahashi, A., Kakimoto, M., Tsurumi, T. et al. 2005. High dielectric ceramic nano particle and polymer composites for embedded capacitor. *Journal of Reinforced Plastics and Composites* 18: 297–300.
113. Gu, A. J., Liang, G. Z. 2007. Novel high performance copper clad laminates based on bismaleimide/aluminium borate whisker hybrid matrix. *Journal of Applied Polymer Science* 103: 1325–1331.
114. Chao, F., Liang, G. Z., Kong, W. F. et al. 2008. Dielectric properties of polymer/ceramic composites based on thermosetting polymers. *Polymer Bulletin* 60: 129–136.
115. Wu, J., Yang, S., Gao, S. et al. 2005. Preparation, morphology and properties of nano-sized Al_2O_3/polyimide hybrid films. *European Polymer Journal* 41: 73–81.
116. Kim, H. J., Yoon, Y. J., Kim, J. H. et al. 2009. Application of Al_2O_3-based polyimide composite thick films to integrated substrates using aerosol deposition method. *Materials Science and Engineering: B* 16: 104–108.

117. Li, H., Liu, G., Liu, B. et al. 2007. Dielectric properties of polyimide/Al_2O_3 hybrids synthesized by in-situ polymerization. *Materials Letters* 61: 1507–1511.
118. Alias, A., Ahmad, Z., Ismail, A. B. 2011. Preparation of polyimide/Al_2O_3 composite films as improved solid dielectrics. *Materials Science and Engineering B* 176: 799–804.
119. Wang, S. F., Wang, Y. R., Cheng, K. C. et al. 2009. Characteristics of polyimide/barium titanate composite films. *Ceramics International* 35: 265–268.
120. Köytepe, S., Seçkin, T., Kıvrılcım, N., et al. 2008. Synthesis and dielectric properties of polyimide-titania hybrid composites. *Journal of Inorganic and Organometallic Polymers and Materials* 18: 222–228.
121. Panomsuwan, G., Ishida, H., Manuspiya, H. 2007. Dielectric properties of barium (strontium) titanate/polybenzoxazine composites with 0-3 connectivity for electrical applications. *Materials Research Society Symposium Proceedings* 993: 25–30.
122. Panomsuwan, G., Kaewwata, S., Manuspiyal, H. et al. 2007. Synthesis of polybenzoxazine and nano-barium titanate for a novel composite. *Proceedings of the 2nd IEEE International Conference on Nano/Micro Engineered and Molecular Systems* 497–501.
123. Selvaraj, V., Jayanthi, K. P., Lakshmikandhan, T. et al. 2015. Development of a polybenzoxazine/TSBA-15 composite from the renewable resource cardanol for low-*k* applications. *RSC Advances* 5: 48898–48907.
124. Sharma, R. A., D'Melo, D., Chaudhari, L. et al. 2012. The effect of absorbed moisture on dielectric behavior of silica (micro)-unsaturated polyester composites. *Journal of Applied Polymer Science* 125: 3788–3793.
125. Hanemann, T., Schumacher, B. 2012. Realization of embedded capacitors using polymer matrix composites with barium titanate as high-k-active filler. *Microsystem Technologies-Micro-and Nanosystems-Information Storage and Processing Systems* 18: 745–751.
126. Sharma, R. A., Melo, D. D., Bhattacharya, S. et al. 2012. Effect of nano/micro silica on electrical property of unsaturated polyester resin composites. *Transactions on Electrical and Electronic Materials* 13: 31–34.
127. Hanemann, T., Gesswein, H., Schumacher, B. 2011. Dielectric property improvement of polymer-nanosized strontium titanate-composites for applications in microelectronics. *Microsystem Technologies-Micro-and Nanosystems-Information Storage and Processing Systems* 17: 1529–1535.
128. Gorelov, B. M., Polovina, O. I., Gorb, A. M. et al. 2012. Fumed silica concentration effect on structure and dielectric properties of a styrene-cross-linked unsaturated polyester resin. *Journal of Applied Physics* 112: 094321.
129. Zhang, X., Shen, Y. Zhang, Q. H. et al. 2015. Ultrahigh energy density of polymer nanocomposites containing $BaTiO_3$@TiO_2 nanofibers by atomicscale interface engineering. *Advanced Materials* 27: 819–824.
130. Ma, D., Hugener, T. A., Siegel, R. W. et al. 2005. Influence of nanoparticle surface modification on the electrical behaviour of polyethylene nanocomposites. *Nanotechnology* 16: 724–731.
131. Singha, S. Thomas, M. J. 2008. Dielectric properties of epoxy nanocomposites. *Dielectrics and Electrical Insulation, IEEE Transactions on* 15: 12–23.
132. Tuncer, E., Sauers, I. 2010. Industrial Applications Perspective of Nanodielectrics. In Nelson, J. K. (Ed.), *Dielectric Polymer Nanocomposites*. Springer: Boston, MA, 321–338.
133. Winey, K. I., Vaia, R. A. 2007. Polymer nanocomposites. *MRS bulletin* 32: 314–322.
134. Fukushima, Y. Inagaki, S. 1987. Synthesis of an intercalated compound of montmorillonite and 6-polyamide. *Journal of Inclusion Phenomena* 5: 473–482.
135. Lewis, T. J. 2005. Interfaces: nanometric dielectrics. *Journal of Physics D: Applied Physics* 38: 202–212.

136. Nelson, J. K. 2010. Background, principles and promise of nanodielectrics. *Dielectric Polymer Nanocomposites*. Springer: Boston, MA, 1–30.
137. Nelson, J. Hu, Y. 2005. Nanocomposite dielectrics—properties and implications. *Journal of Physics D: Applied Physics* 38: 213–218.
138. Yang, T. I., Kofinas P. 2007. Dielectric properties of polymer nanoparticle Composites. *Polymer* 48: 791–798.
139. An, N., Liu, H., Ding, Y. et al. 2011. Preparation and electroactive properties of a PVDF/nano-TiO_2 composite film. *Applied Surface Science* 257: 3831–3835.
140. Fothergill, J. C., Nelson, J. K., Fu, M. 2004. Dielectric properties of epoxy nanocomposites containing TiO_2, Al_2O_3 and ZnO fillers. In: *Electrical Insulation and Dielectric Phenomena, CEIDP'04. 2004 Annual Report Conference on:* IEEE), pp. 406–409.
141. Murugaraj, P., Mainwaring, D., Mora-Huertas, N. 2005. Dielectric enhancement in polymer-nanoparticle composites through interphase polarizability. *Journal of Applied Physics* 98: 1–6.
142. Chu, B., Lin, M., Neese, B. et al. 2007. Large enhancement in polarization response and energy density of poly (vinylidene fluoridetrifluoroethylene-chlorofluoroethylene) by interface effect in nanocomposites. *Applied Physics Letters* 91: 122909.
143. Kim, P., Jones, S. C., Hotchkiss, P. J. et al. 2007. Phosphonic acid—modified barium titanate polymer nanocomposites with high permittivity and dielectric strength. *Advanced Materials* 19: 1001–1005.
144. Tanaka, T. 2005. Dielectric nanocomposites with insulating properties. *Dielectrics and Electrical Insulation* 12: 914–928.
145. Singha, S. Thomas, M. J. 2008. Permittivity and tan delta characteristics of epoxy nanocomposites in the frequency range of 1 MHz-1 GHz. *Dielectrics and Electrical Insulation* 15: 2–11.
146. Nelson, J. K., Fothergill, J. C. 2004. Internal charge behaviour of nanocomposites. *Nanotechnology* 15: 586–595.
147. Nelson, J. K., Utracki, L. A., MacCrone, R. K. et al. 2004. Role of the interface in determining the dielectric properties of nanocomposites. In: *Electrical Insulation and Dielectric Phenomena, 2004. CEIDP '04. 2004 Annual Report Conference on*, pp. 314–317.
148. Roy, M., Nelson, J. K., MacCrone, R. K. et al. 2005. Polymer nanocomposite dielectrics-the role of the interface. *Dielectrics and Electrical Insulation* 12: 629–643.
149. Dang, Z. M., Wang, H. Y., Zhang, Y. H. 2005. Morphology and dielectric property of homogenous $BaTiO_3$/PVDF nanocomposites prepared via the natural adsorption action of nanosized $BaTiO_3$. *Macromolecular Rapid Communications* 26: 1185–1189.
150. Dou, X., Liu, X., Zhang, Y. et al. 2009. Improved dielectric strength of barium titanate-polyvinylidene fluoride nanocomposite. *Applied Physics Letters* 95: 132904.
151. Tuncer, E., Sauers, I., James, D. R. et al. 2008. Nanodielectric system for cryogenic applications: Barium titanate filled polyvinyl alcohol. *Dielectrics and Electrical Insulation* 15: 236–242.
152. Cao, Y., Irwin, P. C., Younsi, K. 2004. The future of nanodielectrics in the electrical power industry. *Dielectrics and Electrical Insulation* 11: 797–807.
153. Tuncer, E., Sauers, I., James, D. R. et al. 2007. Electrical properties of epoxy resin based nano-composites. *Nanotechnology* 18: 025703.
154. Khalil, M. S. 2000. The role of $BaTiO_3$ in modifying the dc breakdown strength of LDPE. *Dielectrics and Electrical Insulation* 7: 261–268.
155. Tanaka, T., Kozako, M., Fuse, N. 2005. Proposal of a multi-core model for polymer nanocomposite dielectrics. *Dielectrics and Electrical Insulation* 12: 669–681.
156. Tanaka, T., Montanari, G. C., Mulhaupt, R. 2004. Polymer nanocomposites as dielectrics and electrical insulation-perspectives for processing technologies, material characterization and future applications. *Dielectrics and Electrical Insulation, IEEE Transactions on* 11: 763–784.

157. Lewis, T. J. 2005. Interfaces: nanometric dielectrics. *Journal of Physics D: Applied Physics* 38: 202–212.
158. Morgan, A. B., Wilkie, C. A. 2007. *Flame Retardant Polymer Nanocomposites*. John Wiley & Sons, Inc., Hoboken, New Jersey.
159. Manias, E. 2007. Nanocomposites: Stiffer by design. *Nature Materials* 6: 9–11.
160. Alexandre, M., Dubois, P. 2000. Polymer-layered silicate nanocomposites: preparation, properties and uses of a new class of materials. *Materials Science and Engineering: R: Reports* 28: 1–63.
161. Smith, R., Liang, C., Landry, M. 2008. The mechanisms leading to the useful electrical properties of polymer nanodielectrics. *Dielectrics and Electrical Insulation* 15: 187–196.
162. Khodaparast, P. Ounaies, Z. 2013. On the impact of functionalization and thermal treatment on dielectric behavior of low content TiO_2 PVDF nanocomposites. *Dielectrics and Electrical Insulation* 20: 166–176.
163. Andou, Y., Jeong, J. M., Nishida, H. et al. 2009. Simple procedure for polystyrene-based nanocomposite preparation by vapor-phase-assisted surface polymerization. *Macromolecules* 42: 7930–7935.
164. Guo, N., DiBenedetto, S. A., Tewari, P. et al. 2010. Nanoparticle, size, shape, and interfacial effects on leakage current density, permittivity, and breakdown strength of metal oxide–polyolefin nanocomposites: experiment and theory. *Chemistry of Materials* 22: 1567–1578.
165. Thomas, P., Dwarakanath, K., Varma, K. 2009. In situ synthesis and characterization of polyaniline–$CaCu_3Ti_4O_{12}$ nanocrystal composites. *Synthetic Metals* 159: 2128–2134.
166. Wang, J. W., Shen, Q. D., Yang, C. Z. et al. 2004. High dielectric constant composite of P (VDF-TrFE) with grafted copper phthalocyanine oligomer. *Macromolecules* 37: 2294–2298.
167. Maliakal, A., Katz, H., Cotts, P. M. et al. 2005. Inorganic oxide core, polymer shell nanocomposite as a high K gate dielectric for flexible electronics applications *Journal of the American Chemical Society* 127: 14655–14662.
168. Xie, L., Huang, X., Wu, C. et al. 2011. Core-shell structured poly (methyl methacrylate)/ $BaTiO_3$ nanocomposites prepared by in situ atom transfer radical polymerization: a route to high dielectric constant materials with the inherent low loss of the base polymer. *Journal of Materials Chemistry* 21: 5897–5906.
169. Lu, J., Wong, C. 2008. Recent advances in high-k nanocomposite materials for embedded capacitor applications. *Dielectrics and Electrical Insulation, IEEE Transactions on* 15: 1322–1328.
170. Yogo, T., Yamamoto, T., Sakamoto, W. et al. 2004. In situ synthesis of nanocrystalline $BaTiO_3$ particle–polymer hybrid. *Journal of Materials Research* 19: 3290–3297.
171. Li, J., Seok, S. I., Chu, B. et al. 2009. Nanocomposites of ferroelectric polymers with TiO_2 nanoparticles exhibiting significantly enhanced electrical energy density. *Advanced Materials* 21: 217–221.
172. Barber, P., Balasubramanian, S., Anguchamy, Y. et al. 2009. Polymer composite and nanocomposite dielectric materials for pulse power energy storage. *Materials* 2: 1697–1733.
173. Xie, S. H., Zhu, B. K., Wei, X. Z. et al. 2005. Polyimide/$BaTiO_3$ composites with controllable dielectric properties. *Composites Part A: Applied Science and Manufacturing* 36: 1152–1157.
174. Thakur, V. K., Kessler, M. R. 2014. Polymer Nanocomposites: New Advanced Dielectric Materials for Energy Storage Applications. *Advanced Energy Materials* 207–257.
175. Lin, Y., Zhou, B., Shiral Fernando, K. et al. 2003. Polymeric carbon nanocomposites from carbon nanotubes functionalized with matrix polymer. *Macromolecules* 36: 7199–7204.

176. Huang, X., Zhi, C., Jiang, P. 2013. Polyhedral oligosilsesquioxane-modified boron nitride nanotube based epoxy nanocomposites: An ideal dielectric material with high thermal conductivity. *Advanced Functional Materials* 23: 1824–1831.
177. Dang, Z. M., Wang, H.Y., Xu, H.P. 2006. Influence of silane coupling agent on morphology and dielectric property in $BaTiO_3$/polyvinylidene fluoride composites. *Applied Physics Letters* 89: 112902.
178. Li, J., Claude, J., Norena-Franco, L. E. et al. 2008. Electrical energy storage in ferroelectric polymer nanocomposites containing surface-functionalized $BaTiO_3$ nanoparticles. *Chemistry of Materials* 20: 6304–6306.
179. Choi, S. H., Kim, I. D., Hong, J. M. et al. 2007. Effect of the dispersibility of $BaTiO_3$ nanoparticles in $BaTiO_3$/polyimide composites on the dielectric properties. *Materials Letters* 61: 2478–2481.
180. Wang, S., Wang, M., Lei, Y. et al. 1999. Anchor effect in poly (styrene maleic anhydride)/TiO_2 nanocomposites. *Journal of materials science letters* 18: 2009–2012.
181. Li, C., Gu, Y., Xiaobo, L. 2006. Synthesis and dielectric property of polyarylene ether nitriles/titania hybrid films. *Thin Solid Films* 515: 1872–1876.
182. Ling, W., Hongxi, L., Peihai, J. et al. 2016. The effects of TiC@AlOOH core–shell nanoparticles on the dielectric properties of PVDF based nanocomposites. *RSC Advances* 6: 25015–25022.

9 Modeling of Polymer/Ceramic Composites Properties

9.1 INTRODUCTION

Experimental results on polymer composite properties are usually modeled using different finite element and micromechanical models to provide more insights into the experimental findings. These models are also applied in predicting the properties of same or similar composite materials, thus avoiding the need for producing each and every composite first to evaluate its properties. Numerous precautions are, however, required to avoid discrepancies in the model outcome; for example, the model applied should not have unrealistic assumptions, and the experimental results should be enough to fit accurately the model.

The mechanical, thermal, and electrical properties of polymer/ceramic composites depend strongly on the filler size, particle–matrix interface adhesion, and the filler ratio. Particle size has an obvious effect on the mechanical properties, thermal conductivity, and electrical performance. The particle dispersion within the matrix can also affect these properties. There are several empirical, semi-empirical, and numerical approaches which were applied to predict the mechanical, thermal, and electrical properties of polymer/ceramic composites. For a fixed ceramic volume ratio, by decreasing the particle size, the effect of the reinforcement of the polymeric phase is remarkably observed. Recently many computational works have shown that the ceramic particle size can influence various material properties such as the Young's and shear moduli [1–3], the thermal expansion coefficient (CTE) [4], the thermal conductivity [5, 6], dielectric constant, and the electrical conductivity [7].

This chapter gives some examples of modeling and prediction of polymer/ceramic micro- and nanocomposite thermal, electrical, and mechanical parameters using micromechanical, finite element techniques. This critical review deals with the theory, modeling, and numerical analysis by providing the various equation models with their theories and scrutinizing their applicability on the different types of polymer/ceramic composite materials. The numerical modeling includes molecular dynamics modeling and finite element modeling.

9.2 MODELING OF MECHANICAL PARAMETERS

Over the past few years, numerous micromechanical models have been developed to predict the mechanical elastic parameters of polymer/ceramic composites. Various empirical and semi-empirical models were used to accurately reproduce the elastic properties of polymeric composites. Prediction of Young's modulus of the polymer/ceramic composites was achieved using the micromechanical models explained in the following text.

9.2.1 Prediction of Elastic Modulus

The elastic modulus of a polymer/ceramic composite is usually determined by the elastic properties of its constituents, ceramic filler and polymeric matrix, filler loading and the aspect ratio [8, 9–11]. As the aspect ratio of ceramic particles approaches unity, such as in the case of spherical particles, the resulting composite modulus will be calculated for the moduli of their components by taking into consideration the particle ratio or particle size. Since the modulus of ceramic fillers is generally much higher than that of polymers, the composite modulus is easily increased by incorporating ceramic particles into a matrix. Serval empirical or semi-empirical equations have been used to predict the modulus of polymer/ceramic composites as summarized below.

Einstein's equation was derived for the effective shear viscosity for dilute suspensions of rigid spheres and was used to study the effective viscosity of concentrated suspensions of mono-sized spheres [12]. The below *Einstein* Eq. (9.1) was used to predict the Young's modulus of polymer/particulate composites is only valid at low filler ratios and consider

The adhesion between filler/matrix and the dispersion of individual filler particles are perfect [13]. This equation implies that elastic modulus of the composite is not related to particle size and establishes a linear trend between the composite elastic modulus and filler ratio. Therefore, it is applied for low filler loadings but not fitted for large filler ratios due to the strong interaction of the strain fields around these fillers. Einstein's equation was established to predict the Young's modulus of polymer/particulate composites, which is stated as:

$$\frac{E_c}{E_m} = 1 + 2.5V_f \tag{9.1}$$

where E_m and E_c are the tensile moduli of the matrix and composite, respectively.

The values of V_f are calculated from Eq. (9.2) as follows:

$$V_f = \frac{W_f/\rho_f}{W_f/\rho_f + (1 - W_f)/\rho_m} \tag{9.2}$$

where W_f is the weight fraction of the ceramic fillers; ρ_f and ρ_m are the filler and the matrix densities.

The parallel model and series model equation are also used to predict the tensile modulus of the polymer composites, and they are expressed as follows.

$$E_c = V_m E_m + V_f E_f \tag{9.3}$$

$$\frac{1}{E_c} = \frac{1}{E_m} + \frac{1}{E_f} \tag{9.4}$$

The elastic modulus of particle-filled composites can also be estimated by *Guth's* equation, which is expressed by Eq. (9.5).

$$E_C = E_m\left(1+2.5V_f+14.1V_f^2\right) \tag{9.5}$$

Concerning expansible polymers incorporating rigid, spherical particles featuring some adhesion, the Kerner equation can be applied to calculate the modulus. The *Kerner* form is written as:

$$E_c = E_m\left[1+\frac{15(1+\nu)}{8-10\nu}\frac{V_f}{1-V_f}\right] \tag{9.6}$$

where ν is poison ratio of the matrix.

An example is shown in Figure 9.1 of SiO_2-reinforced polyamide 6 nanocomposites [14], where the particle size changed in a narrow range of 12, 25, and 50 nm. The experimental data lies within these two bounds. And Kerner's equation provides much better agreement with the tested data, where the matrix Poisson ratio was taken as 0.3 [14].

The semi-empirical model given by Halpin-Tsai is a new model widely used to accurately predict the elastic properties of polymer composite. It is announced by the famous two Eqs. (9.7) and (9.8):

$$E_c = \frac{E_m\left(1+\eta\xi V_f\right)}{1-\eta V_f} \tag{9.7}$$

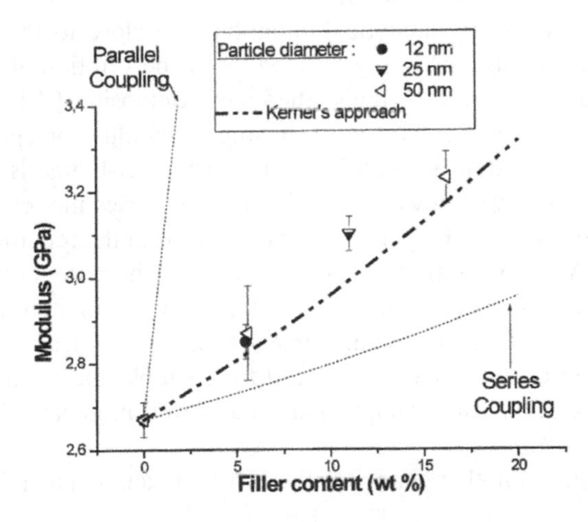

FIGURE 9.1 Variation of the Tensile Modulus of Silica-filled PA6 Nanocomposites with Respect to the Filler Content for Various Mean Particle Sizes: 12, 25, and 50 nm. Adapted from [14].

where

$$\eta = \frac{\frac{E_f}{E_m} - 1}{\frac{E_f}{E_m} + \xi} \quad (9.8)$$

where ξ is the shape factor that equals to 2 for the spherical ceramic particle.

By doing some corrections of the *Halpin-Tsai* equation, *Nielsen* introduced two factors related to the dispersion of fillers and their abilities to form aggregates into the matrix defined as: A is a constant related to the generalized *Einstein* coefficient k_E; ψ is a function related to the maximum-packing fraction ϕ_m of the filler. The modified equations are:

$$\frac{E_C}{E_m} = \frac{1 + ABV_f}{1 - B\psi V_f} \quad (9.9)$$

$$A = \frac{7 - 5\nu}{8 - 10\nu};\ B = \frac{(E_f / E_m - 1)}{(E_f / E_m + A)};\ \psi \approx 1 + \left[(1 - \phi_m)/\phi_m^2\right]V_f$$

The constants $A = 1.088$ and $\phi_m = 0.603$ (for random close packing, non-agglomerated).

Aiming to give more details, the elastic moduli of the studied nanocomposites were also calculated using several analytical equations extensively adopted for traditional composites, namely the *Halpin–Tsai* equation, e.g. [15] and the *Lewis–Nielsen* equation [16]. The comparison between all these results is shown in Figure 9.2, which is necessary to improve the understanding of the nanocomposite behavior. The calculated values using analytical equations are close to the modeling ones. This signifies that models commonly utilized for the description of composites with micro-sized loading are not fitted the studied nanocomposites [17].

Predictions and experimental data of Young's modulus for epoxy/Al_2O_3 nanocomposites are compared in Figure 9.3. The highest upper-bound is afforded by parallel model while the lowest lower-bound is given by series model. The predictions obtained from parallel model significantly deviated from the recorded experimental data because the Al_2O_3 was only the dispersed phase in the system, thus the contribution of Al_2O_3 in the nanocomposites was obviously overestimated in parallel model due to t each constituted phase being subject to the same strain [18]. In addition, the parallel and series models are only valid for a simple and idealized system. On the other hand, series and the Halpin-Tsai model well-predicted the experimental mechanical data.

Mooney [19] made another change to the Einstein equation as follows:

$$E_c / E_m = \exp\left(\frac{2.5V_f}{1 - sV_f}\right) \quad (9.10)$$

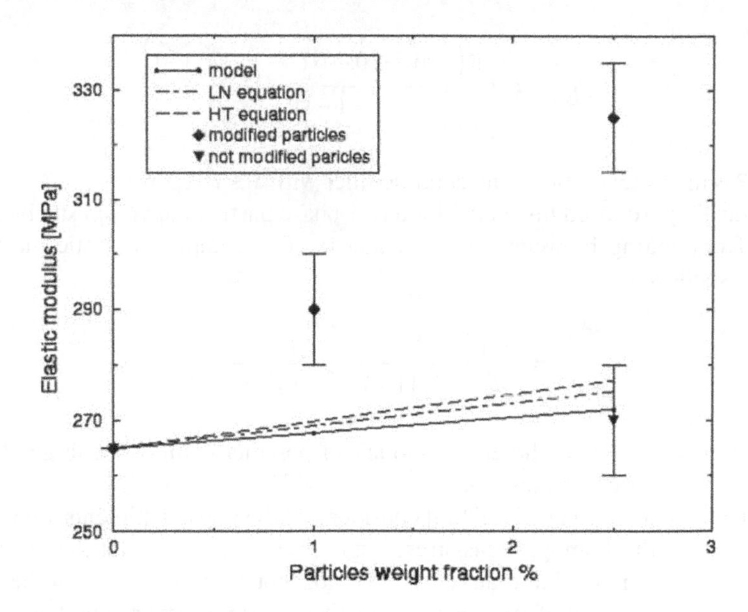

FIGURE 9.2 Comparison of the Modeling Results (Two-phase Model) with Experimental Data and Analytical Equations. The Error Bars Refer to Experimental Data. LN, Lewis–Nielsen Equation; HT, Halpin–Tsai Equation [17].

Where, s is a crowding parameter for the ratio of the apparent volume occupied by the particle to its own true volume, and its value varies between 1.0 and 2.0. This equation is reduced to Einstein's equation at low-volume fractions of spherical particles and represents test data at high-volume fractions. For non-spherical particles, Mooney's equation is further modified to [20].

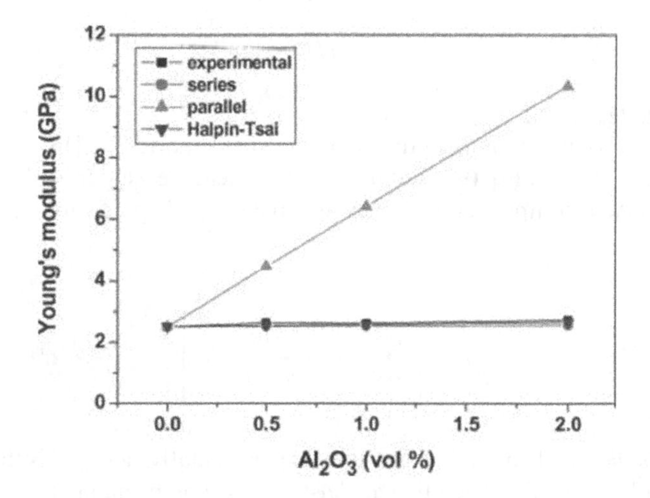

FIGURE 9.3 Young's Modulus of Epoxy/Al_2O_3 Nanocomposites [18].

$$E_c/E_m = \exp\left(\frac{\left(2.5V_f + 0.407(1-P)^{1.508} V_f\right)}{1 - sV_f}\right) \quad (9.11)$$

where P is the aspect ratio of the ceramic filler with $1 \leq P \leq 15$.

Counto [21] proposed his model for a two-phase particulate composite by assuming perfect bonding between filler and matrix. The composite elastic modulus is written as follows:

$$\frac{1}{E_c} = \frac{1 - V_f^{1/2}}{E_m} + \frac{1}{\left(1 - V_f^{1/2}\right)/V_f^{1/2} E_m + E_f} \quad (9.12)$$

This model well-predicts the elastic moduli of polymer composites in good agreement with a wide range of test data.

Ishai and Cohen [22] and Paul [23] consider that the two constituents are in a state of macroscopically homogeneous stress, and adhesion is perfect at the interface of a cubic inclusion in a cubic matrix. As a homogenous stress is used on the boundary, the elastic modulus of the particulate polymer composite is stated by which is another upper-bound solution.

$$E_c/E_m = 1 + \frac{1 + (\delta - 1)V_f^{2/3}}{1 + (\delta - 1)\left(V_f^{\frac{2}{3}} - V_f\right)} \quad (9.13)$$

Using the same model, with uniform strain utilized at the boundary, Ishai and Cohen [22] obtained a lower-bound solution

$$E_c/E_m = 1 + \frac{V_f}{\delta/(\delta - 1) - V_f^{\frac{1}{3}}} \quad (9.14)$$

where $\delta = E_c/E_m$.

Other models can be found in the famous review reported by [18, 24–27].

Ramdani et al. used Eq. (9.13) to fit the experimental elastic moduli data of polybenzoxazine/AlN composites and they estimate the shape factor of Halpin-Tsai model [28].

$$\sigma_E = \left[\sum_{i=1}^{n} \frac{\left(E_c^{exp} - E_c^{cal}\right)^2}{n - m}\right]^{\frac{1}{2}} \quad (9.15)$$

where E_c^{exp} is the tensile modulus measured experimentally, and E_c^{cal} is the calculated tensile modulus using the predictive models; n and m are the number of experimental points and the number of adjustable parameters, respectively.

9.2.2 Prediction of Microhardness

The microhardness of polymer/ceramic composite has been predicted by serval model including the modified rule of mixtures (ROM), Halpin-Tsai, and Nielsen semi-empirical models; the equation models are given below. The modified ROM is applied to predict the experimental microhardness data of some particle/polymer composites by adding the β factor on the role of mixture as given in the Eq. (9.16). This adjusted factor is called the strengthening efficiency factor and is related to the aspect ratio and dispersion sate of the filler into the matrix [29].

$$H_c = \beta H_f . V_f + H_m . V_m \tag{9.16}$$

where H_c, H_f, and H_m are the microhardness of composite, particles, and matrix, respectively. The V_f and β are the filler volume fraction and the strengthening efficiency factor.

The model proposed by Halpin-Tsai is an efficient model to predict the microhardness values of polymer/particulate composites. It is given by the Eqs. (9.17) and (9.18):

$$H_c = \frac{H_m\left(1 + \eta\xi V_f\right)}{1 - \eta V_f} \tag{9.17}$$

where

$$\eta = \frac{\frac{H_f}{H_m} - 1}{\frac{H_f}{H_m} + \xi} \tag{9.18}$$

where ξ is the shape factor that equals to 2 for a spherical particle.

Nielsen modified the Halpin-Tsai equation by introducing two factors characterizing the dispersion state of fillers and their abilities to form aggregates within the matrix, those are: A is a constant related to the generalized Einstein coefficient k_E; ψ is a function related to the maximum packing fraction ϕ_m of the filler which equal to 0.601 (for random loose packing, non-agglomerated), 0.632 (for random close packing, non-agglomerated), and 0.37 (for random close packing, agglomerated) [30]. The modified equations are:

$$\frac{H_C}{H_m} = \frac{1 + ABV_f}{1 - B\psi V_f} \tag{9.19}$$

$$A = \frac{7 - 5\nu}{8 - 10\nu} \tag{9.20}$$

$$B = \frac{\left(H_f / H_m - 1\right)}{\left(H_f / H_m + A\right)} \tag{9.21}$$

$$\psi \approx 1 + \left[\left(1 - \phi_m\right) / \phi_m^2\right] V_f \tag{9.22}$$

The Halpin–Tsai model was used to correlate the microhardness values of PEEK reinforced with micro- or nano-AlN composites. It was detected that the adjustable parameter, ξ, is different for both fillers. The value of ξ is higher for nanocomposites compared to microcomposites [29]. As drawn in Figure 9.4, the microhardness data of the PEEK/AlN c microcomposites and nanocomposites were well-reproduced by modified-rule of mixture with $\beta = 0.065$ and 0.12, respectively. The Halpin–Tsai equation with $\xi = 0.5$ and 3 fit well the data for microcomposites and nanocomposites, respectively.

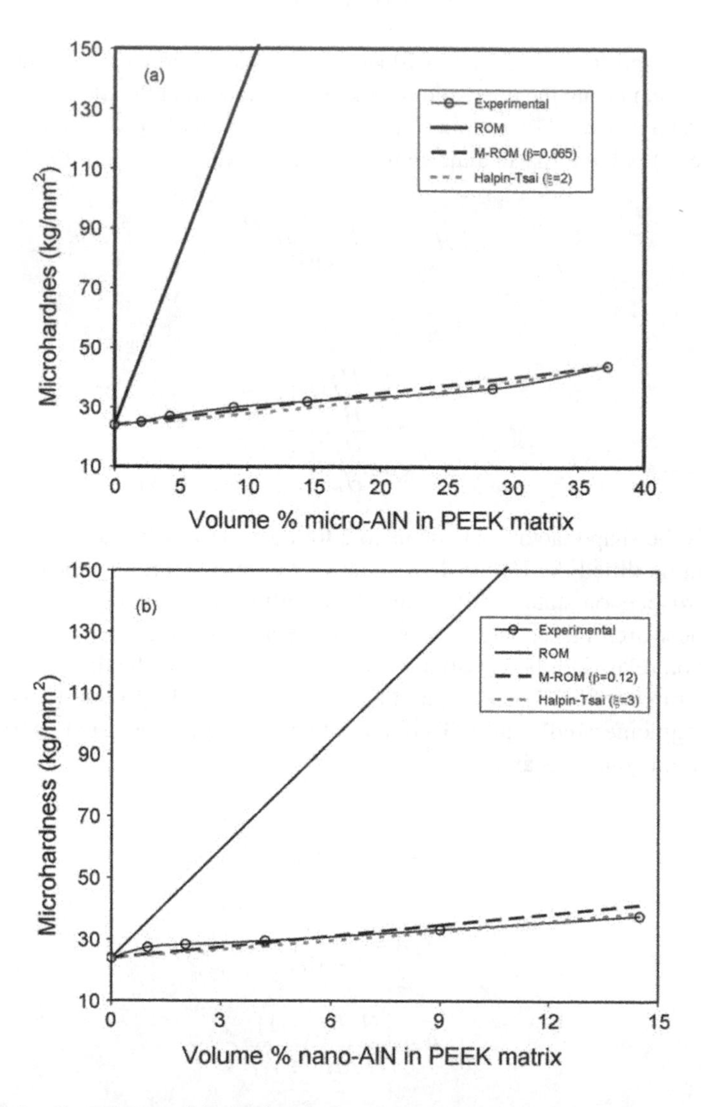

FIGURE 9.4 Correlation of Experimental and Predicted Microhardness for PEEK/AlN (a) Microcomposites and (b) Nanocomposites [29].

9.2.3 Finite Element Modeling

Finite element method (FEM) is considered the most useful and powerful numerical analysis tool, and it is used to predict mechanical properties of polymer composite materials. Since the 1970s, numerous FEMs have been developed to fit the different kinds of composite materials. In the past decade or so, there have been extensive experimental and analytical works, as well as FEM work on developing, analyzing, and characterizing ceramic-reinforced polymer composites and other types of nanocomposites.

9.3 MODELING THERMAL CONDUCTIVITY IN POLYMERIC/CERAMIC COMPOSITES

Due to the almost infinite number of possible thermally-conductive polymer/ceramic composite material compositions (the selection of polymer matrix, the type of fillers, the filler loadings, etc.), it is imperative to limit the numerous possible experiments and to be able to predict the thermal conductivity of the targeted materials. Therefore, theory and modeling play a crucial role in the process of screening potential compositions and choosing the most promising ones. Modeling with molecular dynamic simulations, for example, can estimate "ideal" thermal conductivity of many pure systems, including carbon nanotubes, graphite, ceramic particles, and various polymers. In addition, when applying models, experimenters can predict qualitatively, semi-quantitatively, or even quantitatively the dependency of composite thermal conductivity on the used matrix, type of ceramic filler, and filler loadings [32–34]. However, it is difficult to predict the effect of some factors like morphology and surface treatment, from first principles, thus modeling should be based on the experimental data to afford more explanation and more accuracy.

9.3.1 Micromechanical Modeling

Several key parameters control the prediction of the thermal conductivity of polymer/ceramic composites, this includes the inherent thermal conductivity of each phase, the filler ratios, the filler dimensions and forms, and the degree of dispersion state of ceramic fillers within matrix, etc. The easiest approach that usually fits as the first approximation is to consider two limiting cases: linear mixing rule (parallel model) and inverse mixing rule (series model). The parallel model is hypothesized that the temperature gradient is homogenous, and the heat flux is the weighted sum of heat fluxes through the matrix and filler domains. However, the series model is based on the assumption that the heat flux is uniform, and the temperature gradient is the weighted sum of temperature gradients through the matrix and filler domains. In the parallel model, filler and matrix participate independently in the final thermal conductivity, in proportion to their relative volume ratios [see Eq. (9.23)]. The parallel model maximizes the contribution of the filler phase and assumes perfect contact between particles in a fully percolating network implicitly.

$$\lambda_c = \left(1 - V_f\right)\lambda_m + V_f\lambda_f \tag{9.23}$$

Here, λ_c, λ_{cm} and λ_f are the thermal conductivity value of the composite, polymer matrix and filler respectively, and V_f is the volume fraction of fillers.

The series model is well-fitted for polymer/ceramic composites if the fillers are well-dispersed into the polymeric matrix. For this model, the composite thermal conductivity is expressed by the Eq. (9.24):

$$\lambda_c = \left[\frac{(1 - V_f)}{\lambda_m} + \frac{V_f}{\lambda_f} \right]^{-1} \quad (9.24)$$

The majority of the experimental data were located somewhere between these two models. But, in the majority of cases, the series model has well-fitted the experimental data of composites compared to the parallel model [35], resulting in the development of various models issued from the basic series model. These models added some more complex weighted averages on thermal conductivity values and volume fractions of ceramic filler, and those of polymeric matrix. These models usually include semi-theoretical fitting factors and are based on so-called effective-medium calculations. Two generally used effective-medium approximation methods are the *Maxwell–Garnett* (MG) model [36] and the *Bruggeman* model [37].

Maxwell–Garnett derived the expression for the conductivity λ for suspension of spherical non-interacting particles, as defined in Eq. (9.25):

$$\lambda_c = \lambda_m \left(1 + \frac{3V_f(\delta - 1)}{2 + \delta - V_f(\delta - 1)} \right) \quad (9.25)$$

By deriving the equation of the thermal conductivity taking into his account a solid spherical filler loading in a dilute suspension, Bruggeman established his model. The *Bruggeman model* is given as:

$$1 - V_f = \frac{(\lambda_f - \lambda_c)\left(\lambda_m / \lambda_c\right)^{\frac{1}{3}}}{\lambda_f - \lambda_m} \quad (9.26)$$

Kanari revised Bruggerman's model and developed his new equation [38].

$$1 - V_f = \frac{\lambda_f - \lambda_c}{\lambda_f - \lambda_m} \left(\frac{\lambda_m}{\lambda_c} \right)^{\frac{1}{X}} \quad (9.27)$$

where X is a constant characterizing the spherical shape of the filler. For AlN, 3.5 should be selected for X value in Eq. (9.27) [39]. The results depicted in Figure 9.5 revealed that this model can be accurately used to calculate the thermal conductivity of polyimide/AlN composites within all the ranges of AlN loading [40]. For higher filler ratios, AlN filler start to touch each other and construct conductive pathways in the direction of heat flow, provoking an exponential increase in the thermal conductivity of these composites.

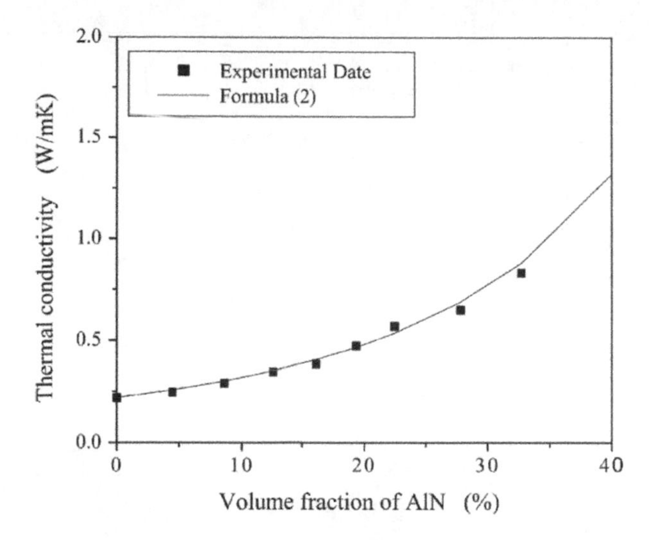

FIGURE 9.5 Effect of the Volume Fraction on Thermal Conductivity of the Composite, since Formula 2 is the *Kanari* Equation [40].

Maxwell also developed his model, which considers a random dispersion of small spheres in a continuous matrix to predict the effective thermal conductivity of polymer/ceramic composites. Lewis and Nielsen modified the role of the fillers' shapes and their orientations or their packing types. Therefore, for a two-constituent polymer composite, the best alternative is to consider these materials ordered in either parallel or series with respect to heat flow. All these equations are in function with the volume fraction of fillers V_f.

The *Maxwell model* is written as:

$$\lambda_c = \lambda_m \frac{\left[2\lambda_m + \lambda_f + 2V_f\left(\lambda_f - \lambda_m\right)\right]}{\left[2\lambda_m + \lambda_f - 2V_f\left(\lambda_f - \lambda_m\right)\right]} \tag{9.28}$$

The *Nielsen-Lewis model* is expressed as:

$$\frac{\lambda_C}{\lambda_m} = \frac{1 + ABV_f}{1 - B\psi V_f} \tag{9.29}$$

where, $B = \frac{\left(\lambda_f / \lambda_m - 1\right)}{\left(\lambda_f / \lambda_m + A\right)}$; $\psi \approx 1 + \left[\left(1 - V_{max}\right)/V_{max}^2\right]V_f$

In the case of spherical filler, A, which is related to geometry of particles, is taken 2, while V_{max} is selected as 0.637, by considering these fillers are randomly packed spheres, respectively.

Generally, it is reported that the experimental thermal conductivity data of polymer/ceramic composites can be accurately predicted by a *Nielsen model* at low-filler

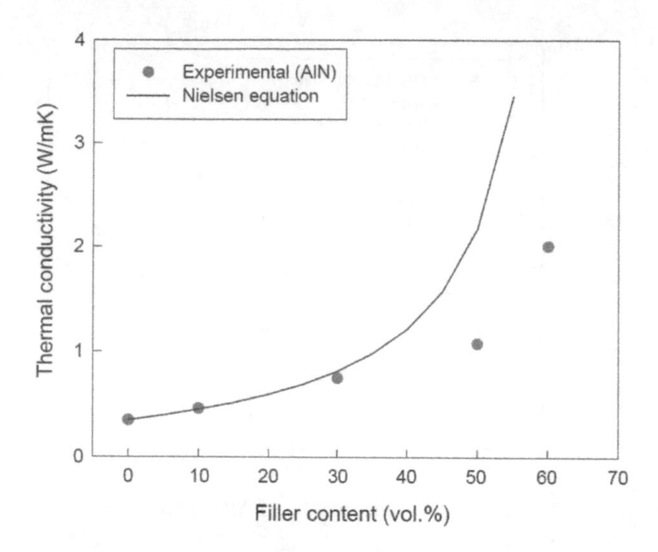

FIGURE 9.6 Thermal Conductivity of the Sample as a Function of Filler Content [42].

loadings [41]. However, as ceramic-filler content increases then particles touch each other, resulting in an overestimation of thermal conductivity by Nielsen equation as revealed in Figure 9.6 [42].

The *Russell model* is expressed as:

$$\lambda_c = \lambda_m \frac{\left[V_f^{2/3} + \frac{\lambda_m}{\lambda_f}\left(1 - V_f^{2/3}\right)\right]}{\left[V_f^{\frac{2}{3}} - V_f + \frac{\lambda_m}{\lambda_f}\left(1 + V_f - V_f^{2/3}\right)\right]} \tag{9.30}$$

The *Agari model* is a semi-empirical model used to estimate the thermal conductivity of two phase polymer/ceramic composite and it is announced as:

$$\log(\lambda_c) = C_f.V_f.\log(\lambda_f) + (1 - V_f).\log\left(C_p.\lambda_p\right) \tag{9.31}$$

since, C_p is a factor related to the feature of the polymer structure, such as crystalinity and physical state, whereas C_f is a factor applied to quantify ease of forming thermally-conductive chains by these fillers. The values of these two parameters should lie between 0 and 1, and as Agari stated, as the C_f values approach 1, the conductive-chains are more likely constructed within the matrix.

Zhou et al. [43] found the experimental data of polyethylene/BN do not fit well for either Maxwell-Eucken and Bruggeman models, although the Bruggeman model is more appropriate for reproducing the thermal conductivity of the composites filled with more than 20 vol.% ratios. From Figure 9.7, the experimental data always are above those predicted because these two models assume that the filler shape of filler

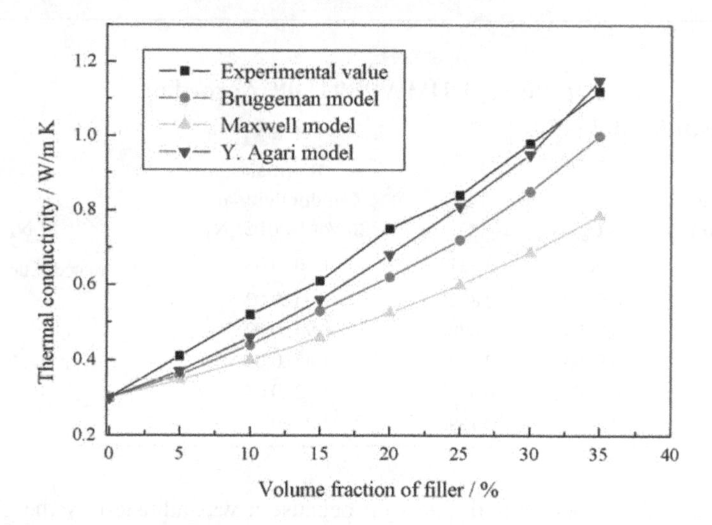

FIGURE 9.7 Thermal Conductivity of the Composites as a Function of Filler Content [43].

is spherical, and the dispersion state of filler is confined to common dispersion state of filler in the composites that is filler mixed into melted matrix.

Figure 9.8 showed a comparison between theoretical models and experimental data for a series of epoxy/AlN composites [44]. As the Maxwell model did not consider the particle's contacts inside the epoxy matrix, so it predicted the lowest thermal conductivity. However, Bruggeman's equation estimated the thermal conductivity was relatively lower than those of the experimental data because this model is based on spherical particles suspended in a dilute matrix. Comparatively, the Agari and

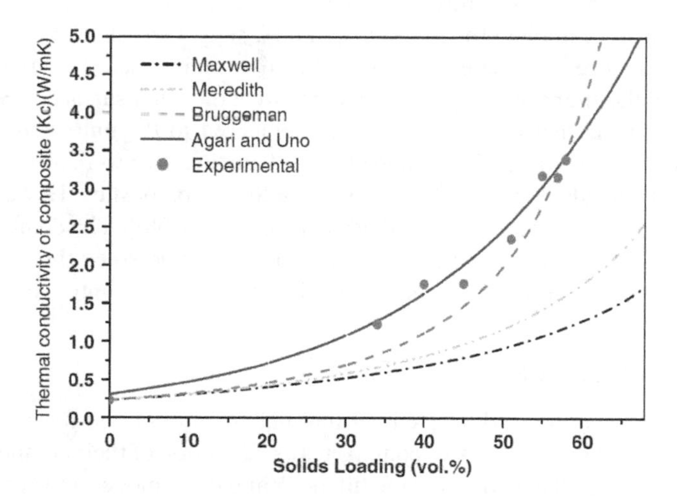

FIGURE 9.8 Comparisons of the Measured Thermal Conductivity of Aluminum Nitride-filled Epoxy Composite with the Calculated Thermal Conductivity by Various Theoretical Models [44].

TABLE 9.1
Values of C_p and C_f for Si_3N_4/UHMWPE/LLDPE According to the Agari Model [45]

Particle Size of Filler (μm)	C_p	C_f	Thermal Conductivity at 20 Vol.% of Si_3N_4	Si_3N_4
0.2	0.9018	0.9213	1.20±0.03	Untreated Filler
3.0	0.9217	0.8729	1.05±0.02	
35	0.9309	0.6282	0.94±0.02	
0.2	0.9283	0.9621	1.85±0.08	Silane (KH-550) treated
3.0	0.9402	0.9037	1.42±0.04	
35	0.9481	0.7395	1.23±0.06	

Uno model fit the experimental data well because it was adjusted by the C_p and C_f factors determined using experimental data. It was reported that the Agari and Uno model fit the experimental data well, not only for AlN fillers, but for many other ceramic fillers, such as SiO_2 and Al_2O_3. In this system, C_p was 41, which confirms the absence of any secondary structure effects of polymer. In addition, the C_f value in this epoxy/AlN composite system reached 0.5758, which was very low compared with other systems, and thus it is suggests that the AlN particles cannot easily contact each other in epoxy resins while still maintaining good fluidity.

By fitting the experimental data of polyethylene composite filled with Si_3N_4 into the Agari model, C_p and C_f for the above three systems are collected in Table 9.1. From this table, it can appear that the Si_3N_4 particles sizes can affect the C_f value more significantly than the C_p value, which reflects that filler particle sizes never influence the secondary structure of matrix. The composites filled with the smallest Si_3N_4 particles exhibited the highest C_f values, suggesting that the smaller filler particles could more easily construct the conductive chains than the larger particles do. The use of the silane agent coupling agent to treat the outer surface of Si_3N_4 obviously improved the thermal conductivity as compared to the untreated filler reinforced system (Table 9.1), and the same effects of particle size were also detected.

This trend was also detected for polystyrene/Si_3N_4 composites. However, C_f for those composites using larger PS particles is much larger [46]. The combining condition at the interface can markedly influence the C_f value since the C_f for silane treated-Si_3N_4 filler-reinforced composites is higher than that of untreated ones.

9.3.2 Finite Element Modeling (FEM)

In the previous section, we described various analytical or closed-form expressions evaluating thermal conductivity of composites as a function of their composition and the characteristics of the matrix and the fillers. While these models are applicable for rapid estimation of the trends expected upon variation in filler loading or filler type, they fail to reproduce the experimental data for some specific cases, such as monodisperse spherical fillers in a homogeneous matrix or perfectly aligned ellipsoidal

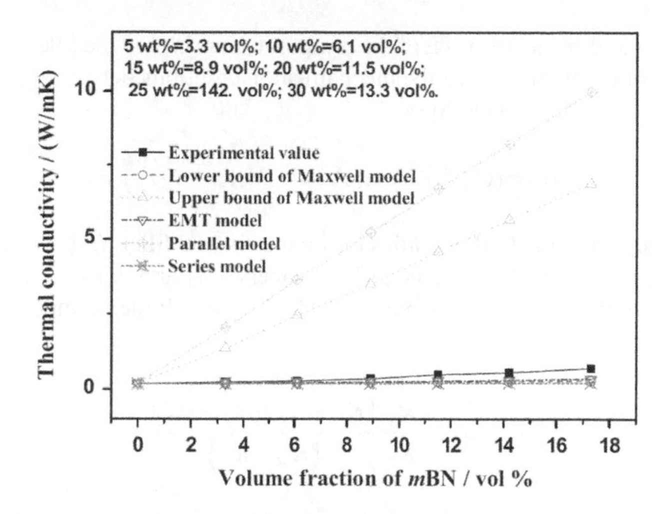

FIGURE 9.9 Thermal Conductivities' Comparison between Experimental and Theory-predicted Values for the mBN/PI Composites [48].

fillers in a homogeneous matrix. In addition, using these models to estimate polymer/ceramic composite morphology could be very complicated, thus numerical approaches, like finite element analysis (FEA) or finite difference (FD) modeling, could be better for resolving such problems [47].

Experimentally measured λ values and those predicted of polyimide/modified-BN by series and parallel conduction models, and *Maxwell* and *Effective Medium Theory* (EMT) models, are given in Figure 9.9 [48]. There is a large discrepancy between the experimental data and those predicted from the *parallel conduction* and upper bound of *Maxwell*, while they are slightly higher than those estimated by the *series conduction*, lower bound of the *Maxwell* and *EMT* models. This difference in the accuracy of theoretical models can be attributed to the low degree of dispersion and the presence of some defects in these composites.

9.4 THERMAL EXPANSION COEFFICIENT MODELS

The thermal expansion coefficient (CTE) models of polymer/ceramic composites are various, and they can be divided into theoretical and semi-empirical models, which generally reproduce better the experimental results. These models are partly based on the theory of elasticity or are based on the mechanics approach [49]. Recently, the most useful models, including the rule of mixtures (ROM) and Turner models, are applied to predict the CTE experimental data of several polymer/ceramic composites. As expressed in Eq. (9.32), the ROM model assumes the absence of any interaction between the host matrix and the ceramic reinforcing agent [50]. In contrast, the Turner model [49] considers the mechanical interactions between the two composites' constituents and its expression is written Eq. (9.11).

$$\alpha_c = \alpha_m \left(1 - V_f\right) + V_f \alpha_f \tag{9.32}$$

where α_c, α_m, and α_f are the CTE of the composite, matrix, and the filler, respectively.

The Turner model introduced the mechanical interactions between the fillers and the matrix and it is stated as follows:

$$\alpha_c = \left(\alpha_m V_m Y_m + \alpha_f V_f Y_f\right) / \left(V_m Y_m + V_f Y_f\right) \tag{9.33}$$

Where, Y_m and Y_f are the bulk modulus of the matrix and filler, respectively

According to thermoelastic principles, Schapery [51] announced his own model to predict the upper and lower bounds of the CTE of a polymer composite. The two bounds are given by:

$$\alpha_c^l = \alpha_m + \frac{K_f}{K_c^u} \frac{\left(K_m - K_c^u\right)\left(\alpha_f - \alpha_m\right)}{\left(K_m - K_f\right)} \tag{9.34}$$

$$\alpha_c^u = \alpha_m + \frac{K_f}{K_c^l} \frac{\left(K_m - K_c^l\right)\left(\alpha_f - \alpha_m\right)}{\left(K_m - K_f\right)} \tag{9.35}$$

where K_f, K_m and K_c are the bulk modulus of the filler, the matrix, and that of the composite. The subscripts u and l refer to the upper and lower bounds, respectively. It can be seen that the upper and lower bounds as calculated from the Hashin-Shtrikman model are used to calculate the lower and upper bounds in the Schapery model (Figure 9.10).

The experimental CTE values of Epoxy/AlN-SiC composites were predicted as drawn in Figure 9.11 [53]. The experimental values are located between those calculated by the models of Turner and Kerner. As the geometry of the binary filler system (particles + whiskers) is complicated, it is difficult to predict the theoretical CTE values that are approaching the experimental ones.

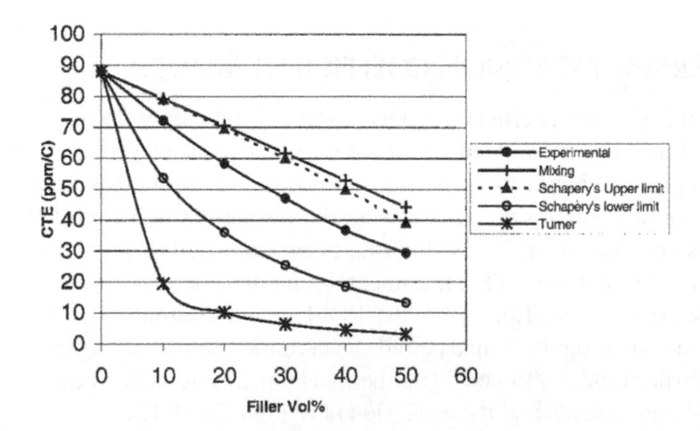

FIGURE 9.10 Comparison of CTE of Silica-filled Composites with Theoretical Predictions [52].

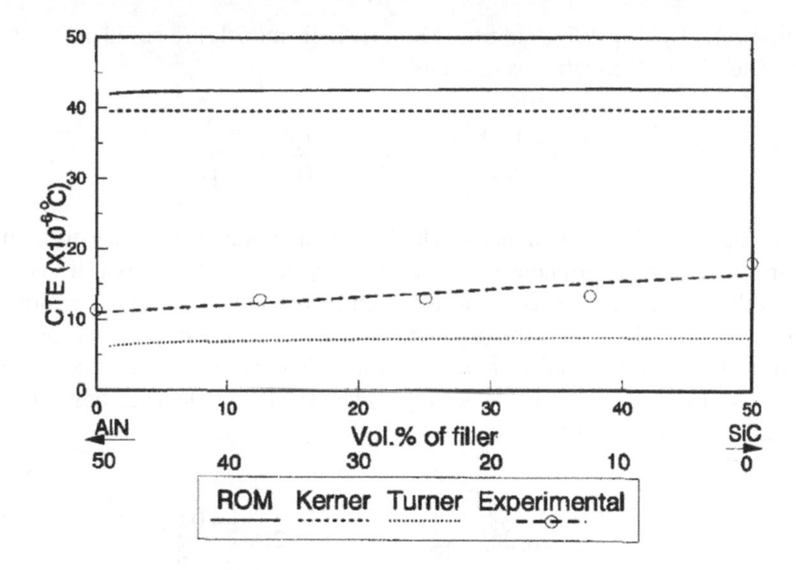

FIGURE 9.11 The Experimental and Measured Results of the CTE of Composites with Various AIN:SiC Volume Ratios [53].

9.5 PREDICTION OF EFFECTIVE DIELECTRIC CONSTANT

The difference in accuracy to estimate the dielectric properties by various models is significantly determined by the numerical formulation and the parameters considered during the formulation. The dielectric properties of the polymer/ceramic composites are directly affected not only by the relative permittivity but also by additional factors like the morphology, dispersion, and the interactions between the two phases [54]. Therefore, it could be a priority to predict the relative permittivity of these composites from the inherent permittivity of their constituents, apart from the volume fraction of the filler. Several theoretical methods have been used to reproduce the experimental effective dielectric constants data of polymer/ceramic composite hybrids. The volume-fraction average is a simple, but inaccurate equation to predict the effective dielectric constant of such composite material, and it stated:

$$\varepsilon_c = \varepsilon_f V_f + \varepsilon_m \left(1 - V_f\right) \tag{9.36}$$

where the subscripts ε_c, ε_f, and ε_m represent the dielectric constant of composite, ceramic phase, and polymer and respectively, and V_f is the volume fraction of the filler.

By analyzing the volume-fraction average model (Eq. 9.37), the effective ε_c of the composite enhances rapidly at low concentrations of the ceramic filler. Many studies based on both experimental data [55] and theory [56, 57] demonstrated that the predicted trend by (Eq.) 9.7 is not true. More realistic models are based on mean field theory. The *Maxwell* equation is given as:

$$\varepsilon_c = \varepsilon_m \frac{\varepsilon_f + 2\varepsilon_m - 2(1 - V_f)(\varepsilon_m - \varepsilon_f)}{\varepsilon_f + 2\varepsilon_m - (1 - V_f)(\varepsilon_m - \varepsilon_f)} \tag{9.37}$$

This equation is based on a mean field approximation of a single spherical inclusion surrounded by a continuous matrix of the polymer [58]. So, Maxwell's equation is only valid when the filler concentration tends to zero, i.e., infinite dilution. The *Bruggeman model* is another model of mean field theory which treats the binary mixture as being constituted of repeated unit cells composed of the matrix phase with spherical inclusions in the center [58]. The effective dielectric constant (ε_c) of the composite is written as:

$$V_f\left(\frac{\varepsilon_m - \varepsilon_c}{\varepsilon_m + 2\varepsilon_c}\right) + (1 - V_f)\left(\frac{\varepsilon_f - \varepsilon_c}{\varepsilon_f + 2\varepsilon_c}\right) = 0 \tag{9.38}$$

The evolution of the dielectric constant of composite predicted using three different models for composite of ceramic nanosphere $(\varepsilon_f = 1{,}000)$ dispersed in a polymer matrix $(\varepsilon_m = 2.3)$ was studied. Many investigations stated that the effective dielectric constant of composite (ε_c) reproduced by the Bruggeman equation increased significantly for filler ratios above 20 vol.% and can be very high for ceramic particle loadings higher than 50 vol.% by volume [58].

Maxwell-Garnett model [59]

$$\varepsilon_c = 1 + \frac{3V_f\left[(\varepsilon_f - \varepsilon_m)/(\varepsilon_f + 2\varepsilon_m)\right]}{1 - V_f\left[(\varepsilon_f - \varepsilon_m)/(\varepsilon_f + 2\varepsilon_m)\right]} \tag{9.39}$$

where ε_m, ε_f, and V_f denote the dielectric constant of the polymer, dielectric constant of the filler, and volume fraction of the filler, respectively. This was the first proposed model to theoretically estimate the dielectric parameters of polymer composites and it is still widely applied today. The spherical ceramic fillers are considered to be randomly dispersed within a polymer matrix without any interaction between these two phases.

Furukawa model is as follows [60]:

$$\varepsilon_c = 1 + \frac{1 - 2V_f}{1 + V_f}\varepsilon_m \tag{9.40}$$

where, ε_m denotes the dielectric constant of the polymer, and V_f volume fraction of the filler. The *Furukawa* model also considers that ceramic filler must have a spherical form, and they are homogenously dispersed into a polymeric matrix. This dielectrically homogeneous composite is assumed to be related to the dielectric behavior of the matrix.

The *Maxwell-Wagner equation* is given as follows [61]

$$\varepsilon_c = \varepsilon_m \frac{2\varepsilon_m + \varepsilon_f + 2V_f\left(\varepsilon_f - \varepsilon_m\right)}{2\varepsilon_m + \varepsilon_f - V_f\left(\varepsilon_f - \varepsilon_m\right)} \tag{9.41}$$

where ε_c, ε_f, and ε_m are the permittivity of the composites, filler, and matrix, respectively, and V_f is the volume fraction of the ceramic. The *Maxwell-Wagner* equation is applicable only when the properties of the two composite phases are identical [61].

Rayleigh's model is written by the Eq. (9.42) below [62]:

$$\varepsilon_c = \varepsilon_m \frac{2\varepsilon_m + \varepsilon_f - 2V_f\left(\varepsilon_f - \varepsilon_m\right)}{2\varepsilon_m + \varepsilon_f + V_f\left(\varepsilon_f - \varepsilon_m\right)} \tag{9.42}$$

here, ε_m, ε_f, and V_f denote the dielectric constant of the polymer, dielectric constant of the filler, and volume fraction of the filler, respectively. This model was issued by Rayleigh according to his theory taken from inferences of both the *Maxwell–Garnett* and *Furukawa* models for biphasic composite materials filled with low spherical-filler loadings.

Bhimasankaram model (BSP model) is given by the Eq. (9.43)[62, 63].

$$\varepsilon_c = \varepsilon_m \frac{\varepsilon_m\left(1 - V_f\right) + \varepsilon_f V_f\left[\dfrac{3\varepsilon_m}{\varepsilon_f + 2\varepsilon_m}\right]\left[1 + \left[\dfrac{3V_f\left(\varepsilon_f - \varepsilon_m\right)}{\varepsilon_f + 2\varepsilon_m}\right]\right]}{\left(1 - V_f\right) + V_f\left[\dfrac{3\varepsilon_m}{\varepsilon_f + 2\varepsilon_m}\right]\left[1 + \left[\dfrac{3V_f\left(\varepsilon_f - \varepsilon_m\right)}{\varepsilon_f + 2\varepsilon_m}\right]\right]} \tag{9.43}$$

where ε_c, ε_m, and ε_f are the dielectric constant of the composites, polymer matrix, and filler respectively, and V_f is the volume fraction of the ceramic.

This model considers that the spherical piezoelectric ceramics are randomly dispersed in a continuous polymer matrix. The major difference between the previous models and this one is that each dielectric sphere is polarized, and the dipoles get aligned in the direction of the applied electric field. These dipoles locally changed the electric field in the neighboring region of fillers and this effect will be considerably increased as higher fraction of piezoelectric ceramic particles are used. This model is also referred to as *Jayasundere–Smith* model.

Series mixing formula is written as follows [64]:

$$\frac{1}{\varepsilon_c} = \frac{V_f}{\varepsilon_f} + \frac{\left(1 - V_f\right)}{\varepsilon_m} \tag{9.44}$$

where ε_c, ε_f, and ε_m are the permittivity of the composites, filler, and matrix, respectively, and V_f is the volume fraction of the ceramic.

A plot showing the difference between all the above models is illustrated in the Figure 9.2. The *Furukawa model* (1979) plotted in the graph is not suitable for V_f 51. Another *Furukawa model* has also been announced, which takes into account the role of both the phases. The equation is similar to that of the Maxwell–Wagner. The

Rayleigh's model, the *Maxwell–Wagner* equation as well as the *Furukawa* model assumes that the influence of both the phases are identical. The *Jaysundere–Smith* equation is similar to the *BSP* model presented by Bhimasankaram et al. [62]. All these models trace the same trajectory at lower ceramic-filler loadings.

Generally, the theoretical predictions are accurate at low ceramic-filler loadings and deviations from predictions grow with rising filler ratios. This is attributed to the low-dispersion state of ceramic filler at higher-filler ratios and is also related to porosity or air enclosed by these composites. All the above cited models do not consider the shape of the ceramic filler or its dimension or the adhesion between the filler and matrix. A general theory is made that the filler has spherical shape and the interactional effect between the particles is overlooked. However, in the next models these effects are taking into consideration.

Modified Lichtnecker model is given by the Eq. 9.45 below [65]:

$$\log \varepsilon_c = \log \varepsilon_m + V_f (1-n) \log \frac{\varepsilon_f}{\varepsilon_m} \tag{9.45}$$

where ε_c, ε_f, and ε_m are the permittivity of the composites, filler, and matrix, respectively, and V_f is the volume fraction of the ceramic.

The *Lichtnecker logarithmic* rule postulated that any composite is a random mixture of nearly spherical inclusions. The theoretical estimation models are valid for composites reinforced with low-filler contents while they exhibit deviation at higher-filler ratios. This is attributed to the inferior dispersion state of the ceramic filler at higher-filler loadings and equally to the presence porosity or air entrapped in such composite. These models are only accurate for the composites based on ceramic filler and matrices having similar relative permittivity values. The revised *Lichtnecker* equation takes into account a fitting parameter n, which somehow measures the interaction between the filler and the host matrix.

Effective-medium theory (EMT) is postulated using Eq. 9.46 [65]:

$$\varepsilon_c = \varepsilon_m \left[1 + \frac{(\varepsilon_f - \varepsilon_m) V_f}{\varepsilon_m + n(1-V_f)(\varepsilon_f - \varepsilon_m)} \right] \tag{9.46}$$

where n presents the fitting parameter or the morphology factor.

The value of n parameter is related to both the polymer and the ceramic filler, thus decreasing the feasibility of the *Lichtnecker* equation for numerous materials [66]. In the *EMT model*, the dielectric property of the polymer/ceramic composite is taken as an effective medium whose relative permittivity is given by averaging the permittivity values of the composite components. This model is a self-consistent model that considers a random unit cell consisting of each filler particle covered by a concentric matrix layer.

Yamada model is written in the equation 9.47 [67]:

$$\varepsilon_c = \varepsilon_m \left[1 + \frac{n.V_f (\varepsilon_f - \varepsilon_m)}{n.\varepsilon_m + (1-V_f)(\varepsilon_f - \varepsilon_m)} \right] \tag{9.47}$$

Yamada et al. [67] described the dielectric behavior of polymer composites in his model. It is based on the properties of the individual constituents and also includes a factor ($n = 4\pi/m$), that depends on the shape and relative orientation of the filler. The shape factor or fitting factor varies from one composite to another, and it is also related to the particle sizes. At higher-volume fractions, the deviation between the predicted and experimental data becomes more significant. The *Jayasundere–Smith* equation is only valid for filler loading up to 0.3 vol.%, as it is based on the interactions between the neighboring spheres. The *Maxwell–Wagner* equation holds good only when the properties of the two phases in the composite are identical [68].

The behavior of the complex dielectric constant was estimated using the theoretical models previously introduced. The results recorded for a-PVDF/PZT and b-PVDF/PZT composites are depicted in Figure 9.12 [69].

In two-phase models, both constituents of the composite system are regarded as different phases rather than considering one phase of the composite. Rao et al. made numerical predictions of the ε_c for polymer/ceramic composites according to the effective-medium theory (EMT) [70]. They reported that as the ceramic-filler particle size is reduced compared to that of the composite, the dielectric permittivity of the composite could be calculated according to the effective medium whose dielectric permittivity can be given by averaging over the dielectric permittivity of the two components. They also included an arbitrary fitting parameter in their model to account for the irregular morphology of the ceramic fillers. The EMT model equation fit the experimental data for lead magnesium niobate-lead titinate/epoxy composites with an error of 10%. However, they found that values of the introduced arbitrary fitting parameter in such models are difficult to optimize.

Much theoretical work has been conducted to include the role of the interface between the matrix phase and the ceramic phase as the interface had a critical effect in determining the dielectric performance of such composite materials. The majority of these models include the role of the "interphase" as a separate phase in addition to the ceramic-filler and organic phases.

Todd et al. created a model known as the "interphase power law" to investigate the complex permittivity of the composite system [71]. The model considers the permittivity and the volume fractions of the matrix, filler, and the interface region. The effective permittivity calculated by this model was compared with experimental data. The model also affords insight into the contributions of the interface towards the dielectric permittivity and considers the interface as the reason for deviations of the predicted permittivity from the predictions of the standard mixture models.

The model evolved by Vo et al. takes into account the participation of molecular polarizability at the interface to the effective dielectric constant of composite [72]. Their model is based on the dielectric constant ratio between that of filler and that of the matrix as well as the degree of filler/matrix interaction. They added a parameter called the interface volume constant, which represents the matrix/filler interaction strength. A value of zero means an insignificant matrix/filler interaction, whereas high-positive values reflected the presence of strong interactions. They also concluded that the interface volume constant should be related to the size of the filler.

Later, Murugaraj et al. utilized the Vo-Shi model to rationalize the high dielectric constant values recorded for polyimide/Al_2O_3 nanocomposite films [73].

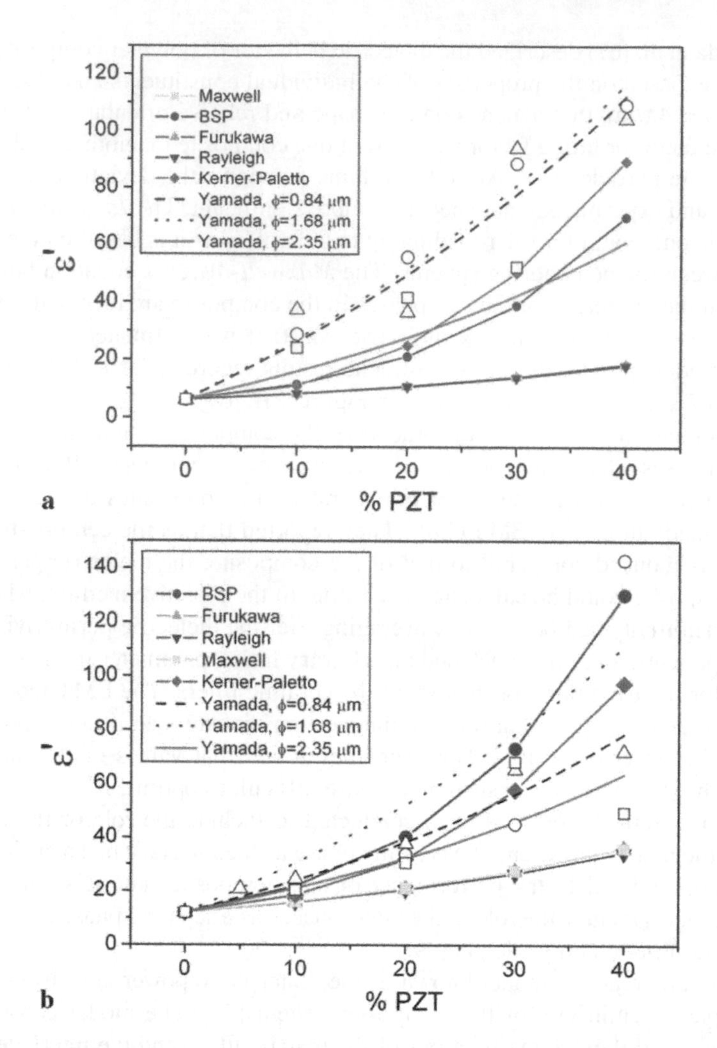

FIGURE 9.12 Variation of ε' as a Function of the PZT Concentration and Particle Size in the (a) a-PVDF and (b) b-PVDF Matrix (the Points Are the Experimental Results and All Fittings Were Realized for Room Temperature and 1 kHz). Open Triangles: 0.84 lm Grain Size; Open Circles: 1.68 lm Grain Size; Open Squares: 2.35 lm Grain Size [69].

The experimental dielectric constant values were well above those estimated by the *Maxwell model*. The predicted values by the Vo-Shi equation agreed with the experimental data after fitting the data by large positive values for the interface-volume constant, signifying the existence of a strong interaction between the polymer and the ceramic filler. They thought that the formation of interfacial dipoles with high-molecular polarizability is the main cause of attaining improved dielectric performance.

In addition to the mathematical modeling, numerical simulations have been also undertaken to estimate the effective dielectric constant data of polymer/ceramic composites. Using the finite element method, Tuncer et al. executed several numerical calculations to scrutinize the frequency-dependent dielectric properties of polymer/ceramic composites. They reported that the dielectric properties exhibited by each composite were ascribed to interfacial polarization. Moreover, they found that the dielectric relaxation was markedly related to the conductivity of the phases and the topology of the dielectric mixtures [74, 75]. Myroshnychenko et al. used the FEM approach to predict the complex effective permittivity of two-phase random statistically isotropic hetero-structures [76]. They considered the composite as a distribution of one dielectric phase randomly dispersed into a continuous matrix of another dielectric phase. The numerical data for the effective complex permittivity were found to situate between the curves of the Maxwell and Bruggeman equations.

Theoretical models have also widely been applied to investigate the effect of interface on the polymer/ceramic nanodielectrics. Tanaka et al. established a multi-core model to study the dielectric properties of several types of polymer nanodielectrics [77]. They stated that the interface of a spherical ceramic filler included into a polymer matrix contains three different regions: a bonded layer (first layer), a bound layer (second layer), and a loose layer (third layer), with an electric double layer covering the three layers. The first one represents a contact layer where the organic matrix is in intimate contact with the ceramic-filler outer surface, whereas the second one corresponds to the interfacial region. However, the third layer determines the properties of the bulk polymer. The second layer accounts for the reduction in the permittivity by disturbing the motion of dipoles resulting from the presence of polar groups.

Experimentally and theoretically dielectric values using various correlative models [78, 79] of the polymethyl-vinyl siloxane rubber/hBN composites are illustrated in Figure 9.13. It appears that experimental ε values of these composites were in

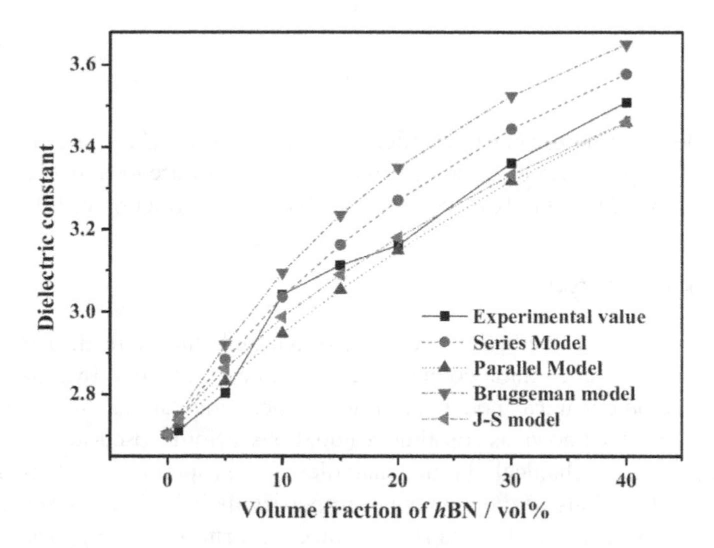

FIGURE 9.13 Experimentally and Theoretical ε Values from Correlative Models of the hBN/VMQ Composites [80].

good agreement with those predicted by J-S equation, higher than those estimated using parallel model, but lower than those given by the series and Bruggeman models. This different evolutions are caused by the random dispersion of hBN fillers into polymethyl-vinyl siloxane rubber matrix.

9.6 PREDICTING PIEZOELECTRIC PROPERTIES

There are few models available for predicting the piezoelectric properties of the polymer/ceramic composite. Two famous models to estimate the piezoelectric properties are stated below:

The *Yamada model* is given here:

$$g_{33} = \frac{V_f n a d33_2 / \varepsilon_0}{\varepsilon_{33_2} + (n-1)\varepsilon_{33_1}\left\{1 + \frac{nV_f\left(\varepsilon_{33_2} - \varepsilon_{33_1}\right)}{n\varepsilon_{33_1} + \left(\varepsilon_{33_2} - \varepsilon_{33_1}\right)\left(1 - V_f\right)}\right\}} \tag{9.48}$$

where α is the poling ratio of the added ceramic, ε_{33_i} is the dielectric constant of the ceramic inclusions in the ith poling direction, and n is a dimensionless factor related to the shape and orientation of the ceramic fillers. By fitting experimental data for ε_{33} and d_{33} values of composites and changing the structural parameter n, d_{33} and g_{33} values for volume fractions up to the theoretical maximum can be predicted. The values are found to be in good agreement up to a volume fraction of 0.7.

The *Furukawa model* is given as

$$d = V_f L_T L_E d_2 \tag{9.49}$$

and

$$g = V_f L_T L_D g_2 \tag{9.50}$$

where d and g are piezoelectric coefficients of composites; d_2, g_2 are piezoelectric constants of the phase 2, i.e., filler material. L_T, L_D, and L_E are local field coefficients with respect to T, D, and E, respectively, and V_f is volume fraction of filler.

9.7 CONCLUSIONS

The models used for predicting the various thermal, mechanical, and electrical properties of polymer/ceramic composite can be categorized into two main classes: analytical micromechanical models and finite-element calculations. Analytical micromechanical models, known as constitutive equations, afford closed-form expressions for predicting the mechanical, thermal, and dielectric properties of polymer/ceramic composite as functions of their inherent properties their basic constituents. These models consider an ideal filler-matrix interface uniform morphology (homogenous dispersion of ceramic filler). Thus, these models can give a quick estimation of the dependency of elastic modulus, dielectric constant, and thermal conductivity of the polymer/ceramic composite based on the inherent filler and matrix constants and the filler ratios. However, they usually overestimate or underestimate these parameters especially at high-filler loadings because they do not consider the interface matrix/filler. FEMs generally provide better opportunity with a good accuracy as they count the realistic composite morphology and the possible interfacial resistance; but they are usually time-consuming while they have been utilized mainly to verify and test predictions of constitutive models.

REFERENCES

1. Fu, S-Y., Feng, X-Q., Lauke, B., Mai, Y-W. 2008. Effects of particle size, particle/matrix interface adhesion and particle loading on mechanical properties of particulate–polymer composites. *Composites: Part B* 39: 933–961.
2. Odegard, G. M., Clancy, T. C., Gates, T. S. 2005. Modeling of the mechanical properties of nanoparticle/polymer composites. *Polymer* 46(2): 553–562.
3. Luo, J. J., Daniel, I. M. 2003. Characterization and modeling of mechanical behavior of polymer/clay nanocomposites. *Composites Science and Technology* 63: 1607–1616.
4. Sideridis, E., Papanicolaou, G. C. 1988. A theoretical model for the prediction of thermal expansion behaviour of particulate composites. *Rheologica Acta* 27(6): 608–616.
5. Progelhof, R. C., Throne, J. L., Ruetsch, R. R. 1976. Methods of predicting the thermal conductivity of composite systems. *Polymer Engineering Science* 16: 615–625.
6. Cheng, S. C., Vachon, R. I. 1969. The prediction of the thermal conductivity of two and three phase solid heterogeneous mixtures. *International Journal of Heat and Mass Transfer* 12: 249–264.
7. Aribou, N., Nioua, Y., Bouknaitir, I., El Hasnaoui, M., Achour, M. E., Costa, L. C. 2017. Prediction of filler/matrix interphase effects on AC and DC electrical properties of carbon reinforced polymer composites. *Polymer Composites.* doi.org/10.1002/pc.24657
8. Fu, S.Y., Lauke, B. 1997. Analysis of mechanical properties of injection molded short glass fibre (SGF)/calcite/ABS composites. *Journal of Material Science and Technology* 13: 389–396.
9. Fu, S. Y., Hu, X., Yue, C. Y. 1999. The flexural modulus of misaligned short fiber reinforced polymers. *Composites Science and Technology* 59: 1533–1542.
10. Fu, S. Y., Lauke, B. 1998. Strength anisotropy of misaligned short-fib rereinforced polymers. *Composites Science and Technology* 58: 1961–1972.
11. Fu, S. Y., Lauke, B. 1996. Effects of fibre length and orientation distributions on the tensile strength of short fibre reinforced polymers. *Composites Science and Technology* 56: 1179–1190.
12. Hsueh, C. H., Becher, P. F. 2005. Effective viscosity of suspensions of spheres. *Journal of the American Ceramic Society* 88: 1046–1049.
13. Tavman, I. H. 1996. Thermal and mechanical properties of aluminum powder-filled high-density polyethylene composites. *Journal of Applied Polymer Science* 62: 2161–2167.
14. Reynaud, E., Jouen, T., Gauthier, C., Vigier, G., Varlet, J. 2001. Nanofillers in polymeric matrix: a study on silica reinforced PA6. *Polymer* 42: 8759–8768.
15. Danusso, F., Tieghi, G. 1986. Strength versus composition of rigid matrix particulate composites. *Polymer* 27: 1385–1390.
16. Levita, G., Marchetti, A., Lazzeri, A. 1989. Fracture of ultrafine calcium carbonate/polypropylene composites. *Polymer Composites* 10: 39–43.

17. Cannillo, V., Bondioli, F., Lusvarghi, L., et al. 2006. Modeling of ceramic particles filled polymer–matrix nanocomposites. *Composites Science and Technology* 66(7-8): 1030–1037.
18. Naous, W., Yu, X. Y., Zhang, Q. X., Naito, K., Kagawa, Y. 2006. Morphology, tensile properties, and fracture toughness of epoxy/Al_2O_3 nanocomposites. *Journal of Polymer Science Part B: Polymer Physics* 44(10): 1466–1473
19. Mooney, M. 1951. The viscosity of a concentrated suspension of spherical particles. *Journal of Colloid Science* 6: 162–170.
20. Brodnyan, J. G. 1959. The concentration dependence of the Newtonian viscosity of prolate ellipsoids. *Transactions of the Society of Rheology* 3: 61–68.
21. Counto, U. J. 1964. Effect of the elastic modulus, creep and creep recovery of concrete. *Magazine of Concrete Research* 16: 129–138.
22. Ishai, O., Cohen, I. J. 1967. Elastic properties of filled and porous epoxy composites. *International Journal of Mechanical Sciences* 9: 539–546.
23. Paul, B. 1960. Prediction of constants of multiphase materials. *Transactions of the American Institute of Mining and Metallurgical Engineer* 218: 36–41.
24. Halpin, J. C. 1969. Stiffness and expansion estimates for oriented short fiber composites. *Journal of Composites Materials* 3: 732–734.
25. Mori, T., Tanaka, K. 1973. Average stress in matrix average elastic energy of materials with misfitting inclusions. *Acta Metallialia* 21: 571–574.
26. Fornes, T. D., Paul, D. R. 2003. Modeling properties of nylon 6/clay nanocomposites using composite theories. *Polymer* 44: 4993–5013.
27. Philip, M. A., Natarajan, U., Nagarajan, R. 2012. Sono-synthesis of polystyrene/alumina nanocomposites. *Proceedings of the Institution of Mechanical Engineers, Part N: Journal of Nanoengineering and Nanosystems* 226: 157–164.
28. Ramdani, N., Derradji, M., Wang, J., Liu, W., Mokhnache, E. O. 2016. Effects of aluminum nitride silane-treatment on the mechanical and thermal properties of polybenzoxazine matrix. *Plastics, Rubber and Composites* 45: 72–80.
29. Goyal, R.K., Tiwari, A.N., Negi, Y.S. 2008. Microhardness of PEEK/ceramic micro- and nanocomposites: Correlation with Halpin–Tsai model. *Materials Science and Engineering: A* 491(1–2): 230–236.
30. Kajohnchaiyagual, J., Jubsilp, C., Dueramae, I., Rimdusit, S. 2014. Thermal and mechanical properties enhancement obtained in highly filled alumina-polybenzoxazine composites. *Polymer Composites* 35(11): 2269–2279.
31. Smit, R.J.M., Brekelmans, W.A.M., Meijer, H.E.H. 1998. Prediction of the mechanical behavior of nonlinear heterogeneous systems by multi-level finite element modeling. *Computer Methods in Applied Mechanics and Engineering* 155(1–2): 181–192
32. Progelhof, R. C., Throne, J. L., Ruetsch, R. R. 1976. Methods for predicting the thermal conductivity of composite systems: A review. *Polymer Engineering and Science* 16(9): 615–625.
33. Agari, Y., Ueda, A., Nagai, S. 1993. Thermal conductivity of a polymer composite. *Journal of Applied Polymer Science* 49(9): 1625–1634.
34. Hill, R. F., and Supancic, P. H. 2002. Thermal Conductivity of Platelet-Filled Polymer Composites. *Journal of the American Ceramic Society* 85(4): 851–857.
35. Bigg, D. M. 1995. Thermal conductivity of heterophase polymer compositions. *Advances in Polymer Science* 119: 1–30.
36. Garnett, J. C. M. 1904. Colours in metal glasses and in metallic films. *Proceedings of the Royal Society of London A* 73: 443–445.
37. Bruggeman D. A. G. 1935. Berechnung Verschiedener Physikalischer Kon-stanten von Heterogenen Substanzen, I. Dielektrizitatskonstantenund Leitfahigkeiten der Mischkorper aus Isotropen Substanzen. *Annalen der Physik* 24: 636–679.

38. Yu, S., Hing, P., Hu, X. 2002. Thermal conductivity of polystyrene–aluminum nitride composites. *Composites Part A: Applied Science and Manufacturing* 33: 289–292.
39. Nagai, Y., Lai, G. C. 1997. Thermal conductivity of epoxy resin filled with particulate aluminum nitride powder. *Journal of the Ceramic Society of Japan* 105(3): 197–200.
40. Xie, S-H., Zhu, B-K., Li, J-B., Wei, X-Z., Xu, Z-K. 2004. Preparation and properties of polyimide/aluminum nitride composites. *Polymer Testing* 23: 797–801.
41. Xu, Y., Chung, D. D. L., Mroz, C. 2001. Thermally conducting aluminum nitride polymer–matrix composites. *Composites Part A* 32(12): 1749–1757.
42. Lee, G.W., Park, M., Kim, J., Lee, J. I., Yoon, H. G. 2006. Enhanced thermal conductivity of polymer composites filled with hybrid filler. *Composites Part A: Applied Science and Manufacturing* 37(5): 727–734.
43. Zhou, W., Qi, S., An, Q., Zhao, H., Liu, N. 2007. Thermal conductivity of boron nitride reinforced polyethylene composites. *Materials Research Bulletin* 42: 1863–1873.
44. Lee, E-S., Lee, S-M. 2008. Enhanced thermal conductivity of polymer matrix composite via high solids loading of aluminum nitride in epoxy resin. *Journal of the American Ceramic Society* 91(4): 1169–1174.
45. Zhou, W., Wang, C., Ai, T., Wu, K., Zhao, F., Gu, H. 2009. A novel fiber-reinforced polyethylene composite with added silicon nitride particles for enhanced thermal conductivity. *Composites Part A: Applied Science and Manufacturing* 40: 830–836.
46. He, H., Fu, R., Shen, Y., Han, Y., Song, X. 2007. Preparation and properties of Si_3N_4/PS composites used for electronic packaging. *Composites Science and Technology* 67: 2493–2499.
47. Zhang, X., Wen, H., Wu, Y. 2017. Computational thermomechanical properties of silica–epoxy nanocomposites by molecular dynamic simulation. *Polymers* 9: 430–447.
48. Gu, J., Meng, X., Tang, Y., Li, Y., Zhuang, Q., & Kong, J. 2017. Dielectric thermally conductive boron nitride/polyimide composites with outstanding thermal stabilities via in-situ polymerization electrospinning-hot press method. *Composites Part A: Applied Science and Manufacturing* 94: 209–216.
49. Vo, H. T., Todd, M., Shi, F. G., Shapiro, A. A., Edwards, M. 2001. Towards model based engineering of underfill materials: CTE modeling. *Microelectronics Journal* 32: 331–338.
50. L. Li, D. D. L. Chung. 1994. Thermally conducting polymer–matrix composites containing both AlN particles and SiC whiskers. *Journal of Electronic Materials* 23: 557–564.
51. Schapery, R. A. 1968. Thermal expansion coefficients of composite materials based on energy principles. *Journal of Composite Materials* 2: 380.
52. Wong, C. P., Bollampally, R. S. 1999. Thermal conductivity, elastic modulus, and coefficient of thermal expansion of polymer composites filled with ceramic particles for electronic packaging. *Journal of Applied Polymer Science* 74: 3396–3403.
53. Li, L., Chung, D. D. L. 1994. Thermally conducting polymer-matrix composites containing both AIN particles and SiC whiskers. *Journal of Electronic Materials* 23(6): 557–564.
54. Sihvola, A. 1999. *Electromagnetic Mixing Formulas and Applications*. The Institute of Electrical Engineering, London, 68.
55. Brosseau, C. 2006. Modellilng and simulation of dielectric heterostructures: A physical survey from an historical perspective. *Journal of Physics D: Applied Physics* 39: 1277–1294.
56. Rao, Y., Qu, J., Marinis, T. et al. 2000. A precise numerical prediction of the effective dielectric constant for polymer-ceramic composite based on effective-medium theory. *IEEE Transactions on Components, Packaging, and Manufacturing Technology: Part A* 23: 680–683.

57. Ying, K. L., Hsieh, T. E. 2007. Sintering behaviors and dielectric properties of nanocrystalline barium titinate. *Materials Science & Engineering. B, Solid-state Materials for Advanced Technology* 138: 241–245.
58. Barber, P., Balasubramanian, S., Anguchamy, Y. et al. 2009. Polymer composite and nanocomposite dielectric materials for pulse power energy storage. *Materials* 2: 1697–1733
59. Nan, C.W. 2001. Comment on "Effective dielectric function of a random medium." *Physical Review B* 63: 176201.
60. Furukawa, T., Ishida, K., Fukada, E. 1979. Piezoelectric properties in the composite systems of polymers and PZT ceramics. *Journal of Applied Physics* 50: 4904.
61. Sun, Y., Zhang, Z., and Wong, C.P. 2005. Influence of interphase and moisture on the dielectric spectroscopy of epoxy/silica composites. *Polymer* 46: 2297–2305.
62. Bhimasankaram, T., Suryanarama, S. V., Prasad, G. 1998. Piezoelectric polymer composite materials. *Current Science* 74: 967–976.
63. Jayasundere, N., Smith, B.V. 1993. Dielectric constant for binary piezoelectric 0-3 composites. *Journal of Applied Physics* 73: 2462.
64. Luo, B., Wang, X., Wang, Y., Li, L. 2014. Fabrication, characterization, properties and theoretical analysis of ceramic/PVDF composite flexible films with high dielectric constant and low dielectric loss. *Journal of Materials Chemistry A* 2: 510–519.
65. Rao, Y., Wong, C. P., Qu, J., and Marinis, T. 2000. A precise numerical prediction of effective dielectric constant for polymer-ceramic composite based on effective-medium theory. *IEEE Transactions on Components and Packaging Technologies* 23: 680–683.
66. Teirikangas, M., Juuti, J., Jantunen, H. 2009. Multilayer BST-COC composite with enhanced high frequency dielectric properties. *Ferroelectrics* 387: 210–215.
67. Yamada, T., Ueda, T., Kitayama, T. 1982. Piezoelectricity of a high-content lead zirconate titanate/polymer composite. *Journal of Applied Physics* 53: 4328.
68. Sun, Y., Zhang, Z., and Wong, C.P. 2005. Influence of interphase and moisture on the dielectric spectroscopy of epoxy/silica composites. *Polymer*, 46: 2297–2305.
69. Firmino Mendes, S., Costa, C. M., Sencadas, V. et al. 2009. Effect of the ceramic grain size and concentration on the dynamical mechanical and dielectric behavior of poly(vinilidene fluoride) $Pb(Zr_{0.53}Ti_{0.47})O_3$ composites. *Applied Physics A* 96: 899–908.
70. Rao, Y., Wong, C. P. 2004. Material characterization of a high-dielectric constant polymer-ceramic composite for embedded capacitor for RF applications. *Journal of Applied Polymer Science* 92: 2228–2231.
71. Todd, M. G., Shi, F.G. 2005. Complex permittivity of composite systems: A comprehensive interphase approach. *IEEE Transactions on Dielectrics and Electrical Insulation* 12: 601–611.
72. Vo, H. T., Shi, F. G. 2002. Towards model-based engineering of optoelectronic packaging materials: Dielectric constant modeling. *Microelectronics* 33: 409–415.
73. Murugaraj, P., Mainwaring, D., Mora-Huertas, N. 2005. Dielectric enhancement of polymer nanoparticle through interphase polarizability. *Journal of Applied Physics* 98: 054304.
74. Tuncer, E. 2004. Signs of low frequency dispersions in disordered binary dielectric mixtures (50-50). *Journal of Physics D: Applied Physics* 37: 334–342.
75. Tuncer, E., Nettelblad, B., Gubanski, S. M. 2002. Non-Debye relaxation of binary dielectric mixtures (50-50): Randomness and regularity in mixture topology. *Journal of Applied Physics* 92: 4612–4624.
76. Myroshnychenko, V., Brosseau, C. 2005. Finite-element modeling method for the prediciton of the complex effective permittivity of two-phase random statistically isotropic heterostructures. *Journal of Applied Physics* 97: 044101.

77. Tanaka, T., Kozaka, M., Fuse, N. et al. 2005. Proposal of a multi-core model for polymer nanocomposite dielectrics. *IEEE Transactions on Dielectrics and Electrical Insulation* 12: 669–681.
78. Ngo, I. L., Vattikuti, S. V. P., Byon, C. 2016. Effects of thermal contact resistance on the thermal conductivity of core-shell nanoparticle polymer composites. *International Journal of Heat and Mass Transfer* 102: 713–722.
79. Jayasundere, N., Smith, B. V. 1993. Dielectric constant for binary piezoelectric 0–3 composites. *Journal of Applied Physics* 73: 2462–2466.
80. Gu, J., Meng, X., Tang, Y., Li, Y., Zhuang, Q., Kong, J. 2017. Hexagonal boron nitride/polymethyl-vinyl siloxane rubber dielectric thermally conductive composites with ideal thermal stabilities. *Composites: Part A* 92 27–32.

10 Barrier Properties of Polymer/Ceramic Composites

10.1 INTRODUCTION

Barrier properties of polymer composites are considered a critical physical property of the system and have to be understood and optimized to ensure potential use in various industrial applications. To improve the barrier performance of composite materials, researchers incorporate high-aspect ratio nanofillers, such as carbon-based nanofillers, ceramics, clays, and biofillers. This strategy led to reducing the material costs and can keep the transparency of polymeric matrices, as these nanofillers could not scatter light. The barrier properties can be controlled by the microstructure of the composites and can be influenced by the low compatibility at the interface between the matrix and inorganic phases. In addition, the barrier properties can be negatively affected by the presence of small amounts of low molecular-weight compatibilizers in the composite system, which are usually included in the non-polar polymer composites to ease the nanofiller dispersion in the high molecular-weight polymer matrix.

Polymer/ceramic composites have been widely used as gas, solvent, and water separation membranes, where their barrier properties are evaluated. The ceramic fillers decreased the gas or liquid permeability according to the type, shape, and content of ceramic filler as well as their interactions with polymeric matrices. Permeation capacity does not depend on pressure, whereas by raising temperature generally leads to a reduction in the penetrant solubility. To optimize the barrier properties of polymer/ceramic micro- and nanocomposites, a wide range of parameters have been studied. In this chapter we will summarized the various critical factor influencing the barrier performance of these kinds of composite systems along with several examples issued from the extensive literature on the subject.

10.2 WATER BARRIER PROPERTIES

The rate of degradation in any polymer/ceramic composite directly depends on the amount of water it absorbs. Plasticization and swelling are among the undesirable consequences of absorbed water by polymer composites. This effect will intensify if such materials are used under severe conditions. A series of polyimide/SiO_2 nanocomposite films were produced. The addition of the silica-modified by oleic acid (OA) SiO_2-OA nanoparticles to a polyimide matrix decreased water absorption [1]. As drawn in Figure 10.1, the water diffusion becomes more difficult in the polyimide nanocomposite films, especially those consisting of SiO_2-OA-1 nanoparticles due to the hydrophobicity of the SiO_2-OA-1 nanoparticles.

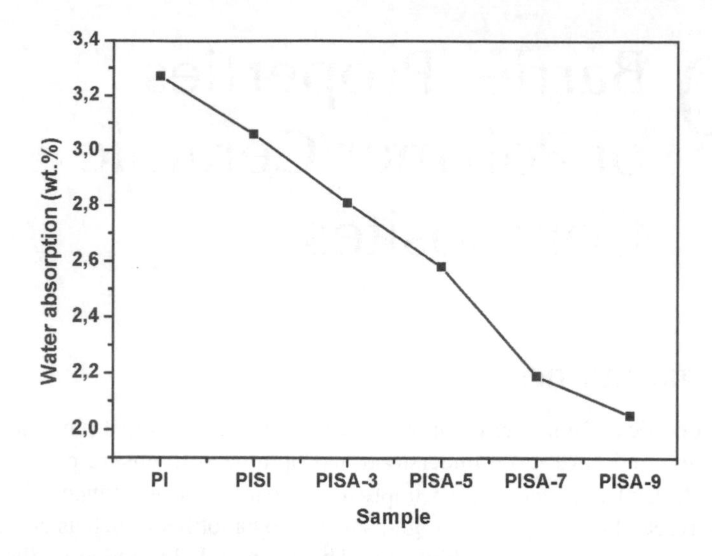

FIGURE 10.1 Water Diffusion in Polyimide/SiO_2 Nanocomposites [1].

As shown in Figure 10.2, the incorporation of Al_2O_3 nanoparticles at various ratios did not change the basic water-diffusion mechanism in the epoxy matrix, while a reduction in water-uptake was observed in the case of the nanocomposite. This is related to two main factors: the volume of epoxy for water diffusion was decreased with enhancing nano-alumina content, and second, the presence of the nanoparticles can enhance the path length for moisture diffusion. On the other hand, both dielectric constant and dielectric loss of the Al_2O_3/epoxy nanocomposites increased markedly after water-absorption treatment [2].

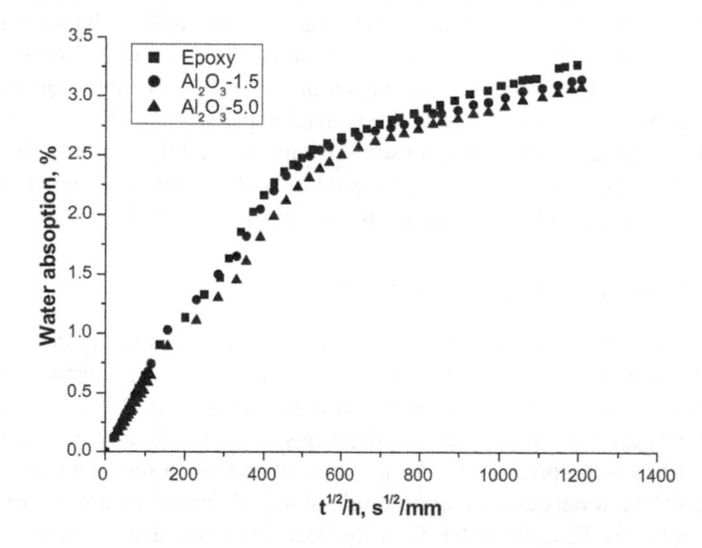

FIGURE 10.2 Curves of Water Absorption Versus Square Root Time per Specimen Thickness, $t^{1/2}/h$ ($s^{1/2}$/mm), for Epoxy and Selected Alumina-filled Nanocomposites [2].

Alamri and Low [3] reported that the inclusion of nano-SiC into epoxy resin decreased both water uptake and diffusivity compared to those of the neat epoxy. The inclusion of 1, 3, and 5 wt.% n-SiC reduced the water uptake by almost 21.8, 28.6, and 33.3%, respectively, comapred to the unfilled matrix. Flexural properties of epoxy-based nanocomposites were decreased due to the water absorption. Surprisingly, fracture toughness and impact strength of these nanocomposites were enhanced after exposure to water [3]. The influence of water absorption on the dielectric properties of epoxy resin and their microcomposites and nanocomposites filled with SiO_2 were evaluated [4]. Nanocomposites were found to absorb water more significantly than the pure epoxy did, while the microcomposites absorbed less water than unfilled epoxy due to the reduced proportion of the epoxy in this composite. The T_g decreased as the water absorption increased and, in all cases, corresponded to a drop of approximately 20 K as the humidity was increased from 0% to 100%.

Greenberg and Kamel [5] found that bound water in porous poly(acrylic acid) PAA/Al_2O_3 can inhibit the onset temperature of anhydride formation. The degree of conversion to anhydride was correlated with the equilibrium swelling level attained by the composite in water. The swelling behavior of optimum superabsorbent Salep-g-poly(sodium acrylate)/Al_2O_3 composite was measured in various pH solutions. In addition, swelling kinetics, swelling in various NaCl concentrations, swelling in various solvents, and the absorbency underload were investigated [6]. The incorporation of SiO_2 into a poly(trimethylolpropane triacrylate) (PTPT) generates a tortuous path and absorbs a water molecule, the PTPT/SiO_2 nanocomposite showed a greatly reduced water-vapor transmission rate relative to the pure matrix [7].

Abenojar et al. [8] studied the effect of moisture and temperature on the mechanical properties of an epoxy reinforced with different B_4C particle sizes (7 and 23 μm) at concentration of 6 wt.%. A general degradation of properties was recorded after water absorption for these composites. However, after the drying process, recovery was observed according to the amount of water entrapped within these composites, which played the role of a plasticizer for these composites, enhancing their strength values. Ramdani et al., [9] also reported that the increasing of nano-B_4C ratio from 0 to 20 wt.%, into polybenzoxazine matrix, the hot water-uptake of their composite was significantly reduced and these nanocomposites follow the Fickian diffusion law (Figure 10.3). The water-swelling value of polyurethane/silica is also gradually reduced with increasing the nano-SiO_2 content. This trend can be attributed to an increase in the water resistance of the nanocomposite films with the inclusion of nano-SiO_2. This resistance was ascribed to the presence of the hydrophobic Si–O–Si groups of the cross-link structure. Such network structures of nano-SiO_2 enhanced the density of the film and thus can prevent water penetration [10].

Figure 10.4 also reveals the water solubility of starch/PVA/SiO_2 nanocomposite films, clearly demonstrating that the water uptake and the water solubility of these films decreased with increasing the nano-SiO_2 loading [11]. This result is attributed to the formation of an intermolecular hydrogen bond and a chemical bond C-O-Si in the nano-SiO_2 and starch/PVA matrix. To ameliorate the barrier properties of poly(ethylene terephthalate) (PET), different weight ratios of untreated and polystyrene-treated nano-SiO_2 were added to this matrix. The water absorption testing for PET/SiO_2 specimens showed that both the maximum water-uptake and the

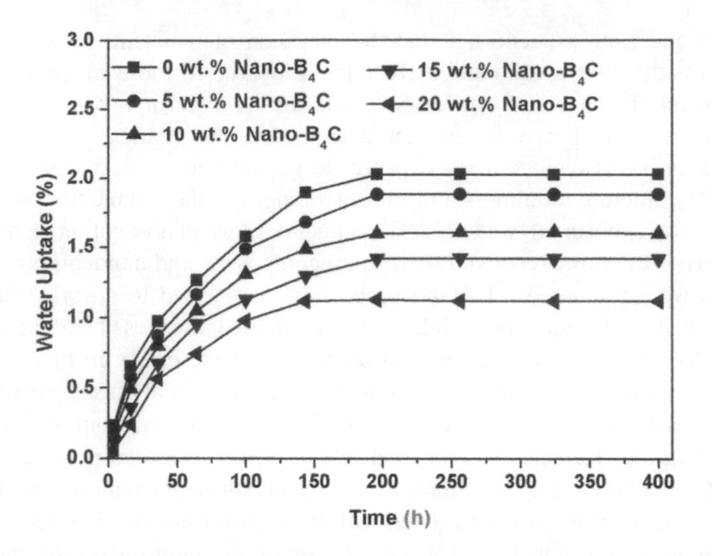

FIGURE 10.3 Water Sorption Versus Time of the Polybenzoxazine/B_4C Nanocomposites [9].

pseudo-diffusion coefficient (*D*) of water decreased with decreasing the SiO_2 particle size from 440 nm to 40 nm, and the water-resistance of these nanocomposites were higher than those recorded for the unfilled PET matrix. At a SiO_2 size of 40 nm, the core-shell SiO_2/polystyrene-reinforced PET nanocomposite compared to those produced from unmodified nano-SiO_2 which were more effective on keeping PET from up-taking water [12].

New polybenzoxazine/SiO_2 nanocomposite films were prepared by a two-step process. The achievable range of the water contact angle of the nanocomposite films was between 167° and 88°, meaning from superhydrophobicity to hydrophilicity.

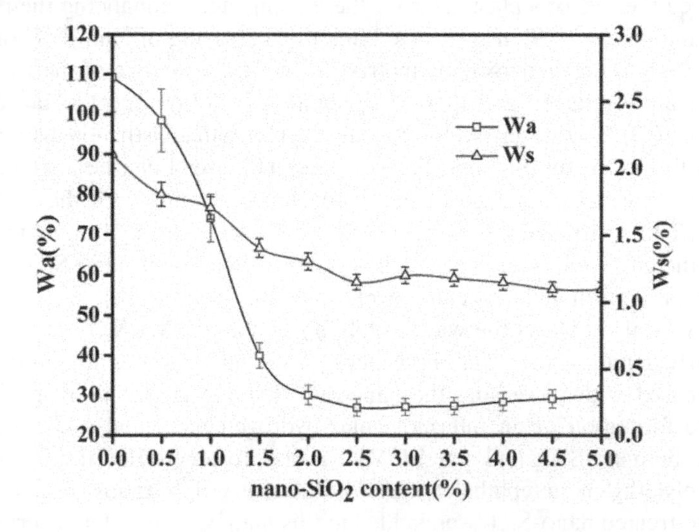

FIGURE 10.4 The Water Absorption and Water Solubility of the Films [11].

The water contact angle of these hybrids can be changed by simply controlling the immersion and drying time [13]. Maheshwari et al., [14] added nano-SiO_2 in order to improve the barrier property of the UPE matrix against water diffusion and thus prevent it from mechanical degradation. The presence of nanoparticles significantly decreased the maximum water-uptake and moisture diffusivity in the composites, and as the salinity of seawater increased, and temperature decreased, degradation caused by water-absorption was reduced. In another work, nylon 11–based coatings reinforced with 0 to 15 vol.% of nanosized SiO_2 were developed using the high-velocity oxy-fuel combustion spray process [15]. Permeability tests revealed that the water vapor transmission rate through nanoreinforced coatings was reduced by almost 50% relative to the unfilled polymer.

Khelifa et al., [16] produced an acrylic-based polymer reinforced with glycidoxypropyl trimethoxysilane (GPTS)-treated nano-SiO_2 prepared by in situ acid hydrolysis and subsequent condensation of tetraethoxysilane (TEOS). It was found that an optimum fraction of nano-SiO_2 having a precise morphology and composition in terms of TEOS/GPTS ratio is required to keep efficient coating barrier properties.

10.3 CORROSION-RESISTANT PROPERTIES

The marine environment is considered to be a highly aggressive environment for metallic materials. Steels and aluminium are the most common materials being used for shipbuilding, which exhibit a high attitude to oxidation. Corrosion is a major cause of structural deterioration in marine and offshore structures. Due to restrictions on using heavy metals and chromates in coatings, as they impact our environment, organic coating approaches for corrosion protection with inherently conducting polymers have been widely explored [17].

Epoxy coatings modified with polyethyleneimine (PEI)-treated mesoporous-TiO_2 nanoparticles (meso-TiO_2) were developed and showed improved nanofillers dispersion into the matrix. The coatings with meso-TiO_2/PEI (600 molecular weight) exhibited the best corrosion resistance among the other samples. The EIS results showed that the resistance value of coating with meso-TiO_2/PEI (600 molecular weight) was above 9.87×10^7 $\Omega.cm^2$ due to chemical interactions between epoxy matrix and meso-TiO_2/PEI, which caused high-barrier properties and a high degree of cross-linking [18]. Bakhshandeh et al. [19], reported on the development and fabrication of organic–inorganic hybrid coatings based on epoxy resin pursuing hydrolyzation of tetraethoxysilane (TEOS) via a sol–gel technique. The EIS results corresponding to the produced coatings confirmed that their corrosion resistance characteristics were greatly enhanced as the nano-SiO_2 loading increased.

Ramezanzadeh et al. [20] synthesized and characterized new silica nanoparticles-decorated graphene oxide (SiO_2-GO) nanohybrids by a two-step in situ and sol–gel process from a mixture of 3-aminopropyl triethoxysilane and tetraethylorthosilicate in water–alcohol mixed solvent. The different tests confirmed that SiO_2-GO nanohybrids significantly improved both the barrier and corrosion-protection properties of the epoxy matrix. Also, the anticorrosion performance of the epoxy coating was improved by the inclusion of 4 to 6 wt.% of treated nano-SiO_2 to the polymeric matrix, as evaluated by Wang et al., [21]. The surface modification of nano-SiO_2 with 3-glycidoxypropyltrimethoxysilane (GPTMS) was found to ameliorate the dispersion of nanofillers.

New coatings based on epoxy composites filled with a fixed concentrations of graphene oxide (GO), Al_2O_3, and GO–Al_2O_3 hybrids sheets were prepared by Yu et al., [22]. Among these coatings GO–Al_2O_3 hybrids achieved a uniform dispersion, improved compatibility in the epoxy matrix, and showed superiority in reinforcing the anti-corrosion performance. Similarly, the corrosion-resistant performance of a 2 wt.% of 3-aminopropyltriethoxysilane-treated TiO_2–GO sheet hybrids filled epoxy resin was significantly increased [23]. At a fixed concentration, results showed that the hybrids of TiO_2–GO had offered more efficiency in the corrosion resistant ability of the epoxy coating relative to other nanofillers. This finding was related to the exfoliation, homogenous dispersion and improved plugging micro-pore property arising from the laminated structure of TiO_2–GO hybrids. In another study, transparent epoxy/3-glycidoxypropyl-trimethoxysilane-modified SiO_2 hybrid-containing various loadings of nano-SiO_2 were produced via a sol–gel method. Superior anticorrosion performance on cold-rolled steel was recorded for the nanocomposites containing low nanofiller loading [24].

The silica/epoxy hybrid coatings were found to be suitable for protection against the corrosion of AA2024 alloy [25]. On the other hand, polystyrene composite coatings reinforced with dodecylbenzenesulfonic acid (DBSA)-doped SiO_2@polyaniline (SP) core–shell microspheres were produced and showed an improved corrosion resistance due to the formation of a dense passive metal oxide layer induced by the redox catalytic effect of the polyaniline shells [26]. In addition, the polystyrene composite coating filled with 10 wt.% of the SP core–shell microspheres exhibited an electrical resistance of nearly $3.65 \times 10^9\ \Omega\ cm^{-2}$, which satisfied the requirements for antistatic applications. Weng et al. [27] prepared novel, efficient anticorrosion coatings for cold-rolled steel (CRS) with synergistic effects of super-hydrophobicity properties (contact angle ca ~ 161°) and redox catalytic capability from fluoropolyaniline incorporated with methyl nano-SiO_2 spheres.

The addition of SiO_2 nanoparticles into the organo-soluble fluorinated polyimide (SFPI) matrix can effectively improve the corrosion-protection performance on (CRS electrodes against corrosive species in saline condition when compared with that of neat polymeric coatings based on a series of standard electrochemical corrosion measurements [28]. Based on a several electrochemical corrosion measurements in saline, electroactive polyimide/SiO_2 composites in the form of coating on CRS electrodes were found to exhibit improved corrosion protection over those of non-electroactive polyimide and EPI materials [29]. The superior corrosion-resistance of EPIS coatings on CRS electrodes are traced to the redox catalytic property of organic EPI inducing the formation of a passive metal oxide layer and the barrier property of well-dispersed treated-SiO_2 nanofillers existing within EPI matrix.

The poly(vinyl carbazole) (PVK) reinforced with 3-(trimethoxysilyl)propyl methacrylate (MSMA)–treated-silica (PVK/SiO_2) hybrid-coated on CRS electrodes at a nano-SiO_2 content 10 wt.% was found to be superior in anticorrosion property over those of neat PVK [30]. This was mainly ascribed to the stronger adhesion strength of such nanocomposite coatings on CRS electrodes, which can be associated with the formation of Fe–O–Si covalent bonds (995 cm^{-1}) at the interface of hybrid coating

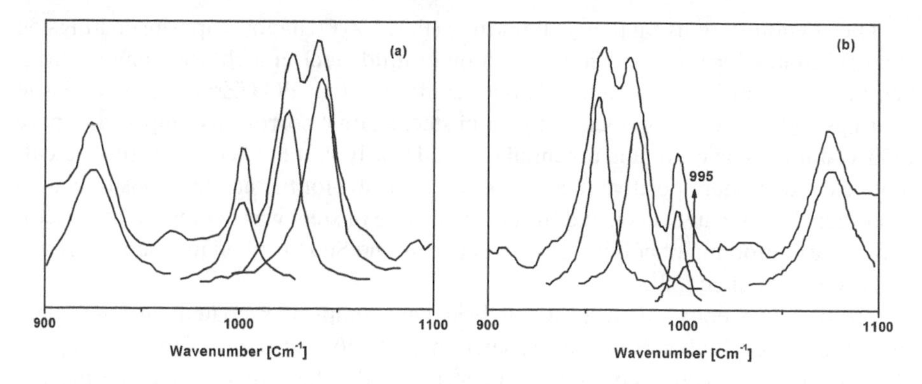

FIGURE 10.5 ATR-FTIR Spectra of PVK/SiO_2 Containing 10 Wt.% Nano-SiO_2 by Curve-fitting Process [30]. Reprinted with permission from Wiley & Sons.

and the CRS electrode as illustrated in the FTIR–RAS (reflection absorption spectroscopy) studies in Figure 10.5. Kumar et al. [31] developed polyaniline/SiO_2 composites for protecting the mild steel from corrosion in a highly-aggressive medium. The PANI/SiO_2 coating revealed a great decrease in the corrosion current density that reflects the improved protection of mild steel in an acidic environment. Higher protection efficiency up to 99% was achieved by using PANI/SiO_2-coated mild steel at 6.0 wt.% loading in epoxy resin. In an acidic medium, the nanocomposite-coated mild steel showed superior anticorrosion performance compared to pure PANI as shown in Figure 10.6.

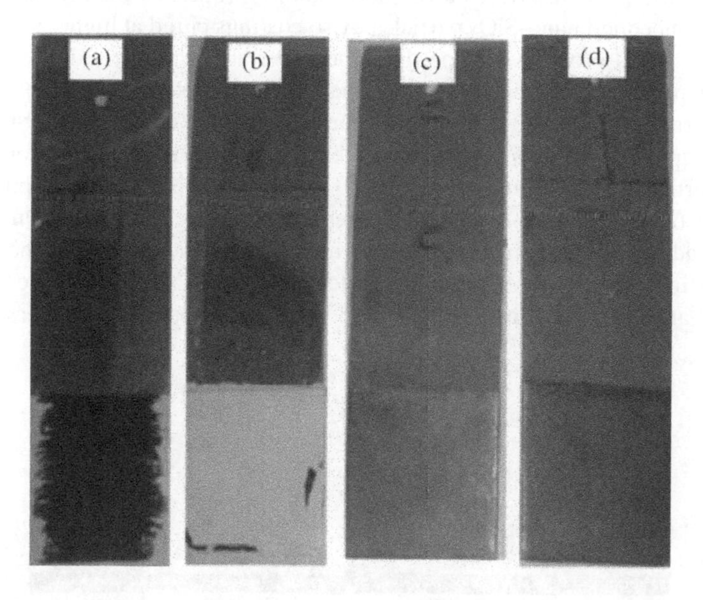

FIGURE 10.6 Weight-loss Images of (a) a Blank Mild Steel Electrode, (b) a Blank Epoxy-resin-coated Electrode, (c) a PANI-coated Electrode and (d) a PANI/SiO_2-coated Electrode after Immersion in 1 mol L^{-1} HCl for 60 days [31].

The possibility of using poly(*o*-toluidine) (POT)/ZrO_2 nanocomposite coatings on steel substrates for the corrosion protection of mild steel in a chloride environment has been studied [32]. The analysis data evidenced that POT/ZrO_2 nanocomposite coatings had improved protection for mild steel against corrosion compared to pure POT coatings. The corrosion potential reached nearly 0.312 V versus a saturated calomel electrode, more positive in aqueous 3 wt.% NaCl for the nanocomposite-coated steel than the uncoated steel, and the corrosion rate of steel was decreased by a factor of 51. The reinforcing of polyimide with both clay and SiO_2 resulted in much-improved corrosion resistance [33].

Electroactive epoxy/50 nm-SiO_2 hybrid nanocomposites coated on CRS electrodes exhibited higher corrosion protection over thoat of non-electroactive epoxy (NEE) and electroactive epoxy (EE) materials [34]. The possible involved mechanisms for these enhancements in corrosion resistance include: (1) electroactive epoxy coatings may act as physical barrier coating; (2) redox catalytic capabilities of ACAT units existed in electroactive epoxy may induce the formation of passive metal-oxide layers on CRS electrodes, as further evidenced by SEM analysis (See Figure 10.7); (3) well-dispersed treated nano-SiO_2 in epoxy matrix could act as effective hinderance to increase the oxygen barrier property of electroactive epoxy matrix.

The pull-off adhesion strength to mild steel substrates and the water contact angle of epoxy coatings were significantly increased by adding SiO_2-GO nanohybrid filler systems [35]. The potentiodynamic polarization test, electrochemical impedance spectroscopy (EIS), and salt-spray test results showed that corrosion-protection performance of epoxy coatings was markedly increased by adding well-distributed SiO_2-GO nanohybrids compared to filling with only GO nanosheets. Electrochemical data revealed a positive effect on the permeability properties for PU coatings with 5 wt.% of embedded nano-SiO_2 particles. Also coatings cured at higher temperatures showed improved protective properties. The Taber abrasion test indicates that both nano-SiO_2 particles and higher-curing temperatures increase abrasion resistance [36].

The corrosion tests showed that the corrosion protection of epoxy/SiO_2 hybrid coatings depends mainly on the silane content, type of the silane precursor, and type of nanoparticles. The coating-protective effect improved by increasing polarization resistance (R_p) for about one decade by replacing silane precursors. On the other hand, the addition of TiO_2 in comparison with AlOOH nanoparticles in the GPTMS-based coatings demonstrated increasing effects on polarization resistance. However, the simultaneous embedding of TiO_2 and AlOOH nanoparticles generated much

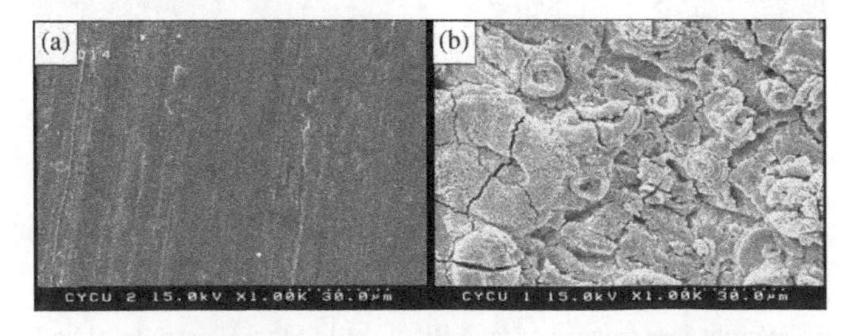

FIGURE 10.7 SEM Image for (a) Polished Cold-rolled Steel (CRS), and the Surface of the (b) Epoxy/SiO_2 Coating and the CRS Metal at 3 Wt.% Nano-SiO_2 [34].

TABLE 10.1
Corrosion Parameters Calculated from the Potentiodynamic Measurements [39]

Material	E_{corr} (mV)	I_{corr} (nA)	R_p (MΩ)	E_b (mV)
DSS 2205	–280	666	0.03	1050
DSS 2205 + epoxy	–400	29	2.2	1150
DSS 2205 + epoxy	–200	18	2.7	1200

higher protective coatings [37]. The anticorrosion behaviors of the annealed ZrO_2/PDA nanocomposite multilayers were significantly improved compared to that of the annealed homogeneous ZrO_2 film. The hybrid microstructure of the annealed ZrO_2/PDA nanocomposite multilayers may largely contribute to the enhanced nano-hardness and excellent corrosion resistance [38].

The effects of adding the SiO_2 nanoparticles on the corrosion resistance of the 50-μm thick SiO_2/epoxy coating-coated steel were investigated in a 3.5 wt.% NaCl solution. Results demonstrated that the filling with SiO_2 significantly affected the microstructure of the coating matrix, since it generated an improved damage resistance, lower degree of delamination, improved surface roughness, and induced hydrophobicity. These hybrid coatings were an efficient barrier in a chloride-ion-rich medium with an improved anticorrosive resistance due to the reduced corrosion rate [39].

During the repetitive cyclic voltammetry scans the presence of TiO_2 nanoparticles inhibits the PANI oxidation process. Good adhesion of PANI/SiO_2 nanocomposite coatings on the steel electrode, with superior electroactive characteristics compared to pure polyaniline, was demonstrated [40]. Due to compact-coating formation, the corrosion behavior of nanocomposites-coated steel exhibited superior corrosion-resistance than the neat PANI coating did. The electrochemical monitoring of the coated steel over 28 days of immersion in both 0.3 wt.% and 3 wt.% NaCl solutions reflected the beneficial role of SiO_2 nanoparticles in significantly improving the corrosion resistance of the coated steel, with the Fe_2O_3 and halloysite clay nanoparticles being the best [41]. The SiO_2 nanoparticles were found to significantly improve both the microstructure and the anticorrosive performance of the epoxy coating.

A 50-μm-thick, well-dispersed epoxy/SiO_2 coating containing 2 wt.% of 130-nm silica ceramic filler was coated on austenitic stainless steel of the type AISI 316L [42]. The silica/epoxy coating plays the role of a successful barrier in a chloride-ion-rich environment with an enhanced anticorrosive performance, which was confirmed by the reduced corrosion rate and the enhancement coating resistance due to zigzagging of the diffusion path available to the ionic species. The corrosion behavior of polypyrrole/silicon nitride (PPy/Si_3N_4) nanocomposite coatings on St-12 steel was studied by electrochemical methods such as electrochemical impedance spectroscopy (EIS) in NaCl 3.5% solution [43]. The recoded data showed that the incorporation of Si_3N_4 nanoparticles enhanced the corrosion-resistance of coatings and ameliorate the surface morphology. As shown in Figure 10.8, SEM images displays that the nanocomposite coatings formed a compact, homogenous and dense films on the St-12 steel outer surfaces.

FIGURE 10.8 SEM Images of: (a) PPy Coating, ×15,000 and (b) PPy/Si_3N_4 Coating, ×15,000 and (c) PPy/Si_3N_4 Coating on St-12 Steel, ×30,000 [43].

A novel anticorrosive material poly(2,3-dimethylaniline)/nano-SiO_2 composite [44]. The results indicated that there was a certain interaction existed between poly(2,3-dimethylaniline) and nano-SiO_2 particles, the thermal stability and electrochemical performances of poly(2,3-dimethylaniline)/nano-SiO_2 was much better than that of P(2,3-DMA). Epoxy coating containing poly(2,3-dimethylaniline)/nano-SiO_2 were coated on steel which showed improved anticorrosion feature compared to that of poly(2,3-dimethylaniline) matrix.

New anticorrosion electroactive PI/10 nm-TiO_2 hybrid nanocomposites were prepared. Higher loading of TiO_2 filler in the as-prepare corresponding PI/TiO_2 nanocomposites provided better corrosion protection performance on CRS electrode, according to the sequential electrochemical corrosion measurements in 5 wt.% NaCl electrolyte [45]. These improvements were attributed to three main reasons, including: (1) electroactive polyimide (EPI) provided a physical barrier coating; (2) the redox catalytic capabilities of ATs units that exist in EPTs may provoke the formation of passive metal-oxide layers on CRS electrode; (3) the homogenous dispersion of nano-TiO_2 particles in PI matrices could act as effective hinderance to increase the oxygen barrier property.

Homogenously-dispersed CeO_2 and ZrO_2 colloidal were used to reinforce the epoxy/SiO_2-based hybrid nanocomposite [46]. The corrosive behavior of the coatings on the 1050 AA substrate was studied by potentiodynamic measurements. The results demonstrated that by incorporating CeO_2 nanoparticles in 1:1 molar ratio to TEOS in coating formulation, corrosion protection was enhanced. But, co-existence of two nanoparticles (CeO_2 and ZrO_2 in 1:1 molar ratio) in such coatings boosted the corrosion-protection efficiency up to 99.8%. This improvement was related to good network interaction between these ceramic nanoparticles and epoxy-molecular chains, which generated better corrosion protection of these coatings. A 0, 1, 2 wt.% of triazine core silane-treated TiO_2 nanoparticles filled polyurethane matrix were produced [47]. Fog test and electrochemical polarization studies suggest that the corrosion resistance increases with increasing the modified TiO_2 content in the coating composition. The coatings substantially acquire hydrophilic features symbiotically with TiO_2 content suggesting its potential application as self-cleanable material.

Surface modification of TiO_2 nanoparticles by amino propyl trimethoxy silane (APS) were synthesized. The results showed that surface treatment of TiO_2 nanoparticles with APS improves nanoparticles dispersion, mechanical properties, and UV protection of the urethane clear coating [48]. An investigation of the corrosion-protection properties of an epoxy/polyamide coating reinforced with various ratios of polysiloxane- modified nano-SiO_2 (see Figure 10.9) coated on several steel substrates was conducted by Matin et al., [49]. It was shown that the coatings' corrosion-protection properties were significantly improved in the presence of 5 wt.% SiO_2 nanoparticles. Insignificant degradation appeared on the surface of the coatings loaded with 5 wt.% nanoparticles.

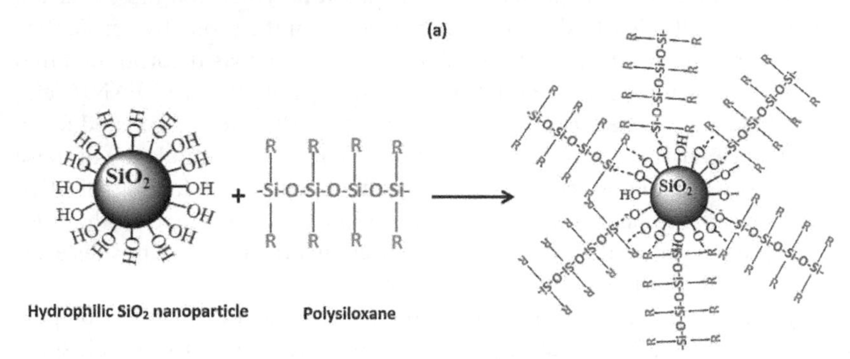

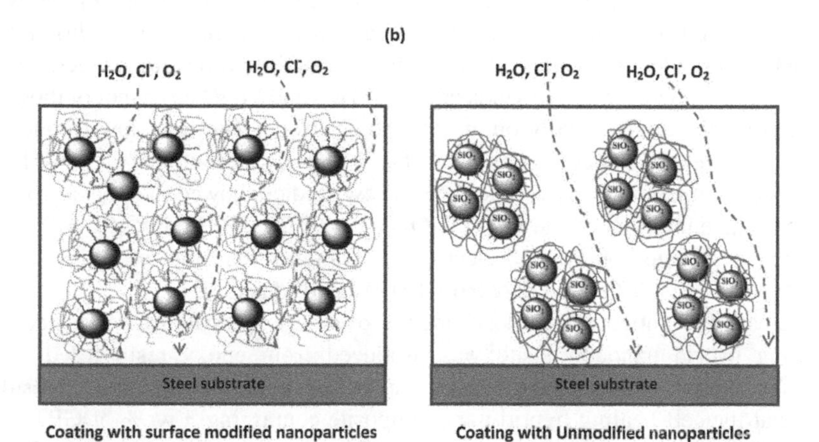

FIGURE 10.9 Schematic Illustration of Surface Grafting of Polysiloxane on the Surface of Silica Nanoparticle (a) and the Effects of Surface Modification on the Coating-barrier Performance (b) [49].

Caldona et al. reported that the rubber-modified polybenzoxazine coating with the optimum SiO_2 loading exhibited improved anti-wettability and anticorrosion performance even for a prolonged exposure to corrosive environment [50]. These coatings also had promising anti-icing ability, preventing ice/snow from adhering to the surface and delaying icing of water upon striking the surface. The fluorinated polyacrylate–silica, having a contact angle of about 153.2°, was coated onto the surface of CRS using the spin-coating technology. These coatings exhibited superior corrosion protection for CRS compared to that of the hydrophobic organic coating as they serve as an effective barrier against aggressive species [51]. Electrochemical-impedance spectroscopy results revealed that a high-performance bisphenol-A phthalonitrile resin reinforced with different amounts of silane treated-TiO_2 nanoparticles offered excellent corrosion protective properties [52].

The effect of including SiO_2 nanoparticles on the corrosion resistance of fluoropolymer-coated steel were studied by potentiodynamic polarization and electrochemical impedance spectroscopy (EIS) [53]. The results shown that nanocomposite particles can be dispersed better in fluoropolymer coatings, and the electrochemical results clearly shown the enhancement of the protective properties of the nanocomposite coatings composed from 4 wt.% SiO_2 was incorporated into the fluoropolymer coatings. The inhibition properties of synthesized PANI/CeO_2 nanocomposites on mild steel (MS) corrosion in 0.5 M HCl were estimated using weight loss and electrochemical techniques [54]. The inhibition efficiency of these nanocomposites was found to increase almost monotonically with loading of filler. In addition, an enhancement in the water contact-angle was detected as the filler content increased in the matrix. It was also reported that the nanocomposites are a potential inhibitor for mild steel in HCl medium.

The polyurethane/3-amino propyl trimethoxy silane-treated Cr_2O_3 nanocomposites were used on the St-37 steel substrates. Results obtained from EIS and salt-spray analyses showed that the surface-treated nanoparticles markedly increased the corrosion protection performance of the polyurethane coating. The improvement was more important for the coating filled with 0.43 g silane/5 g pigment. In addition, the adhesion loss degraded in the presence of surface-treated ceramic nanoparticles with 0.43 silane/5 g pigment [55]. Two kinds of PANI/TiO_2 and PPy/TiO_2 nanocomposites were prepared electrochemically on an Al1050 electrode. The best, low-frequency capacitance values were recorded as C_{LF} = 60.76 and 12.8 mF cm^{-2} for PANI/TiO_2 and PPy/TiO_2 nanocomposites, respectively. These studies showed that the corrosion protection efficiency (PE) of the PANI/TiO_2 (PE = 97.2%), PPy/TiO_2 (PE = 97.4%) nanocomposites coated on Al1050 electrode was higher than that of PANI (PE = 96.4%), PPy (PE = 94.9%) and uncoated Al1050 electrodes [56].

To increase the anticorrosion performance of an epoxy-coated CRS surface, a self-healing hybrid nanocomposite was produced from epoxy resin filled with self-healing microcapsules and organosilane treated nano-SiO_2 [57]. The results showed that the self-healing hybrid nanocomposite containing 3 wt.% of GPTMS-modified silica and 10 wt.% perfluorooctyl tri-ethoxysilane microcapsules (POT) had the highest corrosion performance with the corrosion rate of 0.09 mm/year (I_{corr} of 0.01 mA/cm^2), corrosion rating number of 9, and oxygen permeability about 0.14 Barrer. The addition of modified SiO_2 and self-healing agent microcapsules

(POT) effectively enhanced the length of the diffusion pathways, as well as lowered the O_2 permeability of the nanocomposite film.

Ejenstam et al., found that the most significant result was recorded for the coating having 20 wt.% hydrophobic SiO_2 nanoparticles, where it was possible to attain protection for almost 80 days in 3 wt.% NaCl solution [58]. The protective properties offered by this coating were attributed synergistic effect of the hydrophobicity of the polydimethylsiloxane matrix and the prolonged diffusion path caused by addition of hydrophobic silica particles. Meso-TiO_2 nanoparticles was modified with polyethylenimine (PEI) of various molecular weights were used to reinforce waterborne epoxy coated on mild steel coated specimens. The EIS results revealed that the resistance value of a coating with meso-TiO_2/PEI (600 molecular weight) was above $9.87 \times 10^7\ \Omega.cm^2$, which was higher than that afforded by unfilled epoxy coating [59]. They justified these improvements by the strong chemical interactions between the polymeric matrix and treated meso-TiO_2 providing good barrier properties and increased degree of cross-linking. The Fe_3O_4@SiO_2-filled epoxy nanocomposites not only gave a uniform dispersion and compatibility in the polymeric matrix but also showed a great enhancement in the anticorrosion properties of epoxy coatings [60].

New SiO_2/F-PF hybrid films were developed for enhancing the anti-corrosion performance of Cu materials. The anticorrosive efficiencies of the functionalized FPF and SiO_2/F-PF coating films were superior than those related to the PF and PF/SiO_2 coverings [61].

A series of PBZ-TMOS/SiO_2 nanocomposite coatings were produced for anti-corrosion protection of mild steel. The presence of the covalent bond between nano-SiO_2 and PBZ-TMOS significantly improved the interfacial interactions at the matrix/filler interfaces reflecting the enhanced corrosion resistance ability [62]. The effects of TiO_2 particle loadings on the microstructure and properties of the epoxy coatings of 40-μm thicknesses were studied by Xu et al. [63]. The homogenous dispersion of TiO_2 nanoparticles was observed at concentrations bellow 40 g/l. Results showed that the inclusion of nano-TiO_2 did not affect the thickness of the epoxy coatings and did not degraded the magnetic properties of NdFeB substrates, while could significantly increase the corrosion resistance of the epoxy coatings.

The EIS results corresponding to the epoxy/SiO_2 nano-coating confirmed that the corrosion resistance of these hybrid coatings enhanced with increasing the nano-SiO_2 phase content [64]. The effects of adding nano-Al_2O_3 particles on the anticorrosion performance of an epoxy/polyamide coating applied on the AA-1050 metal substrate was evaluated by Golru et al., [65]. Results showed that even at high nano-Al_2O_3 loadings extent, the nano-Al_2O_3 were uniformly dispersed in the coating matrix at a particle size lower than 100 nm. As illustrated in Figure 10.10, the filling with Al_2O_3 nanofillers significantly ameliorated the corrosion-resistance of the virgin epoxy coating through decreasing the water permeability and increasing its resistance against hydrolytic degradation.

Balaskas et al., [66] recorded a continuous increase of the total impedance value of epoxy/8-hydroxyquinoline-TiO_2 coatings with the time of exposure caused by a possible self-healing effect resulting from the release of the 8-hydroxyquinoline inhibitors. The loading of nano-TiO_2 into the coatings increased the barrier properties of the coatings,

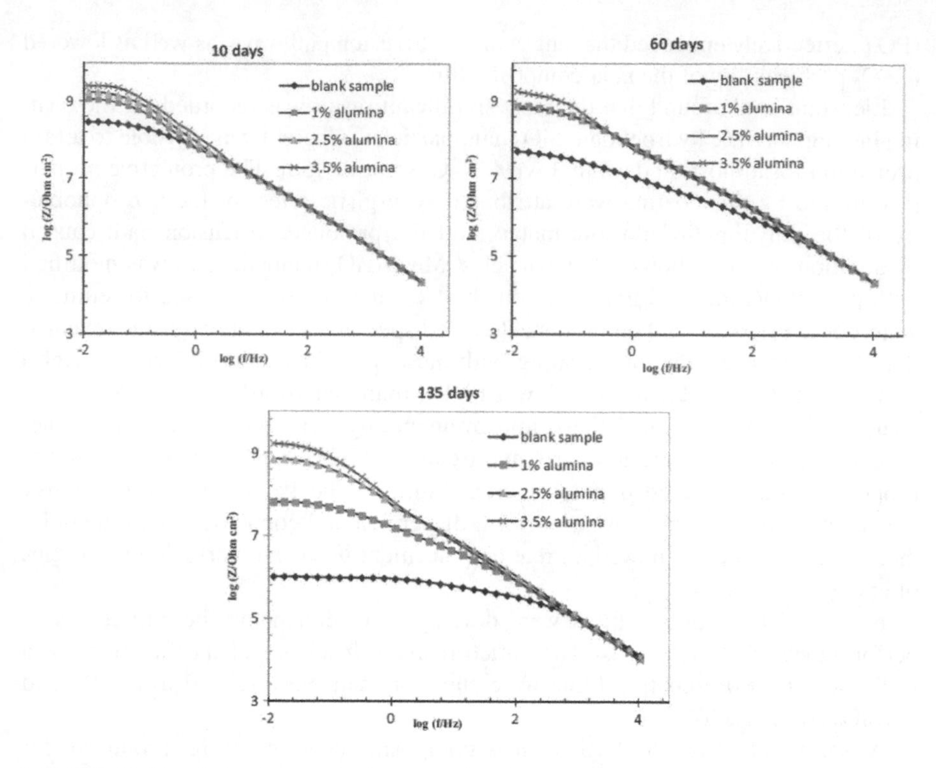

FIGURE 10.10 Bode Plots of the Samples Containing Different Contents of Nano-alumina Exposed to 3.5 wt% NaCl Solution for Different Immersion Times [65].

enhancing the anticorrosion performance. Corrosion performance of epoxy/SiO_2 nano-coatings was studied by Palraj et al. The composite coatings withstood 720 h in a salt-spray test, whereas the micro-SiO_2 coatings withstood up to 650 h. After 30 days' immersion in 3 wt.% NaCl solution, EIS results of epoxy/SiO_2 nanocomposite coatings showed a resistance of 2.36×10^6 Ωcm², whereas micro-composites ones showed a resistance of only 5.41×10^4 Ωcm² (Table 10.2) [67].

TABLE 10.2
Impedance Data of Micro- and Nano-silica–incorporated Epoxy Polyamide Coatings on Mild steel in 3 wt.% NaCl Solutions for Different Durations [67]

Duration	Microsilica-incorporated Epoxy polyamide Coatings		Nanosilica-incorporated Epoxy polyamide Coatings	
	R_{ct} (Ω)	C_{dl} (F/cm²)	R_{ct} (Ω)	C_{dl} (F/cm²)
1 h	7.498×10^6	1.024×10^{-10}	1.920×10^8	1.63×10^{-10}
1 day	3.619×10^5	1.026×10^{-10}	4.446×10^7	2.002×10^{-10}
7 days	2.375×10^5	1.21×10^{-10}	1.863×10^7	3.931×10^{-10}
15 days	1.311×10^5	1.034×10^{-10}	1.757×10^7	3.008×10^{-10}
30 days	5.412×10^4	1.951×10^{-9}	2.361×10^6	3.543×10^{-10}

Epoxy/SiO_2 composites modified by PDMS-OH in order to use them as protective agents for stone surfaces. The addition of PDMS-OH into GPTMS and ATS accelerated the curing process and increased the viscosity of solid phase and it was chemically bonded into epoxy matrix via Si-O-Si bonds. The micro-cracks of epoxy/SiO_2/PDMS-OH nanocomposite was significantly reduced by adding 30% w/w of PDMS-OH [68]. Yu et al. [69] synthesized TiO_2–GO sheet and treated them using ATS silane, and used to reinforce epoxy resin at 2 wt.% fraction. The EIS test revealed that the corrosion-resistant performance improved with the addition of TiO_2–GO hybrids to epoxy compared to adding only GO or TiO_2 as filler alone, due to their exfoliation, dispersion, and excellent plugging micro-pore property arising from their laminated structure. In addition, the corrosion-resistant mechanisms were tentatively proposed for the TiO_2–GO/epoxy coatings.

The failure of epoxy/nano-SiO_2 nanocomposite filled with 5 wt.% nano-SiO_2 on mild steel under high hydrostatic pressure of 3.5 MPa was studied by Liu et al. [70]. The results revealed that high-hydrostatic pressure accelerated the failure of the epoxy/nano-SiO_2 coating by enhancing the diffusion of water in the coating, which increased the water spread and electrochemical reactions at the interface. Anticorrosion performance of an epoxy/GPTMS-g-SiO_2 coating enhanced by improving the dispersion of SiO_2 nanoparticles grafted on GPTMS silane molecules. Corrosion performance of the coated mild-steel specimens was evaluated employing EIS and the salt-spray test. Inclusion of 4 to 6 wt.% nano-SiO_2 had the best corrosion performance [71].

The dodecyl benzene sulphonic acid-doped polyaniline-TiO_2 (DBSA-doped PANI-TiO_2) nanocomposite were used as protective coating for mild steel [72]. The coating resistance of epoxy/DBSA-doped PANI and epoxy/DBSA-doped PANI-TiO_2 are in the order of 1×10^4 Ωcm^2 and 1×10^5 Ωcm^2, respectively. Under different cathodic potentials, the protective performance of epoxy/DBSA-doped PANI coating is improved in the presence of TiO_2 nanoparticles. Bhat and Ahmad [73] reported on synthesis and physico-mechanical and anticorrosive properties on carbon steel in 3.5 wt.% HCl, NaCl and NaOH solutions of nano-TiO_2-filled hyperbranched poly(ester amide) nanocomposite coatings. The TiO_2-dispersed hyperbranched poly(ester amide) coatings demonstrated far superior physico-mechanical, hydrophobic (contact angle values from 98° to 107°) and anticorrosive properties (i_{corr} 9.89×10^{-7} Acm^{-2}, E_{corr} -0.192 V, impedance ~10^5 Ωcm^2, phase angle 77° and R_{pore} 3.7×10^6) than those of unfilled coatings and many previous reported coating systems.

The effects of surface modification of nano-TiO_2 on the adhesion and anticorrosion properties of epoxy-coated mild steel were studied by Pour et al. [74]. Ghiyasi et al. [75] demonstrated that, while including untreated nano-SiO_2 could not markedly curing process, the addition of poly(ethyleneimine)-treated SiO_2 enabled the cross-linking of epoxy/amine system due to the abundance of amine reactive groups on the nano-SiO_2 particle outer surface that make the poly(ethyleneimine)-modified SiO_2 nanofillers a promising reinforcing agent for producing highly-curable epoxy-based nanocomposite coatings for anticorrosion applications. Torrico et al. [76] found that the uniform dispersion of the quasi-spherical sub-nonmetric SiO_2-siloxane nodes into epoxy coating formed a dense nanostructure showing high thermal stability (>300 °C) and constructed strong adhesion to steel substrate. The good dispersion also were responsible for superior barrier property in saline solution and improved corrosion-resistance in the order of $G\Omega.cm^2$.

10.4 GAS BARRIER PROPERTIES

The separation process of gaseous mixtures depends directly on energy-intensive and expensive processes, such as chemical looping of amines. This has concentrated research to explore less energy-intensive, passive methods of performing separations, especially the utilization of polymeric membranes. Although pure polymer membranes have showed good separation performance, they failed to offer a compromise between permeability and selectivity, which decreased their overall performance. Recently, intense efforts have concentrated on the inclusion of a secondary phase, constituted of inorganic fillers, to the polymeric membranes to improve the permeability or selectivity. These hybrid organic/inorganic systems have been recently studied in terms of the size, shape, and poor dispersion of the inorganic fillers, the type of matrix, the transport phenomena, and the interface.

The gas-barrier properties of polyimide/TiO_2 nanocomposites were evaluated by Moghadam et al. [77]. The H_2 and O_2 permeability, as well as H_2/N_2 and O_2/N_2 selectivity, were enhanced with respect to the pure polyimide matrix. A higher gas permeability was recorded due to the formation of void in the TiO_2/polyimide interfaces and the agglomeration of nanoparticles. Thus, a low selectivity of the nanocomposite relative to the unfilled polyimide matrix was obtained. Lua and Shen [78] successfully developed polyimide/SiO_2 composite membranes and tested them using binary mixed gases. The result revealed that the mixed-gas selectivity initially enhanced with increasing the proportion of silicon to carbon with a maximum value of 7.01 was recorded for the C–SiO_2 11% composite membrane.

Permeability rates of O_2 and N_2 gases through films of polypropylene/SiO_2 were also evaluated. Lower rates were found compared to the neat polypropylene, most notably for O_2, which decreased with rising SiO_2 loading. This was related to the increase of tortuous paths, which were covered by the gas molecules, since nano-SiO_2 are considered impenetrable by them [79]. Permeability rates of O_2 and H_2O through films of PMMA-*grafted*-SiO_2/PVC nanocomposites were evaluated [80]. Lower rates were recorded for these nanocomposites when compared with t unfilled PVC matrix. This result was justified by the fact that SiO_2 nanoparticles are impenetrable by gas molecules which generated more tortuous paths covered by the gas molecules on the surface of nanocomposites.

A new series of poly(iso-propylene)/SiO_2 nanocomposites, reinforced with 1 to 10 wt.% fumed nano-SiO_2, was produced by melt mixing in a twin screw co-rotating extruder [81]. The permeability of all the studied gases was decreased to a high extent with the increasing of nano-SiO_2 loadings to attain a maximum reduction of 38% for N_2, 33% for O_2, and 61% for CO_2 at 10 wt.% filler content, as it is depicted in Figure 10.11. Nanocomposite membranes based on a poly(ether-imide) were formed by adding three types of hydrophobically treated nano-SiO_2, since they contained voids or defects at the polymer/particle interface or within aggregates, this enhanced the gas permeability while reducing selectivity of these membranes [82]. The authors

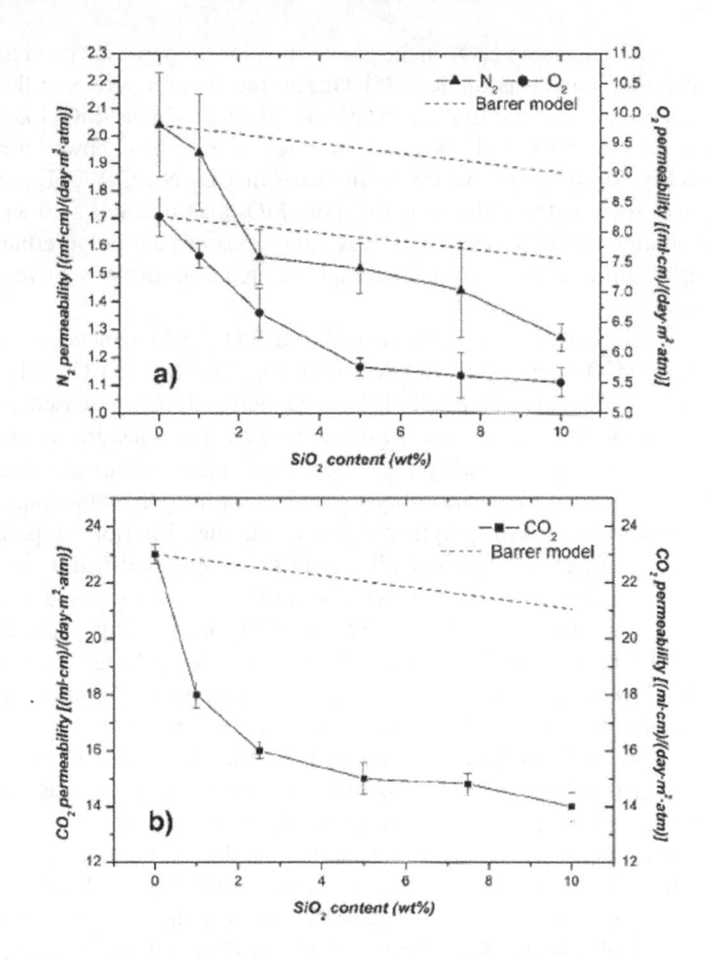

FIGURE 10.11 Permeability of (a) N_2 and O_2 and (b) CO_2 of iPP/SiO_2 Nanocomposites as a Function of Filler Content and Predicted Permeability Using the Barrer Model [81].

try to improve dispersion and enhance the selectivity where they deduced treatments of nano-SiO_2 minimize void formation and particle agglomeration.

Novel well-dispersed HDPE/Al_2O_3 nanocomposite membranes were fabricated, characterized, and tested for their gas barrier performance [83]. The diffusion coefficient of fabricated nanocomposite membranes can be decreased to half with the addition of 7.29 vol.% Al_2O_3 flakes. However, this improvement in permeability was also detected due to the formation of the voids at the interface of HDPE/Al_2O_3 that generally are ascribed to the low wettability of nanoceramics with the matrix, as evidenced by electron microscopy. By using 3-aminopropyltriethoxysilane (APS) to modify the outer surface of nano-Al_2O_3, the diffusion coefficient and permeability were reduced by 20%, relative to the untreated case. Other new membranes were fabricated from polysulfone/mesoporous SiO_2 spheres (MPS) at the range of 0 and 32 wt.% [84]. An optimum content of 8 wt.% was found in terms of H_2/CH_4 separation performance. In addition, the optimum membrane was achieved for CO_2/N_2 separation.

The effect of adding nano-SiO_2 on the gas permeation properties of polyether-based polyurethane membrane was studied [85]. Gas permeation properties of these nanocomposite membranes containing 2.5, 5, 10 and 20 wt.% of nano-SiO_2 loading were investigated for pure CO_2, CH_4, N_2, and O_2 gases. The results showed a reduction in permeability for all gases, while an increase in CO_2/N_2, CO_2/CH_4, and O_2/N_2 selectivity was recorded by enhancing the nano-SiO_2 loadings. For a 20 wt.% membrane, the obtained CO_2/N_2 selectivity was 1.65 times of neat polyurethane, while the CO_2 permeability decrease of nanocomposites reached 35.6% relative to that of pure matrix.

The poly(2,6-dimethyl-1,4-phenylene oxide) ($BPPO_{dp}$)/SiO_2 nanocomposite membranes increased CO_2 permeability while keeping the CO_2/N_2 and CO_2/CH_4 selectivities of the pure $BPPO_{dp}$ membranes [86]. The CO_2 permeability increased as the SiO_2 content rose in the membrane. The 10-nm nano-SiO_2 particles are more effective at enhancing the CO_2 permeability than the 30-nm nanoparticles did. Mechanism revealed that the improved permeability were resulted from the nano-gaps between the SiO_2 nanoparticles and the polymer chains due to their inferior compatibility.

The presence of nano-SiO_2, especially when they were modified by silane molecules, accounting for a greatly enhancement in the crystallinity and in lowering permeability to O_2 and CO_2; values of O_2 and CO_2 permeability were decreased up to 80% and 50%, respectively, respected to the unfilled poly(lactic acid) [87]. In addition, the H_2O vapor permeability is significantly influenced by the presence of the nano-SiO_2 but not by their shapes or by silane-treatments.

The role of adding nano-SiO_2 particles on the permeability of CO_2, CH_4, and N_2 gases in polybenzimidazole (PBI) membranes has been studied [88]. By increasing the silica content in the PBI matrix, gas permeation tests revealed an improvement in the solubility and a corresponding reduction in the diffusivity of the gases through the membranes; thus, the permeability of the condensable CO_2 and CH_4 gases were increased, whereas that of non-condensable N_2 gas was significantly reduced upon increasing nano-SiO_2 loadings. The permeability of CO_2 and its selectivity over N_2 was enhanced from 0.025 Barrer and 3.5 in pure PBI to 0.11 Barrer and 71.3 in the nanocomposite containing 20 wt.% nano-SiO_2.

The relative gas permeability of the poly(ether-imide)/SiO_2 nanocomposite with chemical coupling to matrix was reduced in the presence of nanoparticles [90]. Diffusion coefficients computed from time lag data also decreased with SiO_2 loading. However, solubility coefficients computed by dividing the experimental permeability by the diffusivity resulted from the detected time lag enhanced with SiO_2 loading contrary to simple composite theory. The minimum oxygen permeation rate for poly(ethylene terephthalate) films deposited on SiO_x film is 0.10 $cm^3 m^{-2}$ day^{-1} atm^{-1}, which corresponds to an oxygen permeability coefficient of 1.4×10^{-17} cm^3-cm cm^{-2} s^{-1} cm^{-1} Hg for the SiO_x film itself [91].

The oxygen and water vapor permeation rates of poly(ethylene naphtholate)/$SiOx$ reached 0.08 to 0.13 cm^3 m^{-2} day^{-1} atm^{-1} at 30°C at 90% RH and 0.244 to 0.276 g m^{-2} day^{-1} at 40°C at 90% RH, respectively [92]. From these results, it can be concluded that the ion-assisted plasma polymerization is a potential technique for deposition of gas-barrier $SiOx$ thin films. Polyvinyl alcohol/nano-SiO_2 particles with improved hydrophilicity and polarity were synthesized to reinforce polyurethane [93].

The membranes showed an enhancement in the solubility and a corresponding decrease in the diffusivity of the gases through the membranes by rising the SiO_2 ratios in the polymer matrix; consequently, the permeability of the condensable and polar CO_2 gas was improved, whereas that of other gases reduced. In the membrane containing10 wt.% nano-SiO_2 content an increase of CO_2/CH_4 ($\alpha \approx 10.1$) and CO_2/N_2 ($\alpha \approx 70.7$) selectivities was recorded.

Gas-permeation properties of the polycaprolactone-polyurethane/SiO_2 membranes with different SiO_2 contents, was studied for pure CO_2, CH_4, N_2, and O_2 gases. The data revealed a decrease in gas permeability, but enhancement in CO_2/N_2, CO_2/CH_4 and O_2/N_2 ideal selectivity [94]. Aiming to increase the gas-blocking properties, the metal alloy (Invar) has been mixed with lab-made organic/inorganic nanocomposites, markedly not only increasing the gas resistance but also decreasing the shrinkage after UV curing. The experimental data revealed that using nanocomposite IV can stop the penetration of moisture as well as oxygen in the air into the devices and thus promotes the lifetime of organic solar cells [95].

The oxygen permeability of the epoxy/SiO_2 meso-composites rose significantly at contents ≥5 wt.%, while the compositions made from the calcined form of the meso-structure showed unpermeation dependence on SiO_2 loading [96]. Results showed that the oxygen permeability in the meso-composites filled with 12 wt.% as-made silica reached 6-fold higher than that of the unfilled epoxy membrane. On the other hand, meso-composites filled with calcined nano-SiO_2 do not allow for curing-agent partitioning, and the oxygen permeability is not remarkably affected by the increasing nano-SiO_2 content as revealed in Figure 10.12.

To reduce hydrophilicity, nanosized precipitated SiO_2-coated $CaCO_3$ fillers were treated by an alkyl- and a fluoro-alkoxysilane derivative, respectively. The applied

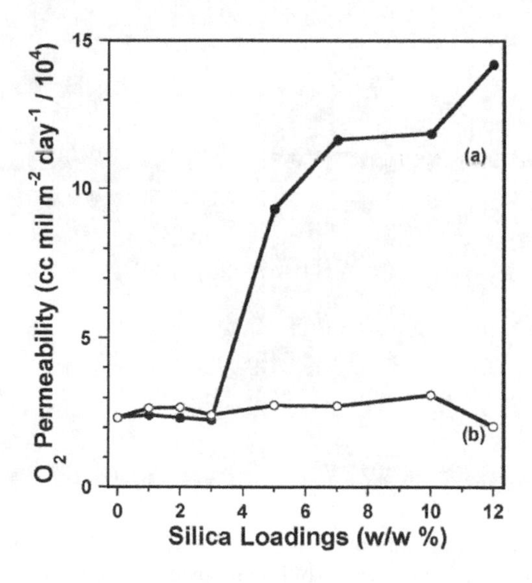

FIGURE 10.12 Dependence of Oxygen Permeability on Silica Loadings for Epoxy Mesocomposites Prepared from (a) As-made SiO_2 and (b) Calcined SiO_2 [96].

surface-treatment led to a similar grafting density of 3.2 μmol. m^{-2} for the two alkoxysilane derivatives. The oxygen permeability of PVDF nanocomposites filled with 10 wt.% of raw SiO_2-coated $CaCO_3$ was reduced, and it agreed with the Maxwell model. The increase in the permeability was more evident for the nanocomposite reinforced with the octyltriethoxysilane-modified SiO_2-coated $CaCO_3$ due to the formation of weak interfaces in such system [97].

Another study examined the effect of adding nano-SiO_2 on the permeation of methane, ethane, and propane gases through two types of polyurethane (PU) membranes: one based on polyether and the second based on polyester [98]. Permeation tests showed that in polyether-based PU, permeability at first stage increased by increasing nano-SiO_2 loading up to 2.5%, and then decreased. However, the permeability of these gases in polyester-based PU was constantly decreased by enhancing nano-SiO_2 contents. The selectivity for C_3H_8 over methane increased with the inclusion of ceramic nanoparticles in the polyether-based PU membranes, while it reduced in polyester-based PU membranes. Results also indicated the high propane permeability and propane/methane selectivity of polyether-based mixed matrix membranes containing 12.5 wt.% nano-SiO_2 at 2 bar pressure up to, 118 Barrer and 7.01, respectively. As illustrated in Figure 10.13, the SEM photos of most of the aggregated SiO_2 particles are smaller than 200 nm in size, which reflects that the prepared membranes have a nonporous, dense structure and there are no pinholes, connected pores, or cracks.

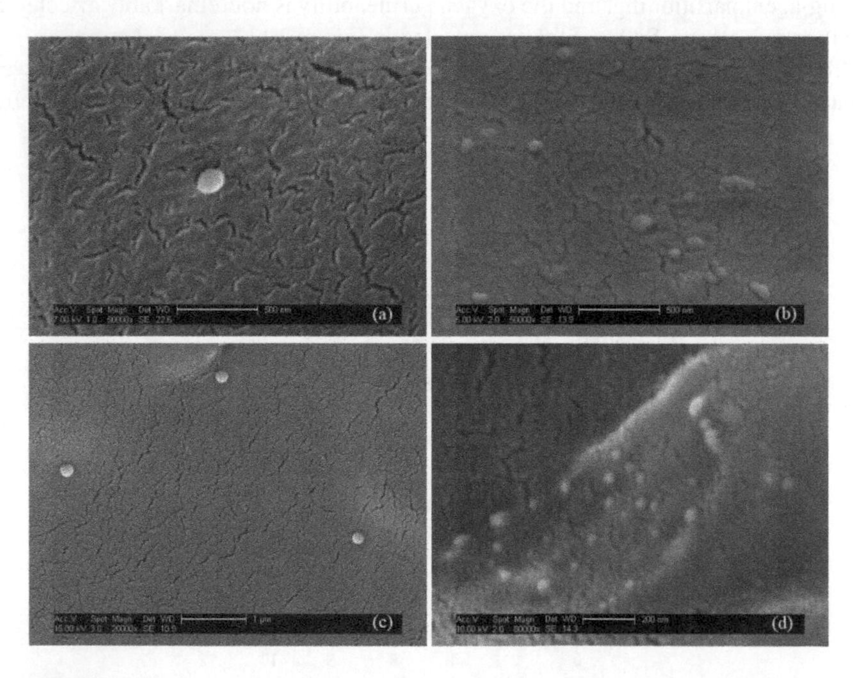

FIGURE 10.13 SEM Micrographs of Cross Section of Hybrid Membranes: (a) Polypropylene Glycol-based Polyurethane/SiO_2-2wt.%; (b) Polypropylene Glycol-based Polyurethane/SiO_2-10wt.%; (c) Polycaprolactone-based Polyurethane/SiO_2-5wt.%; (d) Polycaprolactone-based Polyurethane/SiO_2-20wt.% [98].

The insertion of the stress-relaxation layer drastically reduced the water-vapor permeability coefficient value. The lowest Q value was observed especially when it was embedded between poly(*tert*-butyl acrylate-*co*-2-hydroxyethyl methacrylate)/SiO_2 nanocomposite layer and the PVA substrate. Morphological analysis shows any crack both in the composite layer and near the interface between layers. It would be attributed to the improved compatibility of the TBAH stress relaxation layer to the TBAH/SiO_2 nanocomposite layer and to the PVA substrate [99].

Yanaka et al., [100] studied O_2 permeation through four various SiO_x/poly(ethylene terephthalate) specimens, which were previously strained to give homogenous distribution of cracks, where they showed good agreement as compared with predicted results by analytical model. They also concluded that due to the logarithmic dependence of transmission on the width of a crack, for a fixed strain it is better to have a small number of large cracks rather than a large number of small cracks. Lee et al. [101] reported that the sulfonated poly(arylene ether sulfone)–filled SiO_2 particles having a high surface area and small particle size showed the higher proton conductivity, long membrane life time under oxidative conditions, good dimensional stability, outstanding single cell performance, and reduced methanol crossover. Results showed that the optimal SiO_2 content for maximizing the fuel cell performance of the studied nanocomposite membranes was filled with 2 wt.% nan-SiO_2.

Thermo-sensitive polyurethane (TSPU) reinforced with different *in situ*-generated TiO_2 nanoparticles based nanocomposite membranes were fabricated by Chen et al., [102]. According to the membrane formation temperature (T_{mf}) during casting process, completely-opposite gas transport behaviors of these membranes were found. When T_{mf} was located bellow the melting point (T_m) of the TSPU soft segment, the gas permeability coefficients of the resulted nanocomposite were observed to enhance greatly with increasing nano-TiO_2 loading. However, when T_{mf} was higher than the T_m of the soft segment, flexible TSPU chains tend to pack efficiently around the TiO_2 nanoparticles, where the gas transport behavior of these nanocomposites follows the prediction of conventional composite theory.

The effects of the ATPS-modified SiO_2 nanoparticles modification and the SiO_2 loadings on the gas separation capability of the resulting polyimide nanocomposite membranes were evaluated by Shen and Lu [103]. The permeability of all gases (CO_2, O_2, N_2 and He) enhanced with enhancing the concentrations of the ceramic nanofiller. A decrease in selectivity was detected with increasing modified SiO_2 volume ratios. At a fixed nano-SiO_2 loading the nanocomposite based on the treated nano-SiO_2 showed improved selectivity respect to that based on raw nano-SiO_2.

TiO_2 nano-ceramics were added via solution processing technique in poly(1-trimethylsilyl-1-propyne) (PTMSP) to produce new types of nanocomposite membranes [104]. At low nano-TiO_2 content, light gas permeability was inferior to that of the neat matrix, whereas at higher nano-TiO_2 ratios, light gas permeability (i.e., CO_2, N_2, and CH_4) enhanced to more than four times higher than in pure PTMSP. New PU/Al_2O_3 nanocomposite membranes were developed to study the permeability of CO_2, CH_4, O_2, and N_2 gases in these hybrid membranes. The recorded results revealed a decrease in the gas permeability, but a marked increase in the CO_2/N_2, CO_2/CH_4, and O_2/N_2 selectivity was obtained with increasing Al_2O_3 loading [105]. Sadeghi et al. [106] studied the effect of nano-TiO_2 on the gas separation properties of PU membranes. Gas permeation tests of PU/TiO_2 membranes filled with TiO_2 concentrations up to 30 wt.% were studied for N_2, O_2, CH_4 and CO_2 gases at 1 MPa and 25°C, showing a reduction in the permeability of these gases and an enhancement in gas selectivity as TiO_2 ratio increased.

New types of poly(ether-block-amide)/*cis*-9-Octadecenoic acid-silica (PEBA/-OA-SiO_2) nanocomposite membranes for the separation of CO_2 from syngas and natural gas streams were produced [107]. The inclusion of treated SiO_2 nanoparticles into the PEBA matrix ameliorated the separation performance of the resulted nanocomposite membranes. For example, at 25°C and 2 bar, when the nano-SiO_2 loading ranged from 0 wt.% to 8 wt.%, the best selectivity values of CO_2/H_2, CO_2/CH_4, and CO_2/N_2 were enhanced from 9, 18, and 61 to 17, 45, and 137, respectively.

Su et al., [108] reported on the gas transport properties of tailored hybrid membranes composed of cross-linked poly(ethylene glycol) and SiO_2 nanoparticles by controlling the nanoparticle size, their content, and dispersion. The permeability deviations from Maxwell's model increased as the size of nano-SiO_2 particle decreased and ratios enhanced. These size-dependent deviations from Maxwell's model were ascribed to the interfacial interactions, which scale with surface area and acted to restrain segmental chain mobility.

The simulation techniques have been used to evaluate the structural, physical, and separation properties of penetrant gases including O_2, N_2, CO_2 and CH_4 through pure and nano-SiO_2 particles filled polysulfone (PSF) membranes [109]. Molecular dynamics (MD) and grand canonical Monte Carlo (GCMC) simulations were performed by employing the COMPASS force field to estimate the diffusivity and solubility of the gases into the membrane. The parameters including fractional-free volume, average cavity size, and cavity-size distributions of unfilled and SiO_2-filled PSF were determined using an energetic based cavity-sizing algorithm. These parameters for the SiO_2-filled membrane were higher than those of neat PSF and increased with increasing the amount of the filler loading, and as a result, the diffusion coefficient, solubility, and permeability of penetrant gases in PSF/SiO_2 membranes were more pronounced than that of matrix-based membranes.

Morphological analysis showed that the mean size and the distribution of nano-SiO_2 particles in the thermal-sensitive polyurethane (TSPU) matrix is depended on the SiO_2 loading. As illustrates in Figure 10.14, FTIR study revealed that two chemical bonds of Si–C and Si–O–C occurred at 930 cm^{-1} and at 453 cm^{-1}, 1034 cm^{-1}, respectively, between the organic and the inorganic phases [110]. The water-vapor permeability (WVP) of TSPU nanohybrid membrane relied on the size scale of SiO_2 particles, which depended on SiO_2 ratio. By increasing the temperature over the T_s, from 50°C to 60°C, the WVP of pure matrix increased from 1026 g/m^2/day to 2200 g/m^2/day, while the WVP of the nanocomposites TSPU containing 5.0 wt.% SiO_2 loading increased from 995 g/m^2/day to 2927 g/m^2/day.

Gas permeation properties of PU/TiO_2 nanocomposite membranes filled with up to 30 wt.% of nano-TiO_2 were studied for N_2, O_2, CH_4 and CO_2 gases at 10 bar and 25°C. Results showed a reduction in permeability of the studied gases and enhancement in gas selectivities as TiO_2 loading rose [111]. PMMA/ZrO_2 nanocomposite membranes were prepared by the incorporation of nano-ZrO_2 particles in various loading 2, 4, 6, 8, and 10 wt.% into PMMA matrix by in situ emulsifier-free emulsion

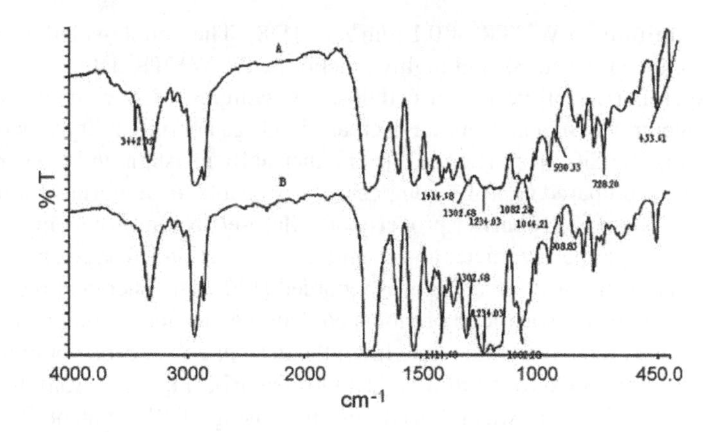

FIGURE 10.14 FTIR Spectra for (a) 1 wt.% Nano-SiO_2/TSPU and (b) Unfilled TSPU [110].

polymerization technique [112]. The oxygen barrier properties of PMMA/ZrO_2 nanocomposites were evaluated by using a gas permeater. The gas barrier property of polyacrylonitrile/zirconium dioxide (PAN/ZrO_2) nanocomposites was studied by using gas permeaters where results revealed that gas barrier properties decreased by about 10 times with increasing of ZrO_2 concentrations [113]. This trend was attributed to the presence of a tortuous path in these nanocomposites due to the addition of nano-ZrO_2 which was responsible for preventing the oxygen permeation.

Erlat and Spontak [114] used non-parametric response surface methods optimization to identify the magnetron-PECVD conditions responsible for superlative SiO_x barrier coatings on poly(ethylene terephthalate) (PET). The thermal E_a for water-vapor permeation, unlike that for oxygen permeation, relied on barrier performance and increases of as much as 20 kJ/mol with an increase in barrier efficacy. Another correlation between SiO_x morphology (including defects) and barrier performance was established. Roberts et al. [115] applied the model to interpret permeability and activation energies measured for the inert gases (He, Ne, and Ar) in evaporated SiO_x films of varying thickness (13 to 70 nm) coated on a polymer substrate. Although no defects could be observed by microscopy, the permeation data revealed that macro-defects (>1 nm), nano-defects (0.3 to 0.4 nm) and the lattice interstices (<0.3 nm) all accounted for to the total permeation.

Single Al_2O_3 atomic-layer deposition (ALD) films on polymers have demonstrated excellent gas diffusion barrier properties. In a Dameron et al. study [116], multilayers of Al_2O_3 ALD and rapid SiO_2 ALD were grown on Kapton and heat-stabilized polyethylene naphtholate substrates. The barrier enhancements recorded for one Al_2O_3/SiO_2 bilayer and two Al_2O_3/SiO_2 bilayers could not be explained using laminate theory. The minimum attained oxygen-transmission rate (OTR) of the gas-barrier layer prepared from nano-SiOx/poly(ether sulfone) substrate via a simple oxygen plasma treatment reached 0.2 $cm^3/(m^2/day)$, which was higher or even comparable to that exhibited by other gas barrier layers developed by plasma-enhanced chemical vapor deposition (PECVD) or sputtering [117]. More significant improvement was attained by depositing a 100-nm-thick SiO_x film on both sides of the PES substrate, which resulted in a minimum WVTR of 0.1 g/m²/day [118]. The double-sided coatings on PES were balanced the stress and highly ameliorate the WVTR data.

The oxygen barrier properties of cellulose/SiC composites were measured using gas permeameter, where a substantial decrease in O_2 gas permeability was observed with increasing the SiC proportion [119]. The thermally resistant and oxygen barrier properties of the prepared nano-biocomposites may enable the materials for the packaging applications. The O_2 barrier property of cellulose/BN nano-biocomposites was measured using a gas permeameter and a significant reduction in oxygen permeability because of increasing of BN loading was recorded [120]. The chemical resistance of the nano-biocomposites was more pronounced than the unfilled biopolymer. At fixed pressures of 2 psi, the oxygen flow rate in all the developed nano-biocomposites was found to be law respected to the unfilled cellulose matrix (Figure 10.15a). It was also observed that the flow rate was reduced with increasing of BN content. The significant decrease in the O_2 permeability was attributed to homogenous nanostructure dispersion of BN into the biopolymer at various pressure up to 2 psi (Figure 10.15b).

Park et al. [121] reported that any change in the siloxane chain length of the polydimethylsiloxane (PDMS) segment in the poly(imide siloxane) (PIS) precursor directly affected the gas permeation and separation properties of the C–SiO_2 membranes. For a constant volume fraction of PDMS moieties, a longer siloxane chain in the PIS led to a drastic increase in gas permeability and a reduction in gas selectivity of the C–SiO_2 membranes. Moreover, the diffusion coefficients of selected gases were also affected by the silica phase, which was embedded in the continuous carbon matrix

Alumina nanosheets are the most efficient fillers among others to restrict the diffusion of water vapor within polyimide (PI) matrix and simultaneously maintain the transparency of PI [122]. With only 0.01 wt.% of nanosheets-Al_2O_3, the PI nanocomposite film exhibited the most significant reduction of 95% in WVTR as compared to that of neat PI matrix. Most importantly, the resultant PI/nanosheets-Al_2O_3-0.01 film had an excellent optical transparency combined with improved mechanical strength and enhanced thermal stability.

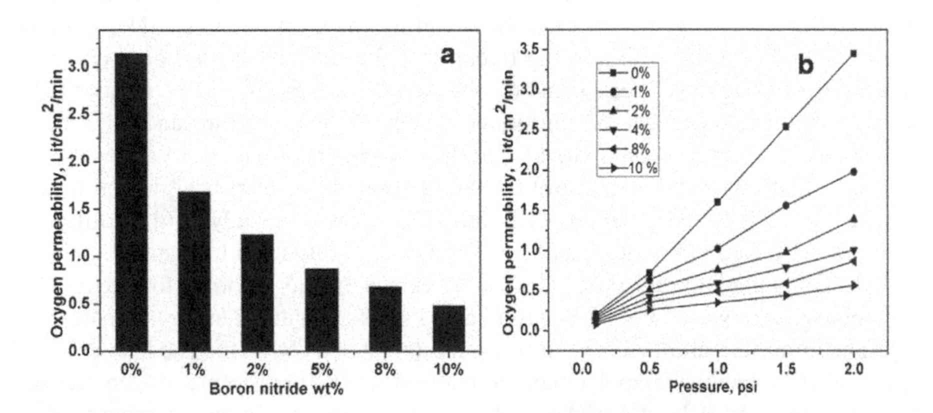

FIGURE 10.15 (a) The Oxygen Permeability of Cellulose/BN Nano-biocomposites with Various Loading of BN at Constant Pressure of 2 psi and (b) the Oxygen Permeability of Cellulose/BN Nano-biocomposites with Various Loading of BN at Difference Pressure [120].

10.5 CONCLUSIONS

In the field of membrane separation technology, polymer/ceramic nanocomposites have attracted both academic and industrial interest due to their exceptional electrical, mechanical, and permeability properties. In this review, we have summarized the state-of-the-art progress on the use of platelet-shaped fillers for the water and gas barrier properties of the various types of polymer/ceramic micro- and nanocomposites. Ceramic nanofillers (such as SiO_2 and TiO_2) appear to be the most promising nanoscale fillers. These nanofillers are able to form a barrier when well-dispersed into a polymer matrix. These barriers do not allow diffusion of small gases or water vapors through them and are able to produce a tortuous path which works as a barrier structure for gases. These membranes also exhibit improved corrosion-resistance.

REFERENCES

1. Tang, J. C., Lin, G. L., Yang, H. C. et al. 2007. Polyimide-silica nanocomposites exhibiting low thermal expansion coefficient and water absorption from surface-modified silica. *Journal of Applied Polymer Science* 104: 4096–4105.
2. Zhao, H., Li, R. K.Y. 2008. Effect of water absorption on the mechanical and dielectric properties of nano-alumina filled epoxy nanocomposites. *Composites Part A: Applied Science and Manufacturing* 39: 602–611.
3. Alamri, H., Low, I. M. 2012. Effect of water absorption on the mechanical properties of nano-filler reinforced epoxy nanocomposites. *Materials & Design* 42: 214–222.
4. Zou, C., Fothergill, J. C., Rowe, S. W. 2008. The effect of water absorption on the dielectric properties of epoxy nanocomposites. *IEEE Transactions on Dielectrics and Electrical Insulation* 15: 12–23.
5. Greenberg, A. R., Kamel, I. 1977. Kinetics of anhydride formation in poly(acrylic acid) and its effect on the properties of a PAA-alumina composite. *Polymer Chemistry* 15: 2137–2149.
6. Bardajee, G. R., Pourjavadi, A., Soleyman, R. et al. 2012. Salep-g-poly(sodium acrylate)/alumina as an environmental-sensitive biopolymer superabsorbent composite: Synthesis and investigation of its swelling behavior. *Advances in Polymer Technology* 31: 41–51.
7. Jo, C., Ko J., Yin, Z. et al. 2016. Solvent-free and highly transparent SiO_2 nanoparticle–polymer composite with an enhanced moisture barrier property. *Industrial & Engineering Chemistry Research* 55: 9433–9439.
8. Abenojar, J., Martinez, M. A., Velasco, F. et al. 2011. Effect of moisture and temperature on the mechanical properties of an epoxy reinforced with boron carbide. *Journal of Adhesion Science and Technology*, 25: 2445–2460.
9. Ramdani, N., Derradji, M., Wang, J. et al. 2016. Improvements of Thermal, Mechanical, and Water-Resistance Properties of Polybenzoxazine/Boron Carbide Nanocomposites. *JOM* 68: 2533–2542.
10. Guo, J. H., Liu, Y. C., Chai, T. et al. 2015. Synthesis and properties of a nano-silica modified environmentally friendly polyurethane adhesive. *RSC Advances* 5: 44990–44997.
11. Tang, S., Zou, P., Xiong, H. et al. 2008. Effect of nano-SiO_2 on the performance of starch/polyvinyl alcohol blend films. *Carbohydrate Polymers* 72: 521–526.
12. Wu, T., Ke, Y. 2006. The absorption and thermal behaviors of PET-SiO_2 nanocomposite films. *Polymer Degradation and Stability* 91: 2205–2212.
13. Lin, H. C., Chang, H. L., Wang, C. F. 2009. Polybenzoxazine-silica hybrid surface with environmentally responsive wettability behavior. *Journal of Adhesion Science and Technology* 23: 503–511.

14. Maheshwari, N., Neogi, S., Praveen, K. M. et al. 2013. Study of the effect of silica nanofillers on the sea-water diffusion barrier property of unsaturated polyester composites. *Journal of Reinforced Plastics and Composites* 32: 998–1002.
15. Petrovicova, E., Knight, R., Schadler, L. S. et al. 2000. Nylon 11/silica nanocomposite coatings applied by the HVOF process. II. Mechanical and barrier properties. *Journal of Applied Polymer Science* 78: 2272–2289.
16. Khelifa, F., Druart, M.E., Habibi, Y. et al. 2013. Sol–gel incorporation of silica nanofillers for tuning the anti-corrosion protection of acrylate-based coatings. *Progress in Organic Coatings* 76: 900–911.
17. Mardare, L., Benea, L. 2017. Development of Anticorrosive Polymer Nanocomposite Coating for Corrosion Protection in Marine Environment. *IOP Conference Series: Materials Science and Engineering* 209: 012056.
18. Wang, N., Fu, W., Zhang, J. et al. 2015. Corrosion performance of waterborne epoxy coatings containing polyethyleneimine treated mesoporous-TiO_2 nanoparticles on mild steel. *Progress in Organic Coatings* 89: 114–122.
19. Bakhshandeh, E., Jannesari, A., Ranjbar, Z. et al. 2014. Anti-corrosion hybrid coatings based on epoxy–silica nano-composites: Toward relationship between the morphology and EIS data. *Progress in Organic Coatings* 77: 1169–1183.
20. Ramezanzadeh, B., Haeri, Z., Ramezanzadeh, M. 2016. A facile route of making silica nanoparticles-covered graphene oxide nanohybrids (SiO_2-GO); fabrication of SiO_2-GO/epoxy composite coating with superior barrier and corrosion protection performance. *Chemical Engineering Journal* 303: 511–528.
21. Ghanbari, A., Attar, M. M. A study on the anticorrosion performance of epoxy nanocomposite coatings containing epoxy-silane treated nano-silica on mild steel substrate. *Journal of Industrial and Engineering Chemistry* 23: 145–153.
22. Yu, Z., Di, H. Ma, Y. et al. 2015. Fabrication of graphene oxide–alumina hybrids to reinforce the anti-corrosion performance of composite epoxy coatings. *Applied Surface Science* 351: 986–996.
23. Yu, Z., Di, H., Ma, Y. 2015. Preparation of graphene oxide modified by titanium dioxide to enhance the anti-corrosion performance of epoxy coatings. *Surface and Coatings Technology* 276: 471–478.
24. Huang, K.Y., Weng, C. J., Lin, S.Y. et al. 2009. Preparation and anticorrosive properties of hybrid coatings based on epoxy-silica hybrid materials. *Journal of Applied Polymer Science* 112: 1933–1942.
25. Tavandashti, N. P. Sanjabi, S., Shahrabi, T. 2009. Corrosion protection evaluation of silica/epoxy hybrid nanocomposite coatings to AA2024. *Progress in Organic Coatings* 65: 182–186.
26. Weng, C. J., Chen, Y. L., Jhuo, Y. S. et al. 2013. Advanced antistatic/anticorrosion coatings prepared from polystyrene composites incorporating dodecylbenzenesulfonic acid-doped SiO_2@polyaniline core–shell microspheres. *Polymer International* 62: 774–782.
27. Weng, C. J., Chang, C. H., Lin, I. L. et al. 2012. Advanced anticorrosion coating materials prepared from fluoro-polyaniline-silica composites with synergistic effect of superhydrophobicity and redox catalytic capability. *Surface and Coatings Technology* 207: 42–49.
28. Hung, W. I., Weng, W. J., Lin, Y. H. et al. 2010. Enhanced anticorrosion coatings prepared from incorporation of well-dispersed silica nanoparticles into fluorinated polyimide matrix. *Polymer Composites* 31: 2025–2034.
29. Chou, Y. C., Lee, P. C., Hsu, T. F. et al. 2014. Synthesis and anticorrosive properties of electroactive polyimide/SiO_2 composites. *Polymer Composites*, 35: 617–625.
30. Kuo, T. K., Weng, C. J., Chen, C. L. et al. 2012. Electrochemical investigations on the corrosion protection effect of poly(vinyl carbazole)-silica hybrid sol–gel materials. *Polymer Composites* 3: 275–281.

31. Kumar, S. A., Bhandari, H., Sharma, C. et al. 2013. A new smart coating of polyaniline–SiO_2 composite for protection of mild steel against corrosion in strong acidic medium. 2: *Polymer Internationals*, 1192–1201.
32. Chaudhari, S., Patil, P. P., Mandale, A. B. et al. 2007. Use of poly(o-toluidine)/ZrO_2 nanocomposite coatings for the corrosion protection of mild steel. *Journal of Applied Polymer Science* 106: 220–229.
33. Huang, T. C., Hsieh, C. F., Yeh, T. C. et al. 2011. Comparative studies on corrosion protection properties of polyimide-silica and polyimide-clay composite materials. *Journal of Applied Polymer Science* 119: 548–557.
34. Huang, T. C., Su, Y. A., Yeh, T. C. et al. Advanced anticorrosive coatings prepared from electroactive epoxy–SiO_2 hybrid nanocomposite materials. *Electrochimica Acta* 56: 6142–6149.
35. Pourhashem, S., Vaezi, M. R., Rashidi, A. Investigating the effect of SiO_2-graphene oxide hybrid as inorganic nanofiller on corrosion protection properties of epoxy coatings. *Surface and Coatings Technology*, 311, 2017: 282–294.
36. Mills, D. J., Jamali, S. S., Paprocka, K. 2012. Investigation into the effect of nano-silica on the protective properties of polyurethane coatings. *Surface and Coatings Technology* 209: 137–142.
37. Abdollahi, H., Ershad-Langroudi, A., Salimi, A. et al. 2014. Anticorrosive coatings prepared using epoxy–silica hybrid nanocomposite materials. *Industrial & Engineering Chemistry Research* 53: 10858–10869.
38. Ou, J., Wang, J., Qiu, Q. et al. 2011. Mechanical property and corrosion resistance of zirconia/polydopamine nanocomposite multilayer films fabricated via a novel non-electrostatic layer-by-layer assembly technique. *Surface and Interface Analysis* 43: 803–808.
39. Conradi, M., Kocijan, A., Zorko, M. et al. 2015. Damage resistance and anticorrosion properties of nanosilica-filled epoxy-resin composite coatings. *Progress in Organic Coatings* 80: 20–26.
40. Karpakam, V., Kamaraj, K., Sathiyanarayananz, S. 2011. Electrosynthesis of PANI-Nano TiO_2 composite coating on steel and its anti-corrosion performance. *Journal of The Electrochemical Society* 158: C416–C423.
41. Shi, X., Nguyen, T. A., Suo, Z. et al. 2009. Effect of nanoparticles on the anticorrosion and mechanical properties of epoxy coating. *Surface and Coatings Technology* 204: 237–245.
42. Conradi, M., Kocijan, A., Kek-Merl, D. et al. 2014. Mechanical and anticorrosion properties of nanosilica-filled epoxy-resin composite coatings. *Applied Surface Science* 292: 432–437.
43. Ashassi-Sorkhabi, H., Bagheri, R. 2014. Sonoelectrochemical synthesis, optimized by Taguchi method, and corrosion behavior of polypyrrole-silicon nitride nanocomposite on St-12 steel. *Synthetic Metals* 195: 1–8.
44. Ma, L., Chen, F., Li, Z. et al. 2014. Preparation and anticorrosion property of poly(2, 3-dimethylaniline) modified by nano-SiO_2. *Composites Part B: Engineering* 58: 54–58.
45. Weng, C. J., Huang, J. Y., Huang, K. Y. et al. 2010. Advanced anticorrosive coatings prepared from electroactive polyimide–TiO_2 hybrid nanocomposite materials. *Electrochimica Acta*, 55: 8430–8438.
46. Ershad-Langroudi, A., Rahimi, V. 2014. Effect of ceria and zirconia nanoparticles on corrosion protection and viscoelastic behavior of hybrid coatings. *Iranian Polymer Journal* 23: 267–276.
47. Arun, M., Kantheti, S., Ranganathan, R. et al. 2014. Surface modification of TiO_2 nanoparticles with 1, 3, 5-triazine based silane coupling agent and its cumulative effect on the properties of polyurethane composite coating. *Journal of Polymer Research*, 21: 600.

48. Sabzi, M., Mirabedini, S. M., Zohuriaan-Mehr, J. et al. 2009. Surface modification of TiO_2 nano-particles with silane coupling agent and investigation of its effect on the properties of polyurethane composite coating. *Progress in Organic Coatings* 65: 222–228.
49. Matin, E., Attar, M. M., Ramezanzadeh, B. 2015. Investigation of corrosion protection properties of an epoxy nanocomposite loaded with polysiloxane surface modified nanosilica particles on the steel substrate. *Progress in Organic Coatings* 78: 395–403.
50. Caldona, E. B., De Leon, A. C., Thomas P. G., et al. 2017. Superhydrophobic rubber-modified polybenzoxazine/sio_2 nanocomposite coating with anticorrosion, anti-ice, and superoleophilicity properties. *Industrial & Engineering Chemistry Research* 56: 1485–1497.
51. Weng, C. J., Peng, C.W., Chang, C. H. et al. 2012. Corrosion resistance conferred by superhydrophobic fluorinated polyacrylate–silica composite coatings on cold-rolled steel. *Journal of Applied Polymer Science* 126: E48–E55.
52. Derradji, M., Ramdani, N., Zhang, T. et al. 2016. Effect of silane surface modified titania nanoparticles on the thermal, mechanical, and corrosion protective properties of a bisphenol-A based phthalonitrile resin. *Progress in Organic Coatings* 90: 34–43.
53. Chen, L., Song, R. G., Li, X. W. et al. 2015. The improvement of corrosion resistance of fluoropolymer coatings by SiO_2/poly(styrene-co-butyl acrylate) nanocomposite particles. *Applied Surface Science* 353: 254–262.
54. Sasikumar, Y., Madhan Kumar, A., Gasem, Z. M. et al. 2015. Hybrid nanocomposite from aniline and CeO_2 nanoparticles: Surface protective performance on mild steel in acidic environment. *Applied Surface Science* 330: 207–215.
55. Palimi, M.J., Rostami, M., Mahdavian, M. et al. 2014. Application of EIS and salt spray tests for investigation of the anticorrosion properties of polyurethane-based nanocomposites containing Cr_2O_3 nanoparticles modified with 3-amino propyl trimethoxy silane. *Progress in Organic Coatings* 77: 1935–1945.
56. Ates, M., Kalender, O., Topkaya, E. et al. 2015. Polyaniline and polypyrrole/TiO_2 nanocomposite coatings on Al1050: Electrosynthesis, characterization and their corrosion protection ability in saltwater media. *Iranian Polymer Journal* 24: 607–619.
57. Kongparakul, S., Kornprasert, S., Suriya, P. et al. 2017. Self-healing hybrid nanocomposite anticorrosive coating from epoxy/modified nanosilica/perfluorooctyl triethoxysilane. *Progress in Organic Coatings* 104: 173–179.
58. Ejenstam, L., Swerin, A., Pan, J. et al. 2015. Corrosion protection by hydrophobic silica particle-polydimethylsiloxane composite coatings. *Corrosion Science* 99: 89–97.
59. Wang, N., Fu, W., Zhang, J. et al. 2015. Corrosion performance of waterborne epoxy coatings containing polyethylenimine treated mesoporous-TiO_2 nanoparticles on mild steel. *Progress in Organic Coatings* 89: 114–122.
60. Zhang, C., He, Y., Xu, Z. et al. 2016. Fabrication of Fe_3O_4@SiO_2 nanocomposites to enhance anticorrosion performance of epoxy coatings. *Polymers for Advanced Technologies* 27: 740–747.
61. Hernández-Padrón, G., Rojas, F., Castaño, V. 2006. Development and testing of anticorrosive SiO_2/phenolic–formaldehydic resin coatings. *Surface and Coatings Technology* 201: 1207–1214.
62. Zhou, C., Lu, X., Xin, Z. et al. 2014. Polybenzoxazine/SiO_2 nanocomposite coatings for corrosion protection of mild steel. *Corrosion Science* 2014; 80: 269–275.
63. Xu, J. L., Huang, Z. X., Luo, J. M. et al. 2014. Effect of titania particles on the microstructure and properties of the epoxy resin coatings on sintered NdFeB permanent magnets. *Journal of Magnetism and Magnetic Materials* 355: 31–36.
64. Bakhshandeh, E., Jannesari, A., Ranjbar, Z. et al. 2014. Anti-corrosion hybrid coatings based on epoxy–silica nanocomposites: Toward relationship between the morphology and EIS data. *Progress in Organic Coatings* 77: 1169–1183.

65. Golru, S. S., Attar, M. M., Ramezanzadeh, B. 2014. Studying the influence of nano-Al_2O_3 particles on the corrosion performance and hydrolytic degradation resistance of an epoxy/polyamide coating on AA-1050. *Progress in Organic Coatings* 77: 1391–1399.
66. Balaskas, A. C., Kartsonakis, I. A., Tziveleka, L. A. et al. 2012. Improvement of anti-corrosive properties of epoxy-coated AA 2024-T3 with TiO_2 nanocontainers loaded with 8-hydroxyquinoline. *Progress in Organic Coatings* 74: 418–426.
67. Palraj, S., Selvaraj, M., Maruthan, K. et al. 2015. Corrosion and wear resistance behavior of nano-silica epoxy composite coatings. *Progress in Organic Coatings* 81: 132–139.
68. Xu, F. G., Wang, C. Y., Li, D. et al. 2015. Preparation of modified epoxy–SiO_2 hybrid materials and their application in the stone protection. *Progress in Organic Coatings* 81: 58–65.
69. Yu, Z., Di, H., Ma, Y. et al. 2015. Preparation of graphene oxide modified by titanium dioxide to enhance the anti-corrosion performance of epoxy coatings. *Surface and Coatings Technology* 276: 471–478.
70. Liu, L., Cui, Y., Li, Y. et al. 2012. Failure behavior of nano-SiO_2 fillers epoxy coating under hydrostatic pressure. *Electrochimica Acta* 62: 42–50.
71. Ghanbari, A., Attar, M. M. 2015. A study on the anticorrosion performance of epoxy nanocomposite coatings containing epoxy-silane treated nano-silica on mild steel substrate. *Journal of Industrial and Engineering Chemistry* 23: 145–153.
72. Hosseini, M. G., Sefidi, P. Y. 2017. Electrochemical impedance spectroscopy evaluation on the protective properties of epoxy/DBSA-doped polyaniline-TiO_2 nanocomposite coated mild steel under cathodic polarization. *Surface & Coatings Technology*. doi:10.1016/j.surfcoat.2017.10.043.
73. Bhat, S. I., Ahmad, S. 2018. Castor oil-TiO_2 hyperbranched poly (ester amide) nanocomposite: a sustainable, green precursor-based anticorrosive nanocomposite coatings. *Progress in Organic Coatings* 123: 326–336.
74. Pour, Z.S., Ghaemy, M., Bordbar S., et al. 2018. Effects of surface treatment of TiO_2 nanoparticles on the adhesion and anticorrosion properties of the epoxy coating on mild steel using electrochemical technique. *Progress in Organic Coatings* 119: 99–108.
75. Ghiyasi, S., Sari, M. G., Shabanian, M., et al. 2018. Hyperbranched poly(ethyleneimine) physically attached to silica nanoparticles to facilitate curing of epoxy nanocomposite coatings. *Progress in Organic Coatings* 120: 100–109.
76. Torrico, R. F.A.O., Har, S.V., Trentin, A., et al. 2018. Structure and properties of epoxy-siloxane-silica nanocomposite coatings for corrosion protection. *Journal of Colloid and Interface Science* 513: 617–628.
77. Moghadam, F., Omidkhah, M., Vasheghani-Farahani, E., et al. 2011. The effect of TiO_2 nanoparticles on gas transport properties of Matrimid5218- based mixed matrix membranes. *Separation and Purification Technology* 77: 128–136.
78. Lua, A. C., Shen, Y. 2013. Preparation and characterization of polyimide–silica composite membranes and their derived carbon–silica composite membranes for gas separation. *Chemical Engineering Journal* 220: 441–451.
79. Vladimirov, V., Betchev, C., Vassiliou, A. et al. 2006. Dynamic mechanical and morphological studies of isotactic polypropylene/fumed silica nanocomposites with enhanced gas barrier properties. *Composites Science and Technology* 66: 2935–2944.
80. Zhu, A., Cai, A., Zhang, J. et al. 2008. PMMA-grafted-silica/PVC nanocomposites: Mechanical performance and barrier properties. *Journal of Applied Polymer Science* 108: 2189–2196.
81. Vassiliou, A., Bikiaris, D., Pavlidou, E. 2007. Optimizing melt-processing conditions for the preparation of ipp/fumed silica nanocomposites: Morphology, mechanical and gas permeability properties. *Macromolecular Reaction Engineering*, 1: 488–501.
82. Takahashi, S., Paul, D. R. 2006. Gas permeation in poly(ether imide) nanocomposite membranes based on surface-treated silica. Part 1: Without chemical coupling to matrix. *Polymer* 47: 7519–7534.

83. Liang, X., King, D. M., Groner, M. D. et al. 2008. Barrier properties of polymer/alumina nanocomposite membranes fabricated by atomic layer deposition. *Journal of Membrane Science* 322: 105–112.
84. Zornoza, B., Irusta, S., Téllez, C. et al. 2009. Mesoporous silica sphere–polysulfone mixed matrix membranes for gas separation. *Langmuir* 25: 5903–5909.
85. Sadeghi, M., Semsarzadeh, M. A., Barikani, M. et al. 2011. Gas separation properties of polyether-based polyurethane–silica nanocomposite membranes. *Journal of Membrane Science* 376: 188–195.
86. Cong, H., Hu, X., Radosz, M. et al. 2007. Brominated Poly(2,6-diphenyl-1,4-phenylene oxide) and Its Silica Nanocomposite Membranes for Gas Separation. *Industrial & Engineering Chemistry Research* 46: 2567–2575.
87. Ortenzi, M. A., Basilissi, L., Farina, H. et al. 2015. Evaluation of crystallinity and gas barrier properties of films obtained from PLA nanocomposites synthesized via "in situ" polymerization of l-lactide with silane-modified nanosilica and montmorillonite. *European Polymer Journal* 66: 478–491.
88. Sadeghi, M., Semsarzadeh, M. A., Moadel, H. 2009. Enhancement of the gas separation properties of polybenzimidazole (PBI) membrane by incorporation of silica nano particles. *Journal of Membrane Science* 331: 21–30.
89. Singha, S., Jana, T. 2014. Structure and Properties of Polybenzimidazole/Silica Nanocomposite Electrolyte Membrane: Influence of Organic/Inorganic Interface. *ACS Applied Materials & Interfaces*, 6 (23): 21286–21296.
90. Takahashi, S., Paul, D. R. 2006. Gas permeation in poly(ether imide) nanocomposite membranes based on surface-treated silica. Part 2: With chemical coupling to matrix. *Polymer* 47: 7535–7547.
91. Inagaki, N., Tasaka, S., Hiramatsu, H. 1999. Preparation of oxygen gas barrier poly(ethylene terephthalate) films by deposition of silicon oxide films plasma-polymerized from a mixture of tetramethoxysilane and oxygen. *Journal of Applied Polymer Science* 71: 2091–2100.
92. Inagaki, N., Cech, V., Narushima, K. et al. 2007. Oxygen and water vapor gas barrier poly(ethylene naphthalate) films by deposition of SiOx plasma polymers from mixture of tetramethoxysilane and oxygen. *Journal of Applied Polymer Science* 104: 915–925.
93. Semsarzadeh M. A., Ghalei, B. 2013. Preparation, characterization and gas permeation properties of polyurethane–silica/polyvinyl alcohol mixed matrix membranes. *Journal of Membrane Science* 432: 115–125.
94. Sadeghi, M., Talakesh, M. M., Behnam Ghalei, B. et al. 2013. Preparation, characterization and gas permeation properties of a polycaprolactone based polyurethane-silica nanocomposite membrane. *Journal of Membrane Science* 427: 21–29.
95. Chen, C. M., Chung, M. H., Hsieh, T. E. et al. 2008. Synthesis, thermal characterization, and gas barrier properties of UV curable organic/inorganic hybrid nanocomposites with metal alloys and their application for encapsulation of organic solar cells. *Composites Science and Technology* 68: 3041–3046.
96. Park, I., Peng, H., Gidley, D. W. et al. 2006. Epoxy–silica mesocomposites with enhanced tensile properties and oxygen permeability. *Chemistry of Materials* 18: 650–656.
97. Morel, F., Bounor-Legaré, V., Espuche, E. et al. 2012. Surface modification of calcium carbonate nanofillers by fluoro- and alkyl-alkoxysilane: Consequences on the morphology, thermal stability and gas barrier properties of polyvinylidene fluoride nanocomposites. *European Polymer Journal* 48: 919–929.
98. Khosravi A., Sadeghi, M., Banadkohi, H. Z. et al. 2014. Polyurethane-silica nanocomposite membranes for separation of propane/methane and ethane/methane. *Industrial & Engineering Chemistry Research* 53: 2011–2021.
99. Saito, R., Nakagawa, D. 2009. Synthesis of water vapor barrier membrane by coating of poly(vinyl alcohol) substrate with poly(tert-butyl acrylate)-silica nanocomposite. *Polymers for Advanced Technologies* 20: 285–290.

100. Yanaka, M. Henry, B. M., Roberts, A. P. et al. 2001. How cracks in SiOx-coated polyester films affect gas permeation. *Thin Solid Films* 397: 176–185.
101. Lee, C. H., Min, K. A., Park, H. B. et al. 2007. Sulfonated poly(arylene ether sulfone)–silica nanocomposite membrane for direct methanol fuel cell (DMFC). *Journal of Membrane Science* 303: 258–266.
102. Chen, Y., Wang, R., Zhou, J. et al. 2011. Membrane formation temperature-dependent gas transport through thermo-sensitive polyurethane containing in situ-generated TiO_2 nanoparticles. *Polymer* 52: 1856–1867.
103. Shen, Y., Lu, A. K. 2012. Structural and transport properties of BTDA-TDI/MDI co-polyimide (P84)–silica nanocomposite membranes for gas separation. *Chemical Engineering Journal* 188: 199–209.
104. Matteucci, S., Kusuma, V.A., Sanders, D. et al. 2008. Gas transport in TiO_2 nanoparticle-filled poly(1-trimethylsilyl-1-propyne). *Journal of Membrane Science* 307: 196–217.
105. Ameri, E., Sadeghi, M., Zarei, N. et al. 2015. Enhancement of the gas separation properties of polyurethane membranes by alumina nanoparticles. *Journal of Membrane Science* 479: 11–19.
106. Sadeghi, M., Afarani, H. T., Tarashi, Z. 2015. Preparation and investigation of the gas separation properties of polyurethane-TiO_2 nanocomposite membranes. *Korean Journal of Chemical Engineering* 32: 97–103.
107. Ghadimi, A., Mohammadi, T., Kasiri, N. 2014. A novel chemical surface modification for the fabrication of PEBA/SiO_2 nanocomposite membranes to separate CO_2 from syngas and natural gas streams. *Industrial & Engineering Chemistry Research* 53: 17476–17486.
108. Su, N. C., Smith, Z.P., Freeman, B. D. et al. 2015. Size-dependent permeability deviations from Maxwell's model in hybrid cross-linked poly(ethylene glycol)/silica nanoparticle membranes. *Chemistry of Materials* 27: 2421–2429.
109. Golzar, K., Amjad-Iranagh, S., Amani, M. et al. 2014. Molecular simulation study of penetrant gas transport properties into the pure and nanosized silica particles filled polysulfone membranes. *Journal of Membrane Science* 451: 117–134.
110. Zhou, H., Chen, Y., Haojun Fan, H. et al. 2008. The polyurethane/SiO_2 nano-hybrid membrane with temperature sensitivity for water vapor permeation. *Journal of Membrane Science* 318: 71–78.
111. Sadeghi, M., Taheri, H., Tarash, A. Z. 2015. Preparation and investigation of the gas separation properties of polyurethane-TiO_2 nanocomposite membranes. *Korean Journal of Chemical Engineering* 32: 97–103.
112. Swain, S. K., Prusty, G., Jena, I. 2013. conductive, gas barrier, and thermal resistant behavior of poly (methyl methacrylate) composite by dispersion of ZrO_2 nanoparticles. *International Journal of Polymeric Materials and Polymeric Biomaterials* 62: 733–736.
113. Prusty, G., Swain, S. K. 2013. Dispersion of ZrO_2 nanoparticles in polyacrylonitrile: Preparation of thermally-resistant electrically-conductive oxygen barrier nanocomposites. *Materials Science in Semiconductor Processing* 16: 2039–2043.
114. Erlat, A. G., Spontak, R. J. 1999. SiO_x Gas Barrier Coatings on Polymer Substrates: Morphology and Gas Transport Considerations. *The Journal of Physical Chemistry B* 103: 6047–6055.
115. Roberts, A. P., Henry, B. M., Sutton, A. P. et al. 2002. Gas permeation in silicon-oxide/polymer (SiO_x/PET) barrier films: role of the oxide lattice, nano-defects and macro-defects. *Journal of Membrane Science* 208: 75–88.
116. Dameron, A. A., Davidson, S. D., Burton, B. B. et al. 2008. Gas diffusion barriers on polymers using multilayers fabricated by Al_2O_3 and rapid SiO_2 atomic layer deposition. *The Journal of Physical Chemistry C* 112: 4573–4580.
117. Kwak, S., Jun, J., Jung, E. S. 2009. Gas barrier film with a compositional gradient interface prepared by plasma modification of an organic/inorganic hybrid sol–gel coat. *Langmuir* 25: 8051–8055.

118. Wu, D. S., Lo, W. C., Chiang, C. C. et al. 2005. Plasma-deposited silicon oxide barrier films on polyethersulfone substrates: temperature and thickness effects. *Surface and Coatings Technology* 197: 253–259.
119. Kisku, S. K., Dash, S., Swain, S. K. 2014. Dispersion of SiC nanoparticles in cellulose for study of tensile, thermal and oxygen barrier properties. *Carbohydrate Polymers* 99: 306–310.
120. Swaina, S. K., Dash, S., Behera, C. et al. 2013. Cellulose nanobiocomposites with reinforcement of boron nitride: Study of thermal, oxygen barrier and chemical resistant properties. *Carbohydrate Polymers* 95: 728–732.
121. Park, H. B., Jung, C. H., Kim, Y. K. et al. 2004. Pyrolytic carbon membranes containing silica derived from poly(imide siloxane): The effect of siloxane chain length on gas transport behavior and a study on the separation of mixed gases. *Journal of Membrane Science* 235: 87–98.
122. Tseng, I. H., Tsai, M. H., Chung, C. W. 2014. Flexible and transparent polyimide films containing two-dimensional alumina nanosheets templated by graphene oxide for improved barrier property. *ACS Applied Materials & Interfaces* 6: 13098–13105.

11 Properties of Polymer/Fiber/Ceramic Composites

11.1 INTRODUCTION

Owing to their superior mechanical properties, good flexibility, and high resistance to both corrosion and fatigue, fiber-reinforced polymer-laminated composites (FRPs) have evoked increased interest from automobile, aeronautical, marine, and construction industrial sectors. In addition, FRP composites are used in engineering materials in commodity applications like, sports and electronics, due to their design flexibility and ease of large processing. Despite their higher attributes, the relatively inferior interlaminar properties of FRPs have limited their uses in practical applications. The inferior interlaminar is the Achilles' heel generated by the relatively low strength and brittle nature of the resins that are conventionally utilized to construct a great majority of structural FRPs. These shortcomings not only decreased the satisfactory performance of FRPs in some applications, but it also influenced the strength of adhesively-bonded joints (ABJs).

Recently, the main properties that have been investigated for improving the delamination and fatigue resistance of FRPs are the interlaminar shear strength, fracture toughness, and fracture energy. The effect of inorganic particle-reinforcements on fracture behavior and interlaminar fracture toughness of fiber/polymer composites was widely studied and discussed in literature. Clays are equally used to mitigate the low fracture energy of several kind of FRPs composites, where great improvements were noticed. To overcome such disadvantages of FRPs composites, the addition of ceramic micro- and nanoparticles like alumina (Al_2O_3), silicon carbide (SiC), and silicon nitride (Si_3N_4), and fumed silica (SiO_2) appear to be one of the most prominent solutions. In order to improve the damping properties of the fiber composites without sacrificing the stiffness, nano-SiO_2 particles/rubber particles were incorporated into the epoxy resin/fiber unidirectional laminates through the sonication process. In this chapter, the various works reporting the usage of ceramic filler to reinforce fiber/polymer laminated composites will be summarized and discussed.

11.2 POLYMER/GLASS FIBER/CERAMIC COMPOSITES

Glass fiber–reinforced polymer (GFRP) composites have demonstrated striking performance combined with high strength, high stiffness, and low moisture absorption, which increases their uses for both structural and functional applications at room temperature (RT) or even at cryogenic temperatures. Polymer/glass fiber composites have been used in many industrial and engineering sectors such as automobile, industrial, marine engineering and others. This is mainly attributed to their lower cost and light-weight properties. Recent researchers further diversified the use of these fiber-based composites by modifying their chemical composition, changing their physical characteristics to provide self-healing properties by means of reinforcements for construction purposes. Use of glass fibers as reinforcements is not only limited to polymer matrices but also to ceramic matrix composites.

However, the interfacial bonding between glass fiber and the polymeric matrix is generally weak, thus interlaminar shear strength (ILSS) is a critical factor for these composites. In addition, the GFRP composites also exhibit low thermal conductivity, which reduces their uses in many aspects. Chemical modifications of glass fiber–reinforced polymers have increased the achievement of enhancing mechanical properties of plastics. Polymer/ceramics/glass fiber hybrid composites are the latest engineering materials showing improved wear- and erosion-resistant features.

A thermosetting epoxy/GF was reinforced by incorporating 9 wt.% of CTBN rubber microparticles and 10 wt.% of nano-SiO_2, where fatigue life of the composites was improved by 6 to 10 times [1]. Suppressed matrix cracking and a reduced crack-propagation rate were detected in the reinforced epoxy matrix, which caused by the different toughening micro-mechanisms induced by the presence of both the CTBN rubber microparticles and nano-SiO_2.

Tsai et al. [2] investigated the compressive strengths of GF/epoxy nanocomposites, containing various loadings ranged from 10 to 30 wt.% of spherical 25 nm-SiO_2 nanofillers. It was revealed that as the SiO_2 contents rose, the compressive strengths of the glass/epoxy composites were accordingly increased, since the enhancing mechanism was explained by the micro-buckling model. By adding 5 to 10 wt.% of 25-nm SiO_2 to epoxy resin/GF composite this increased the in-plane shear strength until the incremental of particle fractions [3]. In addition, results showed that the glass/epoxy composites filled with 20 wt.% nano-SiO_2 content had superior compressive strengths than that of unfilled composites.

In order to enhance the fracture toughness of the GF/epoxy composites without sacrificing their stiffness, Tsai et al. [4] used the nano-SiO_2 in conjunction with the rubber particles were introduced into the epoxy matrix. The fracture tests showed that the incorporation of nano-SiO_2 together with the CSR particle significantly increased the fracture toughness of the glass/epoxy composites by almost 82%. In addition, when the epoxy matrices were filled with CTBN rubber particles and nno-SiO_2, the enhancement of the interlaminar fracture toughness can reach only 48%.

To improve the inferior mechanical performance of glass fiber/polymer composites, ceramic micro- and nanoparticles have been widely introduced into these laminated composites. In this context, the effect of incorporating SiC filler at various ratios on the properties of chopped glass fiber–reinforced epoxy composites has been studied by Agarwal et al. [5]. The resulted data revealed that the physical and mechanical properties of SiC/glass fiber/epoxy composites are better than those of pure glass fiber/epoxy laminated composites. In addition, results showed that adding higher ratios of SiC provoked a degradation in energy-absorption capacity in these composites, which, in turn, affected their overall performance. However, filling with more than 10 wt.% SiC enhanced the elastic properties of the laminated composite.

The effects of adding 0, 3, 5, and 7 wt.% of SiC ceramic filler to unsaturated polyester/chopped glass-fiber composite was investigated by Musa et al. [6] they reported that the tensile, impact strength, and thermal conductivity increased with increasing the SiC content in the laminated composites. Padhi et al. [7] also found that glass fiber (GF) and blast-furnace slag particles (SiO_2 and Al_2O_3) reinforced epoxy-based hybrid composites and exhibited increased hardness, enhanced wear resistance, lower density, and improved dispersion morphology.

The effect of filling with TiC on the erosion behavior of a multiphase-laminated composite consisting of epoxy resin reinforced with E-glass fiber was discussed by Mohapatra et al. [8]. It was noted that with increasing TiC filler content, there was remarkable improvement in hardness and erosion-wear performance. It was also reported that the impact velocity was the most influencing factor followed by TiC ratio and impingement angle, while erodent size had only a slight effect on erosion of the reinforced composite. Glass/epoxy laminates fabricated by the hand lay-up technique and filled with SiC for 5 to 10% with graphite kept constant at 5% resulted in improved dry sliding-wear characteristics of glass [9]. The transfer film formed on the counter surface effectively improved the wear characteristics of filled glass/epoxy composites. In the earlier stage of wear, the fillers' contribution is significant, whereas the transfer film, debris formation, and fiber breakage accounts for the wear at much later steps.

Manjunatha et al. [10] incorporated 10 wt.% of nano-SiO_2 into an epoxy resin and observed an improvement in the tensile fatigue resistance of the resin. They also observed significant improvement of 300% to 400% in the fatigue life of the GF/EP laminate composite made by the reinforced resin compared against the performance of their unfilled laminates. They addressed these enhancements to two main energy-dissipating mechanisms, the first one involves particle debonding and the second deals with the subsequent plastic void growth in the matrix. Similar improvements in the fatigue properties of FRPs by reinforcing with of nano-SiO_2 were also reported by Blackman et al. [11]. However, when Akinyede et al. [12] added nano-Al_2O_3 filler into a polymeric matrix, no remarkable improvement in the fatigue life was detected, although, as discussed earlier, improvement in fracture toughness of resins was obtained by incorporating Al_2O_3 nanofiller.

Under static mechanical and tension–tension fatigue loading, Tate et al. [13] studied the role of adding 6, 7, and 8 wt.% nano-SiO_2 to glass-reinforced epoxy composites. Static mechanical tests such as tensile, flexure and short-beam strength for a 6 wt.% nano-SiO_2 composites revealed the highest increase in tensile strength, percentage elongation, and inter-laminar shear strength compared to the other concentrations and the control. The superhydrophobic glass fiber–reinforced epoxy plastic (GFRP) surfaces were prepared using spray coating a mixture on GFRP surfaces as reported by Sun and Wang [14]. The mixture was prepared from epoxy resin, curing agent, hydrophobic nano-SiO_2 (HSNPs), and acetone. By controlling the ratio of nano-SiO_2 in the solution, a tunable adhesive superhydrophobic surfaces was produced. As given in Figure 11.1, the as-developed specimen not only achieved superhydrophobicity but also offered huge discrepancy in adhesive abilities. For 3 days,

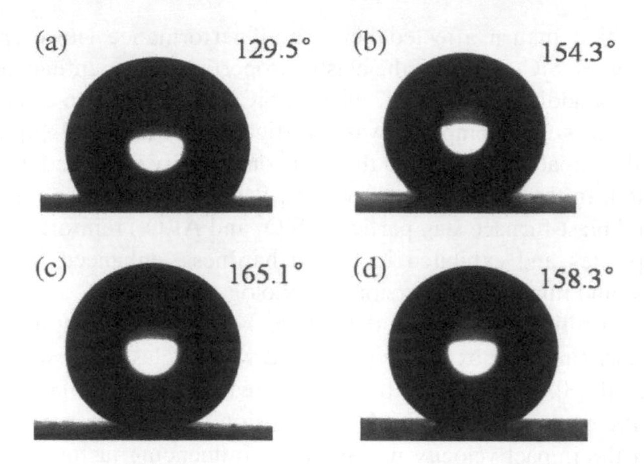

FIGURE 11.1 Water CA Images of Different Sample Surfaces (a) S1, (b) S2, (c) S3, and (d) S4. The Concentrations of Nano-SiO_2 Were Defined as S1 (1 mg·mL–1), S2 (3 mg·mL–1), S3 (5 mg·mL–1), and S4 (7 mg·mL–1) [14].

these superhydrophobic GFRP surfaces exhibited excellent stability in several kinds of solvents, including water (25°C), toluene, acetone, tetrahydrofuran, and ethanol.

The SiO_2/glass fiber/ epoxy nanocomposites were fabricated to investigate the effect of SiO_2 nanofillers on the mechanical properties of the laminated composite [15]. The experimental data revealed that good dispersion of nano-SiO_2 played a key role in promoting the mechanical performance of nanocomposites. The tests on the effects of nano-SiO_2 on the properties of glass-fiber composites revealed that the nano-SiO_2 can usually improve their properties, especially the bending strength that increased by almost 69.4% due to the promoted bonding forces between glass fibers and epoxy in the presence of nano-SiO_2. The epoxy matrix was reinforced with 15 wt.% 20 nm-SiO_2 and E-glass to improve unidirectional strength of epoxy/E-glass-laminated composites. Compression tests showed a great increase (40%) in elastic modulus of the reinforced epoxy. As illustrated in Figure 11.2, the incorporation of nano-SiO_2 markedly enhanced the longitudinal compressive strength and moderately increased the longitudinal and transverse tensile strengths of laminated composites [16].

The hybrid laminates reinforced with surface-functionalized BN showed an increase in tensile and flexural strength and modulus as well as in thermal conductivity compared to those filled with raw BN [17]. SEM photos indicated better compatibility and uniform dispersion of BN nanofiller in epoxy thermoset following either of functionalization route. The effect of adding of 0, 1, 2, 3, 4 and 5 wt.% nano-TiO_2 fillers on the epoxy/Glass fiber hybrid combined with woven roving and chopped strand mat reinforced composite. Inclusion of nano-TiO_2 up to 3% into the epoxy/hybrid glass fiber showed an enhancement in the tensile strength compared to unfilled composites laminates [18]. The improvement in tensile properties of glass fiber/epoxy composite was attributed to the presence of nano-TiO_2 which play the role of an interface material between glass fiber and epoxy thermoset.

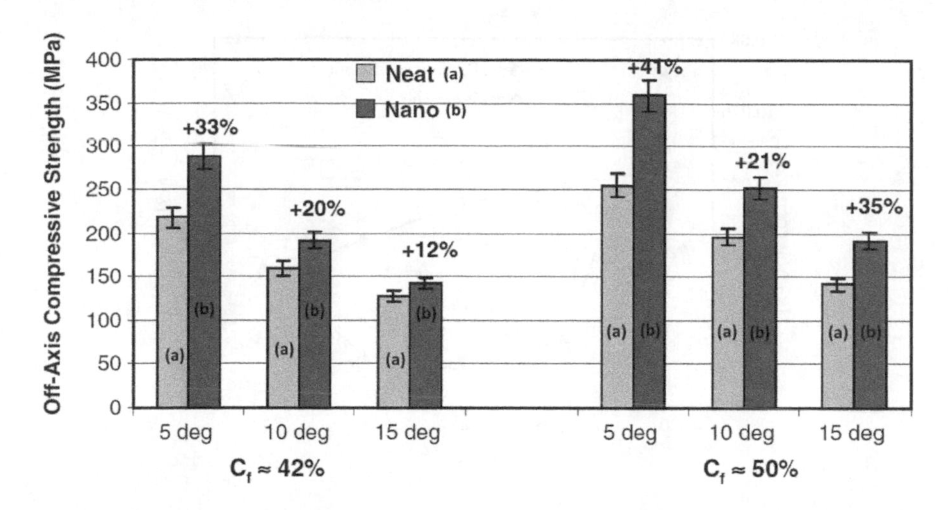

FIGURE 11.2 Off-axis Compressive Strengths of Neat and Nanophased Fiber Composites [16].

Silanized-ZrO_2 nanoparticles were incorporated in glass fiber/epoxy-laminated composites, which showed a marked effect of nano-ZrO_2 on tensile and flexural properties of hybrid multiscale glass fiber/epoxy composites [19]. A significant variation of tensile strength, stiffness, and toughness of ~27%, 62%, and 110% was obtained with respect to unfilled epoxy/glass fiber composites. Strength and modulus under bending were also increased by ~22% and ~38%, respectively, due to reduction in the interfacial delamination for the filled laminates. Epoxy/GF-laminated composites containing 9 wt.% of rubber microparticles and 10 wt.% of SiO_2 nanofillers were produced and tested for their fatigue properties [20]. The fatigue life of the epoxy/GF/SiO_2 composite reached about 4 to 5 times higher than that of unfilled epoxy/GF composite. Under spectrum loads, a fitting correlation was deduced between the experimental and predicted fatigue life data.

The fracture energies of glass fiber composites with an anhydride-cured epoxy matrix modified using core-shell rubber (CSR) particles and SiO_2 nanoparticles were investigated [21]. The composite toughness of epoxy/glass fiber/core-shell rubber/SiO_2 at initiation increased approximately linearly with increasing core-shell rubber/nano-SiO_2 filler loading, from 328 J/m^2 for the control to 842 J/m^2 with 15 wt.% of CSR particles. The effects of adding nano-TiO_2 on the properties of GF-PTFE/PF fabric composites were studied by Su et al. [22]. As shown in Figure 11.3, the wear rate of the nano-TiO_2-filled hybrid glass/PTFE fabric composites gradually reduced with increasing nano-TiO_2 up to 4% and then it was enhanced with the further increase of nano-TiO_2 loading. The improved tribological performance of 4 wt.% nano-TiO_2-filled GF-PTFE/PF hybrid composites were attributed to the improved structural integrity of the composites, the character of transfer film and the special anti-wear action of nano-TiO_2 during friction process.

New GF/unsaturated polyester laminates reinforced by TiO_2 were produced to evaluate the effects of the applied load, sliding velocity, and sliding distance on abrasive wear properties by Moorthy and Manonmani. The load and sliding velocity had

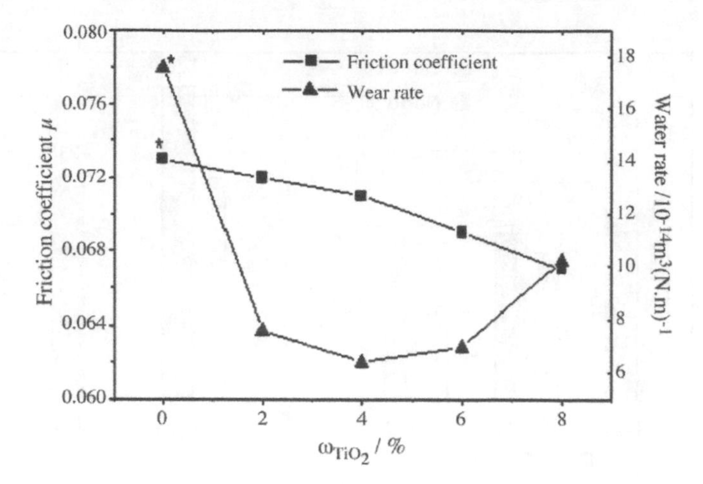

FIGURE 11.3 Effect of Nano-TiO_2 Content on the Friction Coefficient and Wear Rate of Hybrid Glass/PTFE Fabric Composites Filled with Nano-TiO_2. Dry Sliding Condition, 196.0 N, 0.26 m/s and Room Temperature [22].

more pronounced influence on the wear of composites [23]. Mohantyand et al. [24] claimed that the addition of 2 wt.% Al_2O_3 nanoparticles resulted in improved thermal stability and mechanical properties of short GF/CF-reinforced epoxy-based hybrid composites due to the better stress transfer properties from fiber and nanoparticle to the matrix, and to the presence of strong interfacial interactions between both the short GF/CF fibers and epoxy/Al_2O_3, which resisted fiber pull-out from the matrix.

GF/epoxy laminates were reinforced by Al_2O_3, SiO_2 and TiO_2 microparticles for improving their mechanical properties [25]. As depicted in Figure 11.4, both the flexural strength and flexural modulus of the hybrid composite reinforced with microsilica show the best results compared to those reinforced with Al_2O_3 or TiO_2

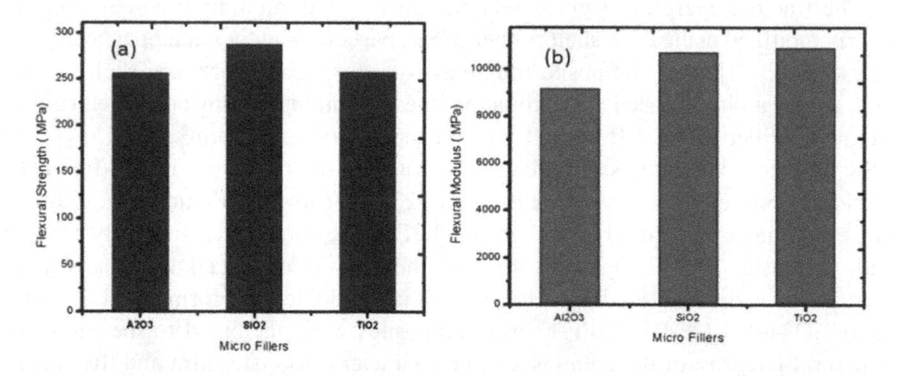

FIGURE 11.4 (a) Comparison of Flexural Strength of Epoxy/GF-hybrid Composites Versus Ceramic Microfillers; (b) Comparison of Flexural Modulus of Epoxy/GF-hybrid Composites Versus Ceramic Microfillers [25].

micro-fillers. Al_2O_3-filled GF/epoxy composites enhanced the hardness and impact energy due to their larger particle size. The mode of failure was identified as a combination of crack in matrix, matrix/fiber debonding, and fiber pull-out for these composites.

The erosion wear behavior and mechanical properties of AlN reinforced GF/epoxy resin composite were studied by Panda et al. [26]. Results revealed that by increasing the AlN loadings, a significant improvement in hardness and erosion wear performance were observed, while the tensile strength was slightly decreased. The erosion of the reinforced aminated composite was mainly governed by the initial impact velocity and filler contents.

The characteristics of GF/UPE composites filled with 10 wt.% TiO_2 filler were explored. Chemical resistance and thermal property of GF/UPE composites were increased due to the addition of TiO_2 reinforcement [27]. A review on the characterization of interfaces in glass fiber and particulate SiO_2-reinforced organic polymers was conducted by DiBenedetto [28].

The fatigue properties of epoxy/GF-laminated composites filled with 0.3 wt.% SiO_2 and MWCNT were evaluated by Böger et al. [29]. Figure 11.5 reveals that the addition of silica nanoparticles lead to increases in inter-fiber fracture strength of up to 16% compared to only 8% when MWCNT were added. More significantly, the high-cycle fatigue life was enhanced by several orders of magnitude in number of load cycles. The apparent density of hybrid epoxy/GF laminates increased by 0.62% when silica was added at the upper-beam side [30]. At the upper-beam side, the inclusions of SiO_2 particle led to an increasing in flexural modulus of 19.6%. The percent increases of 9.8% in impact resistance was found by adding 5 wt.% of SiO_2 particles.

The effect of using poly-diallyl dimethylammonium chloride (PDDA)-treated SiO_2 micro and nanoparticles on the apparent density, tensile and flexural strength

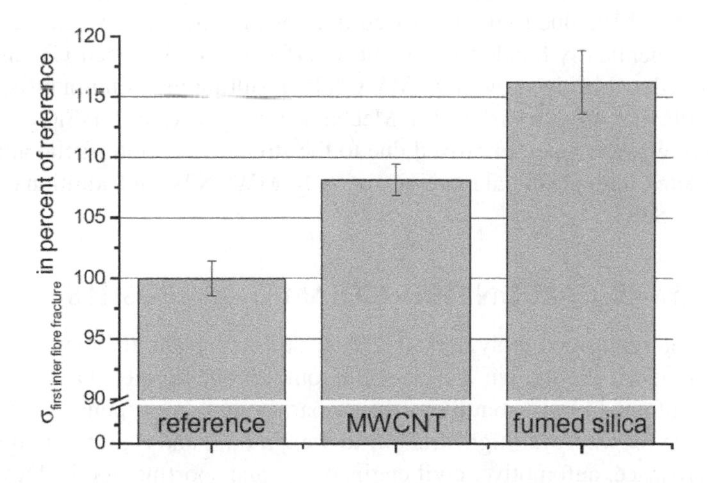

FIGURE 11.5 Stress for the Onset of First Inter-fiber Fracture in 90°-tensile Tests for the Unmodified and Nanoparticle-modified Matrices for GFRP Composites (Normalized) [29].

and modulus of hybrid GF-laminated composites was reported [31]. The addition of 2 wt.% of functionalized nano-SiO_2 reduced the flexural strength and density of the glass fiber laminated composites, while significantly increased both their tensile, flexural modulus and tensile strength. The ceramic particle infusions usually led to an enhanced apparent shear strength and adherent strength of the hybrid glass fibre reinforced composite. The addition of 5 wt.% ceramic particles into 600 g/m^2 cross-ply glass fiber composites increased both adherent and apparent shear strengths [32]. Detomi et al. [33] described the effect generated by adding microceramic particles on the flexural behavior of glass-fiber composites. The study identifies the use of 10 wt.% of silica microparticles in the upper side of the specimen as the best micromechanical configuration to give the highest mechanical performance in these composites.

In order to enhance the fracture toughness of the glass fiber/epoxy composites without losing their stiffness, the SiO_2 nanoparticles together with the rubber particles were concurrently included into the epoxy matrix to give a hybrid nanocomposite [34]. The fracture tests showed that the inclusion of nano-SiO_2 in conjunction with the CSR particle increased the fracture toughness of the glass/epoxy composites up to 82%. In addition, the interlaminar fracture toughness reached 48% as the epoxy matrices were simultaneously filled with CTBN rubber particles and SiO_2. Improvement in energy absorption and interfacial shear strength of epoxy/GF-laminated composites due to the presence of the nano-scale SiO_2 were quantified [35]. Inspection of the failure modes showed that the presence of colloidal silica increased the crack propagation along a more tortuous path within the interphase that generated progressive failure and rising the energy dissipation.

This research work investigates the effect of adding 0, 1, 2, and 3% nano-TiO_2 on the strength, from wear-out resistance and hardness of hybrid E-glass/epoxy composite [36]. The flexural properties of the glass fiber/epoxy increased with increasing the filling with TiO_2 nanofiller. At 3 wt.% of nano-TiO_2, the flexural strength of reached 203.36 MPa due to the improved interfacial bonding between the fiber and epoxy. The chemically-bonded nanoscale interfacial area between GF and matrix was generated and bridged by SiO_2–MWCNTs, resulting in strong interfacial adhesion and effective stress transfer [37]. Mechanical properties of GF/SiO_2–MWCNT/EP composites were much improved due to the strengthened interfacial adhesion in the composites, high chemical reactivity of SiO_2–MWCNTs, and additional reinforcing effect of SiO_2.

11.3 POLYMER/CARBON FIBER/CERAMIC COMPOSITES

Carbon fiber–reinforced polymers (CFRPs) still represent the best strength-to-weight ratio of all the known construction composite materials. They are stronger than GF-reinforced plastic, but they are comparatively more expensive. This makes them suitable for wherever high strength-to-weight ratio and rigidity are demanded, such as aerospace, automotive, civil engineering, and sporting goods. Despite their high initial strength-to-weight ratio, CFRP composites are not suitable for some structural applications due to their lack of a fatigue endurance and their weak out-of-plane properties, particularly the delamination resistance. Therefore, when using

CFRPs for critical cyclic-loading applications, engineers may need to include more strength safety margins to ensure suitable component reliability over a sufficiently long service life. One potential strategy to overcome this problem is the incorporation of micro- and nanoscale inorganic reinforcements, such as carbon nanotube, graphene, clays, and ceramic to these laminated composites [38].

The counterface topography plays an important role on the tribological film structure, and thereby influences significantly the tribological behavior of the PEEK hybrid nanocomposites, i.e. PEEK filled with nano-SiO_2/short carbon fibers and nano-SiO_2/short carbon fibers/PTFE/graphite [39]. Sintering of nanoparticles, oxidation of counterface steel and compaction of wear debris were identified to be competing mechanisms dominating the formation of tribofilm. Tribological behavior of epoxy/short carbon fibers/TiO_2-nanocomposite was also studied in [40].

Polyphenylene sulfide (PPS) composites filled with short carbon fibers (SCFs) (up to 15 vol.%) and sub-micro-scale TiO_2 particles (up to 7 vol.%) were prepared by extrusion and subsequently injection-molding. A synergistic effect of the incorporated short carbon fibers and sub-micro TiO_2 particles was resulted [41]. The lowest specific wear rate was obtained for the composition of PPS with 15 vol.% SCF and 5 vol.% TiO_2. The SEM photos in Figure 11.6 show that this hybrid reinforcement

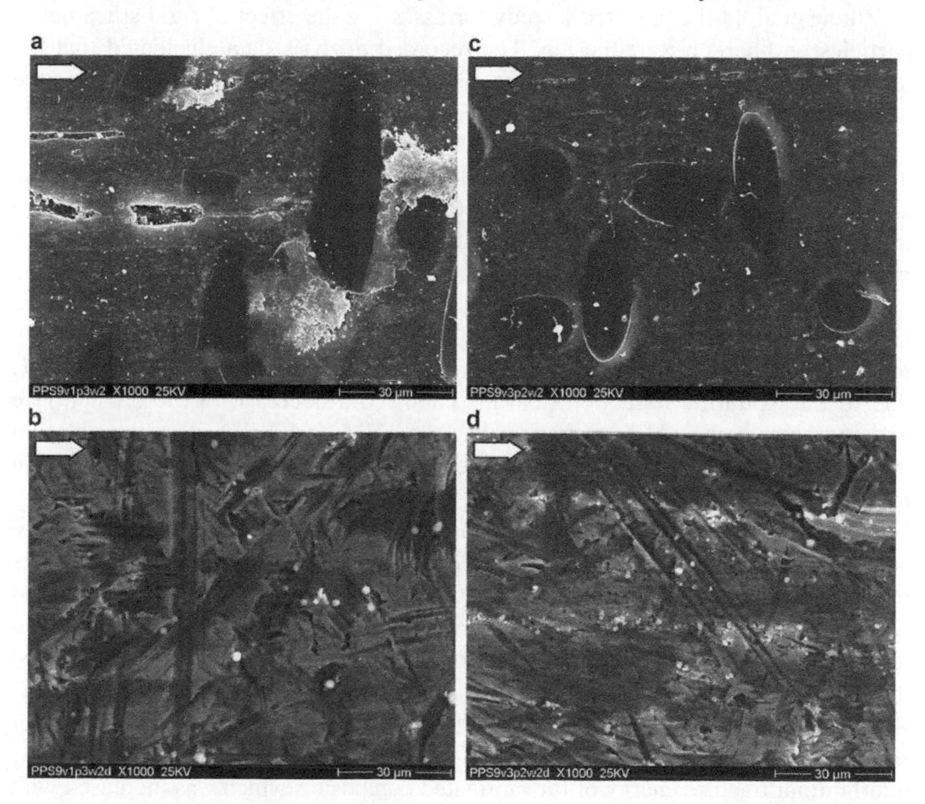

FIGURE 11.6 SEM Micrographs of PPS with 15 vol.% SCF and 5 vol.% TiO_2: (a) Worn Surface and (b) Wear Track at 3 × 1 MPa m/s; (c) Worn Surface and (d) Wear Track at 2 × 3 MPa m/s. The Arrows on the Top-left Corner Indicate the Sliding Direction of the Pin [41].

could be argued in terms of a positive rolling effect of the TiO_2 particles between the two sliding surfaces, which protected the short carbon fibers from being pulled out of the PPS matrix.

The tribological performances of epoxy/TiO_2 nanocomposites combined with short carbon fiber were studied by Chang et al. [42]. The experimental results indicated that the addition of nano-TiO_2 could apparently reduce the μ, and consequently reduced the contact temperature of short carbon fiber–reinforced epoxy composites, which significantly enhanced wear resistance of laminated composites, especially at high contact pressures and sliding velocities.

In a similar work [43], epoxy matrices filled with hybrid filler composed of copolymer of styrene and maleic anhydride-grafted-nano-SiO_2 particles and short pitch-based carbon fiber were produced. The composite containing 4 wt.% nano-SiO_2 and 6 wt.% CF afforded the most pronounced improvement of wear-resistance performance. The use of lower filler loading eased the processing of these composites in practice. The addition of nano-SiO_2 was responsible for the wear-resisting, friction-reducing results; the increase in nanofillers improved the surface hardness and enhanced the lubricating effect of the sheet-like wear debris, which quickly formed the protective transfer film.

Zhang et al. [44] conducted a study for assessing the effect of rigid silica nanoparticles on fiber–epoxy adhesion. The recorded analysis data elucidated that the reinforcing epoxy/carbon fiber laminates using nano-SiO_2 had a slight effect on the interfacial bonding behavior between fibers and matrix.

The heat-resistant properties of methylphenyl silicone/carbon fiber composites were enhanced significantly by adding SiO_2-CNT between SiO_2-CNT. The addition of binary filler system of SiO_2-CNT to the fiber/matrix composites also improved the mechanical stress transfer from methylphenyl silicone matrix to CF reinforcement and mitigated stress concentrations [45]. The degradation kinetic parameters showed that thermal stability of the phenolic resin/carbon fiber increased with increasing of nano-ZrO_2 loading [46]. The degradation enthalpy for these composite filled with 7 wt.% of nano-ZrO_2 was lowered by almost 30%, due to the oxidation reduction, and also showed a decrease in the erosion rate and increase in the ablation resistance. The SEM and XRD after the ablation test proved the existence of a new ceramic phase on the CF due to the reduction reaction of nano-ZrO_2.

The effect of adding nano-SiC combined with chopped carbon fiber to reinforce epoxy thermoset has been investigated [47]. The mechanical and physical tests revealed significant improvements when fine SiC/carbon fiber were used compared with those of the pure carbon fiber/epoxy composite. The loading of 25 wt.% SiC also reduced the capacity of the energy absorption for these composites and increased their elastic behavior characteristics.

By filling the epoxy/carbon fiber composites with combined silica and core-shell rubber (CSR) nanoparticles resulted in a steady-state propagation value of the interlaminar fracture energy of the laminated composites which was increased with enhancing the nanofillers loading [48]. The toughening effects of adding both the nano-SiO_2 and CSR nanoparticles on the laminated epoxy/carbon fiber composites, compared with the unmodified epoxy thermoset, were still evident even at the lower temperature. Indeed, the toughening effect of the nano-SiO_2 was higher at −80°C than at room temperature.

The microstructure and fracture performance of carbon fiber–reinforced epoxy resin filled with nnao-SiO_2 and/or polysiloxane CSR particles were studied in the work of Carolan et al. [49]. Results revealed that the values of toughness of the epoxy/CF laminates, compared to that of bulk epoxy resin, were further increased by extra fiber-based toughening mechanisms, such as fiber bridging, fiber debonding and fiber pull-out. The compressive property of epoxy/CF modified by different types of nanoparticles was first investigated, and the recorded results confirmed that the effect of various SiO_2 contents on the compressive property of CF/epoxy composites was more pronounced than that in pure epoxy polymers [50]. The incorporation of nano-SiO_2 also had a marked strengthening effects on the compression and flexural responses of CF fabrics/epoxy composite laminates.

Another efficient method for improving the interlaminar properties of FRPs and adhesively-bonded joints (ABJs) is the interleaving technique, in which microsize particles as reinforcing phases were included to polymeric matrices. For instance, Taheri [51] discovered that inclusion of 1 wt.% of SiC whiskers significantly increased the shear strength and energy-absorption capacity of ABJs, especially at subfreezing temperatures. However, the author reported that cracking of the whiskers led to a brittle fracture, which limited the effectiveness of the particles.

The effect of other types of rigid SiO_2 nanoparticles on fracture toughness and strength has also been examined by Zeng et al. [52]. Results revealed that by combining both SiO_2 with rubber nanoparticle types, the margin of improvement fell in between the margins obtained by using the individual nanofiller types. This results were attributed the improvement of interlaminar toughness to the cavitation of nano-rubber particles, void growth, and debonding of nano-SiO_2 from the epoxy matrix as shown in Figure 11.7. Hsieh et al. [53] also used rubber/SiO_2 hybrid nanofiller to reinforce CFRP composite, and observed that the inclusion of rubber NPs led to higher fracture toughness than SiO_2 did. Other similar studies regarding nano-silica particles can be found in references [54–62].

High-resolution 2D photoluminescent maps were providing on-site quality control options, capable of quickly and noninvasively providing feedback on Al_2O_3 nanoparticle dispersion and sedimentation in hybrid carbon fiber composites [63]. Lead zirconate titanite (PZT) piezoelectric ceramic particles were used to reinforce of carbon fiber-reinforced polymer matrix. The mode I fracture toughness of the composites degraded with the including of PZT particles. When both PZT and CNT binary fillers were incorporated into the laminated matrix, the mode I fracture toughness slightly enhanced [64]. However, the mode II fracture toughness of the reinforced composites was increased in all three cases. The effect of chemically functionalized nano-Al_2O_3 on the performance of CFRP composite materials has been investigated by Shahid et al. [65].

The transverse tensile properties of SiO_2 reinforced carbon fiber/epoxy (CF/EP) laminates and mode I interlaminar fracture toughness (G_{IC}) increases with increasing

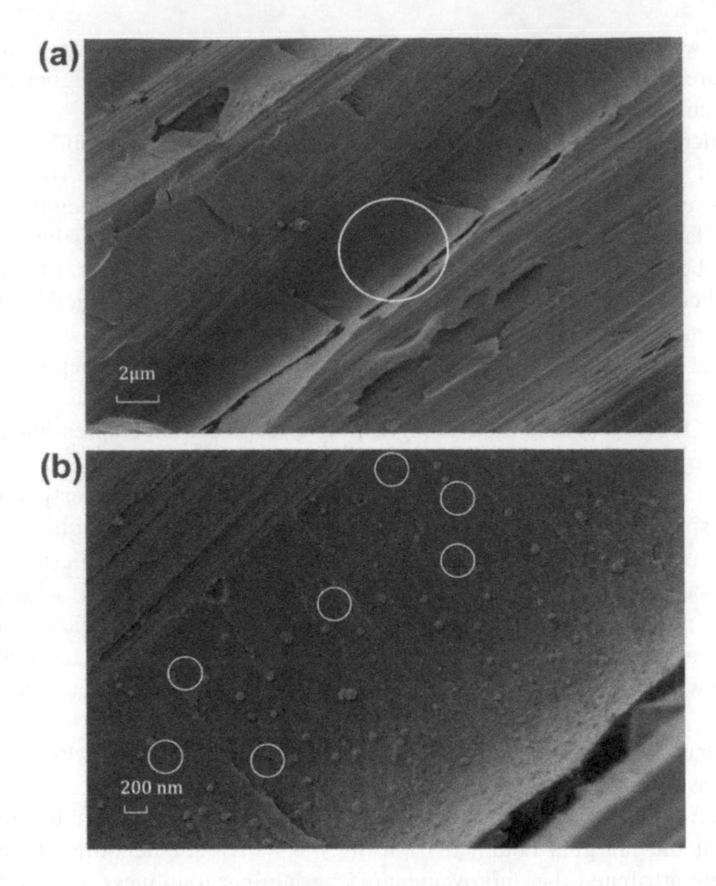

FIGURE 11.7 Scanning Electron Micrographs of Fracture Surface of CF laminates with a 6 wt.% Nano-silica Modified Epoxy Matrix. (a) 10,000 × and (b) 50,000 × [52].

nano-SiO_2 loading in the matrix resins as depicted in Figure 11.8. However, the ILSS and the mode II interlaminar fracture toughness (G_{IIC}) of SiO_2 reinforced carbon fiber/epoxy (CF/EP) decreases as the content of SiO_2 increases [66]. The reduced G_{IIC} value was related to two major competing mechanisms: formation of zipper-like pattern associated with matrix micro-cracks aligned 45° ahead of the crack tip, while the second was simply the shear failure of matrix. The ratio of G_{IIC}/G_{IC} was reduced with rising nano-SiO_2 contents in the CF/EP matrix.

The thermal properties of a CF/epoxy composite filled with nanosized B_4C based nanocomposites was evaluated [67]. The presence of the ceramic filler did not change the mechanical properties, viscosity, or the workability of the CF/epoxy laminates. The recorded data reflected that B_4C allowed keeping a residual structural integrity of the composite after burning due to the fact that chemical reactions occurred in these ceramic fillers at high temperatures; the presence of boron carbide reduced the peak of heat release rate especially at higher heat-fluxes and enhanced the thermal stability of the composite protecting and decelerating the thermal oxidation of the CF (Figure 11.9). The compressive strength and modulus of the nano-Al_2O_3

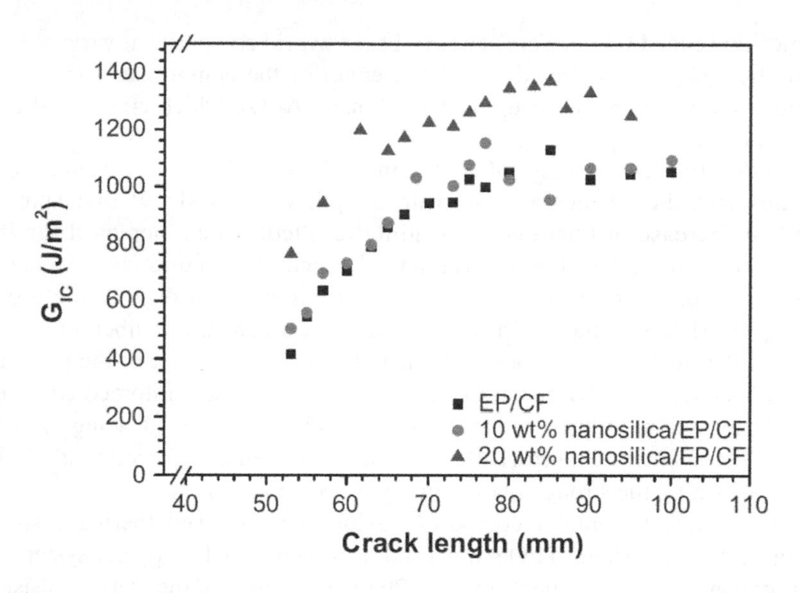

FIGURE 11.8 Typical R-curves for Mode I Interlaminar Fracture of CF/EP Laminates [66].

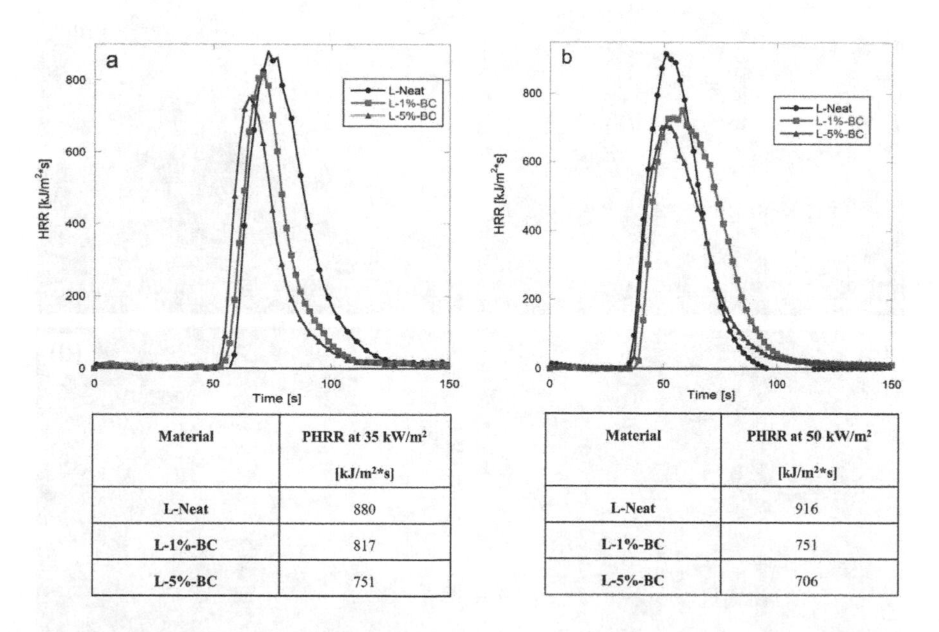

Material	PHRR at 35 kW/m² [kJ/m²*s]
L-Neat	880
L-1%-BC	817
L-5%-BC	751

Material	PHRR at 50 kW/m² [kJ/m²*s]
L-Neat	916
L-1%-BC	751
L-5%-BC	706

FIGURE 11.9 Heat Release Rate (HRR) of L-Neat (●), L-1%-BC (■) and L-5%-BC (▲) Subjected to a Heat-flux of 35 kW/m² (a) and HRR of L-Neat (●), L-1%-BC (■) and L-5%-BC (▲) Subjected to a Heat-flux of 50 kW/m² (b). L-Neat: Epoxy/CF; L-1%-BC: epoxy/CF/1wt.%B_4C; L-5%-BC: Epoxy/CF/5wt.%B_4C [67].

ceramics-embedded CF-reinforced epoxy-based hybrid composite at various rations was studied [68]. It was found that the strength of the compressive strength and modulus both were enhanced up to 2 wt.% nano-Al_2O_3, which considered as the optimum ratio.

The effect of incorporating TiC to the mesophase pitch-matrix precursor of CF laminate composites on interlaminar shear strength was studied [69]. In the presence of TiC, an increase of fiber/matrix bonding resulted, which increased the interlaminar shear strength of the material and changed the fracture mode. As shown in Figure 11.10, the neat materials reveal a strong pull-out on the fracture surface (Figure 11.10a), reflecting a weak bonding between carbon fiber and matrix. Figure 11.10c shows several holes constructed by matrix because of the pull-out of carbon fibers. On the other hand, the fracture surface of the reinforced composites demonstrated a joint failure of fiber and matrix (Figure 10b), evidencing the strong bonding between them. Figure 11.10d shows a greater detail of the joint failure fiber/matrix confirming the strong bonding in TiC-doped composites.

CF/phenolic/SiO_2 ceramer composites exhibiting improved thermal-resistance were fabricated at various TEOS was used as a monomer for sol–gel system. The incorporation of SiO_2 ingredients into the PF matrix increased the thermal resistance of the composites without sacrificing the flexural strength of the CF/PF/SiO_2 ceramer composites up to 60 wt.% [70].

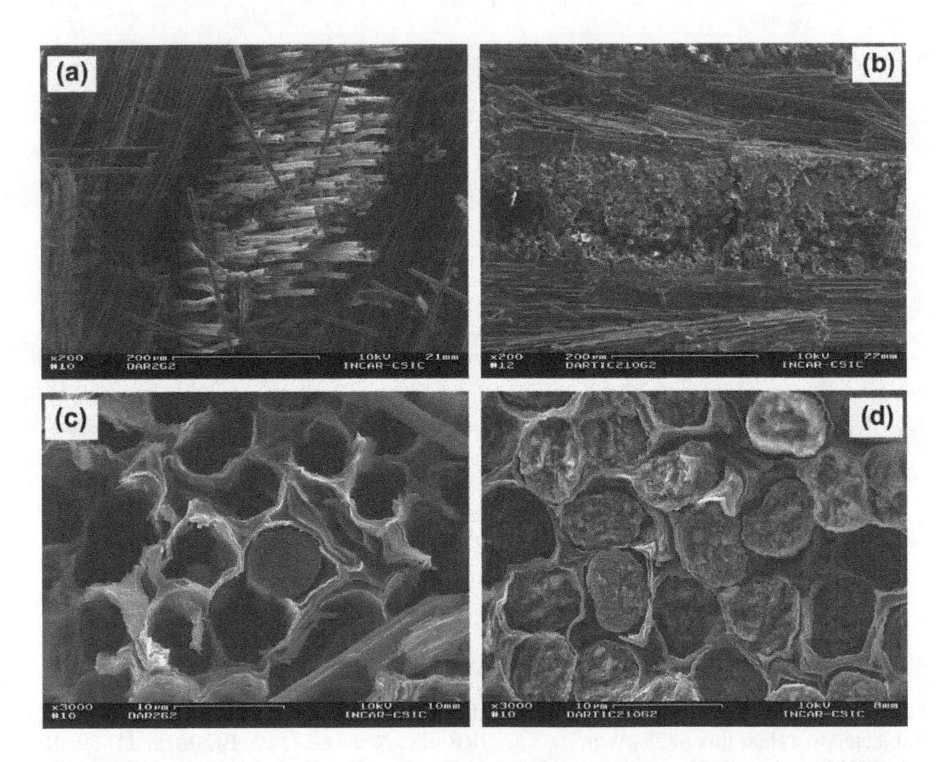

FIGURE 11.10 SEM Micrographs of Fracture Surfaces by Flexure (XZ Plane): (a) and (c) Epoxy/CF Composites; (b) and (d) Epoxy/CF/TiC Composites [69].

The wear properties of carbon fabric reinforced phenolic (CFRPh) composites filled with nano-Al_2O_3 were investigated by Zhang [71]. The author found that the addition of nano-Al_2O_3 effectively decreased the friction coefficient and wear rate of the laminated composites. In addition, the wear resistances of the neat and filled CFRPh composites were suggested to be dependent on the stability of the transfer tribo-film on the counterface.

The thermal conductivity of the SiC/pitch-CF/epoxy composite was studied by Muna et al. [72] and reported that due to the increase in the contact points of pitch-CF, stemming from SiC particulates, this increased the thermal conductivity of SiC/pitch-CF by 38% compared with that of uncoated pitch-based CF at 60 wt.% filler loading.

In a study, He et al. [73] fabricated the PI/SiO_2 nanocomposite coated onto CF by using electrophoretic deposition (EPD) to protect against the thermal oxidation of carbon fiber (Figure 11.11). The fabricated CF/PI/SiO_2 composites can be used in several potential applications as functional material, such as photocatalysis, self-cleaning coatings, microwave absorption, and anti-bacteria coatings.

The effects of nanoparticles on the mechanical and wear properties of the PEEK/CF composites filled with nano-ZrO_2 were studied [74]. The results showed that the addition of ZrO_2 nanoparticles PEEK/CF laminates had noticeably improved the tensile performance of these composites (Figure 11.12). Other synergistic improvements were recorded for the wear-resistant properties under water conditions, in particular when high pressures were used. The addition of nano-ZrO_2 effectively reduced the CF failures by decreasing the stress concentration on these fibers interface and the shear stress between two sliding surfaces.

Hussain et al. [75] reported the improvement of the mechanical properties of CFRPs by the addition of nano- and micro-sized Al_2O_3 ceramics. The inclusion of these particles provided a higher fracture toughness by enhancing significantly the

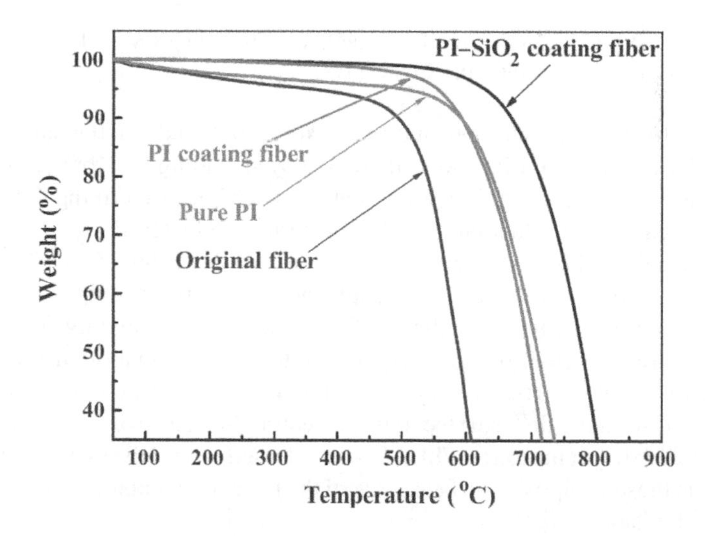

FIGURE 11.11 TGA Curves of Original Carbon Fiber, Pure Polyimide, Polyimide-coating Carbon Fiber, and the PI–silica Hybrid Nanocomposite-coating Carbon Fiber [73].

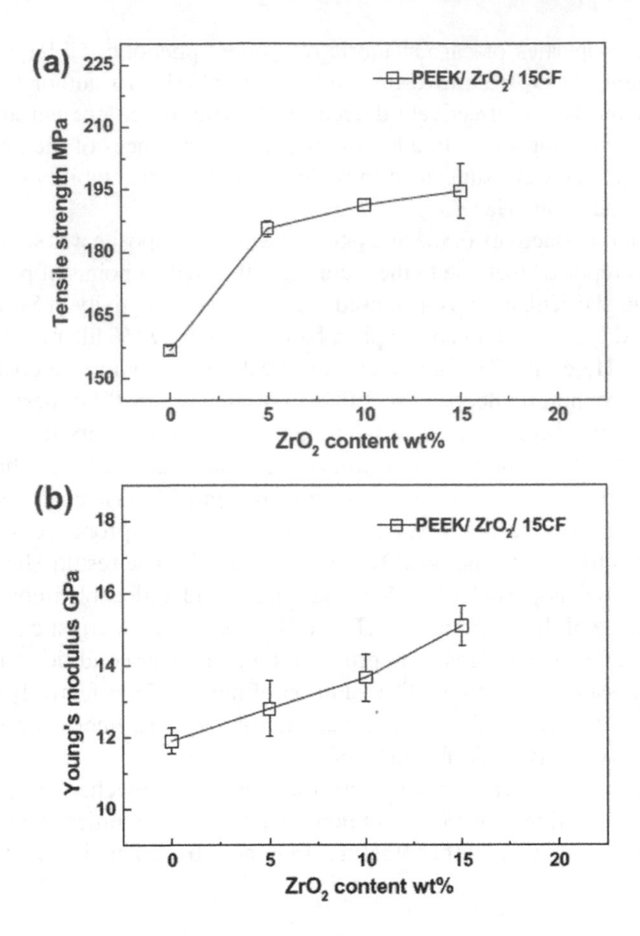

FIGURE 11.12 Variation of (a) Tensile Strength, and (b) Young's Modulus of the 15% CF Composites as a Function of the ZrO_2 Content [74].

toughness of the matrix. Tanimoto and co-workers [76] found that the damping properties and fatigue life of CFRPs were increased by spreading PZT particles between their interlayers. This concept is a prominent approach because mixing multiple constituents can ease the production of higher-performance materials.

Under mixed and boundary lubrication conditions, the filling of EP/CF laminates with SiO_2 nanoparticles reduced the friction and wear of their composites [77]. The recorded wear improvements of the CF/EP composites was mainly related to the high abrasion resistance of the SiO_2/CF and the tribofilm formation. It was reported that the released SiO_2 particles can be mixed with other wear products and produce more stable films at the disk surface, thus preventing further severe oxidational wear of polymer/CF/SiO_2 composites [78]. Also the released wear debris can be embedded within CF at these composite surface to avoid the fiber fragmentation and subsequent third body abrasion.

The tribological behaviors of SiO_2 and ZrO_2 ceramic oxide nanoparticles incorporated into PEEK/CF composites were evaluated [79]. When sliding took place at a low-load and -speed condition, the addition of ZrO_2 nanoparticles led to the formation of patch-like tribofilms enhancing friction and wear. However, at when a load ranged from 30 to 300 N m/s, the hard nanoparticles, i.e., SiO_2 and ZrO_2, resulted in dramatic improvement of the tribological properties. Moreover, nano-ZrO_2 was significantly more effective than nano-SiO_2 for improving the tribological performance. A synergistic effect was obtained for the combination of nano-Si_3N_4 and SCF and graphite, which lead to the best tribological properties of the PI matrix [80]. These composites had good tribological properties under oil lubrication and lower-wear properties under water lubrication compared to those tested under dry sliding condition.

The best wear-resistance performance was conducted by a combination of nano-TiO_2 with conventional fillers; as an example, epoxy + 15 vol.% graphite + 5 vol.% nano-TiO_2 + 15 vol.% short-CF exhibited w_s of 3.2×10^{-7} mm³/Nm, which is about 100 times lower when compared to that of unfilled epoxy [81]. Worn surfaces analysis confirmed that a mechanism of nanoscale rolling controlled this improvement.

It was found that nano-TiO_2 could effectively reduce the μ and wear rate of polyamide 66/SCF/graphite composites, especially under extreme conditions [82]. A positive rolling effect of the nano-TiO_2 within the material pairs can contribute to this remarkable enhancement of the load carrying capacity in these composites. As displayed in Figure 11.13 (b, d, f, h), a positive rolling effect of the TiO_2 nanoparticles is clearly detected. This rolling played the role of decreasing the shear stress and thus the contact temperature, especially at high sliding pressure and speed situations.

The wear behavior of polyetherimide (PEI)/nano-TiO_2 /SCF/ graphite was investigated [83]. Results showed that as nano-TiO_2 ratio increased, μ and the contact temperature of the composite were significantly reduced, especially under higher testing conditions. The roles of low-loading 1 vol.% 13 nm-SiO_2 particles (13 nm) on the tribological behavior of short carbon fiber (SCF)/PTFE/graphite (microsized) was studied [84]. Under 1 MPa, the abrasiveness generated by nano-SiO_2 agglomerates seems to accelerate SCF destructions while under 2 MPa, the nano-SiO_2 markedly decreased the wear rate of these composites. This effect is more remarkable under high pressures and especially at high sliding velocities. The protection of SCF/matrix interface by the SiO_2 nanoparticles was found to be the major mechanism for the wear resistance improvement.

New kinds of composite were fabricated by adding both Al_2O_3 particles at 2 wt.% and CF to the epoxy resin [85]. The recorded data revealed that longitudinal modulus has appreciably enhanced by adding Al_2O_3 particles. Both tensile strength and modulus were further improved for hybrid composites containing both Al_2O_3 fillers and CF. Figure 11.14 illustrates mechanisms of energy dissipation and stress transfer in the interphase layer of fiber composites in the absence and in the presence of nano-Al_2O_3. In the presence of uniformly dispersed nano-Al_2O_3, the rapidly growing cracks can absorb huge energy leading to the delayed mechanical failure of the composite. This mechanism provides larger fracture-surface area during the crack propagation and utilized a greater amount of energy, mainly due to the congested plastic deformation of matrix materials.

Tough CF/epoxy-laminated composites reinforced with liquid rubber and SiO_2 nanoparticles had been produced and studied [86]. The aggregation of nano-SiO_2

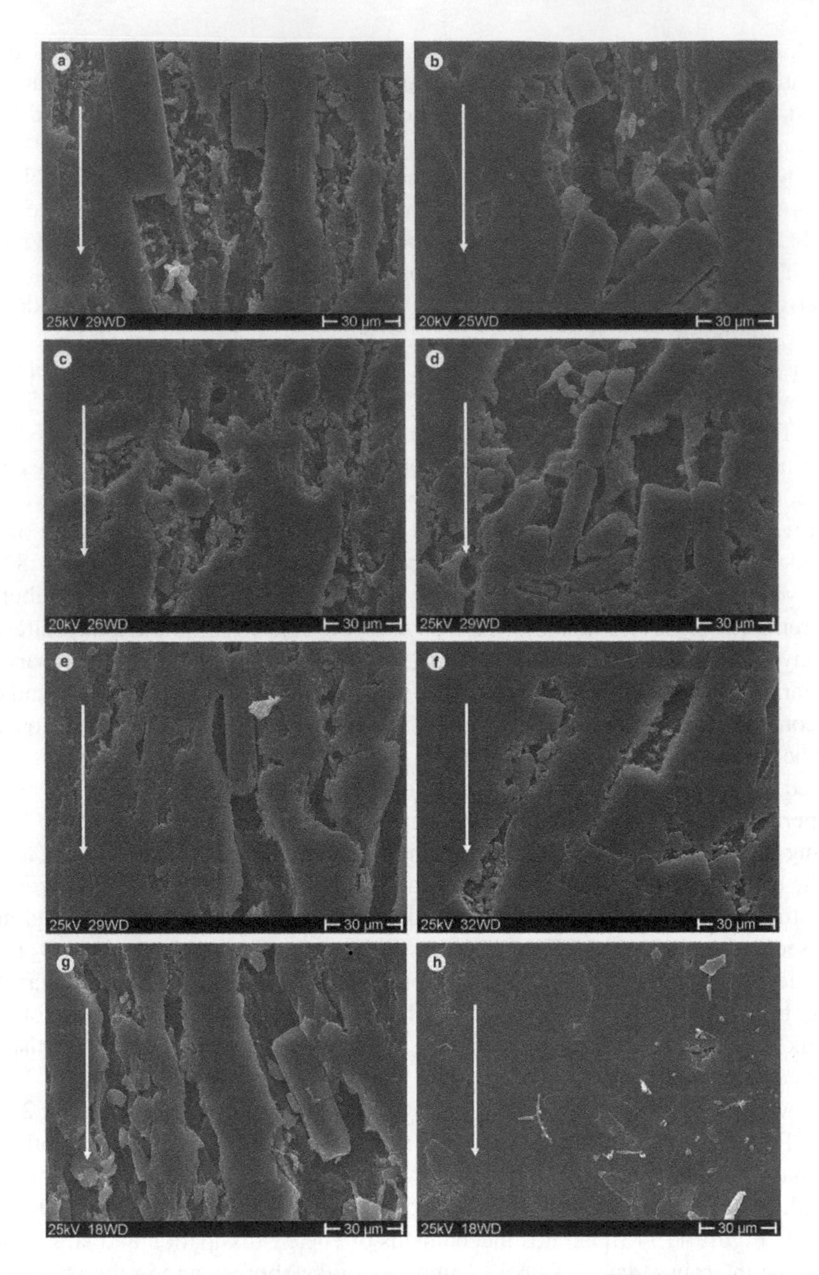

FIGURE 11.13 Comparisons of the Worn Surfaces of Graphite + SCF/PA66 After (a) 10 min (c) 30 min, (e) 60 min (g) 120 min, and that of Nano-TiO_2 + Graphite + SCF/PA66 after (b) 10 min (d) 30 min (f) 60 min (h) 120 min, Respectively [82].

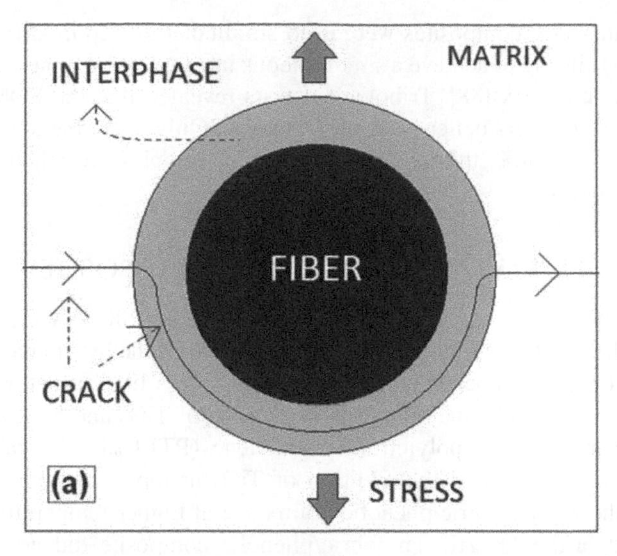

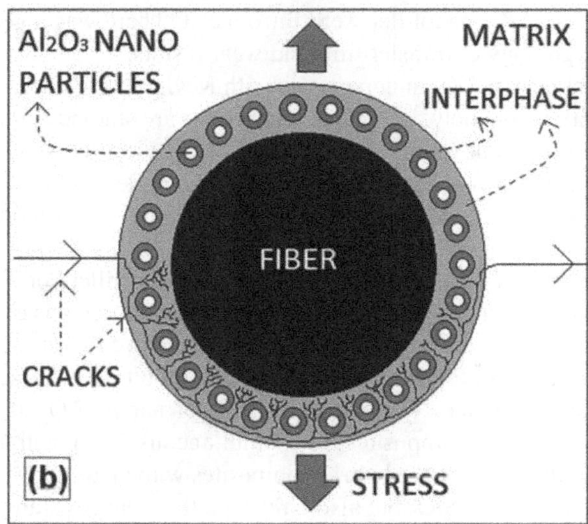

FIGURE 11.14 Schematic Illustration of Energy Dissipation and Stress Transfer in the Interphase Layer of Fiber Composites (a) Without and (b) With Nanoparticle-reinforced Epoxy Composite [85].

was found to negatively influence the mechanical performance of epoxy/CF composites. Low-velocity impact behavior of by 1 to 8 wt.%–treated-nano-SiO_2-reinforced CF/epoxy composites suitable for marine applications was reported by Landowski et al. [87] The most significant effect was the reduced damage; crack branching and deflection, due to the filling with nano-SiO_2, were also noticed. Results equally revealed a decrease in permanent deformation, by ~15%, and absorbed energy, by ~8%, as well as improved fiber/matrix interfacial strength in the presence of the silica nanofiller.

The CF/nano-SiO_2 composites were used simultaneously to modify polyoxymethylene (POM) aiming to achieve a simultaneous increase of toughness and modulus of the polymeric matrix [88]. Tribological tests revealed that the POM filled with 3 vol.% nano-SiO_2 offers better wear performance. Only when the loading of nano-SiO_2 was below 3 vol.%, the CF/POM/nano-SiO_2 samples presented lower μ and smaller wear volumes.

11.4 POLYMER/KEVLAR FIBER/CERAMIC COMPOSITES

Aramid fibers/SiO_2/shape-memory polyurethane (Aramid/SiO_2/SMPU) hybrid were produced [89]. The IFSS of the aramid fiber coated with the hybrid nanocomposites increased by 45%, which benefited from a special "pizza-like" structure on the fiber surface. Zhang et al. [90] studied the incorporation of TiO_2 and SiO_2 nanoparticles into hybrid composites of polytetrafluoroethylene (PTFE)/Kevlar fabric/phenolic systems. The results showed the addition of TiO_2 nanoparticles can decrease the wear rate of the Kevlar fabric/phenolic composite at higher temperatures, however the wear of hybrid PTFE/Kevlar fabric/phenolic composite did not significantly modified as TiO_2 or SiO_2 nanofiller were introduced. There was a good correlation between the morphology of transfer film and wear results.

The stab-resistance of SiO_2 nanoparticles with Kevlar fabric composites and SiO_2 nanoparticles with nylon, polyester fabric composites are studied and results showed a significant improvement over unmodified fabrics having the same areal density [91]. These results also revealed that this SiO_2 nanoparticles and Kevlar fabric composites can be used to prepare stab-resistant clothes, and it was more pronounced, lighter, flexible and lower cost than pure Kevlar fabric.

The combined effect of nano-SiO_2 and nano-Al_2O_3 as fillers on the tribological properties of Kevlar fabric/phenolic laminate (KFPL) in water was evaluated by Liu et al. [92]. Results showed that incorporating only nano-SiO_2 at a proper content decreased the friction coefficient of composites to an ultralow value, but it unfortunately increased wear rate. The single addition of nano-Al_2O_3 at a proper ratio decreased the wear rate of composites. The simultaneous filling with nano-SiO_2 and nano-Al_2O_3 not only provided the hybrid composites with an ultralow friction coefficient originated from nano-SiO_2 but also enhanced the wear resistance.

Alsaadi et al. [93] reported that the inclusion of SiO_2 nanoparticle in the Kevlar/carbon/epoxy laminated composites results in a significant enhancement for tensile and flexural strength, and this was related to the perfect adhesion of nano-SiO_2 with epoxy/fiber composite resulting in an enhancement in load transfer between particle and matrix. Combination of phenolic resin with poly (vinyl butyral), (PVB), is usually used as impregnation of aramid fabrics in prepregs for the ballistic protection armors. In their study, Simić et al. [94] scrutinized the effect of ceramic fullerene-like tungsten disulfide nanoparticles (IF-WS_2) in aramid/phenolic resin/PVB composite for enhancing its viscoelastic and mechanical properties. The presence of WS_2 particle in these composites increased the T_g and significantly improved mechanical properties.

Haro et al. [95] found that addition of ceramic micro- and nanofillers coating on Kevlar fabrics laminates is a prominent way for strengthening the interfacial bonding between the matrix and fibers in hybrid composite laminates, which improve the mechanical and ballistic properties. The modal characterization of aramid/CF- hybrid composites with SiO_2 nanoparticles was investigated through the analytical-experimental transfer-function method [96]. Results indicated that the addition of nano-SiO_2 filler enhanced the stiffness of the hybrid laminates, although the damping response of the reinforced composites was slightly improved.

The dry sliding-wear properties of hybrid PTFE/Nomex fabric/phenolic composite filled with aluminum diboride (AlB_2) was studied by a pin-on-disc tribo-meter [97]. The results showed that only 3 wt.% AlB_2 filled hybrid fabric composite demonstrated excellent wear-resistance properties. In addition, the transfer film formed on the counterpart pin surfaces was showed to be more effective in improving the tribological properties of the reinforced hybrid fabric composites. Hybrid PTFE/Kevlar fabric composite samples were reinforced with nano-Si_3N_4 and/or submicron size WS_2 as fillers. The tribological behaviors of these composites were evaluated by Li et al. [98]. The addition of single nano-Si_3N_4 fillers effectively decreased the wear rate of the hybrid composites, but they did not reduce the friction coefficient. However, hybrid Si_3N_4 and WS_2 fillers can significantly reduce the wear rate and friction coefficient of composites.

Yang et al. [99] found that the addition of 6 wt.% h-BN into air-plasma modified hybrid PTFE/Nomex fabric had the best high-temperature tribological properties. The morphological analysis on the counterpart pin surface supported the improved high-temperature lubricating role of h-BN. As shown in Figure 11.15, the transfer film on the worn surface of the pin sliding against composite B was rough, discontinuous and reflected a scuffing feature (see Figure 11.15a–c). However, the transfer film of composite A is much smoother and continuous (see Figure 11.15d–f). Nomex fabric composites were reinforced with polyfluo150 wax particles (PFW) and nano-SiO_2 using dip-coating of Nomex fabric in a phenolic resin/nano-SiO_2 and the followed

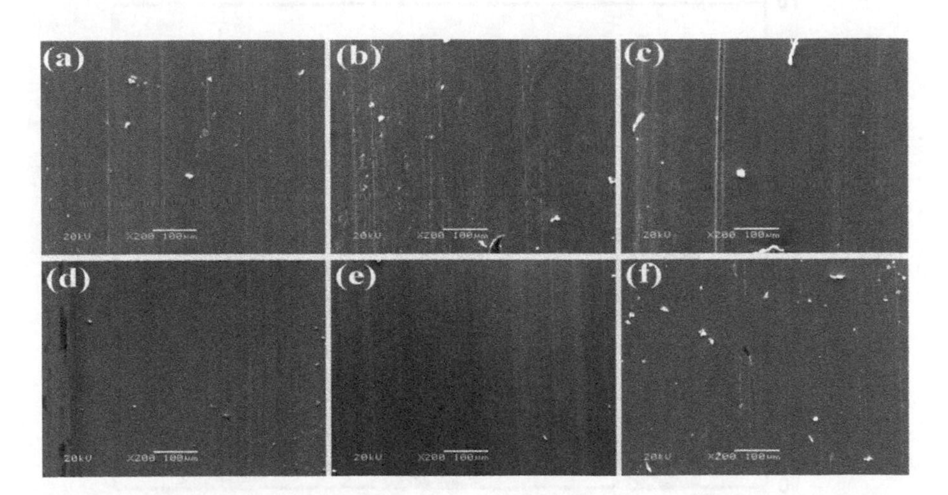

FIGURE 11.15 SEM Images of the Worn Surfaces of the Counterpart Pins Sliding Against 6 Wt.% H-BN-Reinforced Air-Plasma-Treated Hybrid Fabric Composites (Composite A) and Filled Air-Plasma-Treated Fabric Composite (Composite B) Under Different Environmental Temperature, Composite B (a) 240 °C, (b) 270 °C, and (c) 300 °C; Composite A (d) 240 °C, (e) 270 °C, (f) 300 °C (Applied Load 188 N and Sliding Speed 0.26 m/s) [99].

TABLE 11.1
Adhesion and Tensile Strength of the Unfilled and Filled Nomex Fabric Composites [100]

Composite	Tensile strength (MPa)	Adhesion strength (MPa)
Unfilled NFC	41.1	10.3
20%PFW/NFC	69.6	15.9
40%PFW/NFC	57.0	13.2
4%nano-SiO_2/NFC	65.1	13.7
8%nano-SiO_2/NFC	54.7	11.1

PFW: polyfluo150 wax particles; NFC: Nomex fabric composite.

by curing process [100]. The results revealed that incorporating PFW and nano-SiO_2 significantly increased the wear-resistance of fibers/phenolic laminate while reducing the friction coefficient, and these improvements were more substantial compared to the case where only PFW was used (Table 11.1).

Aiming to mitigate the poor interfacial strength and the UV resistance properties of the aramid fiber, TiO_2 nanoparticles were grown on their surfaces using low-temperature hydrothermal technique as reported by Wang et al. [101]. Results confirmed that Anatase TiO_2 nanoparticles were uniformly deposited on the aramid fiber surface while controlling the TiO_2 particle size upon addition of polyethylene glycol (PEG), and this reduced the stress concentration within fiber/matrix interface. Figure 11.16 reveals a

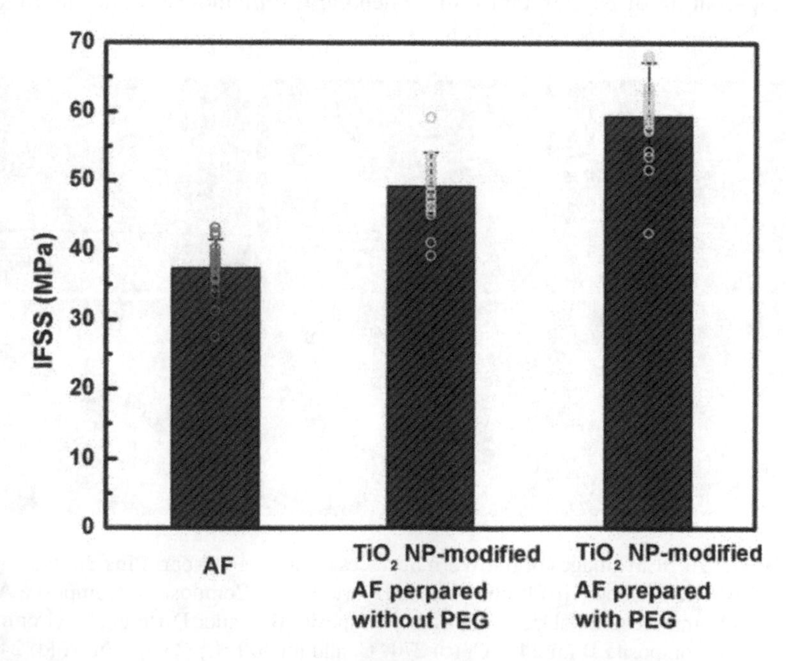

FIGURE 11.16 IFSS Values of AF and TiO_2 NP-modified Aramid Fiber (AF) Composites [101].

good enhancement in the interfacial shear strength (IFSS) upon adding TiO_2 nanofiller (by almost 40% to 67%) whereas retaining the basic strength of the fiber material.

The sol–gel process was used to produce various types of aramid/TiO_2 hybrid materials as reported by Ahmad et al. [102]. Result showed that thin transparent and tough films could be obtained with up to 30 wt.% TiO_2, where the values of the tensile strength in the case of reinforced hybrid composites was enhanced with the loading of nano-TiO_2, and the polyamide system with nonlinear end groupings showed more significant enhancement than did those having the linear chains ends. An advanced stab proof was developed from silica/ethylene glycol, using shear-thickening fluid (STF), and Kevlar fabric [103]. Results reflected that the STF markedly revealed the reversible liquid-solid transition at a certain shear rate. In addition, the STF impregnation significantly enhanced the stab resistance of Kevlar fabric against spike threats as shown in Figure 11.17. This increase the protection performance of Kevlar fabric that can be used a stab proof material.

In their work, Zhang et al. [104] used 3 wt.% SiO_2 nanoparticles to fill an epoxy/Kevlar laminate. The test results confirmed that epoxy reinforced by nano-SiO_2 is an efficient method to improve mechanical properties of such thermosets. The impregnated surface microstate of aramid/nm-SiO_2/epoxy composites are better than that of the matrix unmodified composite. The tensile strength and elongation at break of aramid/nm-SiO_2/epoxy is higher than that of aramid/epoxy by about 10.1% and 7.6%, respectively. The fracture modes of aramid/epoxy-laminated composites have been changed and the properties of composites can be enhanced through the incorporation of nano-SiO_2.

Mahfuz et al. [105] reported that the performance of the shear thickening fluid (STF) strengthened with Kevlar (45 wt.% PEG/55 wt.% silane treated nano-SiO_2/Kevlar)

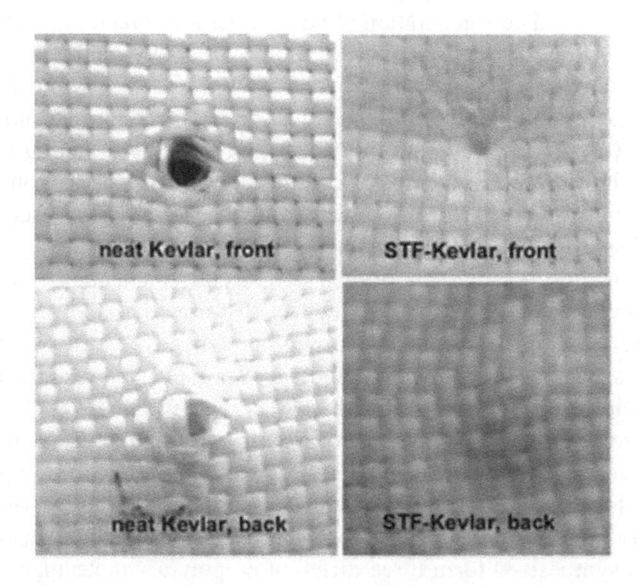

FIGURE 11.17 Photographs of Neat Kevlar Fabric and Fumed Silica/Kevlar Composite Fabric after Quasi-static Stab Testing [103].

armor composites increased significantly. As confirmed by National Institute of Justice spike tests, the energy needed for zero-layer penetration (i.e., no penetration) sharply raised by twofold: from 12 to 25 $J.cm^2/g$ due to the formation of siloxane (Si-O-Si) bonds between SiO_2 and PEG and superior coating of Kevlar filaments with these ceramic nanoparticles. It was also demonstrated that functionalization of SiO_2 nanoparticles followed by direct dispersion into PEG produced a superior Kevlar composites exhibiting much better spike resistance.

Aramid fiber/SiO_2 aerogel composites suitable for thermal insulation were produced under ambient pressure drying [106]. The microstructure showed that the aramid fibers were inlaid in the aerogel matrix, acting as the supporting skeletons, to strengthen the aerogel matrix. The developed aramid fiber/SiO_2 aerogels possessed extremely low thermal conductivity of 0.0227 ± 0.0007 W m^{-1} K^{-1} with the fiber content ranging from 1.5% to 6.6%. The effect of adding SiC particles and their role regarding the mechanical properties of the Kevlar 49 fiber-reinforced polymer composite at different fiber ratios was studied by Potluri et al. [107], using finite element analysis (FEA). The results suggested that SiC particles are a suitable reinforcement for increasing the transverse mechanical properties of the Kevlar fiber-reinforced epoxy composites.

11.5 POLYMER/OTHER FIBERS/CERAMIC COMPOSITES

Basalt fibers are widely used in concrete reinforcing applications due to its excellent mechanical properties and an environmentally friendly manufacturing process. The fibers exhibit a tensile strength superiors than that E-glass fibers, and several times higher than that of steel fibers. A newly developed basalt concrete called minibars (MB) has recently been developed in addition to plain chopped basalt fibers. The minibars are mainly a new generation of basalt fiber reinforced polymer rebar.

Graphite fabric (GrF) were used to reinforce an epoxy matrix as a binder and filled with different ratios h-BN in two sizes of 1.5 μm and 70 nm for producing high-performance composites [108]. Results showed that surface modification of graphite/epoxy composites by nano-hBN was more effective leading to a very low μ and wear rate w_a. The film transfer on the disc which was responsible for this improvement in the wear performance was evidenced by morphological analysis.

The interest in using ecofriendly fibers as reinforcement for the production of lightweight, low-cost polymer composites was recently intensified. One of the most extensively preferred is basalt fiber, which is cost-effective and provides superior properties over glass fibers. There are many prominent advantages of these polymer/basalt composites including high specific mechanic-physico-chemical properties, biodegradability, and non-abrasive qualities to cite a few. Dhand et al. [109] in their recent short review on basalt fibers discussed the use of such fiber in reinforcing composites and evaluate them as an alternative of glass fibers. The improvement in mechanical, thermal and chemical resistant properties achieved by basalt/polymer/ceramic hybrid laminated composites for applications in specific industries.

BMI resin synthesized form three different prepolymer molecular weights were used as matrices for graphite fiber reinforced with nano-SiO_2 [110]. DMA properties and weight-loss tests showed that for the as-prepared composite from prepolymers having lower molecular weight, its activation energy for oxidative degradation was reduced, providing higher total weight loss in short-term under isothermal aging. This was attributed to the unreacted fragments still entrapped within the composites network.

Coatings of epoxy/silane-treated SiO_2 nanocomposites were coated onto the basalt fiber roving. As compared to the pure matrix, the tensile strength and interfacial property of the epoxy/SiO_2/basalt fiber composite coating were significantly improved. The coating modification with silane-treated SiO_2 was effectively enhanced the mechanical properties BF/epoxy resin laminated composites [111]. Figure 11.18 contains the micrographs of fracture surface of the untreated BFRP samples (see Figure 11.18 a) which reveals sample breaks and bursts of basalt fiber into many thin strands and the interfaces are almost totally broken down. Figure 11.18(b–d) all show the effective bonding of the matrix with the fibers in the presence of nano-SiO_2. This result confirms the interface strength approaching the reinforcing basalt fibers, which imply that the comprehensive properties of the composite have been significantly enhanced.

A flax fiber yarn was grafted KH560-TiO_2 nanoparticles to reinforced epoxy plates. With the optimized nano-TiO_2 grafting content (2.34 wt.%), the tensile strength and flexural properties of the flax fibers and the interfacial shear strength to an epoxy resin were enhanced due to formation of Si–O–Ti and C–O–Si bonds

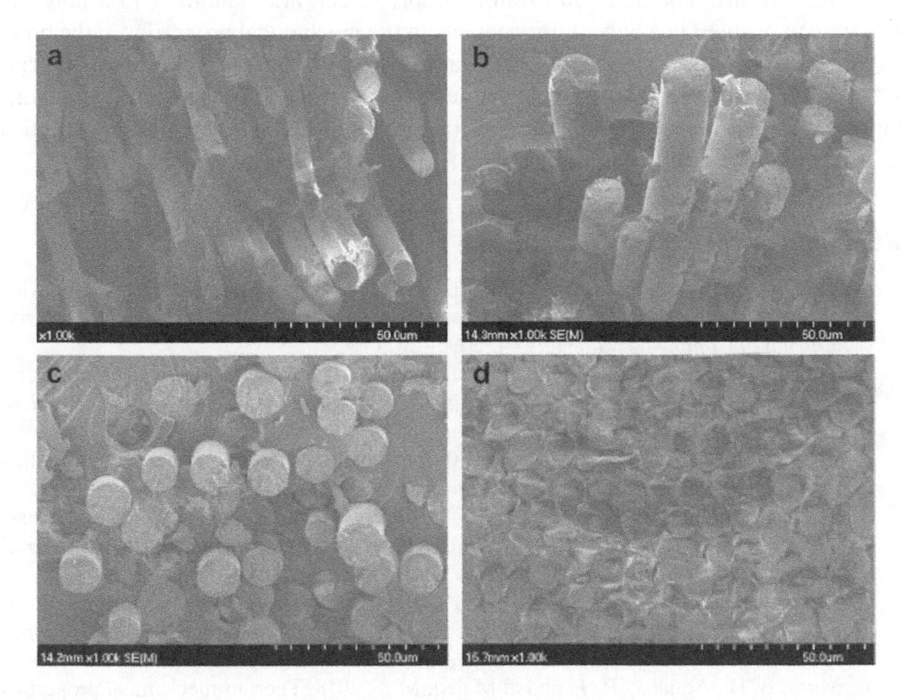

FIGURE 11.18 Fracture Surface Morphologies of Composite Materials Reinforced by Basalt Fibers with Different Coatings: (a) Untreated Basalt Fiber; (b) Basalt Fiber with Coating not Containing SiO_2 Particles; (c) Basalt Fiber with Coating Containing Untreated SiO_2; and (d) Basalt Fiber with Coating Containing Treated SiO_2 [111].

and the presence of the nano-TiO_2 particles on the fiber surfaces [112]. Experimental results showed that with the addition of the SiO_2 nanoparticles together with the rubber, the flexural stiffness of $[90]_{10}$ epoxy/fiber composites was reduced while the damping properties of these laminates were increased. In addition, it was found that a minimum effect of these nanoparticles on the $[0]_{10}$ laminates was obtained [113]. Moreover, vibration-damping responses of composite laminates obtained using the micromechanical analysis together with the modal analysis demonstrated a good agreement with the experimental data.

11.6 CONCLUSIONS

Compressive properties are generally inferior parts in structural application of fiber-reinforced polymer composites. Matrix modification can slightly afford an effective way to enhance compressive performance of such laminated composites. However, FRP composites filled with ceramic micro- and nanoparticles interest many researchers because of the large surface-area-to-volume ratio of nanoscale fillers. The addition of ceramic fillers to polymer/fiber laminates generated much improvements in the thermal and mechanical performances fracture toughness and fatigue endurance of polymers, in turn improving the interlaminar shear strength (ILSS) of the laminate composites and adhesively-bonded joints created by such reinforced resins. The addition of an appropriate ceramic nanofiller to a polymer matrix can resulted in a huge improvements in the mechanical properties of the hosting matrix, so long as these nanofillers have been well-dispersed within the matrix, which can be achieved by surface-treatment. Some numerical approaches and strategies can be also used to predict the role of ceramic filler on the different properties of fiber-reinforced polymer composites.

REFERENCES

1. Manjunatha, C. M., Sprenger, S., Taylor, A.C. et al. 2010. The tensile fatigue behavior of a glass-fiber reinforced plastic composite using a hybrid-toughened epoxy matrix. *Journal of Composite Materials* 44: 2095–2109.
2. Tsai, J. L., Cheng, Y. L. 2009. Investigating silica nanoparticle effect on dynamic and quasi-static compressive strengths of glass fiber/epoxy nanocomposites. *Journal of Composite Materials* 43: 3143–3155.
3. Tsai, J. L., Hsiao, H., Cheng, Y. L. 2010. Investigating mechanical behaviors of silica nanoparticle reinforced composites. *Journal of Composite Materials* 44: 505–524.
4. Tsai, J. L., Huang, B. H., Cheng, Y. L. 2009. Enhancing fracture toughness of glass/epoxy composites by using rubber particles together with silica nanoparticles. *Journal of Composite Materials*. 43: 3107–3123.
5. Agarwal, G., Patnaik, A., Sharma, R. K. 2013. Thermo-mechanical properties of silicon carbide-filled chopped glass fiber-reinforced epoxy composites. *International Journal of Advanced Structural Engineering* 5: 21.
6. Musa, B. H., Aghajan, R. J., abd al Majed, M. F. 2018. Thermo-mechanical properties of unsaturated polyester reinforced with silicon carbide powder and with chopped glass fiber. *Journal of University of Babylon, Pure and Applied Sciences* 26: 141–147.
7. Padhi, P. K., Satapathy A. 2013. Glass fiber and blast furnace slag particles reinforced epoxy-based hybrid composites. *Applied Polymer Composites*, 1: 85–92.

8. Mohapatra, S., Mantry, S., Singh, S. K. 2014. Performance evaluation of glass-epoxy-TiC hybrid composites using design of experiment. *Journal of Composites* 2014: 1–9.
9. Basavarajappa, S., Ellangovan, S. 2012. Dry sliding wear characteristics of glass epoxy composite filled with silicon carbide and graphite particles. *Wear* 296: 491–496.
10. Manjunatha, C. M., Taylor, A. C., Kinloch, A. J. et al. 2010. The tensile fatigue behaviour of a silica nanoparticle-modified glass fibre reinforced epoxy composite. *Composites Science and Technology* 70: 193–199.
11. Blackman, B. R. K., Kinloch, A. J., Sohn Lee, J. et al. 2007. The fracture and fatigue behaviour of nano-modified epoxy polymers. *Journal of Materials Science* 42: 7049–7051.
12. Akinyede, O., Mohan, R., Kelkar, A. et al. 2009. Static and fatigue behavior of epoxy/fiber glass composites hybridized with alumina nanoparticles. *Journal of Composite Materials* 43: 769–781.
13. Tate, J. S., Akinola, A. T., Espinoza, S. et al. 2018. Tension–tension fatigue performance and stiffness degradation of nanosilica-modified glass fiber-reinforced composites. *Journal of Composite Materials* 52: 823–834.
14. Sun, J., Wang, J. 2015. The fabrication of superhydrophobic glass fiber-reinforced plastic surfaces with tunable adhesion based on hydrophobic silica nanoparticle aggregates. *Colloid and Polymer Science* 293: 2815–2821.
15. Zheng, Y., Ning, R., Zheng, Y. 2005. Study of SiO_2 nanoparticles on the improved performance of epoxy and fiber composites. *Journal of Reinforced Plastics and Composite* 24: 223–233.
16. Uddin, M. F., Sun, C. T. 2008. Strength of unidirectional glass/epoxy composite with silica nanoparticle-enhanced matrix. *Composites Science and Technology* 68: 1637–1643.
17. Kelkar, A. D., Tian, Q., Yu, D. et al. 2016. Boron nitride nanoparticle enhanced prepregs: A novel route for manufacturing aerospace structural composite laminate. *Materials Chemistry and Physics* 176: 136–142.
18. Thiagarajan, A., Gilbert, S. A., Govindaraj, A. et al. 2015. Mechanical properties of hybrid glass fibre/epoxy reinforced TiO_2 nanocomposites. *International Journal of Scientific & Engineering Research* 6: 390–392.
19. Halder, S., Ahemad, S., Das, S. et al. 2016. Epoxy/glass fiber laminated composites integrated with amino functionalized ZrO_2 for advanced structural applications. *ACS Applied Materials & Interfaces* 8: 1695–1706.
20. Manjunatha, C. M., Bojja, R., Jagannathan, N. et al. 2013. Enhanced fatigue behavior of a glass fiber reinforced hybrid particles modified epoxy nanocomposite under WISPERX spectrum load sequence. *International Journal of Fatigue* 54: 25–31.
21. Ngah, S. A., Taylor, A. C. 2016. Toughening performance of glass fibre composites with core–shell rubber and silica nanoparticle modified matrices. *Composites Part A: Applied Science and Manufacturing* 80: 292–303.
22. Su, F. H., Zhang, Z. Z., Liu, W. M. 2008. Tribological behavior of hybrid glass/PTFE fabric composites with phenolic resin binder and nano-TiO_2 filler. *Wear* 264: 562–570.
23. Moorthy, S. S., Manonmani, K. 2014. Research on dry sliding wear behaviour of (TiO_2) particulate filled polyester-based polymer composite. *Journal of the Balkan Tribological Association* 20: 217–226.
24. Mohantyand, A., Srivastava, V. K. 2015. Effect of alumina nanoparticles on the enhancement of impact and flexural properties of the short glass/carbon fiber reinforced epoxy based composites. *Fibers and Polymers* 16: 188–195.
25. Nayak, R. K., Dash, A., Ray, B. C. 2014. Effect of epoxy modifiers (Al_2O_3/SiO_2/TiO_2) on mechanical performance of epoxy/glass fiber hybrid composites. *Procedia Material Science* 6: 1359–1364.
26. Panda, P., Mantry, S., Mohapatra, S. et al. 2014. A study on erosive wear analysis of glass fiber-epoxy-AlN hybrid composites. *Journal of Composite Materials* 48: 107–118.

27. Moorthy, S. S., Manonmani, K. 2013. Preparation and characterization of glass fiber reinforced composite with TiO_2 particulate. *Fibers* 69: 154–158.
28. DiBenedetto, A.T. 2001. Tailoring of interfaces in glass fiber reinforced polymer composites: a review. *Materials Science and Engineering A* 302: 74–82.
29. Böger, L., Sumfleth, J., Hedemann, H. et al. 2010. Improvement of fatigue life by incorporation of nanoparticles in glass fibre reinforced epoxy. *Composites Part A Applied Science and Manufacturing* 41: 1419–1424.
30. Bagni, R., Júlio, T., Santos, C. et al. 2017. Hybrid glass fibre reinforced composites containing silica and cement microparticles based on a design of experiment. *Polymer Testing* 57: 87–93.
31. Santos, J. C., Vieira, L. M. G., Túlio H. Panzera, T. H. et al. 2015. Hybrid glass fibre reinforced composites with micro and poly-diallyldimethylammonium chloride (PDDA) functionalized nano silica inclusions. *Materials & Design* 65: 543–549.
32. Santana, P. R. T., Panzera, T. H., Freire, R. T. S., Christoforo, A. L. 2017. Apparent shear strength of hybrid glass fibre reinforced composite joints. *Polymer Testing* 64: 307–312.
33. Detomi, A. C., dosSantos, R. M., Ribeiro Filho, S. L. M. et al. 2014. Statistical effects of using ceramic particles in glass fibre reinforced composites. *Materials & Design* 55: 463–470.
34. Tsai, J. L., Huang, B. H., Cheng, Y. L. 2011. Enhancing fracture toughness of glass/epoxy composites for wind blades using silica nanoparticles and rubber particles. *Procedia Engineering* 14: 1982–1987.
35. Gao, X., Jensen, R. E., McKnight, S. H. et al. 2011. Effect of colloidal silica on the strength and energy absorption of glass fiber/epoxy interphases. *Composites Part A: Applied Science and Manufacturing* 42: 1738–1747.
36. Nallusamy, S. 2016. Characterization of epoxy composites with TiO_2 additives and E-glass fibers as reinforcement agent. *Journal of Nano Research* 40: 99–104.
37. Jia, X., Li, G., Liu, B. et al. 2013. Multiscale reinforcement and interfacial strengthening on epoxy-based composites by silica nanoparticle-multiwalled carbon nanotube complex. *Composites Part A: Applied Science and Manufacturing* 48: 101–109.
38. Lubineau, G., Rahaman, A. 2012. A review of strategies for improving the degradation properties of laminated continuous-fiber/epoxy composites with carbon-based nanoreinforcements. *Carbon* 50: 2377–2395.
39. Zhang, G., Wetzel, B., Jim, B. et al. 2015. Impact of counter face topography on the formation mechanisms of nanostructured tribofilm of PEEK hybrid nanocomposites. *Tribology International* 83: 156–165.
40. Xian, G. J., Walter, R., Haupert, F. 2006. Friction and wear of epoxy/TiO_2 nanocomposites: Influence of additional short carbon fibers, Aramid and PTFE particles. *Composites Science and Technology* 2006; 66: 3199–3209.
41. Jiang, Z., Gyurova, L. A., Schlar, A. K. et al. 2008. Study on friction and wear behavior of polyphenylene sulfide composites reinforced by short carbon fibers and sub-micro TiO_2 particles. *Composites Science and Technology* 68: 734–742.
42. Chang, L, Zhang, Z., Breidt, C. et al. 2005. Tribological properties of epoxy nanocomposites I. Enhancement of the wear resistance by nano-TiO_2 particles. *Wear* 258: 141–148.
43. Guo, Q. B., Rong, M. Z., Jia, G. L. et al. 2009. Sliding wear performance of nano-SiO_2/short carbon fiber/epoxy hybrid composites, *Wear* 266: 658–665.
44. Zhang, J., Deng, S., Wang, Y. et al. 2013. Effect of nanoparticles on interfacial properties of carbon fibre–epoxy composites. *Composites Part A: Applied Science and Manufacturing* 55: 35–44.
45. Wu, G. S., Ma, L. C., Wang, Y. W. et al. 2016. Improvements in interfacial and heat-resistant properties of carbon fiber/methylphenylsilicone resins composites by

incorporating silica-coated multi-walled carbon nanotubes. *Journal of Adhesion Science and Technology* 30: 117–130.
46. Naderi, A., Mazinani, S., Ahmadi, S. J. et al. 2014. Modified thermo-physical properties of phenolic resin/carbon fiber composite with nano zirconium dioxide. *Journal of Thermal Analysis and Calorimetry* 117: 393–401.
47. Nassar, A., Nassar, E. 2014. Thermo and mechanical properties of fine silicon carbide/chopped carbon fiber reinforced epoxy composites. *Universal Journal of Mechanical Engineering* 2: 287–292.
48. Carolan, D., Ivankovic, A., Kinloch, A. J. et al. 2017. Taylor Toughened carbon fibre-reinforced polymer composites with nanoparticle-modified epoxy matrices. *Journal of Materials Science* 52: 1767–1788.
49. Carolan, D., Kinloch, A.J., Ivankovic, A. et al. 2016. Mechanical and fracture performance of carbon fibre reinforced composites with nanoparticle modified matrices. *Procedia Structural Integrity* 2: 096–103.
50. Liu, F., Deng, S., Zhang, J. 2017. Mechanical properties of epoxy and its carbon fiber composites modified by nanoparticles. *Journal of Nanomaterials* 2017: 1–9.
51. Taheri, F. 1997. Improvement of Strength and Ductility of Adhesively Bonded Joints by Inclusion of SiC Whiskers. *Journal of Composites, Technology and Research* 19: 86–92.
52. Zeng, Y., Liu, H. Y., Mai, Y. W. et al. 2012. Improving interlaminar fracture toughness of carbon fibre/epoxy laminates by incorporation of nano-particles. *Composites Part B: Engineering* 43: 90–94.
53. Hsieh, T. H., Kinloch, A. J., Masania, K. et al. 2010. The toughness of epoxy polymers and fibre composites modified with rubber microparticles and silica nanoparticles. *Journal of Materials Science* 45: 1193–1210.
54. Kinloch, A. J., Mohammed, R. D., Taylor, A. C. et al. 2005. The effect of silica nano particles and rubber particles on the toughness of multiphase thermosetting epoxy polymers. *Journal of Materials Science* 40: 5083–5086.
55. Johnsen, B. B., Kinloch, A. J., Mohammed, R. D. et al. 2007. Toughening mechanisms of nanoparticle-modified epoxy polymers. *Polymer* 48: 530–541.
56. Ma, J., Mo, M. S., Du, X. S. et al. 2008. Effect of inorganic nanoparticles on mechanical property, fracture toughness and toughening mechanism of two epoxy systems. *Polymer* 49: 3510–3523.
57. Kinloch, A. J., Mohammed, R. D., Taylor, A. C. et al. 2006. The interlaminar toughness of carbon-fibre reinforced plastic composites using "hybrid-toughened" matrices. *Journal of Materials Science* 41: 5043–5046.
58. Chisholm, N., Mahfuz, H., Rangari, V. K. et al. 2005. Fabrication and mechanical characterization of carbon/SiC-epoxy nanocomposites. *Composite Structures* 67: 115–124.
59. Kinloch, A. J., Masania, K., Taylor, A. C. et al. 2008. The fracture of glass-fibre-reinforced epoxy composites using nanoparticle-modified matrices. *Journal of Materials Science* 43: 1151–1154.
60. Zheng, Y., Zheng, Y., Ning, R. 2003. Effects of nanoparticles SiO_2 on the performance of nanocomposites. *Materials Letters* 57: 2940–2944.
61. Rosso, P., Ye, L., Friedrich, K. et al. 2006. A toughened epoxy resin by silica nanoparticle reinforcement. *Journal of Applied Polymer Science* 100: 1849–1855.
62. Liu, H. Y., Wang, G. T., Mai, Y. W. et al. 2011. On fracture toughness of nano-particle modified epoxy. *Composites Part B: Engineering* 42: 2170–2175.
63. Hanhan, I., Selimov, A., Carolan, D. et al. 2017. Quantifying alumina nanoparticle dispersion in hybrid carbon fiber composites using photoluminescent spectroscopy. *Applied Spectroscopy* 71: 258–266.
64. Kostopoulos, V., Karapappas, P., Loutas, T. et al. 2011. Interlaminar fracture toughness of carbon fibre-reinforced polymer laminates with nano- and micro-fillers. *International journal of experimental mechanics* 47: E269–E282.

65. Shahid, N., Villate, R. G., R. Barron A. R. 2005. Chemically functionalized alumina nanoparticle effect on carbon fiber/epoxy composites. *Composites Science and Technology* 65: 2250–2258.
66. Tang Y., Ye, L., Zhang, D. et al. 2011. Characterization of transverse tensile, interlaminar shear and interlaminate fracture in CF/EP laminates with 10 wt.% and 20 wt.% silica nanoparticles in matrix resins. *Composites Part A: Applied Science and Manufacturing* 42: 1943–1950.
67. Rallini, M., Natali, M., Kenny, J.M. et al. 2013. Effect of boron carbide nanoparticles on the fire reaction and fire resistance of carbon fiber/epoxy composites. *Polymer* 54: 5154–5165.
68. Mohanty, A., Srivastava, V. K. 2012. Compressive failure analysis of alumina nano particles dispersed short glass/carbon fiber reinforced epoxy hybrid composites. *International Journal of Scientific & Engineering Research* 3:1–7.
69. Centeno, A., Viña, J. A., Blanco, C. et al. 2011. Influence of titanium carbide on the interlaminar shear strength of carbon fibre laminate composites. *Composites Science and Technology* 71: 101–106.
70. Lin, J. M., Ma, C. C. M., Tai, W. H. et al. 2000. Carbon fiber reinforced phenolic resin/silica ceramer composites-processing, mechanical and thermal properties. *Polymer Composites* 27: 305–311.
71. Zhang, X. 2017. Study on the tribological properties of carbon fabric reinforced phenolic composites filled with nano-Al_2O_3. *Journal of Macromolecular Science, Part B* 56: 568–577.
72. Muna, S. Y., Lima, H. M., Lee, D. J. 2015. Thermal conductivity of a silicon carbide/pitch-based carbon fiber-epoxy composite. *Thermochimica Acta* 619: 16–19.
73. He, S., Zhang, S., Lu, C. 2011. Coating carbon fiber using polyimide–silica nanocomposite by electrophoretic deposition. *Colloids and Surfaces A: Physicochemical and Engineering Aspects* 387: 86–91.
74. Lin, G., Xie, G., Sui, G. et al. 2012. Hybrid effect of nanoparticles with carbon fibers on the mechanical and wear properties of polymer composites. *Composites Part B* 43: 44–49.
75. Hussain, M., Nakahira, A., Nishijima, S. et al. 2000. Evaluation of mechanical behavior of CFRC transverse to the fiber direction at room and cryogenic temperature. *Composites Part A* 31: 173–179.
76. Tsantzalis, S., Karapappas, P., Vavouliotis, A. et al. 2007. On the improvement of toughness of CFRPs with resin doped with CNF and PZT particles. *Composites Part A* 38: 1159–1162.
77. Gao, C. P., Guo, G. F., Zhao, F. Y. et al. 2016. Tribological behaviors of epoxy composites under water lubrication conditions. *Tribology International* 95: 333–341.
78. Österle, W., Dmitriev, A. I., Wetzel, B. et al. 2016. The role of carbon fibers and silica nanoparticles on friction and wear reduction of an advanced polymer matrix composite. *Materials & Design* 93: 474–484.
79. Guo, L., Qi, H., Ga Zhang, G. et al. 2017. Distinct tribological mechanisms of various oxide nanoparticles added in PEEK composite reinforced with carbon fibers. *Composites Part A: Applied Science and Manufacturing* 97: 19–30.
80. Wang, Q., Zhang, X., Pei, X. 2010. Study on the synergistic effect of carbon fiber and graphite and nanoparticle on the friction and wear behavior of polyimide composites. *Materials & Design* 31: 3761–3768.
81. Zhang, Z., Breidt, C., Chang, L. et al. 2004. Enhancement of the wear resistance of epoxy: short carbon fibre, graphite, PTFE and nano-TiO_2. *Composites Part A: Applied Science and Manufacturing* 35: 1385–1392.
82. Chang, L., Zhang, Z., Zhang, H. et al. 2006. On the sliding wear of nanoparticle filled polyamide 66 composites. *Composites Science and Technology* 66: 3188–3198.

83. Chang, L., Zhang, Z., Zhang, H. et al. 2006. Effect of nanoparticles on the tribological behaviour of short carbon fibre reinforced poly(etherimide) composites. *Tribology International* 38: 966–973.
84. Zhang, G., Chang, L., Schlarb, A. K. 2009. The roles of nano-SiO_2 particles on the tribological behavior of short carbon fiber reinforced PEEK. *Composites Science and Technology* 69: 1029–1035.
85. Mohanty, A., Srivastava, V. K., Sastry, P.U. 2014. Investigation of mechanical properties of alumina nanoparticle-loaded hybrid glass/carbon-fiber-reinforced epoxy composites. *Journal of Applied Polymer Science* 131: 39749.
86. Sprenger, S., Kothmann, M. H., Altstaedt, V. 2014. Carbon fiber-reinforced composites using an epoxy resin matrix modified with reactive liquid rubber and silica nanoparticles. *Composites Science and Technology* 105: 86–95.
87. Landowski, M., Strugała, G., Budzik, M. et al. 2017. Impact damage in SiO_2 nanoparticle enhanced epoxy–Carbon fibre composites. *Composites Part B: Engineering* 113: 91–99.
88. Xu, Z., Li, Z., Li, J. et al. 2014. The effect of CF and nano-SiO_2 modification on the flexural and tribological properties of POM composites. *Journal of Thermoplastic Composite Materials* 27: 287–296.
89. Chen, J., Zhu, Y., Ni, Q. et al. 2014. Surface modification and characterization of aramid fibers with hybrid coating. *Applied Surface Science* 321: 103–108.
90. Zhang, H. J., Zhang, Z. Z., Guo, F. 2010. A Study on the Sliding Wear of Hybrid PTFE/Kevlar Fabric/Phenolic Composites Filled with Nanoparticles of TiO_2 and SiO_2. *Tribology Transactions* 53: 678–683.
91. Wang, Z., Zhou, L. Zhu, J. 2008. Study of the SiO_2 nanoparticle/fabric composites of stab-resistance. *Technical Textiles* 10.
92. Liu, N., Wang, J., Yang, J. et al. 2016. Combined effect of nano-SiO_2 and nano-Al_2O_3 on improving the tribological properties of Kevlar fabric/phenolic laminate in water. *Journal Tribology Transactions* 59: 163–169.
93. Alsaadi, M., Bulut, M., Erkliğ, A. et al. 2018. Nano-silica inclusion effects on mechanical and dynamic behavior of fiber reinforced carbon/Kevlar with epoxy resin hybrid composites. *Composites Part B*. doi: 10.1016/j.compositesb.2018.07.015.
94. Simić, D. M., Stojanović, D. B., Brzić, S. J. et al. 2017. Aramid hybrid composite laminates reinforced with inorganic fullerene-like tungsten disulfide nanoparticles. *Composites Part B: Engineering* 123: 10–18.
95. Haro, E. E., Odeshi, A. G., Szpunar, J. A. 2016. The energy absorption behavior of hybrid composite laminates containing nano-fillers under ballistic impact. *International Journal of Impact Engineering* 96: 11–22.
96. Mansour, G., Tsongas, K., Tzetzis, D. 2016. Modal testing of epoxy carbon–aramid fiber hybrid composites reinforced with silica nanoparticles. *Journal of Reinforced Plastics and Composites* 35: 1401–1410.
97. Yang, M., Zhang, Z., Yuan, J. et al. 2017. Synergistic effects of AlB_2 and fluorinated graphite on the mechanical and tribological properties of hybrid fabric composites. *Composites Science and Technology* 143: 75–81.
98. Li, H. L., Wei, Z., Jiang, Y. D. et al. 2014. Tribological behavior of hybrid PTFE/Kevlar fabric composites with nano-Si_3N_4 and submicron size WS_2 fillers. *Tribology International* 80: 172–178.
99. Yang, M., Zhu, X., Ren, G. et al. 2015. Influence of air-plasma treatment and hexagonal boron nitride as filler on the high temperature tribological behaviors of hybrid PTFE/Nomex fabric/phenolic composite. *European Polymer Journal* 67: 143–151.
100. Su, F. H., Zhang, Z. Z., Liu, W. M. 2007. Tribological and mechanical properties of Nomex fabric composites filled with polyfluo 150 wax and nano-SiO_2. *Composites Science and Technology* 67: 102–110.

101. Wang, B., Duan, Y., Zhang, J. 2016. Titanium dioxide nanoparticles-coated aramid fiber showing enhanced interfacial strength and UV resistance properties. *Materials & Design* 103: 330–338.
102. Ahmad, Z., Sarwar, M. I., Mark, J. E. 1998. Thermal and mechanical properties of aramid-based titania hybrid composites. *Journal of Applied Polymer Science* 70: 297–302.
103. Kang, T. J., Hong, K. H., Yoo, M. R. 2010. Preparation and properties of fumed silica/Kevlar composite fabrics for application of stab resistant material. *Fibers and Polymers* 11: 719–724.
104. Zhang, S., Yan, M., Cui, H. et al. 2011. Effect of nm-SiO_2 on the matrix and aramid fiber/epoxy composite applied to the repair and reinforcement of oil and gas pipelines. *International Conference on Pipelines and Trenchless Technology* 1305–1315.
105. Mahfuz, H., Clements, F., Rangari, V. et al. 2009. Enhanced stab resistance of armor composites with functionalized silica nanoparticles. *Journal of Applied Physics* 105: 064307.
106. Li, Z., Cheng, X., He, S. et al. 2016. Aramid fibers reinforced silica aerogel composites with low thermal conductivity and improved mechanical performance. *Composites Part A: Applied Science and Manufacturing* 84: 316–325.
107. Potluri, R, Paul, K. J., Babu, B. M. 2018. Effect of Silicon Carbide Particles Embedment on the properties of Kevlar Fiber Reinforced Polymer Composites. *Materials Today: Proceedings* 5: 6098–6108.
108. Kadiyala, A. K., Bijwe, J. 2013. Surface lubrication of graphite fabric reinforced epoxy composites with nano- and micro-sized hexagonal boron nitride. *Wear* 301: 802–809.
109. Dhand, V., Mittal, G., Rhee, K. Y. et al. 2015. A short review on basalt fiber reinforced polymer composites. *Composites Part B: Engineering* 73: 166–180.
110. Geng, D. B., Zeng, L. M., Li, Y. 2009. Thermo-oxidative stability of nano-SiO_2/bismaleimide composite. *4th IEEE International Conference on Nano/Micro Engineered and Molecular Systems* 1–2: 698–702.
111. Wei, B., Song, S. H., Cao, H. 2011. Strengthening of basalt fibers with nano-SiO_2–epoxy composite coating. *Materials & Design* 32: 4180–4186.
112. Wang, H. G., Xian, G. J., Li, H. 2015. Grafting of nano-TiO_2 onto flax fibers and the enhancement of the mechanical properties of the flax fiber and flax fiber/epoxy composite. *Composites Part A: Applied Science and Manufacturing* 76: 172–180.
113. Bashar, M. T., Sun, U., Mertiny, P. 2013. Mode-I interlaminar fracture behaviour of nanoparticle modified epoxy/basalt fibre-reinforced laminates. *Polymer Testing* 32: 402–412.

12 Applications of Polymer/Ceramic Composites

12.1 INTRODUCTION

Since the incorporation of ceramics to polymers not only can improve their physical properties, such as the mechanical and thermal properties, but can also result in some new unique properties, their resulting composites have attracted strong interest in many industrial sectors. Many potential and practical applications of this type of composite have been reported, such as in coatings, flame-retardant materials, optical devices, electronics and optical packaging materials, photoresistant materials, photoluminescent-conducting film, pervaporation membranes, ultra-permeable reverse-selective membranes, proton exchange membranes, grouting materials, sensors, materials for metal uptake, etc. Also, polymer/ceramic nanocomposites demonstrating various morphologies, can generally exhibit improved, even novel, properties compared with the traditional microcomposites. In this chapter, several applications of polymer/ceramic composite materials will be presented, which are mainly based on specific characteristics of these composites.

Polymer/ceramic composites with improved mechanical, thermal, and electrical properties have been widely explored for many potential applications, including sensors and actuators, lightweight armors, capacitors, static-charge dissipation, electromagnetic interference (EMI) shielding, and photovoltaic devices [1–3]. They offer several processing advantages, such as mechanical flexibility and the ability to be molded into intricate configurations; both features make them very suitable candidates to be applied in compact electronics and larger geometrically specific electrical devices [4–6].

Recently, ceramics are added to different polymer matrices, such as polyvinylidene fluoride (PVDF), paraffin, polyethylene, and polystyrene, aiming to use them as dielectrics in capacitors due to their large electric breakdown (E_b) [7–11]. With the incorporation of ceramic particles into these matrices, the effective dielectric constant can be increased. Polymer coatings on the surface of ceramic fillers are primarily controlled by the interaction between the filler-surface groups and polymeric end groups, in addition to the reaction medium. This chapter focuses on the various applications of polymer/ceramic composite materials in major industries, including electronics, armor, coatings, biomedical technology, microwave absorbers, anticorrosion coatings, thermal ablation, etc.

12.2 MICROELECTRONICS

The need for denser and faster microelectronic circuits reduces the utilization of traditional packaging materials. Both polymers and ceramics demonstrate extreme electrical, thermal, and mechanical properties for the packaging of electronic and microelectronic materials. The electrical properties of microelectronic equipment, such as signal attenuation, propagation velocity, and cross talk are affected by the inherent dielectric characteristics of packaging materials.

Heat management is still a real problem for many kinds of industries, such as flexible electronics, displays, and biointegrated and implantable sensors [12–16]. In these applications, composite materials must exhibit high thermal conductivity, electrical insulation, a light weight, and a low thermal expansion coefficient at low-filler contents. Therefore, ceramic-reinforced polymer composites are promising materials to be applied as substrates, thermal interface materials (TIM) [17], and packages [18] of electronic devices. Zeng et al. produced PVA/BN composite films with improved flexibility in a low-matrix fraction (6 wt.% PVA) through vacuum-assisted self-assembly technique. Figure 12.1 shows great improvements in the heat dissipation of the prepared composite compared to pure polyimide substrates when a light-emitting diode (LED) is used. In the case of the LED, external quantum efficiency (EQE) and light-output power are significantly enhanced when temperature is reduced [19]. As a result, such composites with aligned BN fillers can ameliorate the efficiency and reliability of LEDs.

Aligning techniques can be utilized to develop TIM as an adhesive for connecting a device to a substrate or to act as an underfill to occupy void space of a solder joint between the device and the substrate. The TIM of a randomly dispersed ceramic-reinforced polymer composite has relatively low thermal conductivity (in the range of 0.2 to 1.2 W/m.K), while metal or carbon composite has high thermal conductivity

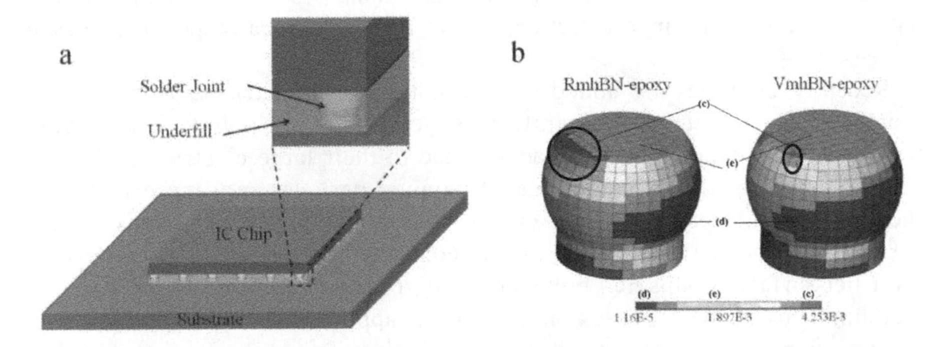

FIGURE 12.1 Practical Examples of Enhanced Heat Dissipation and Minimized Thermal Shrinkage with Aligned Ceramic/Polymer Composites. The Simulation of an Accumulated Plastic Strain to Solder Joints Influenced by the Underfill's CTE of Randomly Dispersed and Vertically Aligned BN-epoxy Composites (Bottom) [22]. Reprinted with Permission from [22], Copyright (2013) American Chemical Society.

of approximately 3 to 5 W/m.K. However, such TIM needs an external layer for electrical insulation [20]. Thus, an aligned ceramic filler composite is the best choice for overcoming these shortcomings. In addition to heat dissipation, aligned fillers can be applied as underfill to lower the CTE value. Lin et al. prepared magnetically-aligned magnetite-coated BN-epoxy composites to decrease vertical CTE so as to diminish thermo-mechanical strain to the solder joint to improve the mechanical reliability of the solder joint interconnection between the substrate and IC chip. The corner of the solder joint with a randomly aligned BN composite showed the highest thermo-mechanical strain because of the *z*-direction CTE of the composite during the temperature cycle.

On the other hand, the vertically aligned BN/epoxy composite was found to decrease accumulated plastic strain on the solder joint. High thermal conductivity, low CTE, and electrical insulation are essentials for dissipating heat from electronic devices [21]. An aligned nano-ceramic/polymer composite is a promising material for single-chip packaging and 3D chip packaging that needs electrical insulation and anisotropic heat flow. By using an aligned polymer/ceramic composite as a glue and insulator layer, stacked heat in the upper chip layers can be anisotropically dissipated to the heat sink [22, 23].

Electronic I–V curves obtained after encapsulation of testing devices indicated that the highly AlN-filled epoxy slip, which exhibited a thermal conductivity as high as 3.39 W/m.K, appeared to be feasible for use in the encapsulation of integrated circuit chips [24]. Agrawal et al. developed AlN/epoxy micro-composites with AlN contents ranging between 0 and 25 vol.%. The incorporation of AlN in resin increased the thermal conductivity, whereas CTE of the composite favorably decreased. By modifying the thermal and dielectric characteristics of these composites, they can be used for several microelectronics applications [25].

Direct encapsulation of organic light-emitting devices (OLEDs) is achieved by using highly transparent, photocurable *co*-polyacrylate/SiO_2 nanocomposite. The feasibility of such a composite for OLED encapsulation was evaluated by physical/electrical property analysis of resins and driving voltage/luminance/lifetime measurement of OLEDs [26]. Electrical tests showed a higher electrical insulation of photocured nanocomposite film at 3.20×10^{12} Ω compared with that of oligomer film at 1.18×10^{12} Ω at 6.15 V to drive the bare OLED. This resulted in a decreased leakage current, and the device-driving voltage was efficiently decreased so that the nanocomposite-encapsulated OLED could be driven at a lower driving voltage of 6.09 V rather than 6.77 V for the oligomer-encapsulated OLED at the current density of 20 mA/cm^2.

Polymer/ceramic nanocomposites based on new concepts were developed for embedded capacitor applications. The dielectric constant of a blended polymer of an aromatic polyamide (PA) and an aromatic bismaleimide (BMI) reinforced with $BaTiO_3$ exceeded 80 at 1 MHz and the specific capacitance was reached at 8 nF/cm^2 [27]. Using this nanocomposite, multilayer printed wiring boards with embedded passive components were produced for prototypes as depicted in Figure 12.2. Thus, these technologies were combined and optimized for embedded capacitor materials.

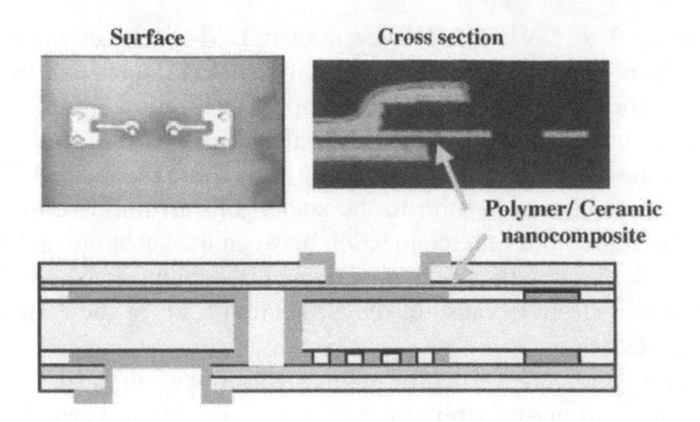

FIGURE 12.2 Surface and Cross Section of Polymer/ceramic Substrate [27].

12.3 SENSORS AND ACTUATORS

Conducting polymers reinforced with nano-ceramic have been widely studied to be used as gas sensors. Because of the large surface-area-to-volume ratio of nanostructure ceramic-reinforced conducting-polymers nanocomposites, they demonstrated improved sensing performance compared to pure conducting polymers. However, the sensitivity and selectivity of these sensors is still a serious concern to be solved. The conductive polymer/ceramic nanocomposites have reached a wide recognition in the sensors' application area due to the synergistic effect of small-size ceramic nanofillers and high electrical conductivity of conducting polymer matrices. Thus, the sensitivity and selectivity of conducting polymer nanocomposites-based sensors have been gradually ameliorated compared to that of the unfilled conductive polymers-based sensors.

The properties of these conjugated polymers can be controlled by incorporating ceramic fillers. In this issue, ceramic nanoparticles are more attractive due to intriguing properties arising from their nano-size and large surface area. The insertion of nanoscale fillers may enhance the electrical and sensing features of these polymers. For example, PAni/TiO_2 nanocomposite has attracted more attention in recent years in the development of gas chemosensors using ultrasonic irradiation [28, 29] and the sol–gel technique [30]. The introduction of TiO_2 ceramic particles improves the conductivity of PAni, while the surfactant improves the dispersibility of TiO_2 nanoparticles in the macromolecules.

Conducting polythiophene/SiO_2 (PT/SiO_2) nanocomposites were developed in the presence of three different surfactants (anionic, cationic, and non-ionic) through a chemical oxidative polymerization in an anhydrous medium to produce an enzyme-immobilized polymeric amperometric biosensor. Glucose oxidase (GOX) was immobilized by crosslinking to the conducting PT/SiO_2 nanocomposites and was utilized for amperometric detection of glucose [31].

It can be seen that the nanocomposite showed a relatively fast response towards aqueous NH_3 in the range of 0.1 to 1 N. The reproducibility of the response of the nanocomposite was also studied, and the response of polyaniline-*p*-toluenesulfonic/TiO_2-5 nanocomposite towards 0.2 N and 0.3 N NH_3 was found to be highly reproducible during the test of cyclic measurements as illustrated in Figure 12.3 [32].

A new method for elaborating polydiacetylene/SiO_2 nanocomposite to be used as chemosensor was reported. The disordered 10, 12-pentacosadiynoic acid (PCDA) aggregates were absorbed on the surfaces of nano-SiO_2 in aqueous medium. After irradiation with UV light, polydiacetylene/SiO_2 nanocomposites took on a blue color. A variety of environmental perturbations, such as temperature, pH, and amphiphilic molecules, could provoke in a colorimetric change of the polydiacetylene/SiO_2 nanocomposites from the blue to the red color [33].

TiO_2/polyaniline nanocomposite synthesized by in-situ chemical polymerization. The conductivity of the nanocomposite layer enhanced upon exposure to O_2 gas (an oxidizing agent) and reduced upon exposure to NH_3 gas (a reducing agent). The *n-p*

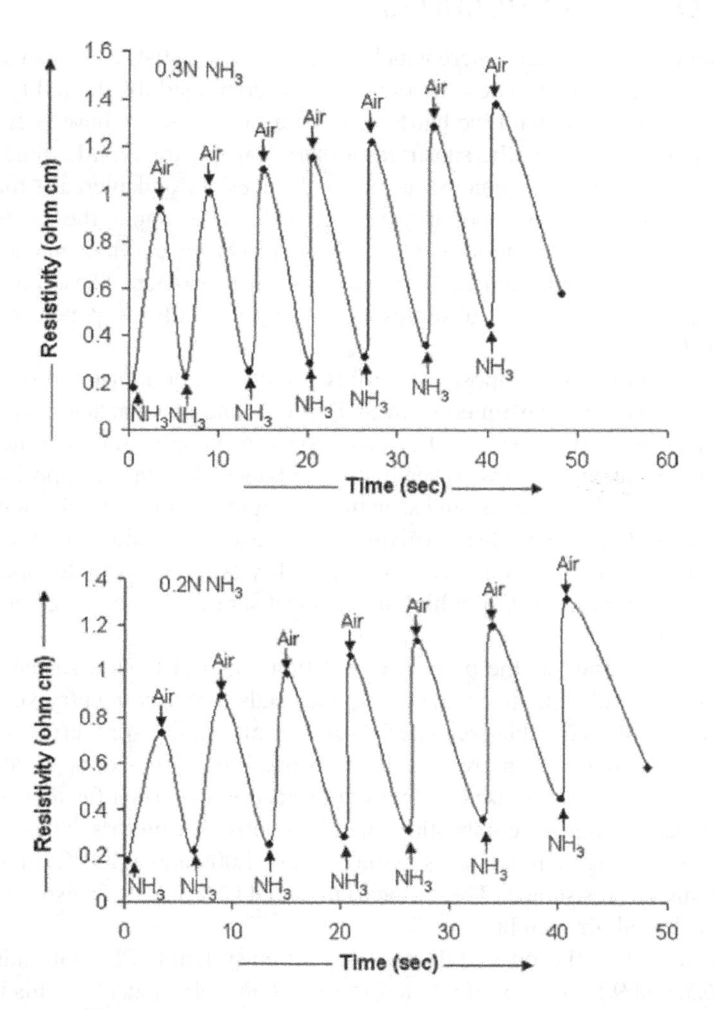

FIGURE 12.3 Variation in Resistivity of Polyaniline-*p*-toluenesulfonic/TiO_2-5 on Intermittent Exposure to 0.2 N NH_3 and 0.3 N NH_3 [32].

contacts between TiO_2 nanoparticles and polyaniline matrix increased the various the physical adsorption sites for oxygen gas molecules resulting in an enhancement in the gas sensitivity [34].

The photoluminescence (PL) spectra of pure polythiophene (PT) and PT/Al_2O_3 nanocomposite samples showed mainly three visible emission peaks at around *462, *490, and *522 nm. The ammonia gas (NH_3)–sensing characteristics have been evaluated at room temperature by changing the NH_3 gas concentration from 25 ppm to 650 ppm. The response is more considerable as Al_2O_3 doping concentrations and the concentrations of ammonia gas are increased. The highest Al_2O_3-doped PT nanocomposite showed its maximum sensitivity at lower concentration of NH_3 gas and rapid response and recovery time [35].

12.4 COATING APPLICATIONS

Polymer/ceramic composites were widely explored by scientists as new materials for various types of coatings. These hybrid materials combined the flexibility and easy processing of polymers with the hardness of ceramic fillers and have been successfully used on various metallic substrates as shown in Figure 12.4. In general, these coatings are transparent, demonstrate a good adhesion, and increase the scratch and abrasion resistance of a polymeric hosts [36]. For example, the reinforcement of acrylates with surface-modified nano-SiO_2 produced acrylate nanocomposite coatings exhibiting enhanced scratch and abrasion resistance. These coatings can be applied on many types of substrates including polymer films, paper, metals, and woody surfaces [37].

Compared with nanocomposite materials, a much better abrasion resistance was recorded for coatings constituted from SiO_2 micro- and nanoparticles. These nano/micro-composites are widely used as clear coats for parquet and flooring applications [38]. For instance, it was revealed that polymer/SiO_2 nanocomposites can be fabricated under various forms and structures through the mini-emulsion polymerization process [39]. These hybrid structures can be used for production of waterborne hybrid coatings, which demonstrate a strong ability of the polymer for spontaneous film formation combined with a high mechanical scratch resistance given by these ceramic nanoparticles.

Ablation is defined as the process of deleting material from a surface or other erosive process and generally is related to materials for space reentry vehicles and rocket nozzles [40–42]. Ablative materials are mainly applied for thermal protection and have been utilized for more than 40 years in a broad range of applications [43]. The ablative materials are used as thermal-protection materials for rocket nozzles, space engines, as well as combustion chambers of rocket motors. These materials can support very high temperatures in the range of 1,000 and 4,000°C, highly thrust and high-impact resistance. These materials should be able to construct complex shapes and have light weight.

Srikanth et al. [44] produced thermal protection system (TPS) materials by adding 3.5, 6.5, and 9.5 wt.% of ZrO_2 to a resol-based phenolic matrix. In this investigation, he found that the presence of ZrO_2 increased the ablation rate due to increased

FIGURE 12.4 Some Applications of Abrasive Coating Polymer/ceramic on Metallic Equipment.

conversion of solid char into zirconium carbide (ZrC) and CO gas resulting in under-protection of the carbon fibers facing the plasma jet used. Chen et al. [45] also enhanced the ablation resistance of carbon-phenolic composites by incorporating zirconium diboride (ZrB_2) particles. They deduced that the selected method for adding ZrB_2 into carbon-phenolic composites could significantly boost the ablation performance. The linear ablation rate of ZrB_2 carbon-phenolic composites was decreased by almost 79% compared to that recorded for carbon-phenolic composites. Kim et al. [46] used polymer/silica/reinforced carbon-fiber composites to join them to aluminum metal with high adhesive-strength levels and good coating properties (Figure 12.5). It was found that the presence of SiO_2 in the surface coating improves the adhesive strength by approximately 20%.

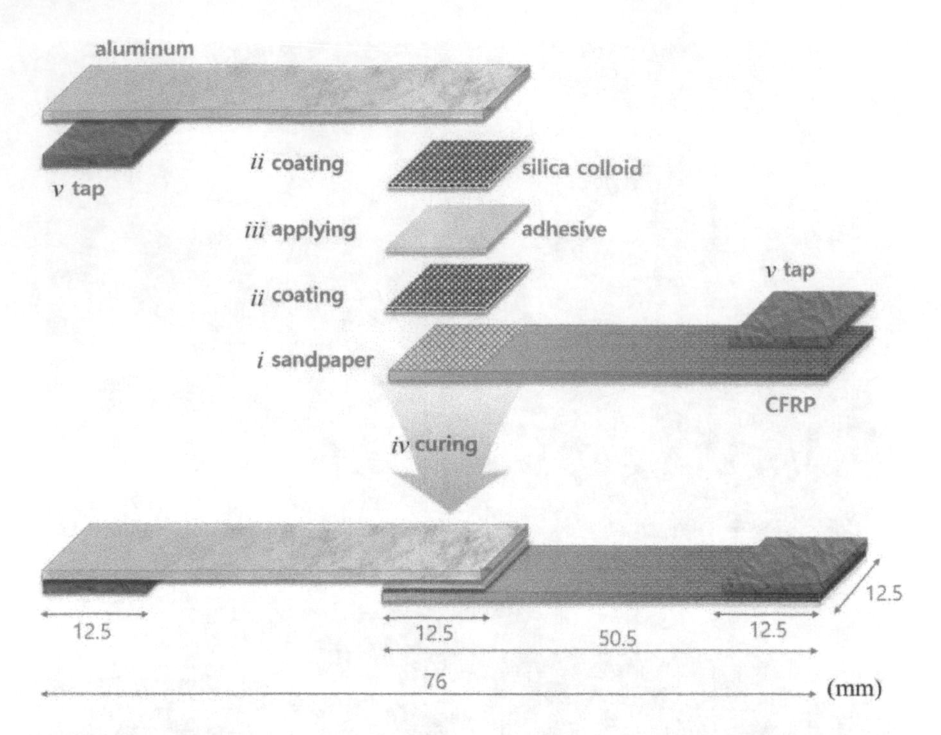

FIGURE 12.5 Specimen Fabrication Sequence of the Silica-coated CFRP/Al Specimen Preparation Process for the Lap Shear Test: (i) Sandpaper to Roughen the Contact Surface, (ii) Drop-coat Colloidal Silica, (iii) Apply Adhesive onto the Silica-coated Surface, (iv) Fix CFRP/Al with Adhesive by a Clamp and Follow Subsequent Curing Steps, v. Tap Specimens by Sandpaper [46].

Polybenzoxazine/ceramic nanocomposites are promising materials for a variety of potential applications in high-temperature filtration, self-cleaning coatings, catalyst carriers, etc., and also provide new insight into the design and development of functional nanofibrous membranes through functionalization of polybenzoxazine. Superhydrophobic polybenzoxazine/SiO_2 membranes exhibiting robust thermal stability and flexibility were prepared by a facile combination of electrospun SiO_2 nanofibers and a novel in situ polymerized matrix. The wettability of prepared membranes could be controlled by tuning the surface composition and the hierarchical structures [47].

Bacterial and fungal stains of polyurethane/APTES-trated-TiO_2 nanocomposites increased with increasing the APTES-TiO_2 loadings as shown in Figure 12.6 [48]. The coatings substantially gain hydrophilic nature symbiotically with TiO_2 content suggesting its potential application as self-cleanable material.

12.5 CORROSION-RESISTANCE COATING

Ceramic fillers including SiO_2, ZrO_2, Al_2O_3, and TiO_2 are known for their good chemical stability and their ability to protect metal substrates from corrosion reactions in a harsh environment, however their resulted coatings on metallic substrates suffer from

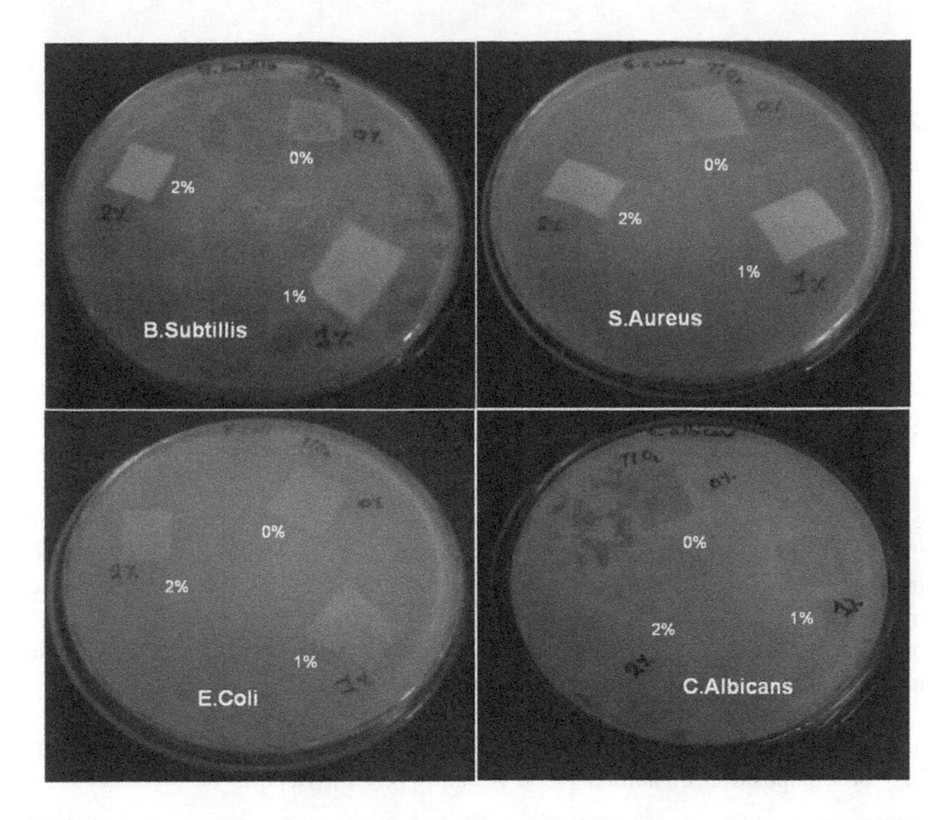

FIGURE 12.6 Antibacterial Activity of 0%, 1% and 2% PTNs. Reprinted from Springer [48].

high brittleness problems. Therefore, it was too difficult to realize thick coatings for protecting metal substrates from corrosion without cracking. Even in that circumstance, ceramics are preferred because they demonstrate several outstanding structural properties. Thus, polymer/ceramic systems have been intended to be more suitable for surpassing the limitations ascribed to the brittle nature of the ceramic fillers.

The basic application areas for such hybrid coatings are abrasion and scratch resistance or corrosion-resistant applications [49]. The most effective corrosion-protective polymer/ceramic coatings on metallic substrates have been collected in Table 12.1. Recently, a series of polybenzoxazine-TMOS/SiO_2 nanocomposite coatings were produced for anti-corrosion protection of mild steel [50].

12.6 BIOMEDICAL APPLICATIONS

The fabricated composite materials containing nano-metric amorphous ceramic directly, or indirectly incorporated into the resorbable polymer matrix are promising techniques for developing strong and bioactive nanohybrids implants suitable for application in bone surgery, e.g., for regeneration of bone defects (Figure 12.7). It is well-known that ceramic-based glass is bioactive due to its ability to join to living bone. From extensive literature, it can be deduced that the crucial condition for glasses and

TABLE 12.1
Corrosion protective coatings on metallic substrates

Polymer/ceramic Coating	Substrate	Thickness (μm)
SiO_2-PMMA	Steel	1.0
ZrO_2-PMMA	Steel	0.2
ZrO_2-PMMA	Steel	0.2–1.0
SiO_2-PVB	Steel	1.0
SiO_2-PMMA	Aluminum	0.1–0.3
SiO_2-PVB	Aluminum	0.1–0.3
SiO_2-vinylpolymer	Aluminum	3–4
Ce-SiO_2-epoxy	Aluminum	2–3
ZrO_2-TiO_2-oil	Aluminum	45–95

glass-ceramics for bonding to living bone is the building of a bone-like apatite layer on the outer surfaces of ceramics, which can be investigated via in vitro studies after soaking the bioactive materials in a simulated body fluid. Thus, bio-ceramic/polymer hybrid materials were recognized as having high potential as they exhibit bioactive properties of glasses or glass-ceramics and demonstrate improved elasticity of their polymeric matrices, thus they are very important materials for orthopedic applications [51–53].

The bioactivity of the polycaprolactone (PCL)/SiO_2 composite system formed a layer of hydroxyapatite on the surface of specimens soaked in a fluid simulating the composition of human blood plasma [54, 55]. Reports on the bioactivity

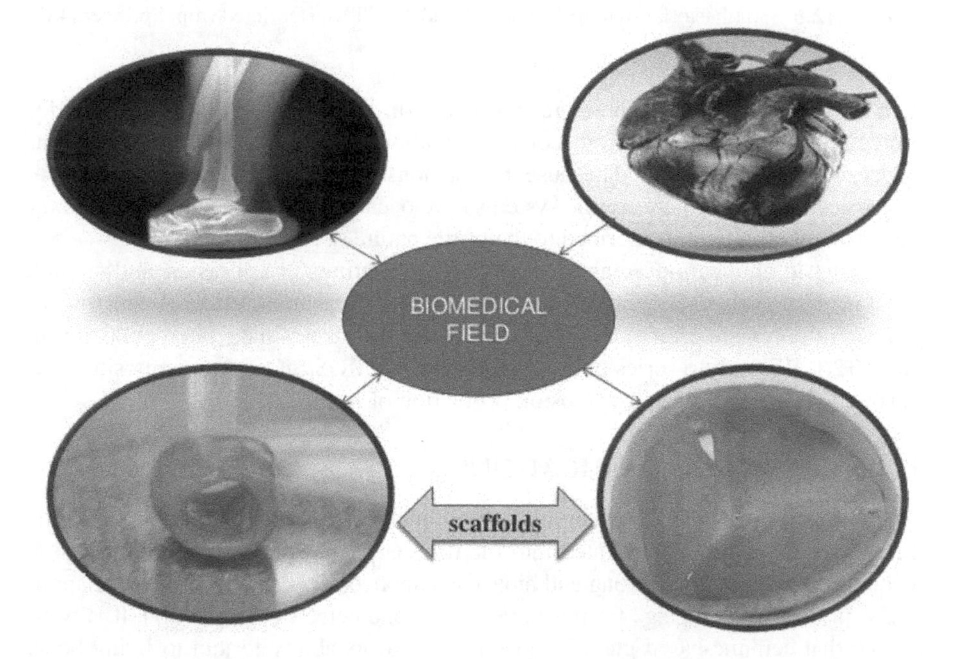

FIGURE 12.7 Biomedical Applications of Bio-ceramic/polymer Composites. [51]

of polymethyl methacrylate (PMMA)/SiO_2 nanocomposites has revealed that they are very suitable systems to be applied as both bioactive bone substitutes and as nanofillers for PMMA bone cement [56]. On the other hand, the PMMA/CaO-SiO_2 nanohybrid composites were found to be suitable for bone cement and dental composite resin applications because of their good bioactivity and enhanced mechanical properties [57]. PDMS/ZrO_2 nanocomposites were regarded as suitable materials for tissue-implant integration functions due to their advantageous effects on the proliferation and viability of human primary osteoblast and fibroblast cells and therefore can be used in the form of coatings for orthopedic trauma implants [58].

Kasuga et al. prepared vaterite-poly(lactic acid) (PLA) composite for bone regeneration [59]. There are three polymorphisms of calcium carbonate, the calcite, the aragonite, and the vaterite. The addition of excess vaterite accelerates the release of calcium ions which in turn is responsible for increasing the deposition of calcium phosphate on the body environment.

12.7 SEPARATION MEMBRANES

Polymer/ceramic nanocomposite membranes are useful methods for enhancing the separation efficiency of membranes because they combine the properties of both polymeric matrices and ceramic fillers, such as good permeability, selectivity, mechanical strength, and thermal and chemical stability [60]. Thus, these nanocomposite materials can boost the application area of membranes. To exemplify, the current spectrum of applications for gas-separation membranes include nitrogen enrichment, oxygen enrichment, hydrogen recovery, acid gas (CO_2, H_2S) removal from natural gas, and dehydration of air and natural gas [61, 62].

Recently, polymer/ceramics have been widely used as gas separation membranes (Table 12.2) [63]. For example, epoxy/porous SiO_2 composites were prepared with the pore surface modified using various silane-coupling agents. The epoxy/porous SiO_2 composite surface modified by APTES exhibited CO_2/N_2 gas selectivity at a lower pressure

TABLE 12.2
Reported Data Related to the Gas-Separation Performance of Polymer/ceramic Membranes

Material	P_{CO_2} (Barrer)	P_{N_2} (Barrer)	P_{O_2} (Barrer)	α_{CO_2/N_2}	α_{O_2/N_2}
PU/SiO_2	120	2.98	7.69	40.26	2.58
PI/SiO_2	2.03	0.1	0.38	20.3	3.8
PI/SiO_2	2.8	0.13	-----	22	
PAN/SiO_2	----	0.17	2.6	----	14.9
PI/SiO_2	41	7.74	-----	5.3	-----
PI/SiO_2	80	5	-----	16	------
PI/SiO_2	19	0.46	----	41	------
PI/SiO_2	15	0.3	2.61	50	8.7
PEG/SiO_2	94.2	2.46	-----	38.3	----
PI/TiO_2	----	0.07	0.72	------	9.5
PVAc/TiO_2	5.26	0.07	0.5	74.32	7.1

drop and gas-separation capability, due to the affinity between amino group of the APTES and CO_2 gas. Gas separation tests were conducted on the prepared membranes at 25°C and showed that the 80% N_2/20% CO_2 mix gas was transformed to 68.2% N_2/31.8% CO_2 gas. The gas separation capability was maintained at high temperatures [64].

12.8 ARMOR SYSTEMS

Lightweight body armors are very important equipment for saving the lives of military soldiers. The ballistic protection system is mainly composed of rigid panels based on ceramic plates such as boron carbide or alumina inserted into an aramid fabric pouch or incorporated in hard formed gear. Such system protects only the head and torso of soldiers while excluding the extremities of their bodies, like hands, arms, necks, etc. (Figure 12.8). The head and torso are extremely critical to protect because these two parts consist of life-supporting organs. The need for lightweight and flexible armor is to maximize maneuverability without compromising the protection of the soldier. In addition to ballistic resistance, attention is given to threats imposed by sharp weapons rather than by guns. The flexible armor that has been fabricated generally provides resistance against ballistics only or sharp cutting impacts only.

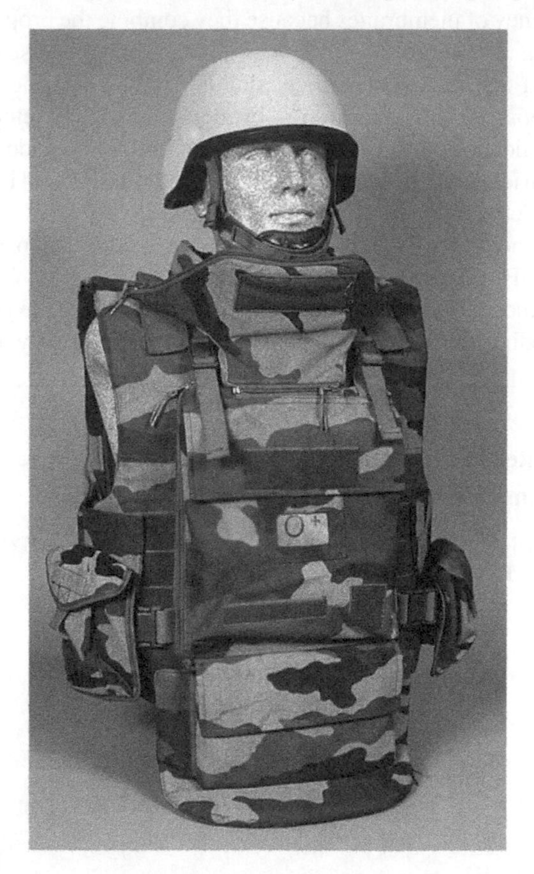

FIGURE 12.8 Futuristic Epoxy/Kevlar Body Armor Reinforced with TiO_2 Nanoparticles.

Armor systems need engineering solutions based on a combination of materials having different natures to achieve adequate ballistic protection and reduced weight. The most known lightweight ballistic armor is constituted from layered composite systems based on monolithic ceramic tiles bonded to polymer/aramid laminated composites. Other systems contained cell ceramic foams, like SiC, of various sizes infiltrated with thermosetting or elastomeric polyurethane (PU). Such a design was proposed as protection solutions for structural and vehicular armors, including blast protection [65].

Shear-thickening fluid systems based on treated-SiO_2 nanoparticles ultrasonically dispersed into a mixture of polyethylene glycol (PEG) and ethanol, reinforced Kevlar fibers has been recently developed as new solution for decreasing the weight of armor. Improvement in the protection was detected due to the formation of siloxane bonds during functionalization [66].

12.9 ABLATION AND FIRE RESISTANCE

Rallini et al. [67] filled epoxy matrices with B_4C nanoparticles at 1 and 5 wt.% ratios, and the nanostructured polymer was used as a matrix to impregnate carbon-fiber fabrics. These composites showed many benefits, which were related to the chemical reactions of B_4C when exposed to an oxidizing hyperthermal environment. TheB_4C can react not only with oxygen to generate liquid boron oxide and carbon dioxide, but also with other combustion products of the epoxy matrix, like CO and H_2O vapor, to produce boron oxide and amorphous carbon. All these reactions reduce the volumetric shrinkage of the charring material.

In addition, the glassy network generated by the liquid B_2O_3 acted as a protective layer for the carbon fibers, preventing them from oxidizing and also serving as a high-temperature adhesive that joined the plies of the laminated composite.

Tibiletti et al. [68] introduced nano-Al_2O_3 and submicron Al_2O_3 trihydrate particles into the UP matrix to produce synergistic effects on the thermal stability of their resulting composites. The best result for fire behavior was recorded for a global loading of 10 wt.% with an equal mass fractions for both kinds of fillers. This improvement was attributed to the formation of a mineral barrier, which enhances the catalytic effects related to the huge specific surface area of oxide nanoparticles.

12.10 WEATHER-RESISTANT AUTOMOTIVE COATINGS

Recently, various ceramic nanoparticles, including cerium oxide (CeO_2), TiO_2, and SiO_2, have been added to several kinds of polymeric coatings to improve their resistance against sunlight. These ceramics exhibited a high-surface area for absorbing the dangerous ultraviolet part of sunlight, protected the coatings from the weathering degradation. Due to their inorganic and particulate natures, they demonstrate a more stable and non-migratory characteristics within an applied coating. Thus, they revealed better effectiveness and longer protection duration.

For example, TiO_2 nanoparticles can effectively handle UV rays and can well-protect the coating against weathering. But, these ceramic nanoparticles can also exert strong oxidizing power and generate highly reactive free radicals and thus deteriorate the coating in which they have been attached. Therefore, the photocatalytic activity of TiO_2 nanoceramics has to be managed.

Aiming to control this phenomena, the surface-treatment of nanoparticles using various methods such as silane-coupling agents not only decreases photocatalytic activity of titinia nanoparticles but also gives more advantages, such as simplicity, low cost, and low-temperature processing. Studies showed that the surface modification of TiO_2 nanofillers by aminopropyltriethoxysilane (APS), significantly reduced photocatalytic activity of these ceramic nanoparticles while increasing the weathering resistance of a polyurethane coating [69].

The UV-cured waterborne polyurethane acrylat/SiO_2 hybrids were elaborated at wide filler dispersion, via in-situ polymerization method at different content TEOS and GPTMS as coupling agents. The UV-cured nanocomposite hybrids materials exhibited improved tensile strength, water-resistance, and thermal properties than those of the WPUA matrix [70]. The nano-hybrids are promising for a number of applications such as for high-performance water-based UV-curable coatings. POSS–TiO_2 hybrid filler system was synthesized and incorporated into epoxy resin to form POSS–TiO_2/EP nanocomposite. The POSS–TiO_2/EP exhibited excellent properties of anti-space ultraviolet radiation [71].

12.11 NUCLEAR-SHIELDING APPLICATIONS

Neutron radiation is generally encountered in many industries, like aerospace, healthcare, and nuclear power plants. It has been a great challenge to shield this neutron radiation for improving the equipment safety and protect human health. Aiming to produce an effective neutron-shielding material, alternating multi-layered composites (high-density polyethylene)/(high-density polyethylene/boron nitride), (HDPE/(HDPE/BN)) and HDPE/$BaSO_4$ were developed by Zhang et al. [72] as illustrated in Figure 12.9.

Jun et al. developed new the B_4C/epoxy composites having good mechanical properties and uniform dispersion of filler to use them as a neutron shield for spent nuclear fuel casks, using a direct ultrasonic dispersion of the B_4C particles in the

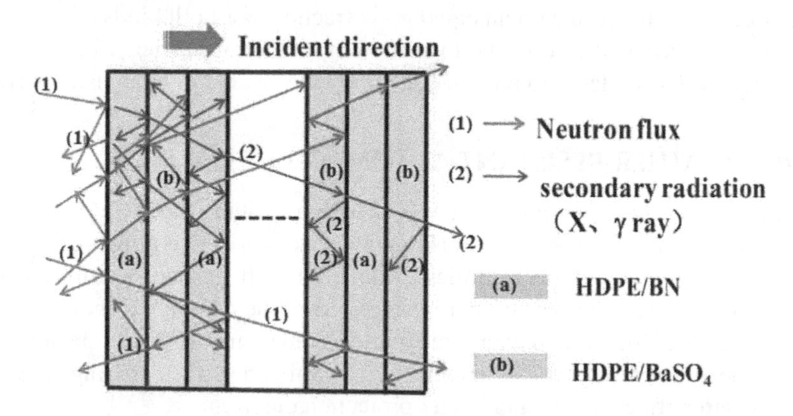

FIGURE 12.9 The Alternating Multilayered Distribution Structure of the Composite (One Layer Was High-density Polyethylene (HDPE)/Boron Nitride (BN), and Another Layer was HDPE/Barium Sulfate ($BaSO_4$) [72].

hardener. The strong adhesion between B_4C and epoxy matrix improved the properties of these materials without any chemical-treatment, and this was by avoiding the formation of the undesired impurities [73]. Epoxy/B_4C nanocomposites were produced by ultrasonic dispersion of B_4C nanoparticles in low-viscosity hardener and a subsequent mixing of a hardener-nanoparticle colloid with epoxy resins to ensure homogenous dispersion and good wetting with enhanced matrix/filler interfacial adhesion. By adding B_4C, PbO, and $Al(OH)_3$ to the epoxy matrix, a synergistic effect resulted in mechanical, thermal, and radiation shielding efficiency [74].

Kim et al. reported on the development of micro- and milled B_4C and BN nano-powders mixed with high-density polyethylene (HDPE) using a polymer mixer followed by hot pressing to fabricate sheet composites. Thermal neutrons' attenuation of the prepared HDPE nanocomposites was evaluated using a monochromatic ~0.025 eV neutron beam. Thermal neutron attenuation of the HDPE nanocomposites was greatly enhanced compared to their micro-counterparts at the same B-10 areal densities [75]. HDPE/modified-BN composites showed a better dispersion state of the filler particles and greater neutron-shielding properties compared with the HDPE/BN and HDPE/B_4C composite [76].

Harrison et al. [77] prepared new composites of HDPE and BN, and evaluated them for space-radiation shielding properties. The authors compared the shielding effectiveness of 2 wt.% BN composite with neat HDPE and aluminum (Al) against neutron-beam energies up to 600 MeV, and against 120 GeV protons. Under high-energy neutrons, both neat HDPE and HDPE/BN composites exhibited similar shielding efficiencies to that of Al (Figure 12.10). However, Al showed the best shielding capability for high-energy protons.

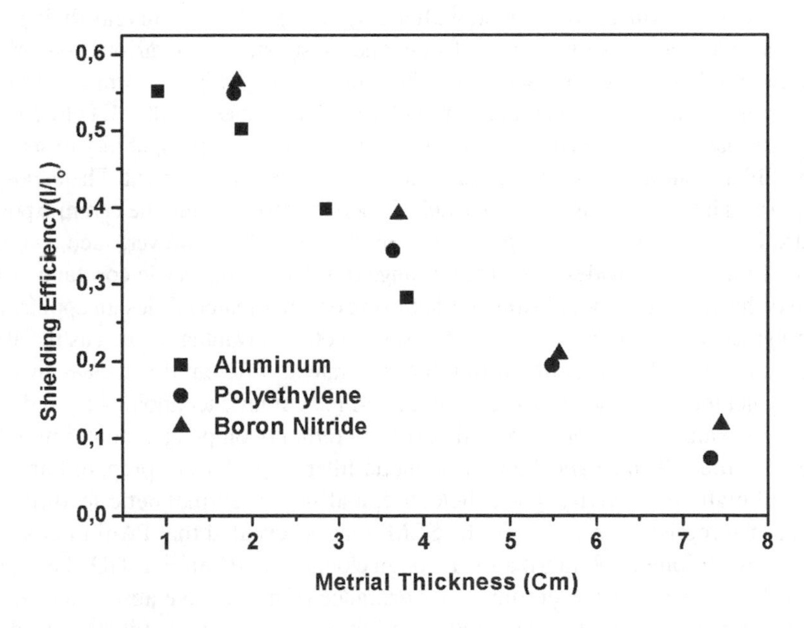

FIGURE 12.10 Comparison of Attenuation Results for Al, PE, and PE/BN Composites. Reprinted with Permission from [77], Copyright 2008 Wiley.

12.12 MICROWAVE ABSORPTION

Polymeric/ceramic composites exhibiting conducting, ferromagnetic, or microwave-absorbing functions are very useful in various electronic systems applications. Dielectric substrates used for microwave applications differ in several ways from the conventional printed circuit boards (PCBs) applied at low frequencies. Microwave packaging technology recommends substrate materials exhibiting high-frequency compatibility, decreased dielectric losses, low coefficient of thermal expansion, and high thermal conductivity. The CTE of polymer/ceramic composites is a critical factor to be considered in packaging technology.

The ceramic-reinforced polyethylene composites have been widely studied using $ZrSiO_4$, $Sm_2Si_2O_7$, CeO_2, $Sr_2Ce_2Ti_5O_{16}$, $Ca_4La_6(SiO_4)_4(PO_4)_2O_2$, $Mg_{095}Ca_{005}TiO_3$, $BaTiO_3$, and Li_2MgSiO_4 ceramic fillers for applying them as microwave substrates. The dielectric characteristics of these composites were studied at radio and microwave frequencies and showed that their densities and relative permittivity values increased with increasing the amount of ceramic fillers. The CTE of these composite is also an important parameter to be considered as it determines the dimensional stability, compatibility with metal electrodes, and mechanical robustness, etc., when applied to microwave substrate applications.

The fabricated composite materials from ceramic powders and polymers generated a high-contrast substrate that was concurrently bendable. In this context, many kinds of polymer/ceramic substrates are produced and applied to investigate the performance of a patch antenna and a coupled-line filter. For example, polymer/ceramic composites were used as substrate materials for a scanning antenna, while other polymer/ceramic hybrid systems were utilized as thin-film capacitors. In all these systems, polymers were functionalized by ceramic fillers whereas their pliable nature permit their uses like flexible free-standing substrates in various shapes.

Ceramic-filled elastic polymer composite substrates for truly conformal microwave applications suitable for a wide range of operating frequencies of 100 MHz to 20 GHz. A key advantage of the polymer/ceramic mixtures is their higher capability to specify a range of high-contrast substrates through tailoring the ceramic content. The ceramic is incorporated into the polymer via a particle dispersion process, and the ceramic powder mixture is kept below a certain percentage of 30% to 40%. However, such a concentration of mixtures provides a significant range of substrate dielectric constants, which can also change within the substrate for texturing or other material-design applications. The polymer/ceramic substrates also offer several other advantages, such as the ability for including metallic structures within the substrate at different thicknesses, which is typically not the case when low-temperature co-fired ceramic technology is used.

Novel hexanoic acid (HA) treated PAni micro/nanocomposites containing TiO_2 (dielectric filler, 30 nm) and Fe_3O_4 (magnetic filler, 1 μm) were prepared and they exhibited high conductivity, good dielectric, and improved magnetic features. The resultant nanorods/tubes shown in the SEM images revealed that PAni micro/nanocomposites exhibited polymerization through elongation. PAni/HA/TiO_2/Fe_3O_4 with 40% of Fe_3O_4 provides the optimum performance of microwave absorption (higher than 99.4% absorption) at a high-frequency range that makes it suitable for microwave absorbing and shielding applications [78].

12.13 PROTON-EXCHANGE MEMBRANES

The proton-exchange membrane (PEM) is considered one of the critical components in solid-type fuel cells, including the proton-exchange membrane fuel cells (PEMFC) and the direct methanol fuel ones (DMFC). Till now, research groups have been working on the fabrication of polymer/ceramic nanocomposites as PEM. Sulfonated poly(phthalazinone ether ketone) (s-PPEK) with a degree of sulfonation of 1.23 was mixed with nano-SiO_2 to produce hybrid materials suitable to be used as PEMs [79]. These hybrid membranes demonstrated better swelling behavior, high thermal stability, and improved mechanical properties. The methanol crossover behavior of the membrane was also reduced, revealing that these membranes were useful for a high methanol concentration in the feed within a cell testing. The membrane with 5 wt.% nano-SiO_2 exhibited an open cell potential of 0.6 V and an optimum power density of 52.9 m.W cm^{-2} at a current density of 264.6 mA cm^{-2}, which was significantly higher than the performance of the pristine sPPEK membrane or Nafion 117.

Sulfonated P(St-co-MA)-PEG/SiO_2 nanocomposite polyelectrolyte membranes were also developed at various nano-SiO_2 contents mixed with PEG of different molecular weights to control spacing between SiO_2 domains, up to a few nanometers by chemically-bound interior polymer chain [80]. These membranes were widely characterized for DMFC applications. In the Devrim et al., [81] study, sulfonated polysulfone (sPS)/TiO_2 composite membranes were produced for applying in proton-exchange membrane fuel cells (PEMFCs). The evolution of various performances showed that sPS/TiO_2 presented a promising membrane material for possible use in proton-exchange membrane fuel cells (Figure 12.11).

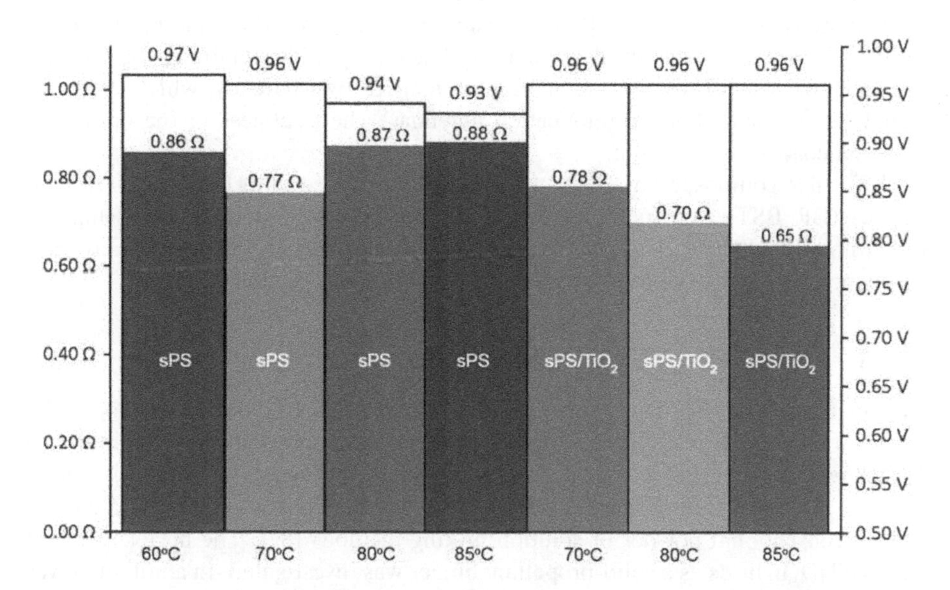

FIGURE 12.11 The Comparison of the Change of Open-circuit Voltage (OCV) and Cell Resistance of the Sulfonated Polysulfone sPS (40% Sulfonation) and sPS/TiO_2 Composite Membranes at Different Temperatures [81].

12.14 FABRICATION OF ANTENNA

The accelerated development in wireless communication systems has made ultrawideband (UWB) an advantageous technology to take the place of the conventional wireless technologies in today's use, including that of Bluetooth and wireless (local area networks, LANs), etc. To implement UWB technology, there are several challenges to surpass. UWB has a marked influence on antenna design. It significantly increases the interest in antenna design by affording new challenges and opportunities for antenna designers as UWB systems recommend antennas with an operating bandwidth covering the entire UWB in the range of 3.1 to 10.6 GHz and capable of receiving associated frequencies at the same time [82].

Recently, Ganguly et al. [83] developed a mushroom-like antenna exhibiting about 137% bandwidth with monopole-like radiation and 6 to 10 dBi gain over the entire band. This antenna was made of polyethylene/$SrTiO_3$ and $SrCuSi_4O_{10}$ ceramic composites at various weight ratios to give varying relative permittivity values.

High-density polyethylene (HDPE)/polystyrene-coated $BaO–Nd_2O_3–TiO_2$ (BNT) ceramic-based composites were fabricated using twin-screw melt extrusion. The microwave dielectric properties of the prepared composites were evaluated systematically. The results revealed that as the volume ratio of BNT ceramic fillers increased from 10 to 50 vol.%, the dielectric constant increased from 3.54 (9.23 GHz) to 13.14 (7.20 GHz), which is beneficial for the miniaturization of microwave devices. The GPS microstrip antennas were therefore developed and produced from the HDPE–PS/BNT composites. They possess good thermal stability, which satisfies requirements of practical antenna applications [84].

The study of Esther et al. [85] indicates that the Al_2O_3 thin coating is able to prevent surface degradation of germanium black polyimide, retaining the thermo-optical properties of the Ge-coated Kapton and RF transparency, which are functional requirements for communication antennas. The thickness of the optimized alumina coating was ~60 nm.

A flexible composite based on butyl rubber reinforced with $Ba_{0.7}Sr_{0.3}TiO_3$ composites (BR–BST) were prepared by sigma mixing followed by hot pressing. By examining the dielectric properties of the composites at both radio and microwave frequencies, this composite is very suitable for electronic applications [86].

12.15 OTHER APPLICATIONS

Reid et al. prepared a polyurethane/TiO_2 nanocomposites with uniform nanoparticle dispersion by the in situ synthesis of TiO_2 nanoparticles in a solution containing the HTPB prepolymer. The in situ TiO_2 nanoparticles demonstrated substantially higher activity as combustion catalysts compared to TiO_2 nanoparticles incorporated into PU by conventional powder or solution-mixing methods [87]. The performance of the PU/TiO_2 hybrids as a solid-propellant binder was investigated. In addition, it was reported that the Al_2O_3/polyimide separators are thus a promising separator candidate for next-generation lithium-ion batteries [88].

12.16 CONCLUSION

Polymer/ceramic micro- and nanocomposites have many striking properties, including mechanical, thermal, electrical, barrier, and optical. Owing to this, they have demonstrated extensive uses in different fields of application. This chapter summarized the different advanced and prospective engineering and industrial applications of polymer/ceramic composite materials. Due to their unique microstructures and property of polymer/ceramic micro- and nanocomposites, as well as their high versatility and ease of processing, polymer/ceramic composites represent an emerging class of multifunctional materials that are expected to be useful for many advanced applications, including catalysis, sensing, optoelectronics, biomedicine, and energy harvesting, conversion, and storage.

REFERENCES

1. Lai, M. B., Yu, S., Sun, R. et al. 2013. Effects and mechanism of graft modification on the dielectric performance of polymer-matrix composites. *Composites Science and Technology* 89: 127–133.
2. Nguyen, H., Navid, A., Pilon, L. 2010. Pyroelectric energy converter using co-polymer P(VDF-TrFE) and Olsen cycle for waste heat energy harvesting. *Applied Thermal Engineering* 30: 2127–2137.
3. Polizos, G., Tomer, V., Manias, E., Randall, C. A. 2010. Epoxy-based nanocomposites for electrical energy storage. II: nanocomposites with nanofillers of reactive montmorillonite covalently-bonded with barium titanate. *Journal of Applied Physics* 108: 074117.
4. Xia, W. M., Xu, Z., Wen, F., Zhang, Z. C. 2012. Electrical energy density and dielectric properties of poly(vinylidene fluoride-chlorotrifluoroethylene)/$BaSrTiO_3$ nanocomposites. *Ceramics International* 38: 1071–1075.
5. Wang, Q., Zhu, L. 2011. Polymer nanocomposites for electrical energy storage. *Journal of Polymer Science Part B: Polymer Physics* 49: 1421–1429.
6. Wu, W., Huang, X. Y., Li, S. T., Jiang, P. K., Toshikatsu, T. 2012. Novel three-dimensional zinc oxide superstructures for high dielectric constant polymer composites capable of withstanding high electric field. *The Journal of Physical Chemistry C* 116: 24887–24895.
7. Li, J. J., Seok, S. I., Chu, B. et al. 2009. Nanocomposites of ferroelectric polymers with TiO_2 nanoparticles exhibiting significantly enhanced electrical energy density. *Advanced Materials* 21: 217–221.
8. Rabuffi, M., Picci, G. 2002. Status quo and future prospects for metallized polypropylene energy storage capacitors. *IEEE Transactions on Plasma Science* 30: 1939–1942.
9. Wang, Y., Zhou, X., Chen, Q., Chu, B. J., Zhang, Q. M. 2010. Recent development of high energy density polymers for dielectric capacitors. *IEEE Transactions on Dielectrics and Electrical Insulation* 17: 1036–1042.
10. Zhang, Q. M., Li, H., Poh, M. et al. 2002. An all-organic composite actuator material with a high dielectric constant. *Nature* 419: 284–287.
11. Zhou, X., Zhao, X., Suo, Z. et al. 2009. Electrical breakdown and ultrahigh electrical energy density in poly(vinylidene fluoride-hexafluoropropylene) copolymer. *Applied Physics Letters* 94: 162901.
12. Liu, Z., Xu, J., Chen, D., Shen, G. 2015. Flexible electronics based on inorganic nanowires. *Chemical Society Reviews* 44: 161–192.
13. Kim, T. I., McCall, J. G., Jung, Y. H. et al. 2013. Injectable, cellular-scale optoelectronics with applications for wireless optogenetics. *Science* 340: 211–216.

14. McCall, J. G., Kim, T., Shin, G. et al. 2013. Fabrication and application of flexible, multimodal light-emitting devices for wireless optogenetics. *Nature Protocols* 8: 2413–2428.
15. Kim, T. I., Jung, Y. H., Song, J. et al. 2012. High-efficiency, microscale GaN light-emitting diodes and their thermal properties on unusual substrates. *Small* 8: 1643–1649.
16. Choi, M. K., Park, I., Kim, D. C. et al. 2015. Thermally controlled, patterned graphene transfer printing for transparent and wearable electronic/optoelectronic system. *Advanced Functional Materials* 25: 7109–7118.
17. Lin, Z., Yao, Y., McNamara, A., Moon, K., Wong, C. P. 2012. Single/few-layer boron nitride-based nanocomposites for high thermal conductivity underfills. *In Proceedings of the 2012 IEEE 62nd Electronic Components and Technology Conference (ECTC)*, San Diego, CA, 1437–1441.
18. Lin, Z., Mcnamara, A., Liu, Y., Moon, K., Wong, C. P. 2014. Exfoliated hexagonal boron nitride-based polymer nanocomposite with enhanced thermal conductivity for electronic encapsulation. *Composites Science and Technology* 90: 123–128.
19. Kim, M. H. 2007. Origin of efficiency droop in GaN-based light-emitting diodes. *Applied Physics Letters* 91: 183507.
20. Sarvar, F., Whalley, D., Conway, P. 2006. Thermal interface materials—A review of the state of the art. In *Proceedings of the 1st Electronics System integration Technology Conference*, Dresden, Germany, 5-7 September: 1292–1302.
21. Cui, Y., Li, Y., Xing, Y., Ji, Q., Song, J. 2017. Thermal design of rectangular microscale inorganic light-emitting diodes. *Applied Thermal Engineering* 122: 653–660.
22. Lin, Z., Liu, Y., Raghavan, S., Sitaraman, S. K., Wong, C. 2013. Magnetic alignment of hexagonal boron nitride platelets in polymer matrix: Toward high performance anisotropic polymer composites for electronic encapsulation. *ACS Applied Materials & Interfaces* 5: 7633–7640.
23. Zeng, X., Ye, L., Yu, S. et al. 2015. Artificial nacre-like papers based on noncovalent functionalized boron nitride nanosheets with excellent mechanical and thermally conductive properties. *Nanoscale* 7: 6774–6781.
24. Lee, E. S., Lee, S. M. 2008. Enhanced thermal conductivity of polymer matrix composite via high solids loading of aluminum nitride in epoxy resin. *Journal of the American Ceramic Society* 91: 1169–1174.
25. Agrawal, A., Satapathy, A. 2015. Epoxy Composites Filled with Micro-Sized AlN Particles for Microelectronic Applications. *Particulate Science and Technology: An International Journal* 33: 2–7.
26. Wang, Y. Y., Hsieh, T. E. 2007. Effect of UV Curing on Electrical Properties of a UV-Curable co-Polyacrylate/Silica Nanocomposite as a Transparent Encapsulation Resin for Device Packaging. *Macromolecular Chemistry and Physics* 208: 2396–2402.
27. Kakimoto, M., Takahashi, A., Tsurumi, T. et al. 2006. Polymer-ceramic nanocomposites based on new concepts for embedded capacitor. *Materials Science and Engineering B* 132: 74–78.
28. Pawar, S. G., Patil, S. L., Chougule, M. A. et al. 2010. Synthesis and characterization of polyaniline: TiO_2 nanocomposites. *International Journal of Polymeric Materials* 10: 777–785.
29. Xia, H., Wang, Q. 2002. Ultrasonic irradiation: A novel approach to prepare conductive polyaniline/nanocrystalline titanium oxide composites. *Chemistry of Materials* 5: 2158–2165.
30. Schnitzler, D. C., Zabin, A. J. G. 2004. Organic/Inorganic hybrid materials formed from TiO_2 nanoparticles and polyaniline. *Journal of the Brazilian Chemical Society* 3: 378–384.
31. Uygun, A., Yavuz, A. G., Sen, S., Omastovác, M. 2009. Polythiophene/SiO_2 nanocomposites prepared in the presence of surfactants and their application to glucose biosensing. *Synthetic Metals* 159: 2022–2028.

32. Ansari, M. O., Mohammad, F. 2011. Thermal stability, electrical conductivity and ammonia sensing studies on p-toluenesulfonic acid doped polyaniline: titanium dioxide (pTSA/Pani:TiO_2) nanocomposites. *Sensors and Actuators B* 157: 122–129.
33. Su, Y. L. 2006. Preparation of polydiacetylene/silica nanocomposite for use as a chemosensor. *Reactive Functional Polymers* 66: 967–973.
34. Huyen, D. N., Tung, N. T., Thien, N. D., Thanh, L. H. 2011. Effect of TiO_2 on the Gas Sensing Features of TiO_2/PANi Nanocomposites. *Sensors* 11: 1924–1931.
35. Tripathi, A., Mishra, S. K., Bahadur, I. et al. 2015. Optical properties of regiorandom polythiophene/Al_2O_3 nanocomposites and their application to ammonia gas sensing. *Journal of Materials Science: Materials in Electronics* 26: 7421–7430.
36. Soloukhin, V. A., Posthumus, W., Brokken-Zijp, J. C. M., Loos, J., de With, G. 2002. Mechanical properties of silica–(meth)acrylate hybrid coatings on polycarbonate substrate. *Polymer* 43: 6169–6181.
37. Bauer, F., Mehnert, R. 2005. UV Curable Acrylate Nanocomposites: Properties and Applications. *Journal of Polymer Research* 12: 483–491.
38. Bauer, F., Flyunt, R., Czihal, K. et al. 2007. UV curing and matting of acrylate coatings reinforced by nano-silica and micro-corundum particles. *Progress in Organic Coatings* 60: 121–126.
39. Tiarks, F., Landfester, K., Antonietti, M. 2001. Silica Nanoparticles as Surfactants and Fillers for Latexes Made by Miniemulsion Polymerization. *Langmuir* 17: 5775–5780.
40. Tate, J. S., Gaikwad, S., Theodoropoulou, N., Trevino, E., Koo, J. H. 2013. Carbon/Phenolic Nanocomposites as Advanced Thermal Protection Material in Aerospace Applications. *Journal of Composites* 2013: 1–9.
41. Ogasawara, T., Aoki, T., Hassan, M. S. A., Mizokami, Y., Watanabe, N. 2011. Ablation behavior of SiC fiber/carbon matrix composites under simulated atmospheric reentry conditions. *Composites Part A: Applied Science and Manufacturing* 42: 221–228.
42. Bahramian, A. R., Kokabi, M., Famili, M. H. N., Beheshty, M. H. 2006. Ablation and Thermal Degradation Behavior of a Composite Based on Resol Type Phenolic Resin: Process Modeling and Experimental. *Polymer* 47: 3661–3673.
43. Pulci, G., Tirillò, J., Marra, F. et al. 2010. Carbon–phenolic ablative materials for re-entry space vehicles: Manufacturing and properties. *Composites Part A: Applied Science and Manufacturing* 41: 1483–1490.
44. Srikanth, I., Padmavathi, N., Kumar, S. et al. 2013. Mechanical, thermal and ablative properties of zirconia, CNT modified carbon/phenolic composites. *Composites Science Technology* 80: 1–7.
45. Chen, Y., Chen, P., Hong, C., Zhang, B., Hui, D. 2013. Improved ablation resistance of carbon– phenolic composites by introducing zirconium diboride particles. *Composites part B: Engineering* 47: 320–325.
46. Kim, K., Jung Y. C., Kim, S. Y., Yang, B. J., Kim, J. 2018. Adhesion enhancement and damage protection for carbon fiber-reinforced polymer (CFRP) composites via silica particle coating. *Composites Part A* 109: 105–114.
47. Yang, L., Raza, A., Si, Y. et al. 2012. Synthesis of superhydrophobic silica nano-fibrous membranes with robust thermal stability and flexibility via in situ polymerization. *Nanoscale* 4: 6581–6587.
48. Arun, M., Kantheti, S., Gaddam, R. R. et al. 2014. Surface modification of TiO_2 nanoparticles with 1,3,5-triazine based silane coupling agent and its cumulative effect on the properties of polyurethane composite coating. *Journal of Polymer Research* 21: 600–611.
49. Kaur S., Gallei M., Ionescu E. 2014. Polymer–ceramic nanohybrid materials. In: Kalia S., Haldorai Y. (Eds.), *Organic-Inorganic Hybrid Nanomaterials. Advances in Polymer Science*, vol. 267. Springer, New York, pp. 143–185.
50. Zhou, C., Lu, X., Xin, Z. et al. 2014. Polybenzoxazine/SiO_2 nanocomposite coatings for corrosion protection of mild steel. *Corrosion Science* 80: 269–275.

51. Stodolak, E., Gadomska, K., Lacz, A. et al. 2009. Polymer-ceramic nanocomposites for applications in the bone surgery. *Journal of Physics: Conference Series* 146: 012026.
52. Mills, K. L., Zhu, X., Takayama, S. et al. 2008. The mechanical properties of a surface-modified layer on poly(dimethylsiloxane). *Journal of Materials Researches* 23: 37–48.
53. Martin, R. A., Yue, S., Hanna, J. V. et al. 2012. Characterizing the hierarchical structures of bioactive sol–gel silicate glass and hybrid scaffolds for bone regeneration. *Philosophical Transactions of the Royal Society A: Mathematical, Physical and Engineering Sciences* 370: 1422–1443.
54. Catauro, M., Raucci, M. G., De Gaetano, F. et al. 2003. Sol–gel synthesis, characterization and bioactivity of polycaprolactone/SiO_2 hybrid material. *Journal of Materials Science* 38: 3097–3102.
55. Camargo, P. H. C., Satyanarayana, K. G., Wypych, F. 2009. Nanocomposites: synthesis, structure, properties and new application opportunities. *Materials Researches* 12: 1–39.
56. Rhee, S. H., Choi, J. Y. 2002. Preparation of a bioactive poly(methyl methacrylate)/silica nanocomposite. *Journal of the American Ceramic Society* 85: 1318–1320.
57. Lee, K. H., Rhee, S. H. 2009. The mechanical properties and bioactivity of poly(methyl methacrylate)/SiO_2-CaO nanocomposite. *Biomaterials* 30: 3444–3449.
58. Thomas, N. P., Tran, N., Tran, P. A. et al. 2013. Characterization and bioactive properties of zirconia based polymeric hybrid for orthopedic applications. *Journal of Materials Science: Materials in Medicine* 25: 347–354.
59. Kasuga, T., Maeda, H., Kato, K. et al. 2003. Preparation of poly(lactic acid) composites containing calcium carbonate (vaterite). *Biomaterials* 24: 3247–3253.
60. Cong, H., Radosz, M., Towler, B. et al. 2007. Polymer–inorganic nanocomposite membranes for gas separation. *Separation and Purification Technologies* 55(3): 281–291.
61. Pandey, P., Chauhan, R. S. 2001. Membranes for gas separation. *Progress in Polymer Science* 26: 853–893.
62. Koros, W. J., Mahajan, R. 2000. Pushing the limits on possibilities for large scale gas separation: Which strategies? *Journal Membrane Science* 175: 181–196.
63. Cong, H., Radosz, M., Towler, B. F., Shen, Y. 2007. Polymer–inorganic nanocomposite membranes for gas separation. *Separation and Purification Technology* 55: 281–291.
64. Isobe, T., Nishimura, M., Takada, Y. et al. 2014. Gas separation using Knudsen and surface diffusion II: Effects of surface modification of epoxy/porous SiO_2 composite. *Journal of Asian Ceramic Societies* 2: 190–194.
65. Colombo, P., Zordan, F., Medvedovski, E. 2006. Ceramic–polymer composites for ballistic protection. *Journal Advances in Applied Ceramics Structural, Functional and Bioceramics* 105: 78–83.
66. Young, S., Lee, E. D., Wetzel, N. J. Wagner. 2003. The ballistic impact characteristics of Kevlar® woven fabrics impregnated with a colloidal shear thickening fluid. *Journal of Materials Science* 38: 2825–2833.
67. Rallini, M., Natali, M., Kenny, J.M. et al. 2013. Effect of boron carbide nanoparticles on the fire reaction and fire resistance of carbon fiber/epoxy composites. *Polymer* 54: 5154–5165.
68. Tibiletti, L., Longuet, C., Ferry, L. 2011. Thermal Degradation and Fire Behavior of Unsaturated Polyester Filled with Metallic Oxides. *Polymer Degradation and Stability* 96: 67–75.
69. Charpentier, P. A., Burgess, K., Wang, L., Chowdhury, R. R., Lotus, A. F., and Moula, G. 2012. Nano-TiO_2/polyurethane composites for antibacterial and self-cleaning coatings. *Nanotechnology* 23: 425606–425615.
70. Qiu, F., Xu, H., Wang, Y. et al. 2012. Preparation, characterization and properties of UV-curable waterborne polyurethane acrylate/SiO_2 coating. *Journal of Coatings Technology and Research* 9: 503–514.

71. Peng, D., Qin, W., Wu, X. 2015. A study on resistance to ultraviolet radiation of POSS–TiO_2/epoxy nanocomposites. *Acta Astronautica* 111: 84–88.
72. Zhang, X., Yang, M., Zhang, X., Wu, H., Guo, S., Wang Y. 2017. Enhancing the neutron shielding ability of polyethylene composites with an alternating multi-layered structure. *Composites Science and Technology* 150: 16–23.
73. Jun, J., Kim, J., Bae, Y. et al. 2011. Enhancement of dispersion and adhesion of B_4C particles in epoxy resin using direct ultrasonic excitation. *Journal of Nuclear Materials* 416: 293–297.
74. Lee, M. K., Lee, J. K., Kim, J. W. et al. 2014. Properties of B_4C–PbO–$Al(OH)_3$-epoxy nanocomposite prepared by ultrasonic dispersion approach for high temperature neutron shields. *Journal of Nuclear Materials* 445: 63–71.
75. Kim, J., Lee, B. C., Uhmd, Y. R., Millere, W. H. 2014. Enhancement of thermal neutron attenuation of nano-B4C, -BN dispersed neutron shielding polymer nanocomposites. *Journal of Nuclear Materials* 453: 48–53.
76. Shin, J. W., Lee, J. W., Yu, S. et al. Polyethylene/boron-containing composites for radiation shielding. *Thermochimica Acta* 585: 5–9.
77. Harrison, C., Weaver, S., Bertelsen, C., Burgett, E., Hertel, N., Grulke, E. 2008. Polyethylene/boron nitride composites for space radiation shielding. *Journal of Applied Polymer Science* 109: 2529–2538.
78. Phang, S. W., Tadokoro, M. Watanabe, J. et al. 2009. Effect of Fe_3O_4 and TiO_2 addition on the microwave absorption property of polyaniline micro/nanocomposites. *Polymers for Advanced Technologies* 20: 550–557.
79. Su, Y. H., Liu, Y. L., Sun, Y. M., Lai, J. Y., Guiver, M. D., Gao, Y. 2006. Using silica nanoparticles for modifying sulfonated poly(phthalazinone ether ketone) membrane for direct methanol fuel cell: A significant improvement on cell performance. *Journal of Power Sources* 155: 111–117.
80. Saxena, A., Tripathi, B. P., Shahi, V. K. 2007. Sulfonated poly(styrene-co-maleic anhydride)–poly(ethylene glycol)–silica nanocomposite polyelectrolyte membranes for fuel cell applications. *Journal of Physical Chemistry B* 111: 12454–12461.
81. Devrim, Y., Erkan, S., Baç, N., Eroğlu, I. 2009. Preparation and characterization of sulfonated polysulfone/titanium dioxide composite membranes for proton exchange membrane fuel cells. *International Journal of Hydrogen Energy* 34: 3467–3475.
82. Joseph, S., Sreekumari Amma, P., Modal, A. K., and Ratheesh, R. 2015. Compact UWB antenna with WiMAX and WLAN rejection. *IEEE International Microwave and RF Conference (IMaRC)*, Hyderabad, India.
83. Ganguly, D., Guha, D., George, S. et al. 2016. New design approach for hybrid monopole antenna to achieve increased ultra-wide bandwidth. *Antennas and Propagation Magazine* 59 (1): 139–144.
84. Zhang, L., Zhang, J., Yue, Z., Li, L. 2016. Thermally stable polymer–ceramic composites for microwave antenna applications. *Journal of Advanced Ceramics* 5(4): 269–276.
85. Esther, A. C. M., Sridhara, N., Sebastian, S. V. et al. 2016. Evaluation of nanoalumina coated germanium black polyimide membrane as sunshield for application on the communication satellite antenna. *Ceramics International* 42: 2589–2598.
86. Chameswary, J., Sebastian, M. 2013. Butyl rubber–$Ba_{0.7}Sr_{0.3}TiO_3$ composites for flexible microwave electronic applications. *Ceramic Internationals* 39: 2795–2802.
87. Reid, D. L., Draper, R., Richardson, D. et al. 2014. In situ synthesis of polyurethane–TiO_2 nanocomposite and performance in solid propellants. *Journal of Materials Chemistry* 2: 2313–2322.
88. Lee, J., Lee, C. L., Park, K. et al. 2014. Synthesis of an Al_2O_3-coated polyimide nanofiber mat and its electrochemical characteristics as a separator for lithium ion batteries. *Journal of Power Sources* 248: 1211–1217.

Index

Note: *Italicized* page numbers refer to figures, **bold** page numbers refer to tables

D

E

L

M

Q

R

S

T

U

V

W

X

Y

Z